T0406852

METHODS IN MOLECULAR BIOLOGY™

For other titles published in this series, go to
www.springer.com/series/7651

Engineered Zinc Finger Proteins

Methods and Protocols

Edited by

Joel P. Mackay

University of Sydney, Sydney, Australia

David J. Segal

University of California, Davis, CA, USA

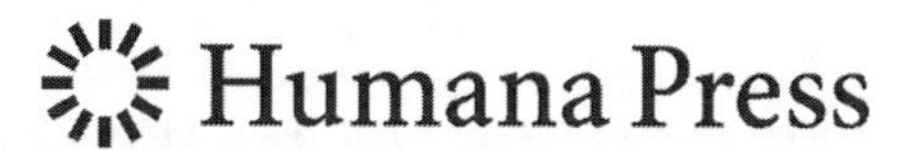

Editors
Joel P. Mackay
School of Molecular Bioscience
University of Sydney
Cnr. Butlin Ave. & Maze Cres.
2006 New South Wales
Australia
joel.mackay@sydney.edu.au

David J. Segal
Department of Pharmacology & Genome Center
University of California, Davis
451 E. Health Sciences Drive
Davis, CA 95616
USA
djsegal@ucdavis.edu

ISSN 1064-3745 e-ISSN 1940-6029
ISBN 978-1-60761-752-5 e-ISBN 978-1-60761-753-2
DOI 10.1007/978-1-60761-753-2
Springer New York Dordrecht Heidelberg London

Library of Congress Control Number: 2010930513

Cover illustration: Structural cartoons of engineered zinc fingers and their applications. Clockwise from upper right: A three-zinc finger protein (fingers 1, 2, and 3 in red, blue, and green, respectively) bound to DNA (orange), viewed along the DNA helical axis (PDB 1AAY). Zinc ions, yellow spheres. Two zinc fingers wrapping through the major groove of DNA (white surface), with DNA-contacting amino acids in positions -1, 2, 3, and 6 of each α-helix (red) shown (PDB 1AAY). Beta strands, green arrows. A model of an artificial transcription factor (ATF), composed of a six-zinc finger protein and a KRAB transcriptional repression domain (red helices at the N-terminus of the extended strand) bound to DNA (white) (based on PDB 1AAY and 1V65). Model of a heterodimeric zinc finger nuclease (ZFN), composed of two three-zinc finger proteins (purple and blue) bound to DNA in opposite orientation to juxtapose the FokI nuclease domains at their C-termini (based on PDB 1AYY and 2FOK). Models were generated by D.J.S. using INSIGHT and ICMBrowserPro.

Printed on acid-free paper

Humana Press is part of Springer Science+Business Media (www.springer.com)

Preface

Among the many types of DNA-binding domains, C2H2 zinc finger proteins (ZFPs) have proven to be the most malleable for creating custom DNA-binding proteins. Most classical ZFPs contain multiple fingers arranged in tandem arrays. The first crystal structure of a ZFP bound to its DNA target revealed a relatively simple mode of recognition, in which 3–4 amino acids from each finger made specific hydrogen bond interactions with 3–4 DNA base pairs in the major grove. It was realized that with each finger recognizing 3–4 bp, a six-finger protein has (in principle) the capacity to bind 18 bp of DNA, enough to specify a unique site in any known genome. These tantalizing observations suggested that ZFPs with new DNA-binding specificities could be created by simply changing the identities of a subset of amino acid side chains and that such 'designer' fingers could be combined to make powerful tools for a range of applications. Seeing this potential, several groups employed a variety of protein engineering methods and a great deal of labor to derive 'first-generation' engineered ZFPs, which were composed of individual modified fingers that could indeed be assembled in a modular fashion to recognize user-defined DNA sequences. Subsequent 'second-generation' methods employed an additional combinatorial selection step to optimize the binding of multi-finger ZFPs to extended DNA target sites. In recent years, we have begun to see the spectacular harvest that has come from all of this engineering work. Designer ZFPs are being used in a broad range of applications carried out in a rapidly increasing number of experimental systems.

This volume, *Engineered Zinc Finger Proteins: Methods and Protocols*, presents methods, guidance, and perspectives from 27 of the most active laboratories in this field. In **Section I**, methods are presented for engineering ZFPs. The first chapter presents first-generation modular assembly methods, which are still the most widely used noncommercial methods for ZFP construction. It also describes the many resources, tools, and organizations that have been developed to help researchers achieve success in these endeavors. Other chapters describe second-generation methods such as OPEN and bipartite selection. The final chapter in the section describes a computational approach to ZFP design.

Section II of this volume presents methods for the creation, evaluation, and delivery of artificial transcription factors (ATFs). There are over 700 endogenous ZFPs in the human genome, most of which are transcription factors. These factors typically consist of a zinc finger DNA-binding domain attached to either a transcriptional activation or a repression domain. In direct analogy, transcriptional activation or repression domains can be appended to engineered ZFPs to create ATFs. Unlike natural factors, ATFs are often designed to activate or repress expression of one single gene in the genome. This early application of engineered ZFPs was fairly successful: the first Phase I Clinical Trial of an ATF was initiated by Sangamo Biosciences in 2005. We now appreciate, however, that genes' expression is dependant not only on *genetic* elements (such as transcription factor-binding sites), but also on *epigenetic* elements, such as DNA and histone methylation. Two chapters in this section describe methods for ATF-directed epigenetic gene regulation. Another chapter describes the construction and use of libraries of ATFs, enabling the selection of ATFs based on the cellular phenotype they produce. These methods offer the potential to extend this technology beyond gene regulation to gene and pathway discovery. The final chapter describes the production and analysis of transgenic mice expressing active ATFs.

Section III presents methods for the creation and evaluation of zinc finger nucleases (ZFNs). Like ATFs, ZFNs are created by appending a cleavage domain to engineered ZFPs. The cleavage domain used most often is that of the non-specific *Fok*I endonuclease, which requires dimerization for activity. Therefore, a ZFN most commonly consists of two ZFPs, each with a *Fok*I cleavage domain appended to its C-terminus (although some chapters describe alternative architectures). Only when the two ZFP–*Fok*I monomers bind simultaneously in the correct spatial orientation will an active nuclease be created and produce a double strand break (DSB) in the DNA. The first chapter in this section describes how custom ZFNs can be used to clone large genomic fragments in vitro. However, it is the ability to induce a DSB at a specific site inside a living cell that has led to a revolution in genetic manipulation. The DSB in a cell can either be repaired by a non-homologous end-joining (NHEJ) pathway, leading to targeted mutagenesis at the cleavage site, or, in the presence of an appropriate donor DNA, by the homologous recombination (HR) pathway, leading to the targeted insertion or replacement of genetic information. ZFNs have been shown to induce NHEJ and HR events in as high as 10–50% of treated cells. The prospect of highly efficient, high-fidelity genomic modification raises the tantalizing possibility for the eventual therapeutic correction of endogenous genetic defects. The first Phase I Clinical Trial of a ZFN was initiated by Sangamo Biosciences in 2009. Several chapters in this section are based on recent exciting studies describing how these capabilities can now be applied to organisms for which targeted gene modification was hitherto impossible, such as fruit flies, zebrafish, and plants. Other chapters describe robust methods for assessing the desired (targeted cleavages) or undesired (off-target cleavages, often manifesting in cytotoxicity) activities of ZFNs. The final chapter describes non-*Fok*I-based ZFNs.

Section IV describes a small collection of the many applications and assays beyond ATFs and ZFNs. Methods for the construction and testing of zinc finger transposases are described. Other chapters explain how the ability of ZFPs to read DNA sequences directly from double-stranded DNA without denaturation or reannealing can be exploited to produce probes that can report on DNA sequence information in vitro or inside living cells. Other chapters in this section were included to help practitioners of engineered ZFPs to better understand and improve their systems. These methods include the incredibly powerful ChIP-seq methodology to identify the actual binding sites of ZFPs in living cells, methods for performing biophysical assays on purifed proteins, and methods for structural characterization of engineered ZFPs. The final chapter in this volume is a review of recent discoveries in the area of RNA-binding ZFPs, with speculation on the potential for creating custom RNA-binding proteins.

The co-editors greatly appreciate the time and effort of all of the contributors to this volume. *Engineered Zinc Finger Proteins: Methods and Protocols* is the first book devoted to the many exciting advances in this emerging field. We hope that both seasoned practitioners and new investigators will find these methods and insights useful as they create the next generation of engineered ZFPs and applications.

Sydney, Australia and Davis, CA, USA
July 2010

Joel P. Mackay
David J. Segal

Contents

Contributors

ANDREU ALIBÉS • *EMBL-CRG Systems Biology Unit, Centre de Regulació Genòmica, PRBB-UPF, Barcelona, Spain*

SHARON L. AMACHER • *Department of Molecular and Cell Biology, University of California Berkeley, Berkeley, CA, USA*

WOJCIECH BAL • *Institute of Biochemistry and Biophysics, Polish Academy of Sciences, Warsaw, Poland*

ADRIANA S. BELTRAN • *Department of Pharmacology and Lineberger Comprehensive Cancer Center, University of North Carolina at Chapel Hill, Chapel Hill, NC, USA*

KELLY J. BEUMER • *Department of Biochemistry, University of Utah School of Medicine, Salt Lake City, UT, USA*

MITAL S. BHAKTA • *Department of Pharmacology, Genome Center, University of California at Davis, Davis, CA, USA*

ALAN BILSLAND • *Cancer Research UK Beatson Laboratories, Centre for Oncology and Applied Pharmacology, University of Glasgow, Glasgow, UK*

PILAR BLANCAFORT • *Department of Pharmacology and Lineberger Comprehensive Cancer Center, University of North Carolina at Chapel Hill, Chapel Hill, NC, USA*

DANA CARROLL • *Department of Biochemistry, University of Utah School of Medicine, Salt Lake City, UT, USA*

TONI CATHOMEN • *Hannover Medical School, Department of Experimental Hematology, Hannover, Germany*

NICOLETTA CORBI • *IBPM-CNR, Regina Elena Cancer Institute, Rome, Italy*

TATJANA I. CORNU • *Institute of Virology (CBF), Charité Medical School, Berlin, Germany; Department of Experimental Hematology, Hannover Medical School, Hannover, Germany*

PEGGY J. FARNHAM • *Department of Pharmacology, Genome Center, University of California at Davis, Davis, CA, USA*

LINDA S. FENG • *Division of Cell Biology and Biophysics, School of Biological Sciences, University of Missouri-Kansas City, Kansas City, MO, USA*

JOSEP FONT • *School of Molecular Bioscience, University of Sydney, Sydney, NSW, Australia*

SETH FRIETZE • *Department of Pharmacology, Genome Center, University of California at Davis, Davis, CA, USA*

ALFRED L. GEORGE, JR. • *Division of Genetic Medicine, Department of Medicine, and the Institute for Integrative Genomics, Vanderbilt University, Nashville, TN, USA*

INDRANEEL GHOSH • *Department of Chemistry, University of Arizona, Tucson, AZ, USA*

MARIA GRAZIA DI CERTO • *INMM-CNR, Fondazione Santa Lucia/Ebri, Rome, Italy*

DMITRY Y. GUSCHIN • *Sangamo BioSciences, Inc., Richmond, CA, USA*

ANDREA HARTWIG • *Fachgebiet Lebensmittelchemie und Toxikologie, Technische Universität Berlin, Berlin, Germany*

MICHAEL C. HOLMES • *Sangamo BioSciences, Inc., Richmond, CA, USA*

NANCY C. HORTON • *Department of Biochemistry and Molecular Biophysics, University of Arizona, Tucson, AZ, USA*

JUSTIN P. HOSHAW • *Department of Genetics, Cell Biology and Development and Center for Genome Engineering, University of Minnesota, Minneapolis, MN, USA*

MIKI IMANISHI • *Institute for Chemical Research, Kyoto University, Kyoto, Japan*

ALBERT JELTSCH • *Biochemistry Lab, School of Engineering and Science, Jacobs University Bremen, Bremen, Germany*

J. KEITH JOUNG • *Molecular Pathology Unit, Center for Cancer Research, and Center for Computational and Integrative Biology, Massachusetts General Hospital, Charlestown, MA; Biological and Biomedical Sciences Program and Department of Pathology, Harvard Medical School, Boston, MA, USA*

RENATA Z. JURKOWSKA • *Biochemistry Lab, School of Engineering and Science, Jacobs University Bremen, Bremen, Germany*

GEORGE E. KATIBAH • *Sangamo BioSciences, Inc., Richmond, CA, USA*

W. NICOL KEITH • *Cancer Research UK Beatson Laboratories, Centre for Oncology and Applied Pharmacology, University of Glasgow, Glasgow, UK*

EUN JI KIM • *Department of Chemistry, Seoul National University, Seoul, South Korea*

JIN-SOO KIM • *Department of Chemistry, Seoul National University, Seoul, South Korea*

SEOKJOONG KIM • *Department of Chemistry, Seoul National University, Seoul, South Korea*

JOHN H. LAITY • *Division of Cell Biology and Biophysics, School of Biological Sciences, University of Missouri-Kansas City, Kansas City, MO*

BEATRICE I. LINDHOUT • *Clusius Laboratory, Department of Molecular and Developmental Genetics, Institute of Biology, Leiden University, Leiden, The Netherlands*

SARAH H. LOCKWOOD • *Department of Pharmacology, Genome Center, University of California at Davis, Davis, CA, USA*

JOEL P. MACKAY • *School of Molecular Bioscience, University of Sydney, Sydney, NSW, Australia*

MORGAN L. MAEDER • *Molecular Pathology Unit, Center for Cancer Research, and Center for Computational and Integrative Biology, Massachusetts General Hospital, Charlestown, MA; Biological and Biomedical Sciences Program, Harvard Medical School, Boston, MA, USA*

ELISABETTA MATTEI • *INMM-CNR, Fondazione Santa Lucia/Ebri, Rome, Italy*

JASMINE M. MCCAMMON • *Department of Molecular and Cell Biology, University of California Berkeley, Berkeley, CA, USA*

TOBIAS MECKEL • *Clusius Laboratory, Department of Molecular and Developmental Genetics, Institute of Biology, Leiden University, Leiden, The Netherlands*

JEFFREY C. MILLER • *Sangamo BioSciences, Inc., Richmond, CA, USA*

MICHAL MINCZUK • *Medical Research Council Mitochondrial Biology Unit, Cambridge, UK*

ALEJANDRO D. NADRA • *EMBL-CRG Systems Biology Unit, Centre de Regulació Genòmica, PRBB-UPF, Barcelona, Spain*

SHIGERU NEGI • *Faculty of Pharmaceutical Sciences, Doshisha Women's University, Kyoto, Japan*

HENRIETTE O'GEEN • *Department of Pharmacology, Genome Center, University of California at Davis, Davis, CA, USA*

ANNALISA ONORI • *IBPM-CNR, Regina Elena Cancer Institute, Rome, Italy*

CHAD K. PARK • *Department of Biochemistry and Molecular Biophysics, University of Arizona, Tucson, AZ, USA*

CLAUDIO PASSANANTI • *IBPM-CNR, Regina Elena Cancer Institute, Rome, Italy*

JASON R. PORTER • *Department of Chemistry, University of Arizona, Tucson, AZ, USA*

EDWARD J. REBAR • *Sangamo BioSciences, Inc., Richmond, CA, USA*

MARIANNE G. ROTS • *Department of Pathology and Medical Biology, University Medical Center Groningen, Groningen, The Netherlands*

TANJA SCHWERDTLE • *Institut für Lebensmittelchemie, Westfälische Wilhelms-Universität, Münster, Germany*

DAVID J. SEGAL • *Department of Pharmacology, Genome Center, University of California at Davis, Davis, CA, USA*

TAKASHI SERA • *Department of Synthetic Chemistry and Biological Chemistry, Graduate School of Engineering, Kyoto University, Kyoto, Japan*

LUIS SERRANO • *EMBL-CRG Systems Biology Unit, Centre de Regulació Genòmica, PRBB-UPF, Barcelona, Spain; Institució Catalana de Recerca i Estudis Avançats, Barcelona, Spain*

JIA-CHING SHIEH • *Department of Biomedical Sciences, Chung Shan Medical University, Taichung City, Taiwan*

CEM SÖLLÜ • *Institute of Virology (CBF), Charité Medical School, Berlin, Germany; Department of Experimental Hematology, Hannover Medical School, Hannover, Germany*

SABINE STOLZENBURG • *Department of Pathology and Medical Biology, University Medical Center Groningen, Groningen, The Netherlands*

YUKIO SUGIURA • *Faculty of Pharmaceutical Sciences, Doshisha Women's University, Kyoto, Japan*

STACEY THIBODEAU-BEGANNY • *Molecular Pathology Unit, Center for Cancer Research, and Center for Computational and Integrative Biology, Massachusetts General Hospital, Charlestown, MA, USA*

ANDRIY TOVKACH • *Department of Molecular, Cellular and Developmental Biology, The University of Michigan, Ann Arbor, MI*

JONATHAN K. TRAUTMAN • *Department of Biochemistry, University of Utah School of Medicine, Salt Lake City, UT, USA*

D. LORNE J. TYRRELL • *Department of Medical Microbiology and Immunology, University of Alberta, Edmonton, AB, Canada*

TZVI TZFIRA • *Department of Molecular, Cellular and Developmental Biology, The University of Michigan, Ann Arbor, MI, USA*

ERICA UNGER-WALLACE • *Department of Genetics, Development, and Cell Biology, Iowa State University, Ames, IA, USA*

BERT J. VAN DER ZAAL • *Clusius Laboratory, Department of Molecular and Developmental Genetics, Institute of Biology, Leiden University, Leiden, The Netherlands*

DANIEL F. VOYTAS • *Department of Genetics, Cell Biology and Development, Center for Genome Engineering, University of Minnesota, Minneapolis, MN, USA*

ADAM J. WAITE • *Sangamo BioSciences, Inc., Richmond, CA; Fred Hutchinson Cancer Research Center, Seattle, WA, USA*

MATTHEW H. WILSON • *Department of Medicine, Michael E. DeBakey VA Medical Center, Baylor College of Medicine, Houston, TX, USA*

VARDIT ZEEVI • *Department of Molecular, Cellular and Developmental Biology, The University of Michigan, Ann Arbor, MI, USA*

FENG ZHANG • *Department of Genetics, Cell Biology and Development, Center for Genome Engineering, University of Minnesota, Minneapolis, MN, USA*

KIMBERLEY A. HOEKSEMA • *Department of Medical Microbiology and Immunology, University of Alberta, Edmonton, AB, Canada*

Section I

Engineering Zinc Finger Proteins

Chapter 1

The Generation of Zinc Finger Proteins by Modular Assembly

Mital S. Bhakta and David J. Segal

Abstract

The modular assembly (MA) method of generating engineered zinc finger proteins (ZFPs) was the first practical method for creating custom DNA-binding proteins. As such, MA has enabled a vast exploration of sequence-specific methods and reagents, ushering in the modern era of zinc finger-based applications that are described in this volume. The first zinc finger nuclease to cleave an endogenous site was created using MA, as was the first artificial transcription factor to enter phase II clinical trials. In recent years, other excellent methods have been developed that improved the affinity and specificity of the engineered ZFPs. However, MA is still used widely for many applications. This chapter will describe methods and give guidance for the creation of ZFPs using MA. Such ZFPs might be useful as starting materials to perform other methods described in this volume. Here, we also describe a single-strand annealing recombination assay for the initial testing of zinc finger nucleases.

Key words: Zinc finger nuclease, single-strand annealing, homologous recombination, zinc finger protein engineering methods, modular assembly.

1. Introduction

ZFPs are among the most common type of DNA-binding protein found in eukaryotes and often comprise arrays of 8–10 fingers (1). The 1991 crystal structure of the three-finger ZFP Zif268 bound to DNA (2) revealed that each finger used amino acids in positions –1, 3, and 6 of its α-helix to contact a 3-bp subsite on the DNA. This seemingly simple mode of recognition inspired an early effort to create custom DNA-binding proteins (3). Through a combination of rational design and combinatorial methods such as phage display, individual zinc fingers were optimized to recognize one of the 64 possible 3-bp DNA subsites (4–8). In principle, these fingers could be assembled in any order

J.P. Mackay, D.J. Segal (eds.), *Engineered Zinc Finger Proteins*, Methods in Molecular Biology 649,
DOI 10.1007/978-1-60761-753-2_1,

to create multi-finger ZFPs to any desired sequence. Since each finger recognized an independent 3-bp subsite, ZFPs containing six fingers were expected to recognize 18 bp of DNA, a site sufficiently large to specify a unique locus in the human genome (9). The construction of ZFPs in this manner is referred to as modular assembly (MA).

Three large sets of MA fingers have been described. A widely used and well-characterized set was developed by Carlos Barbas of The Scripps Research Institute that covers all 3-bp subsites of the type GNN, most ANN, many CNN, and some TNN (*see* **Table 1.1**) (4, 5, 7). Sangamo Biosciences developed a set that covers GNN subsites, often with slightly different protein sequences depending on whether the finger would be used as finger 1, 2, or 3 in a three-finger protein (*see* **Table 1.2**) (6). This set is not widely used by Sangamo Biosciences, and several recent studies found this set was less likely than others to generate highly active ZFNs (10, 11). ToolGen, another company, developed a set that was based on selections of zinc fingers that occurred naturally in the human genome, as opposed to synthetic variants of Zif268 (8). The final set contains more than 50 full-length fingers that collectively recognize about 25 subsites (*see* **Tables 1.3** and **1.4**).

The ability to assemble the fingers in any order is based on the assumption that each finger binds its 3-bp subsite as an independent module. However, evidence soon began to emerge that, in some cases, there were "context-dependent effects," interactions that occur when two particular fingers are assembled next to each other on a particular DNA sequence. For example, strong structural and biochemical evidence demonstrated that when the residue in position 2 of the α-helix was aspartate, its side chain made energetically important contacts that favored G or T as the first bp of the previous finger's subsite (*see* **Fig. 1.1A**) (12). We can refer to this type of context-dependant effect, a helical amino acid (aa) specifying a base pair (bp) in the neighboring subsite, as target site overlap (TSO). However, there is only scattered evidence for other examples of TSO interactions. Other types of context-dependent effects, such as sequence-specific irregularities in the DNA structure due to base stacking, may in fact be more prevalent but are difficult to discover or predict.

An appreciation of these context-dependent effects led to the development of what can be called "second-generation" ZFP engineering methods, in which selections were performed on larger segments of the intended DNA target sequence. Yen Choo and Sir Aaron Klug developed a "bipartite" method (*see* **Chapter 3** by J.-C. Shieh, this volume), allowing for rapid selection of essentially 1.5 fingers in the context of each other (13). Sangamo Biosciences eventually acquired a set of two-finger modules created by this method and has used them as starting

Table 1.1
The Barbas set of zinc finger modules

DNA[a]	Finger[b]	DNA	Finger	DNA	Finger	RA[c]	DNA	Finger
AAA	QRANLRA	CAA	QSGNLTE	GAA	QSSNLVR	0.5	TAA	
AAC	DSGNLRV +	CAC	SKKALTE +	GAC	DPGNLVR	3	TAC	
AAGk	RKDNLKN	CAGk	RADNLTE +	GAGk	RSDNLVR	1	TAGk	REDNLHT +
AAT	TTGNLTV	CAT	TSGNLTE –	GAT	TSGNLVR	3	TAT	
ACA	SPADLTR +	CCA	TSHSLTE +	GCA	QSGDLRR[d]	2	TCA	
ACC	DKKDLTR	CCC	SKKHLAE –	GCC	DCRDLAR	80	TCC	
ACGk	RTDTLRD	CCGk	RNDTLTE +	GCGk	RSDDLVR	9	TCG	
ACT	THLDLIR +	CCT	TKNSLTE +	GCT	TSGELVR[d]	65	TCT	
AGA	QLAHLRA	CGA	QSGHLTE	GGA	QRAHLER	3	TGA	QAGHLAS
AGC		CGC	HTGHLLE +	GGC	DPGHLVR +	40	TGC	
AGGk	RSDHLTN	CGGk	RSDKLTE	GGGk	RSDKLVR	6	TGGk	RSDHLTT
AGT	HRTTLTN –	CGT	SRRTCRA +	GGT	TSGHLVR	15	TGT	
ATA	QKSSLIA	CTA	QNSTLTE	GTA	QSSSLVR	2.5[e]	TTA	
ATC		CTC		GTC	DPGALVR	40	TTC	
ATGk	RRDELNV –	CTGk	RNDALTE –	GTGk	RSDELVR	15	TTG	
ATT	HKNALQN	CTT	TTGALTE –	GTT	TSGSLVR	5	TTT	

[a] The 3-bp subsite to which the finger was designed. A "k" following the triplet indicates that this neighboring subsite should begin with either a G or a T (K in the IUPAC nomenclature). This is due to TSO interactions, as described in the Section 1.
[b] The DNA-recognition portion of the finger, consisting of amino acids –1, 1, 2, 3, 4, 5, and 6 with respect to the α-helix (based on (4, 5, 7)). These amino acids should be inserted into the appropriate positions in an Sp1C or Zif268 zinc finger scaffold (*see* **Fig. 1.1 C, D**). Also indicated is whether the finger was recommended favorably (+) or unfavorably (–) by the large-scale ToolGen study (28). Note that many fingers were not evaluated.
[c] The relative affinity of each finger in nM, as determined in a context as the middle finger of the three-finger Zif268 DNA-binding domain. Note that this information is only currently available for the GNN set of fingers (based on (7)).
[d] As described in Section 1, QSGDLRR also binds the subsite 5′-GCT-3′ with a relative affinity of 10, which is sixfold stronger than the more 5′-GCT-3′-specific finger, TSGELVR.
[e] The original relative affinity of QSSSLVR reported in (7) was determined to be erroneous. The correct value should be closer to 2.5 nM (24).

Table 1.2
The Sangamo set of zinc finger modules

DNA[a]	Finger 1[b]	Finger 2	Finger 3
GAA	QRSNLVR	QSGNLAR	QSGNLAR
GAC	DRSNLTR	DRSNLTR	DRSNLTR
GAGk	RSDNLAR	RSDNLAR	RSDNLTR
GAT	QSSNLAR	TSGNLVR	TSANLSR
GCA	QSGSLTR	QSGDLTR	QSGDLTR
GCC	ERGTLAR	DRSDLTR	DRSDLTR
GCGk	RSDDLTR	RSDDLQR	RSDDLTR
GCT	QSSDLTR	QSSDLTR	QSSDLQR
GGA	QSGHLAR	QSGHLQR	QSGHLQR
GGC	DRSHLTR	DRSHLAR	DRSHLAR
GGGk	RSDHLAR	RSDHLTR	RSDHLSR
GGT	QSSHLTR	TSGHLVR	TSGHLVR
GTA	QSGALTR	QSGALAR	QRASLTR
GTC	DRSALAR	DRSALAR	DRSALAR
GTGk	RSDALRT	RSDALSR	RSDALTR
GTT	TTSALTR	TSGALTR	QSSALTR

[a]The 3-bp subsite to which the finger was designed. A "k" following the triplet indicates that this neighboring subsite should begin with either a G or a T (K in the IUPAC nomenclature). This is due to TSO interactions, as described in the Section 1.
[b]The DNA-recognition portion of the finger, consisting of amino acids –1, 1, 2, 3, 4, 5, and 6 with respect to the α-helix (based on (6)). Note that this set often recommends a slightly different protein sequence depending on whether the finger would be used as finger 1, 2, or 3 in a three-finger protein. These amino acids should be inserted into the appropriate positions in an Sp1C or Zif268 zinc finger scaffold (*see* **Fig. 1.1 C, D**).

materials for their engineered ZFPs (14). Carl Pabo developed a "sequential selection" method, in which each new finger was selected in the context of its neighbor (15). A second method from the Pabo group, dubbed "B2H" in reference to the use of a bacterial two-hybrid system for selections, involved two steps: an enrichment of fingers on out-of-context 3-bp subsites as in classic modular assembly, followed by a final selection of the enriched fingers in the context of the full intended DNA target sequence (16). This method was extremely laborious, but was shown to frequently produce three-finger ZFPs that were superior in both affinity and specificity to three-finger ZFPs assembled by classical MA. Recently, J. Keith Joung developed a more streamlined version of this method called "oligomerized pool engineering" (OPEN) (11). In this method, the first step of the B2H method is performed by the Joung laboratory in high volume. The pools of enriched fingers are then made available to other investigators

Table 1.3
The ToolGen set of zinc finger modules

DNA[a]	Module name[b]	DNA	Module name	DNA	Module name		DNA	Module name
AAA	QSNI + QSNK QSNT QSNV1 QSNV2 QSNV3 QSNV4	CAA	QSNI QSNV1 QSNV2 + QSNV3 QSNV4	GAA	CSNR1 CSNR2 ISNR QGNR QSHR5 QSNK	QSNR1 + QSNR2 QSNR3 QSNR4 QSNV2 QSTV	TAA	QSNK – QSNV2
AAC	HSNK	CAC		GAC	CSNR1 + CSNR2 DSNR +	HSNK – QSTV	TAC	
AAGk		CAGk		GAGk	CSNR1[c] CSNR2[c] KSNR[c] QFNR[c] –	QSTV[c] RDER1 RSNR[c] – SSNR[c]	TAGk	
AAT	ISNV VSNV +	CAT		GAT	ISNR +		TAT	VDYK +
ACA		CCA		GCA	QSSR1 + QSSR2 QSSR3	QSTR – VSTR	TCA	
ACC		CCC		GCC	DSCR +		TCC	
ACGk		CCGk		GCGk	RDER1 + RDER2 + RDER3	RDER4 RDER5 RDER6	TCGk	
ACT		CCT		GCT	VSTR+		TCT	

(Continued)

Table 1.3
(Continued)

DNA[a]	Module name[b]	DNA	Module name	DNA	Module name		DNA	Module name
AGA	QSHR1 QSHR3 QSHT QSHV QTHQ + QTHR1	CGA	QSHT – QSHV + QTHQ QTHR1	GGA	QAHR QSHR1 QSHR2 + QSHR3 QSHR4	QSHR5 QTHR1 – QTHR2 WSNR	TGA	QSHT – QSHV + QTHQ
AGC		CGC		GGC			TGC	
AGGk	RDHT RDKR +	CGGk	RDHT +	GGGk	RDHR1 + RDHR2 RDHT	RDKI RDKR RSHR[c] +	TGGk	RDHT +
AGT		CGT		GGT	WSNR +		TGT	
ATA	QNTQ +	CTA		GTA	QSSR1 + QSSR2 QSSR3	QSTR – VSSR	TTA	
ATC	DSAR	CTC		GTC	DSAR +		TTC	
ATGk		CTGk		GTGk	RDER1 + RDER2 + VSSR[c]		TTGk	
ATT		CTT		GTT	HSSR – VSSR –		TTT	

[a]The 3-bp subsite to which the finger was designed. A "k" following the triplet indicates that this neighboring subsite should begin with either a G or a T (K in the IUPAC nomenclature). This is due to TSO interactions, as described in the Section 1.

[b]The unique ToolGen name for each finger (8). The name corresponds to the amino acids in positions –1, 2, 3, and 6 of the α-helix; however, the full sequence of the finger is provided in **Table 1.4**. Also indicated is whether the finger was recommended favorably (+) or unfavorably (–) by the large-scale ToolGen study (28). Note that some fingers were not evaluated.

[c]These fingers do not contain Asp in position 2 of the α-helix. Therefore, they are not expected to have a TSO interaction (i.e., the neighboring subsite could begin with any base).

Table 1.4
The ToolGen finger sequences and functional data

Module name[a]	Finger sequence[b]			DNA-full[c]	T[d]	DNA-ZFN[e]	RA[f]	DNA-RA[g]
CSNR1	YKCKQCGKAFG	CPSNLRR	HGRTH	GA(ACG)	+	GAC	2.3	GAC
CSNR2	YQCNICGKCFS	CNSNLHR	HQRTH	GA(ACG)				
DSAR	YSCGICGKSFS	DSSAKRR	HCILH	(AG)TC	+	GTC	0.4	GTC
DSCR	YTCSDCGKAFR	DKSCLNR	HRRTH	GCC	+	GCC	3.9	GCC
DSNR	YRCKYCDRSFS	DSSNLQR	HVRNIH		+	GAC		
HSNK	YKCKECGKAFN	HSSNFNK	HHRIH	(AG)AC	–	GAC	178.6	GAC
HSSR	FKCPVCGKAFR	HSSSLVR	HQRTH	GTT	–	GTT	0.8	GTT
ISNR	YRCKYCDRSFS	ISSNLQR	HVRNIH	GA(AT)	+	GAT	0.1	GAT
ISNV	YECDHCGKAFS	IGSNLNV	HRRIH	AAT				
KSNR	YGCHLCGKAFS	KSSNLRR	HEMIH	GAG			1.7	GAG
QAHR	YKCKECGQAFR	QRAHLIR	HHKLH	GGA			8.4	GGA
QFNR	YKCHQCGKAFI	QSFNLRR	HERTH	GAG	–	GAG	66.1	GAG
QGNR	FQCNQCGASFT	QKGNLLR	HIKLH	GAA			1.2	GAA
QNTQ	YTCSYCGKSFT	QSNTLKQ	HTRIH		+	ATA		
QSHR1	YACHLCGKAFT	QSSHLRR	HEKTH	(AG)GA				
QSHR2	YKCGQCGKFYS	QVSHLTR	HQKIH	GGA	+	GGA		
QSHR3	YACHLCGKAFT	QCSHLRR	HEKTH	(AG)GA				
QSHR4	YACHLCAKAFI	QCSHLRR	HEKTH	GGA				
QSHR5	YVCRECGRGFR	QHSHLVR	HKRTH	G(AG)A			2	GGA
QSHT	YKCEECGKAFR	QSSHLTT	HKIIH	(ACT)GA	–	(CT)GA	17.1	AGA
QSHV	YECDHCGKSFS	QSSHLNV	HKRTH	(ACT)GA	+	(CT)GA	64.3	CGA
QSNI	YMCSECGRGFS	QKSNLTI	HQRTH	(AC)AA	+	AAA	51.8	CAA

Table 1.4
(Continued)

Module name[a]	Finger sequence[b]			DNA-full[c]	T[d]	DNA-ZFN[e]	RA[f]	DNA-RA[g]
QSNK	YKCEECGKAFT	QSSNLTK	HKKIH	(AGT)AA	–	TAA	2.7	GAA
QSNR1	FECKDCGKAFI	QKSNLIR	HQRTH	GAA	+	GAA	1.1	GAA
QSNR2	YVCRECRRGFS	QKSNLIR	HQRTH	GAA				
QSNR3	YECEKCGKAFN	QSSNLTR	HKKSH	GAA				
QSNR4	YECVQCGKSYS	QSSNLFR	HQRRH	GAA				
QSNT	YECVQCGKGFT	QSSNLIT	HQRVH	AAA				
QSNV1	YECNTCRKTFS	QKSNLIV	HQRTH	(AC)AA				
QSNV2	YVCSKCGKAFT	QSSNLTV	HQKIH	(ACGT)AA	+	CAA	4.1	CAA
QSNV3	YKCDECGKNFT	QSSNLIV	HKRIH	(AC)AA				
QSNV4	YECDVCGKTFT	QKSNLGV	HQRTH	(AC)AA				
QSSR1	YKCPDCGKSFS	QSSSLIR	HQRTH	G(CT)A	+	G(CT)A	0.8	GTA
QSSR2	YECQDCGRAFN	QNSSLGR	HKRTH	G(CT)A				
QSSR3	YECNECGKFFS	QSSSLIR	HRRSH	G(CT)A				
QSTR	YKCEECGKAFN	QSSTLTR	HKIVH	G(CT)A	–	G(CT)A	0.9	GTA
QSTV	YECNECGKAFA	QNSTLRV	HQRIH	GA(ACG)				
QTHQ	YECHDCGKSFR	QSTHLTQ	HRRIH	(ACT)GA	+	AGA	0.6	CGA
QTHR1	YECHDCGKSFR	QSTHLTR	HRRIH	(ACG)GA	–	GGA	1.6	GGA
QTHR2	HKCLECGKCFS	QNTHLTR	HQRTH	GGA				
RDER1	YVCDVEGCTWKFA	RSDELNR	HKKRH	G(ACT)Gk	+	G(CT)Gk	1.4	GCG
RDER2	YHCDWDGCGWKFA	RSDELTR	HYRKH	GCGk	+	G(CT)Gk		
RDER3	YRCSWEGCEWRFA	RSDELTR	HFRKH	GCGk				
RDER4	FSCSWKGCERRFA	RSDELSR	HRRTH	GCGk				
RDER5	FACSWQDCNKKFA	RSDELAR	HYRTH	GCGk				

(Continued)

Table 1.4 (Continued)

Module name[a]	Finger sequence[b]	DNA-full[c]	T[d]	DNA-ZFN[e]	RA[f]	DNA-RA[g]
RDER6	YHCNWDGCGWKFA RSDELTR HYRKH	GCGk				
RDHR1	FLCQYCAQRFG RKDHLTR HMKKSH	GGGk	+	GGGk	0.9	GGG
RDHR2	CRCNECGKSFS RRDHLVR HQRTH	GGGk				
RDHT	FQCKTCQRKFS RSDHLKT HTRTH	(ACGT)GGk	+	(CT)GGk	6.8	AGG
RDKI	FACEVCGVRFT RNDKLKI HMRKH	GGGk				
RDKR	YVCDVEGCTWKFA RSDKLNR HKKRH	(AG)GGk	+	AGGk	4.5	GGG
RSHR	YKCMECGKAFN RRSHLTR HQRIH	GGG	+	GGG	0.2	GGG
RSNR	YICRKCGRGFS RKSNLIR HQRTH	GAG	–	GAG	0.7	GAG
SSNR	YECKECGKAFS SGSNFTR HQRIH	GAG			1.3	GAG
VDYK	FHCGYCEKSFS VKDYLTK HIRTH		+	TAT		
VSNV	YECDHCGKAFS VSSNLNV HRRIH	AAT	+	AAT	2.5	AAT
VSSR	YTCKQCGKAFS VSSSLRR HETTH	GT(AGT)	–	GTT	2.1	GTG
VSTR	YECNYCGKTFS VSSTLIR HQRIH	GC(AT)	+	GCT	14.5	GCT
WSNR	YRCEECGKAFR WPSNLTR HKRIH	GG(AT)	+	GGT	1.3	GGT
Zif268	FACDICGRKFA RSDERKR HTKIH	GCG			10	GCG

[a]The unique ToolGen name for each finger (8). The name corresponds to the amino acids in positions –1, 2, 3, and 6 of the α-helix; however, the full sequence of the finger is provided in the next column.
[b]The full sequence of the human zinc finger (based on (28, 44)). These amino acids should be inserted between canonical 5-residue linkers (TGEKP) to create new multi-finger ZFPs (*see* **Fig. 1.1E**). The DNA-recognition portion of the finger, consisting of amino acids –1, 1, 2, 3, 4, 5, and 6 with respect to the α-helix, is separated from the scaffold regions by spaces.
[c]The full set of 3-bp subsites to which the finger bound as determined by a yeast one-hybrid assay (8).
[d]Indicates this finger was recommended favorably (+) or unfavorably (–) by the large-scale ToolGen study (28). Note that some fingers were not evaluated.
[e]The 3-bp subsite to which the finger bound when it performed favorably in the large-scale ToolGen study (28). A "k" following the triplet indicates that this neighboring subsite should begin with either a G or a T (K in the IUPAC nomenclature). This is due to TSO interactions, as described in the Section 1.
[f]The relative affinity of each finger in nanomolar, as determined in a context as finger 3 of the three-finger Zif268 DNA-binding domain. Note that these values are based on (8), but have been normalized (multiplied by a factor of 17.86) to set the value of Zif268 equal to 10 nM. This was done for easier comparison with the Barbas relative affinity values, for which Zif268 had a value of 10 nM (7).
[g]The 3-bp subsite used to test the finger in the relative affinity assay (8).

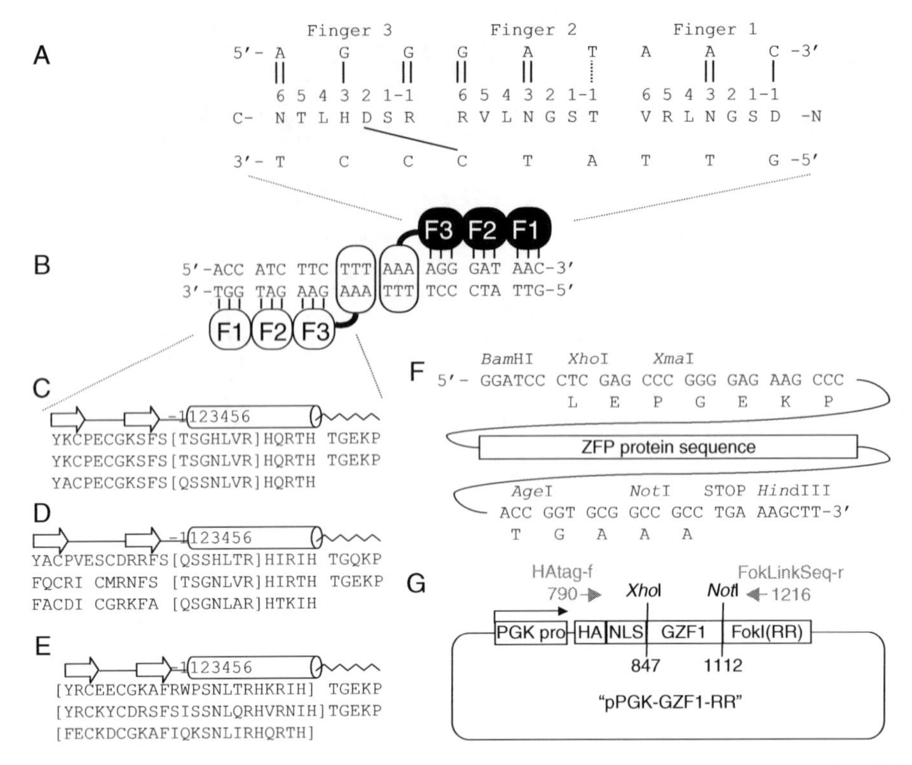

Fig. 1.1. Construction of ZFNs by MA. (**A**) A schematic diagram of the hydrogen bonding between the ZFP GZF3 and its 9-bp DNA target site. The ZFP binds "anti-parallel" (C-terminal to N-terminal) with respect to the DNA strand containing the target sequence (5′ to 3′). *Vertical lines* are presumed hydrogen bonds, dotted lines are Van Der Waal interactions. The *diagonal line* represents a target site overlap (TSO) by finger 3, influencing the 5′ base of the neighboring Finger-2 subsite to be G. Fingers with potential TSO interactions are indicated in **Tables 1.1**, **1.2**, **1.3**, and **1.4**. (**B**) A heterodimeric ZFN. Each monomer binds in an everted orientation so that the *FokI* cleavage domains at their C-termini (*open ovals*) can dimerize and cleave the DNA within the 6-bp spacer (TTTAAA) between the ZFP sites. (**C**) The sequence of the three-finger ZFP GZF1 created by inserting the Barbas modules (*in brackets*) into the Sp1C scaffold. Regions corresponding to beta strands (*arrows*), alpha helix (*tube*), and the interfinger linker (TGEKP, *wavy line*) are indicated. (**D**) The sequence for GZF1 using the Sangamo modules inserted into the Zif268 scaffold. (**E**) The sequence of GZF1 using the ToolGen modules, which consist of full-length fingers (*in brackets*) joined by interfinger linkers. Note that **d** and **e** are for illustrative purposes; they have not been evaluated for function. (**F**) Additional protein sequences must be added to the ZFPs in (**C**, **D**, or **E**), and must be encoded by the specific DNA sequences indicated in order to allow proper cloning into the ZFP cloning vector. (**G**) A schematic diagram of the ZFP cloning vector. A PGK promoter (PGK pro), HA tag (HA), SV40 nuclear localization signal (NLS), ZFP (GZF1), and *FokI* cleavage domain (here shown to contain the obligate heterodimer modification RR) are indicated. PCR primers used for sequencing are shown in *gray*.

for the second step of selection on the intended target (*see* **Chapter 2**).

Currently, there are three major sources of ZFPs for investigators interested in engineered zinc finger technology, such as zinc finger nucleases (ZFNs). Sangamo Biosciences use proprietary methods to create ZFNs that are highly potent and can target an impressively diverse spectrum of DNA sequences. Custom ZFNs from Sangamo are available through Sigma-Aldrich as CompoZr ZFN Technology, albeit at considerable cost (http://www.compozrzfn.com). MA is still one of the most popular methods for engineering ZFPs and has been referred to in several chapters in this volume (*see* **Chapters 6–11**, **17**, **23**, and **24**). Due to its large repertoire of targetable subsites, low cost, and ease of use, MA has been used to produce numerous regulators of endogenous genes, targetable live-cell imaging probes, recombinases, integrases, transposases, and methylases (17–21). Indeed, the first zinc finger nuclease to cleave an endogenous site was created using MA, as was the first artificial transcription factor to enter phase II clinical trial (22, 23). However, many have reported difficulty creating highly active ZFNs using MA. The Zinc Finger Consortium (ZFC), a confederation of academic laboratories, recently found that three-finger MA ZFPs composed of any set of modules could only produce a "positive" binding signal in a B2H reporter assay for 24% of 104 DNA targets (i.e., an "unexpected" failure rate of 76%) (10). The OPEN system has become the method of choice for the ZFC and has been shown consistently to create high-quality ZFNs (11). However, it is important to consider that the current version of OPEN is still a fairly complex procedure and only allows selection of fingers that recognize GNN and a few TNN subsites. If three-finger ZFPs engineered by OPEN were evaluated on the same 104 target sequences as used in the MA study, they would have an "expected" failure rate of 87%. Further advances are therefore required to develop methods for engineering high-quality ZFPs that are simultaneously inexpensive, accessible, comprehensive, and robust.

In that regard, several recent reports offer helpful guidance for practitioners of MA. Many of the Barbas set of fingers were selected based on high specificity rather than high affinity. A more recent study by the ZFC showed that three-finger MA ZFPs containing one or more of the four lowest affinity Barbas fingers (GGC, GTC, GCT, GCC) tended to be poor binders in a B2H assay (24). These results argue against the proposition that MA is fundamentally flawed because it fails to address context-dependent effects (25) and suggest that one approach to improving performance might be to replace low affinity fingers with higher affinity variants. An interesting example supporting this approach is TSGELVR, a Barbas finger engineered to recognize 5′-GCT-3′ with high specificity but relatively low affinity (7). QSGDLRR, a Barbas finger designed to recognize 5′-GCA-3′, also binds 5′-GCT-3′ with sixfold higher affinity than TSGELVR. Several successful ZFNs in the literature actually use fingers essentially identical to QSGDLRR to recognize 5′-GCT-3′ (26, 27). Another approach might be to construct ZFPs containing 4, 5, or 6 fingers, which may have increased affinity compared to the three-finger MA ZFPs examined in the ZFC study (10). In support of this approach, a recent large-scale analysis by ToolGen found that ZFN pairs composed of two four-finger MA ZFPs were more frequently active than those composed of three-finger ZFPs (28). The authors also recommended a subset of 37 ToolGen and Barbas modules to produce active ZFNs based on their study (*see* **Tables 1.1** and **1.3**).

In this chapter we describe methods and design considerations for creating ZFPs by MA (*see* **Section 3.1**). Once constructed, it is often useful to test some performance characteristics of the ZFPs (i.e., affinity, specificity, or activity). Many such assays are described in other chapters of this volume, including in vitro methods such as EMSA (*see* **Chapters 6** and **16**), in vitro ZFN cleavage assays (*see* **Chapter 13**), cell-based transcription reporter assays such as B2H or luciferase (*see* **Chapters 2** and **7**), and expression assays such as westerns (*see* **Chapters 11** and **16**). Here we describe a single-strand annealing (SSA) recombination assay for the initial analysis of ZFNs (*see* **Section 3.2**). Once active ZFN pairs are identified, assays at the endogenous site can be performed, such as the Surveyor mutagenesis detection assay (*see* **Chapter 15**), the restriction site modification assay (*see* **Chapter 19**), the chromatin immunoprecipitation (ChIP) binding assay (*see* **Chapters 11** and **27**). Upon achieving activity at the endogenous site, more laborious characterizations of the ZFN are warranted, such as in vitro target site selection assays (SELEX or Bind-n-Seq (29, 30)), in vivo genome-wide binding site assays (ChIP-seq, *see* **Chapter 27**), or other target-specific activity assays (e.g., *see* **Chapter 18**).

In the SSA recombination assay, a gene encoding firefly luciferase is divided into two segments, separated by a stop codon and a ZFN target site. Both segments contain an 870-bp region of homology in direct repeat orientation (*see* **Fig. 1.2**). A ZFN-induced DSB between the segments allows efficient SSA homologous recombination, resulting in an active luciferase gene. Firefly luciferase activity is therefore proportional to ZFN activity. The assay has several features that make it well suited as an initial test of ZFN activity. First, the assay is performed in the biologically relevant environment of a human cell (or could be any easily transfected cell type). Adverse events, such as cytotoxicity due to off-target cleavages, will result in the death of transfected cells. Loss of firefly luciferase signal due to toxicity-dependant cell death can be distinguished from poor ZFN activity by the inclusion of a Renilla luciferase control plasmid. Second, there are many special challenges to the success of a ZFN at an endogenous target site, including limited chromatin accessibility, competing endogenous binding factors, and unexpected local DNA structures. If ZFN activity is not observed at an endogenous site, knowing that the ZFN was functional in an SSA assay will help to differentiate complications at the endogenous site from problems with the ZFN. Third, because it is difficult to anticipate the challenges at the endogenous site, a pragmatic approach would be to create

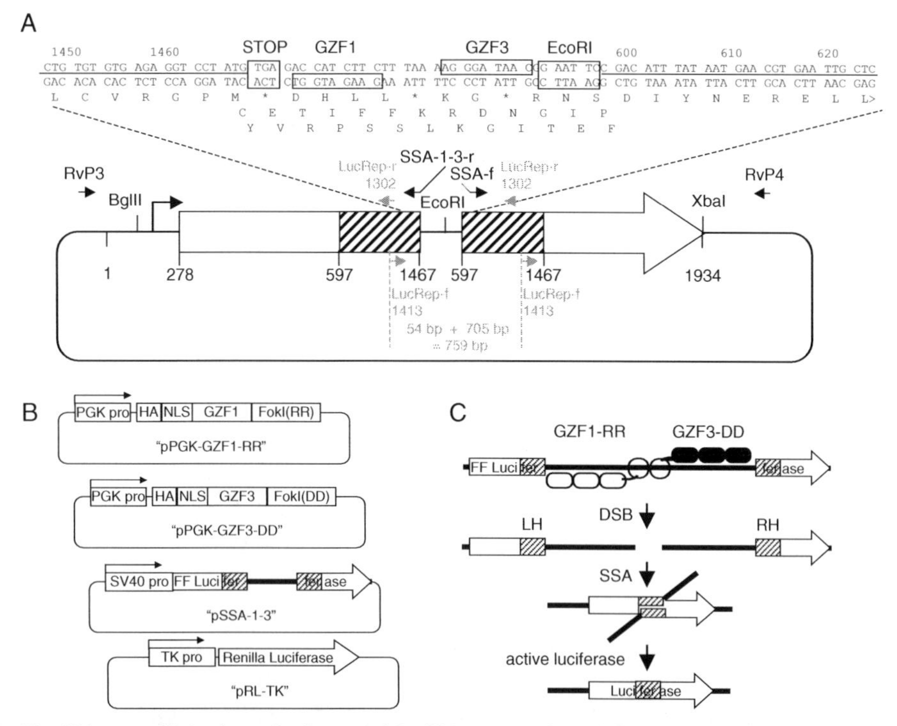

Fig. 1.2. The SSA assay. (**A**) A schematic diagram of the SSA reporter plasmid. The sequence at the junction site between the homologous repeats is shown at the *top*. The GZF1 and GZF3 heterodimer target site is indicated. PCR primers used in construction are shown as *black arrows* on the plasmid diagram below. PCR primers used for amplification of the junction region and sequencing are shown in *gray*. (**B**) Schematics of the four plasmids that will be transfected into the test cells for the SSA assay. (**C**) The expressed ZFN pair will bind at the target site and create a double-strand break (DSB). The break will be efficiently repaired by the SSA pathway to reconstruct an active firefly luciferase reporter gene.

5–10 ZFNs that target different sites in the region of interest. The SSA assay can be easily reconfigured with a new ZFN target site with a single oligonucleotide. Finally, ZFNs are composed of a heterodimer of two ZFP-*Fok*I cleavage domain fusion proteins ((31), and *see* **Fig. 1.1B**). Often only one monomer of the ZFN pair can be the cause of heterodimer failure. It is therefore important that the functional assay identify potential underperforming monomers, which can be easily done by configuring the SSA assay with homodimer ZFN target sites. Once identified, improvements can be directed toward the deficient monomer, such as replacing poorly binding fingers with stronger variants or adding additional fingers.

2. Materials

2.1. The Construction of ZFNs by MA

1. The list of Barbas zinc finger modules (*see* **Table 1.1**).
2. The list of Sangamo zinc finger modules (*see* **Table 1.2**).

3. The list of ToolGen zinc finger modules (*see* **Tables 1.3** and **1.4**).
4. pPGK-GZF1-wt ZFN-expression plasmid: this plasmid contains a phosphoglycerate kinase promoter driving expression of a gene encoding an HA-tag (YPYDVPDYA), the nuclear localization signal of SV40 large T antigen (PKKKRKV), a three-finger ZFP (GZF1), a 6-aa linker (TGAAAR), and a wild-type *Fok*I endonuclease catalytic domain (32). The expressed ZFN is active as a homodimer. All non-commercial plasmids are available from the authors upon request.
5. pPGK-GZF1-RR ZFN-expression plasmid: This plasmid is similar to pPGK-GZF1-wt, except the *Fok*I cleavage domain contains a modification that prevents activity as a homodimer. The expressed ZFN is only active as a heterodimer with a DD-modified ZFN (i.e., GZF3-DD; see below).
6. pPGK-GZF3-DD ZFN-expression plasmid: This plasmid is similar to pPGK-GZF3-wt, except the *Fok*I cleavage domain contains a modification that prevents activity as a homodimer. The expressed ZFN is only active as a heterodimer with an RR-modified ZFN (i.e., GZF1-RR; see above).
7. Oligonucleotide primers for sequencing the pPGK ZFN-expression plasmid:
   a. HAtag-f: 5′-ATGTACCCATACGATGTCCCAGACTA CG-3′
   b. FokLinkSeq-r: 5′-TCTATCCTGAGTGGAATTTCTG GC-3′
8. Standard equipment and reagents for agarose gel electrophoresis.
9. Restriction endonucleases and appropriate buffers: *Xhol*I and *Not*I.
10. T4 DNA ligase.

2.2. The SSA Recombination Assay

2.2.1. Preparation of the SSA Reporter Plasmid

1. Oligonucleotide primers for SSA reporter plasmid construction:
   a. RvP3-f: 5′-CTAGCAAAATAGGCTGTCCC-3′
   b. SSA-1-3-r: 5′- CTC GAATTC C – [GTT ATC CCT TTTAAA GAA GAT GGT] – CTCACATAGGACCTCT-CACACACAG –3′

c. Additional custom SSA-r oligonucleotides, in which the desired ZFN target site replaces the bracketed sequence in SSA-1-3-r.

2. Oligonucleotide primers for sequencing the SSA reporter plasmid:
   a. LucRep-f: 5′- CGAAGGTTGTGGATCTGGATACC-3′
   b. LucRep-r: 5′- TAGCTGATGTAGTCTCAGTGAGC-3′
3. pGL3-Control plasmid: This plasmid expresses full-length firefly luciferase under the control of a SV40 promoter and enhancer (Invitrogen). It serves as the template for construction of the SSA reporter plasmid and can be used as a positive control.
4. pSSA-1-3 reporter plasmid: This plasmid contains the target site for the ZFN target site for the heterodimer of GZF1 and GZF3, flanked by a split firefly luciferase reporter gene. This is the exemplary plasmid that will be constructed in **Section 3.2.1**, but it is also used for the construction of new SSA reporter plasmids. This and all other non-commercial plasmid vectors are available from the investigator upon request.
5. 2× Taq-Pro Complete polymerase mix with dNTPs and 2.0 mM $MgCl_2$ (Denville Scientific, Cat. CB4070-4). Thaw the 3 mL solution on ice, aliquot 250 μL volumes into 1.5 μL tubes, and store at –20°C.
6. Standard equipment and reagents for agarose gel electrophoresis.
7. PCR purification kit (e.g., QIAquick PCR purification kit, Qiagen).
8. Restriction endonucleases and appropriate buffers: *Bgl*II and *Eco*RI.
9. T4 DNA ligase.

2.2.2. Performing the SSA Assay

1. Appropriate ZFN-expression plasmids and SSA reporter plasmids (*see* **Note 1**).
2. pPGK-empty vector: This plasmid is similar to pPGK-GZF1-wt but lacks the ZFP and *Fok*I domains. It can be used as a negative control.
3. pRL-TK vector: This plasmid expresses Renilla luciferase under control of a HSV-thymidine kinase promoter. It can be used to assess cytotoxicity.
4. 24-well polystyrene tissue culture plates.
5. 2× Poly-D-Lysine: Poly-D-Lysine (Poly-K; MP Biomedicals, Cat. 102694) is supplied as a 10 mg desiccant that

is resuspended in 50 mL of filtered autoclaved water. This 2× stock should be stored at 4°C. Fresh 1× Poly-K is made by diluting the 2× stock in filtered autoclaved water.

6. HEK 293-T cells (ATCC, Cat. CRL-11268).
7. 75 cm^2 tissue culture flasks.
8. Phosphate-buffered saline, sterile for tissue culture (e.g., Dulbecco's PBS, Invitrogen, Cat. 14190).
9. Trypsin-EDTA (Invitrogen, Cat. 25300-054).
10. Pen–Strep (Invitrogen, Cat. 15140-122) contains 100 U/mL of penicillin and 100 μg/mL streptomycin. 5 mL aliquots are stored at –20°C.
11. Bovine Calf Serum (JR Scientific, Cat. WBJR050). 50 mL aliquots are stored at –20°C.
12. DMEM Complete Medium: DMEM medium (Invitrogen, Cat. 11995) is supplemented with 5 mL of Pen–Strep and 50 mL of Bovine Calf Serum (10% final). This mixture should be stored at 4°C.
13. Hemocytometer.
14. OPTI-MEM reduced serum medium (Invitrogen, Cat. 31985). This should be stored at 4°C.
15. Lipofectamine2000 (1 mg/mL solution, Invitrogen, Cat. 11668-019). This should be stored at 4°C.
16. 1× Reporter Lysis Buffer. Dilute the 5× stock solution (Promega, Cat. E397A) with filtered autoclaved water to make a 1× working solution. The 1× solution can be stored at room temperature.
17. 25× Protease Inhibitor Cocktail: Suspend one Complete Protease Inhibitor Cocktail Tablet without EDTA (Roche, Cat. 11873580001) in 2 mL of filtered autoclaved water. Prepare small aliquots and store at –20°C.
18. Bright-Glo Luciferase Assay System (Promega, Cat. E2620). Aliquot in 15 mL tubes and store at –80°C. This substrate is light sensitive, keep it wrapped in aluminum foil.
19. Renilla Luciferase Assay System (Promega, Cat. E2820). Aliquot in 15 mL tubes and store at –80°C. This substrate is light sensitive, keep it wrapped in aluminum foil.
20. Opaque white 96-well plates (Denville, Cat. P9725).
21. Microplate luminometer (e.g., Veritas Microplate Luminometer, Turner Biosystems).

3. Methods

3.1. The Construction of ZFNs by MA

1. Find 5–10 appropriate target sites within your DNA region of interest (*see* **Note 2**).
2. Verify that the target sites are reasonably unique (not more than 1–2 mismatches) in the target genome. Usually, this can be achieved using BLAST (*see* **Note 3**).
3. Design the ZFP coding region based on the information in **Tables 1.1**, **1.2**, **1.3**, **1.4** and **Fig. 1.1**. For the Barbas and Sangamo module sets, only the DNA-contacting regions corresponding to positions −1, 1, 2, 3, 4, 5, and 6 of the α-helix are inserted into an Sp1C (*see* **Fig. 1.1C**) or Zif268 (*see* **Fig. 1.1D**) three-finger scaffold. It is not clear if either scaffold has any advantage compared to the other. The ToolGen modules listed in **Table 1.4** are connected with TGEKP linkers to create a finger-specific scaffold (*see* **Fig. 1.1E** and **Note 4**).
4. The coding regions can be physically constructed by commercial synthesis (*see* **Note 5**). Typically this is rapid and relatively inexpensive, especially for new users. In addition, the codons can be optimized to the target organism. Be sure to request that the DNA sequences are appended to the protein coding region as indicated in **Fig. 1.1F** to facilitate cloning into the pPGK-based ZFN-expression plasmids (*see* **Fig.1.1G**).
5. Digest both the ZFP coding region and a pPGK-based ZFN-expression vector with *Xho*I and *Not*I. Subclone the ZFP into the expression vector. To create obligate heterodimer ZFN pairs, one ZFP will replace GZF1 in pPGK-GZF1-RR and the other will replace GZF3 in pPGK-GZF3-DD. For homodimer ZFNs, each ZFP will replace GZF1 in pPGK-GZF1-wt.
6. Verify the new constructs by sequencing using primers HAtag-f and/or FokLinkSeq-r.

3.2. The SSA Recombination Assay

3.2.1. Preparation of the SSA Reporter Plasmid

1. Prepare the left arm of the split luciferase gene in a 50 μL PCR reaction consisting of the following components (*see* **Fig. 1.2** and **Note 6**):
   a. 19 μL of H_2O
   b. 1.0 μL of 10 ng/μL pGL3-Control template

c. 2.5 μL of 5 μM RvP3 primer

d. 2.5 μL of 5 μM SSA-1-3-r primer (or other SSA-r primer with a desired ZFN site)

e. 25 μL of 2× TaqPro polymerase mix

The reactions conditions are as follows: 95°C for 30 s; 25 cycles of 95°C for 30 s, 58°C for 30 s, 72°C for 30 s; 72°C for 2 min; hold at 4°C.

2. Check for correct amplification by analyzing 10 μL of the reaction using agarose gel electrophoresis. Purify the amplicon using a PCR purification kit.
3. Digest both the amplicon and pSSA-1-3 using *Bgl*II and *Eco*RI (*see* **Note 7**).
4. Ligate the amplicon into the pSSA vector using T4 DNA ligase.
5. Because the duplicated region also makes sequencing of the final plasmid difficult, the junction of the repeated region (containing the ZFN target site) should be first amplified in the following 50 μL PCR reaction:

a. 19 μL of H_2O

b. 1.0 μL of 10 ng/μL pSSA reporter plasmid

c. 2.5 μL of 5 μM LucRep-f primer

d. 2.5 μL of 5 μM LucRep-r primer

e. 25 μL of 2× TaqPro polymerase mix

The reactions conditions are as follows: 95°C for 30 s; 25 cycles of 95°C for 30 s, 58°C for 30 s, 72°C for 30 s; 72°C for 2 min; hold at 4°C.

6. Check for correct amplification by analyzing 10 μL of the reaction using agarose gel electrophoresis. Purify the amplicon using a PCR purification kit.
7. The sequence of the newly constructed SSA reporter plasmid can then be verified by sequencing the PCR amplicon using LucRep-f as the sequencing primer.

3.2.2. Performing the SSA Assay

Day 0: *Seed the cells in the* 24*-well plates.*

1. Pretreat the assay tissue culture plates with Poly-K to help loosely adherent cells such as HEK 293-T to remain bound during washing procedures. Add 450 μL of 1× Poly-K to each well of a 24-well plate. Allow the Poly-K solution to incubate at room temperature (RT) for 30 min. Steps 1–19 should be performed in a biosafety cabinet to maintain sterility and biosafety.
2. Prepare the HEK 293-T cells to seed on 24-well plates. Aspirate the medium from the cells growing in a 75 cm^2 flask.

3. Wash the cells with 5 mL of PBS. Swirl the PBS around to wash away remaining medium, then aspirate the PBS.
4. Add 2 mL of Trypsin-EDTA and swirl the flask to ensure the entire surface is covered. Incubate at RT for 1 min.
5. Resuspend the cells using 8 mL of DMEM Complete Medium.
6. Place 15 μL of the above suspension onto a hemocytometer. Count the number of cells using an inverted microscope. After use, decontaminate the hemocytometer in 10% final bleach solution.
7. Aspirate the 1× Poly-K solution from the 24-well plate.
8. Add 100,000–150,000 cells to each well.
9. Incubate for 24 h in a 37°C CO_2 incubator before transfection.

Day 1: Transfection

10. To prepare the DNA for transfection, add an appropriate amount of DNA (25 ng of Renilla reporter plasmid, 25 ng of SSA reporter plasmid, and 100 ng of each ZFN plasmid) to 250 μL of OPTI-MEM in a 1.5 mL tube, and incubate at RT for 5 min.
11. Mix 2 μL of Lipofectamine2000 and 250 μL of OPTI-MEM in a separate 1.5-mL tube. Incubate at RT for 5 min (*see* **Note 8**).
12. Combine the two and mix gently. Incubate at RT for 20 min.
13. To transfect the cells, remove the medium from cells in the 24-well plate from the previous day. Add the DNA-Lipofectmine2000-OPTI-MEM solution to wells.
14. Incubate for 4–5 h in a 37°C incubator.
15. Add 1 mL of DMEM Complete Medium.
16. Incubate the cells in a 37°C incubator, then allow 24 h for ZFN expression and activity (*see* **Note 9**).

Day 2: Harvesting

17. To prepare the cell lysates, remove the media from wells.
18. Wash twice with 0.5 mL of PBS. Carefully aspirate the PBS.
19. Supplement the 1× Reporter Lysis Buffer with 1× Protease Inhibitor Cocktail (4 μL of the 25× Protease Inhibitor Cocktail stock for every 100 μL of 1× Reporter Lysis Buffer). Add 100 μL of supplemented Reporter Lysis Buffer to each well of the 24-well plate.
20. Allow to incubate on ice for 5 min.
21. Harvest the cell lysates by scraping the cells from the bottom of the well with pipette tip. Keep the 24-well plate on

ice to prevent any degradation of proteins. Wear appropriate personal protective equipment, such as a lab coat and eye goggles.

22. Transfer the lysates to 1.5 mL tubes on ice.
23. Centrifuge the tubes at 20,817 × *g* for 5 min at 4°C to pellet cellular debris. The supernatants can be used directly in the luciferase assays (*see* **Note 10**).
24. Thaw the Bright-Glo solution. This luciferase substrate is light sensitive so keep it wrapped in aluminum foil.
25. For each sample, pipette 20 μL of cell lysate into a well of an opaque white 96-well plate. The opaque walls ensure that the luminescence generated by one sample does not interfere with measurements of the neighboring samples.
26. Add 100 μL/well of Bright-Glo Luciferase Assay solution and pipette to mix. Cover the plate with aluminum foil.
28. Read the plate using a microplate luminometer. Integrate the signal of each sample for 1 s.
29. To read the Renilla luciferase signal, pipette another 20 μL of cell lysate into unused wells of the opaque white 96-well plate. Add 100 μL/well of Renilla Luciferase Assay solution, mix, and read using the microplate luminometer (*see* **Note 11**).
30. Calculate and plot the average and standard deviation of the firefly and Renilla luciferase signals for each set of duplicates or triplicates (*see* **Fig. 1.3**). The standard deviation of the replicates will be an indication of well-to-well variation, which should be small. The gain in firefly luciferase signal is an indicator of ZFN activity. The loss of Renilla luciferase signal is an indicator of ZFN toxicity, due to significant off-target cleavage activity (typically leading to apoptosis of the transfected cells). Note that a low firefly luciferase signal may be due to either low ZFN activity or high ZFN toxicity. A good example is the ZFN GZF3-wt (*see* **Fig. 1.3**). The presence of this monomer, either as a GZF3-wt homodimer or as a GZF1-wt/GZF3-wt heterodimer, leads to significant toxicity (reduced Renilla signal) and a corresponding apparent loss of ZFN activity (reduced firefly signal). In fact, the nuclease is active, but is causing the transfected cells to die. This conclusion is supported by the obligate heterodimer modifications, GZF1-RR/GZF3-DD. The GZF3 monomer is no longer active by itself, and the heterodimer pair now displays reduced toxicity and increased activity. It is also advisable to normalize the firefly signal to the Renilla to compensate for variable transfection efficiency and cell recovery.

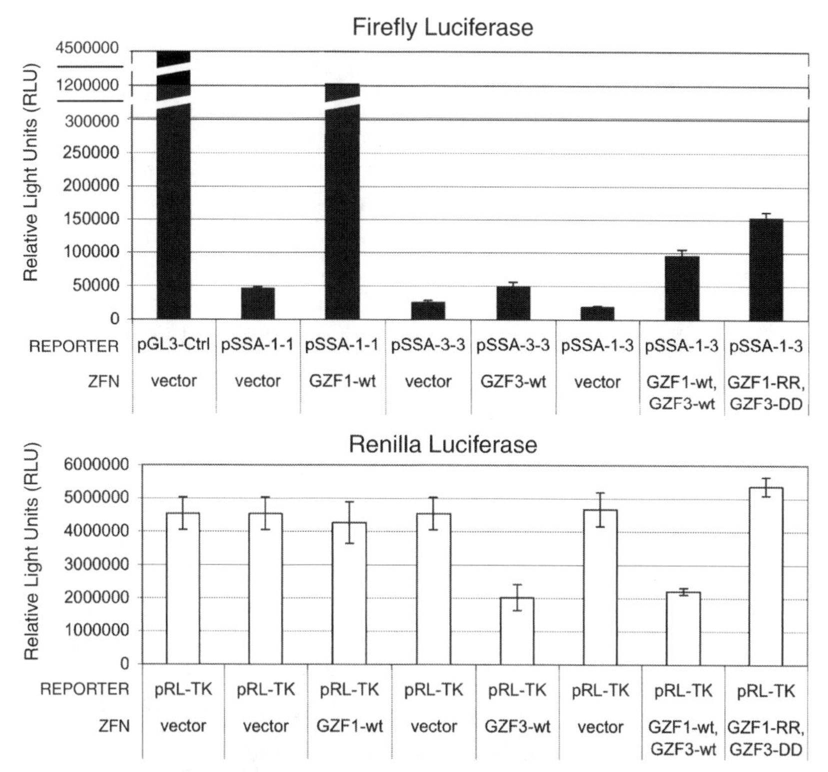

Fig. 1.3. Exemplary data from an SSA assay. Firefly (SSA recombination due to ZFN *activity*) and Renilla (constitutively high in the absence of ZFN *toxicity*) luciferase data are shown for some of the experiments described in **Note 1**, including the analysis of GZF1-wt and GZF3-wt homodimers, as well as GZF1-wt/GZF3-wt and GZF1-RR/GZF3-DD heterodimers. Luciferase activity was measured 24 h post-transfection. *Error bars* indicate the standard deviation of triplicate experiments.

4. Notes

1. As a guide for experimental design, ZFNs and reporters to be analyzed would include some of the following. (Note that each sample below will additionally include a Renilla luciferase expression plasmid to assess toxicity.)
   a. pPGK-GZF1-RR, pPGK-GZF3-DD, pSSA-1-3. This set can serve as a standard and should be included routinely in every assay.
   b. pPGK-empty vector, pSSA-1-3. In the absence of a ZFN-induced double-strand break, the reporter plasmid should not produce a luciferase signal. However, mechanical damage to the plasmid, such as occasionally occurs during the preparation of plasmid from bacteria, can cause strand breaks that enable reporter to recombine in the absence of ZFN cleavage. For this reason, it is important to monitor the background signal of the

reporter plasmid. Preparations that produce a significant signal should be remade. Conversely, precleavage at the unique EcoRI site is a good control for maximum reporter activity.

c. pPGK-empty vector, pGL3-Control. This sample produces a luciferase signal that would be equivalent to every molecule of SSA reporter plasmid recombining to yield an active luciferase gene immediately upon transfection. Since the efficiency and kinetics of recombination will always be less than this idyllic scenario, the pGL3-Control signal will always be higher than that of a ZFN/SSA reporter sample. The standard ZFN/SSA set described in (a) above typically produces a signal that is typically 10–30% of the pGL3-Control signal (somewhat less than 10% under the conditions used in **Fig. 1.3**).

d. A new ZFN pair and the cognate new heterodimer SSA reporter plasmid. The ZFN pair can have either wild-type cleavage domains or obligate heterodimer domains with DD and RR modifications (33). Obligate heterodimers generally reduce the cytotoxicity of the ZFN.

e. pPGK-empty vector and the new heterodimer SSA reporter plasmid.

f. Optionally, it may be useful to test the activity of each monomer of the ZFN pair, especially if the heterodimer activity is low. For testing homodimers, wild-type cleavage domains should be used, since the use of obligate heterodimer versions of the domains will cause 2/3 of the homodimer pairs to be inactive. Therefore, testing of the homodimers may require additional subcloning steps for the ZFNs.

g. Optionally, pPGK-empty vector and the new homodimer SSA reporter plasmid. Note that each ZFN pair consists of two monomers; therefore, testing of the homodimers requires the construction of two new homodimer SSA reporters.

h. For each sample a-g listed above, duplicate or triplicate assays should be performed on the same plate. 24-well plates are useful to accommodate the many samples required for an accurate assessment of ZFN activity.

2. Clearly there are many factors that should be considered in selecting a target site, most of which will be beyond the scope of this methods article. If the intended target is within a living cell, it is often advisable to seek regions with accessible chromatin. In principle, accessibility can be determined experimentally (e.g., in vivo DNaseI

footprinting) or extrapolated from existing genomic data (e.g., the UCSC Genome Browser). Targeting regions near actively transcribed genes is often useful. Depending on the application, chromatin-remodeling drugs can be used to increase chromatin accessibility temporarily (34). However, the most important strategy to ensure a successful outcome is to choose 5–10 different target sites.

The composition of the fingers in the ZFPs is also a critical parameter. Our understanding of the relationship between affinity and the number of fingers in a ZFP is still evolving. Generally, three fingers are the minimum required for biologically relevant binding. Recent studies suggest that fingers with greater relative affinity improve the functional capabilities of three-finger ZFPs (24). The conventional wisdom is that ZFP composed of GNN-binding fingers tend to perform better; however, it is important to remember that not all GNN-binding fingers have high relative affinity. Increasing the number of fingers may be a useful strategy to increase the overall affinity of the ZFP, especially (as is common) if lower affinity fingers are required to bind the target site. In principle, more fingers also provide greater specificity. Even a three-finger ZFP with perfect specificity for a 9-bp site will nominally have about 22,000 perfect binding sites in the human genome. A six-finger ZFP with perfect specificity for an 18-bp site would have 0–1 perfect sites. However, the caveat for multi-finger proteins is that subgroups of three or more fingers may have sufficient affinity by themselves to cause biologically relevant binding (35, 36). If this occurs, the ZFP will use different binding modes to recognize numerous and diverse binding sites. The subgroup phenomenon will be exacerbated by using an excess of fingers with high relative affinity, such as many GNN-binding fingers. The goal should therefore be to create ZFPs with "appropriate" affinity, using fewer but stronger binding fingers, or more but weaker binding fingers.

Some guidance is available to help predict performance. The relative affinities for some Barbas and ToolGen fingers have been measured (*see* **Tables 1.1** and **1.4**). ToolGen has identified a number of Barbas and ToolGen fingers that occurred frequently in successful ZFNs (*see* **Tables 1.1** and **1.3**). Also, since G and A bases often form stronger bonds to their contacting amino acids than C and T, a simple count of the purines (including any TSO interactions that are present) can serve as a rough guide to affinity. Two web-based tools are also available. The site http://www.zincfingertools.org/ is run by the Barbas lab and allows easy searching and coding of the Barbas set of modules (37). The site http://bindr.gdcb.iastate.edu/ZiFiT/ is run by the ZFC (38). It allows searching of Barbas, Sangamo, and ToolGen modules. The output of the search is not protein sequence but rather a set of catalog numbers to their library of pre-cloned fingers, which are available from the Internet distribution service http://addgene.org/. Both sites provide voluminous predictive data, which in most cases will have similar accuracy to the guidance outlined here. Unfortunately, due to the many parameters and significant weaknesses in our understanding, our ability to predict the behaviors of assembled ZFPs is limited. A robust assay, such as the SSA assay described in this chapter (*see* **Section 3.2**), may be the most useful approach to functionally determine the optimal length and composition of the ZFP.

Finally, ZFNs have special target site considerations. The two monomers of the ZFN pair must bind in a tail-to-tail format, such that their C-termini (containing the cleavage domains) face each other. The optimal spacing between the monomer binding sites is 5, 6, or 7 bp. Maximum ZFN activity for a 5, 6, or 7-bp spacer may require different peptide linkers between the last zinc finger and the *Fok*I cleavage domain. The relationship of spacers and linkers is best described in (39). The ZFN constructs described in this chapter are designed to provide maximal activity at target sites with a 6-bp spacer.

3. Unfortunately, BLAST does not work very well for ZFN target sites, because the 6-bp spacer between the two monomer binding sites can have any sequence without affecting cleavage efficiency. A custom search algorithm would be more appropriate. A C-based program for searching any downloaded genome is available from the author upon request.

4. Pay very close attention to the order of the fingers with respect to the DNA (*see* **Fig. 1.1**). The most 5′ 3-bp subsite of the DNA sequence to be recognized is bound by the most C-term finger of the ZFP. The most 3′ 3-bp subsite is bound by the most N-terminal finger (i.e., finger 1). Failure to correctly order the fingers in this "antiparallel" arrangement is the most common technical error when designing ZFPs.

5. For commercial synthesis, we use a Canadian company called BioBasic. At the time of this writing, the cost of synthesis is \$0.50/bp and deliver typically takes 3–6 weeks. Of course there are many other assembly methods. **Chapters 7** and **12** of this volume describe two oligonucleotide-based assembly methods, and several others have been

described (40–42). The ZFC has constructed a library of single fingers cloned into plasmids and has published detailed methods for their assembly (43). The entire library of 141 fingers is available at a modest cost from Addgene.

6. It is often difficult to clone small DNA fragments, such as a 24-bp ZFN target site. Therefore, we usually amplify the whole "left arm" of the split luciferase gene with the ZFN target site appended, then subclone the amplicon into the SSA reporter vector. The forward primer anneals upstream of the *Bgl*II site (RvP3, *see* **Fig. 1.2**). The reverse primer anneals around nucleotide 1,467 (in the numbering of the pGL-Control plasmid), and additionally contains the ZFN target sequence followed by an *Eco*RI site (SSA-1-3-r). Since this primer anneals to both parts of the duplicated region, this PCR must use the original pGL3-Control as a template (i.e., not a SSA plasmid).

7. As an alternative to ligating the "left arm" into an existing pSSA vector, a "right arm" can be prepared, and the left and right arms can be ligated into a digested pGL3-Control plasmid in a three-part ligation. The right arm is prepared by performing a PCR reaction similar to that described in Step 1, **Section 3.2.1**, using primers SSA-f (5′-GAGGAG GAATTC CGACATTTATAATGAAC GTGAATTGCTC-3′) and RvP4-r (5′-GACGATAGTCAT GCCCCGCGC-3′) (*see* **Fig. 1.2**). The right arm is digested with *Eco*RI and *Xba*I, and a pGL3-Control plasmid is digested with *Bgl*II and *Xba*I.

8. For HEK 293-T cells, we normally transfect with 1–2 μL of Lipofectamine2000 and approximately 250 ng of DNA when the cells are about 95% confluent. However, these parameters should be re-optimized for different cell types and conditions, as large variations in transfection efficiency can result.

9. Firefly luciferase signals are often higher after 48 h of incubation, probably due to an increased opportunity for ZFN-induced cleavage and an increased time to express the recombined luciferase.

10. Alternatively, the cell lysates can be transferred to a clean tube and stored at –80°C until further use. Thaw these lysates on ice prior to use.

11. Alternatively, the firefly and Renilla luciferase signals can be read sequentially from a single cell lysate sample using the Dual-Luciferase Reporter Assay System (Promega, Cat. E1960). However, this system requires a microplate luminometer with dual injection pumps.

Acknowledgments

This chapter is based upon work supported by grant CA103651 from the National Cancer Institute, NIH (DJS).

References

1. Emerson, R.O. and Thomas, J.H. (2009) Adaptive evolution in zinc finger transcription factors. *PLoS Genet.* **5**, e1000325.
2. Pavletich, N.P. and Pabo, C.O. (1991) Zinc finger-DNA recognition: crystal structure of a Zif268-DNA complex at 2.1 Å. *Science.* **252**, 809–817.
3. Segal, D.J. and Barbas, C.F., III (2001) Custom DNA-binding proteins come of age: polydactyl zinc-finger proteins. *Curr Opin Biotechnol.* **12**, 632–637.
4. Dreier, B., Beerli, R.R., Segal, D.J., Flippin, J.D., and Barbas, C.F., III (2001) Development of zinc finger domains for recognition of the 5′-ANN-3′ family of DNA sequences and their use in the construction of artificial transcription factors. *J Biol Chem.* **276**, 29466–29478.
5. Dreier, B., Fuller, R.P., Segal, D.J., Lund, C.V., Blancafort, P., Huber, A., Koksch, B., and Barbas, C.F., III (2005) Development of zinc finger domains for recognition of the 5′-CNN-3′ family DNA sequences and their use in the construction of artificial transcription factors. *J Biol Chem.* **280**, 35588–35597.
6. Liu, Q., Xia, Z., Zhong, X., and Case, C.C. (2002) Validated zinc finger protein designs for all 16 GNN DNA triplet targets. *J Biol Chem.* **277**, 3850–3856.
7. Segal, D.J., Dreier, B., Beerli, R.R., and Barbas, C.F., III (1999) Toward controlling gene expression at will: selection and design of zinc finger domains recognizing each of the 5′-GNN-3′ DNA target sequences. *Proc Natl Acad Sci USA.* **96**, 2758–2763.
8. Bae, K.H., Kwon, Y.D., Shin, H.C., Hwang, M.S., Ryu, E.H., Park, K.S., Yang, H.Y., Lee, D.K., Lee, Y., Park, J., Kwon, H.S., Kim, H.W., Yeh, B.I., Lee, H.W., Sohn, S.H., Yoon, J., Seol, W., and Kim, J.S. (2003) Human zinc fingers as building blocks in the construction of artificial transcription factors. *Nat Biotechnol.* **21**, 275–280.
9. Liu, Q., Segal, D.J., Ghiara, J.B., and Barbas, C.F., III (1997) Design of polydactyl zinc-finger proteins for unique addressing within complex genomes. *Proc Natl Acad Sci USA.* **94**, 5525–5530.
10. Ramirez, C.L., Foley, J.E., Wright, D.A., Muller-Lerch, F., Rahman, S.H., Cornu, T.I., Winfrey, R.J., Sander, J.D., Fu, F., Townsend, J.A., Cathomen, T., Voytas, D.F., and Joung, J.K. (2008) Unexpected failure rates for modular assembly of engineered zinc fingers. *Nat Methods.* **5**, 374–375.
11. Maeder, M.L., Thibodeau-Beganny, S., Osiak, A., Wright, D.A., Anthony, R.M., Eichtinger, M., Jiang, T., Foley, J.E., Winfrey, R.J., Townsend, J.A., Unger-Wallace, E., Sander, J.D., Muller-Lerch, F., Fu, F., Pearlberg, J., Gobel, C., Dassie, J.P., Pruett-Miller, S.M., Porteus, M.H., Sgroi, D.C., Iafrate, A.J., Dobbs, D., McCray, P.B., Jr., Cathomen, T., Voytas, D.F., and Joung, J.K. (2008) Rapid "open-source" engineering of customized zinc-finger nucleases for highly efficient gene modification. *Mol Cell.* **31**, 294–301.
12. Isalan, M., Choo, Y., and Klug, A. (1997) Synergy between adjacent zinc fingers in sequence-specific DNA recognition. *Proc Natl Acad Sci USA.* **94**, 5617–5621.
13. Isalan, M., Klug, A., and Choo, Y. (2001) A rapid, generally applicable method to engineer zinc fingers illustrated by targeting the HIV-1 promoter. *Nat Biotechnol.* **19**, 656–660.
14. Urnov, F.D., Miller, J.C., Lee, Y.L., Beausejour, C.M., Rock, J.M., Augustus, S., Jamieson, A.C., Porteus, M.H., Gregory, P.D., and Holmes, M.C. (2005) Highly efficient endogenous human gene correction using designed zinc-finger nucleases. *Nature.* **435**, 646–651.
15. Greisman, H.A. and Pabo, C.O. (1997) A general strategy for selecting high-affinity zinc finger proteins for diverse DNA target sites. *Science.* **275**, 657–661.
16. Joung, J.K., Ramm, E.I., and Pabo, C.O. (2000) A bacterial two-hybrid selection system for studying protein-DNA and protein-protein interactions. *Proc Natl Acad Sci USA.* **97**, 7382–7387.

17. Carroll, D. (2008) Progress and prospects: Zinc-finger nucleases as gene therapy agents. *Gene Ther.* **15**, 1463–1468.
18. Kolb, A.F., Coates, C.J., Kaminski, J.M., Summers, J.B., Miller, A.D., and Segal, D.J. (2005) Site-directed genome modification: nucleic acid and protein modules for targeted integration and gene correction. *Trends Biotechnol.* **23**, 399–406.
19. Gordley, R.M., Gersbach, C.A., and Barbas, C.F., III (2009) Synthesis of programmable integrases. *Proc Natl Acad Sci USA.* **106**, 5053–5058.
20. Camenisch, T.D., Brilliant, M.H., and Segal, D.J. (2008) Critical parameters for genome editing using zinc finger nucleases. *Mini Rev Med Chem.* **8**, 669–676.
21. Sera, T. (2009) Zinc-finger-based artificial transcription factors and their applications. *Adv Drug Deliv Rev.* **61**, 513–526.
22. Bibikova, M., Golic, M., Golic, K.G., and Carroll, D. (2002) Targeted chromosomal cleavage and mutagenesis in Drosophila using zinc-finger nucleases. *Genetics.* **161**, 1169–1175.
23. Price, S.A., Dent, C., Duran-Jimenez, B., Liang, Y., Zhang, L., Rebar, E.J., Case, C.C., Gregory, P.D., Martin, T.J., Spratt, S.K., and Tomlinson, D.R. (2006) Gene transfer of an engineered transcription factor promoting expression of VEGF-A protects against experimental diabetic neuropathy. *Diabetes.* **55**, 1847–1854.
24. Sander, J.D., Zaback, P., Joung, J.K., Voytas, D.F., and Dobbs, D. (2009) An affinity-based scoring scheme for predicting DNA-binding activities of modularly assembled zinc-finger proteins. *Nucleic Acids Res.* **37**, 506–515.
25. Cathomen, T. and Joung, J.K. (2008) Zinc-finger nucleases: the next generation emerges. *Mol Ther.* **16**, 1200–1207.
26. Santiago, Y., Chan, E., Liu, P.Q., Orlando, S., Zhang, L., Urnov, F.D., Holmes, M.C., Guschin, D., Waite, A., Miller, J.C., Rebar, E.J., Gregory, P.D., Klug, A., and Collingwood, T.N. (2008) Targeted gene knockout in mammalian cells by using engineered zinc-finger nucleases. *Proc Natl Acad Sci USA.* **105**, 5809–5814.
27. Minczuk, M., Papworth, M.A., Miller, J.C., Murphy, M.P., and Klug, A. (2008) Development of a single-chain, quasi-dimeric zinc-finger nuclease for the selective degradation of mutated human mitochondrial DNA. *Nucleic Acids Res.* **36**, 3926–3938.
28. Kim, H.J., Lee, H.J., Kim, H., Cho, S.W., and Kim, J.S. (2009) Targeted genome editing in human cells with zinc finger nucleases constructed via modular assembly. *Genome Res.* **19**, 1279–1288.
29. Segal, D.J., Beerli, R.R., Blancafort, P., Dreier, B., Effertz, K., Huber, A., Koksch, B., Lund, C.V., Magnenat, L., Valente, D., and Barbas, C.F., III (2003) Evaluation of a modular strategy for the construction of novel polydactyl zinc finger DNA-binding proteins. *Biochemistry.* **42**, 2137–2148.
30. Zykovich, A., Korf, I., and Segal, D.J. (2009) Bind-n-Seq: high-throughput analysis of in vitro protein-DNA interactions using massively parallel sequencing. *Nucleic Acids Res.* **22**, e151.
31. Bibikova, M., Carroll, D., Segal, D.J., Trautman, J.K., Smith, J., Kim, Y.G., and Chandrasegaran, S. (2001) Stimulation of homologous recombination through targeted cleavage by chimeric nucleases. *Mol Cell Biol.* **21**, 289–297.
32. Alwin, S., Gere, M.B., Guhl, E., Effertz, K., Barbas, C.F., III, Segal, D.J., Weitzman, M.D., and Cathomen, T. (2005) Custom zinc-finger nucleases for use in human cells. *Mol Ther.* **12**, 610–617.
33. Szczepek, M., Brondani, V., Buchel, J., Serrano, L., Segal, D.J., and Cathomen, T. (2007) Structure-based redesign of the dimerization interface reduces the toxicity of zinc-finger nucleases. *Nat Biotechnol.* **25**, 786–793.
34. Beltran, A.S., Sun, X., Lizardi, P.M., and Blancafort, P. (2008) Reprogramming epigenetic silencing: artificial transcription factors synergize with chromatin remodeling drugs to reactivate the tumor suppressor mammary serine protease inhibitor. *Mol Cancer Ther.* 7, 1080–1090.
35. Filippova, G.N., Fagerlie, S., Klenova, E.M., Myers, C., Dehner, Y., Goodwin, G., Neiman, P.E., Collins, S.J., and Lobanenkov, V.V. (1996) An exceptionally conserved transcriptional repressor, CTCF, employs different combinations of zinc fingers to bind diverged promoter sequences of avian and mammalian c-myc oncogenes. *Mol Cell Biol.* **16**, 2802–2813.
36. Imanishi, M., Nakamura, A., Morisaki, T., and Futaki, S. (2009) Positive and negative cooperativity of modularly assembled zinc fingers. *Biochem Biophys Res Commun.* **387**, 440–443.
37. Mandell, J.G. and Barbas, C.F., III (2006) Zinc finger tools: custom DNA-binding domains for transcription factors and nucleases. *Nucleic Acids Res.* **34**, W516–W523.
38. Sander, J.D., Zaback, P., Joung, J.K., Voytas, D.F., and Dobbs, D. (2007) Zinc fin-

ger targeter (ZiFiT): an engineered zinc finger/target site design tool. *Nucleic Acids Res.* **35**, W599–W605.

39. Handel, E.M., Alwin, S., and Cathomen, T. (2009) Expanding or restricting the target site repertoire of zinc-finger nucleases: the inter-domain linker as a major determinant of target site selectivity. *Mol Ther.* **17**, 104–111.
40. Cathomen, T., Segal, D.J., Brondani, V., and Muller-Lerch, F. (2008) Generation and functional analysis of zinc finger nucleases. *Methods Mol Biol.* **434**, 277–290.
41. Carroll, D., Morton, J.J., Beumer, K.J., and Segal, D.J. (2006) Design, construction and in vitro testing of zinc finger nucleases. *Nat Protoc.* **1**, 1329–1341.
42. Mani, M., Kandavelou, K., Dy, F.J., Durai, S., and Chandrasegaran, S. (2005) Design, engineering, and characterization of zinc finger nucleases. *Biochem Biophys Res Commun.* **335**, 447–457.
43. Wright, D.A., Thibodeau-Beganny, S., Sander, J.D., Winfrey, R.J., Hirsh, A.S., Eichtinger, M., Fu, F., Porteus, M.H., Dobbs, D., Voytas, D.F., and Joung, J.K. (2006) Standardized reagents and protocols for engineering zinc finger nucleases by modular assembly. *Nat Protoc.* **1**, 1637–1652.
44. Kim, J.S., Kwon, Y.D., Kim, H., Ryu, E.H., and Hwang, M.S. (2007) Zinc finger domains and methods of identifying same. US Patent US2007087371.

Chapter 2

Engineering Single Cys2His2 Zinc Finger Domains Using a Bacterial Cell-Based Two-Hybrid Selection System

Stacey Thibodeau-Beganny, Morgan L. Maeder, and J. Keith Joung

Abstract

Individual synthetic Cys2His2 zinc finger domains with novel DNA-binding specificities can be identified from large randomized libraries using selection methodologies such as phage display. We have previously demonstrated that a bacterial cell-based two-hybrid system is at least as effective as phage display for selecting zinc fingers with desired specificities from such libraries. In this chapter we provide updated, detailed protocols for performing zinc finger selections using the bacterial two-hybrid system.

Key words: Bacterial two-hybrid, zinc finger selection.

1. Introduction

Artificial Cys2His2 zinc finger domains (C2H2 ZFs) with engineered DNA-binding specificities have shown promise for applications in both biological research and gene therapy (1–7). Selection-based methods for altering the specificities of individual C2H2 ZFs typically involve randomization of amino acid residues in the DNA recognition helix to generate large libraries followed by use of a selection method to identify variants with desired DNA-binding specificities. Early studies utilized phage display for the selection method (8–10) but more recent studies have demonstrated that a bacterial cell-based two-hybrid (B2H) system works as well as phage display and may be, in certain cases, more effective (11, 12). In addition, the B2H system is somewhat

This is a revised version of a protocol that was originally published as Thibodeau-Beganny S, Joung JK. Engineering Cys2His2 zinc finger domains using a bacterial cell-based two-hybrid system, Methods Mol Biol. 2007, 408: 317–34.

J.P. Mackay, D.J. Segal (eds.), *Engineered Zinc Finger Proteins*, Methods in Molecular Biology 649, DOI 10.1007/978-1-60761-753-2_2, © Springer Science+Business Media, LLC 2010

easier to use than phage display because it directly selects for proteins in an in vivo, cellular context whereas phage display requires multiple rounds of in vitro selection.

In this chapter, we describe detailed methods for using the B2H system to identify individual C2H2 ZFs with desired DNA-binding specificities from randomized libraries >10^8 in size. Although we have outlined similar protocols in earlier publications (11, 12), the overall approach has evolved in our laboratory as we have gained experience with the method. The protocol described in detail here is the most up-to-date method currently utilized by our laboratory for selections of individual ZFs.

The B2H system, as used in this protocol, links the occurrence of a protein–DNA interaction to the activation of two reporter genes: the yeast *HIS3* and the bacterial *aadA* genes. These genes code for an enzyme that is essential for histidine synthesis and a gene that confers resistance to streptomycin, respectively. These genes are used to create a "selection strain" harboring a co-cistronic *HIS3/aadA* reporter on a single copy episome. The reporter also contains a target DNA site of interest positioned just upstream of the weak promoter directing *HIS3/aadA* expression. If a zinc finger domain capable of binding the target DNA site of interest (and fused to a fragment of the yeast Gal11P protein) is expressed in the selection strain, this leads to recruitment of RNA polymerase to the weak promoter and subsequent activated expression of *HIS3/aadA* transcription; this occurs because the Gal11P fragment interacts with a yeast Gal4 protein fragment that is fused to a subunit of the *E. coli* RNA

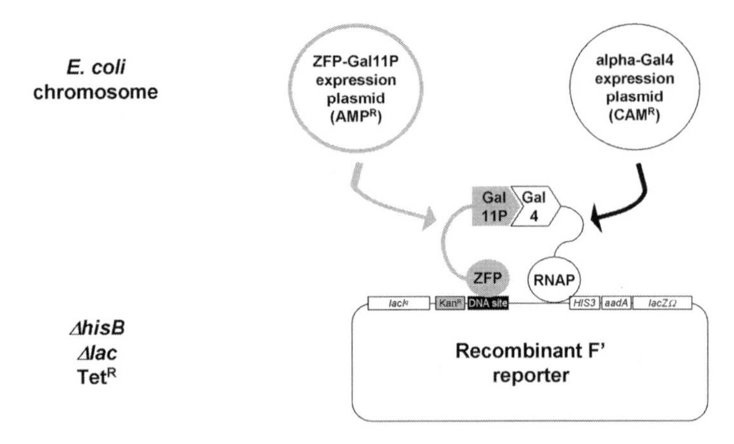

Fig. 2.1. Schematic overview of the bacterial two-hybrid selection system. A selection strain harboring the *HIS3/aadA* reporter and a kanamycin resistance gene on a single copy recombinant F' is transformed with plasmids encoding a hybrid alphaGal4 protein (harboring chloramphenicol resistance [CAM^R]) and a ZF domain-Gal11P hybrid protein (harboring carbenicillin resistance [AMP^R]). If the ZF domain binds to the target DNA site present on the F' reporter (*black box*), transcription of the *HIS3/aadA* reporter gene cistron is activated via recruitment of RNA polymerase to the reporter promoter mediated by interaction of the Gal11P and Gal4 domains. See text for additional details.

polymerase alpha-subunit (a hybrid protein we refer to as the alphaGal4 protein, *see* **Fig. 2.1**) (12). Cells harboring such a ZF domain can be identified on a medium lacking histidine and containing the antibiotic streptomycin. The stringency of the *HIS3 or aadA* selections can be increased by adding higher concentrations of 3-AT (3-aminotriazole, a competitive inhibitor of the HIS3 enzyme) or streptomycin, respectively, to the medium.

We first describe methods for engineering "selection strains" harboring target DNA sequences of interest (**Section 3.1**). We then describe methods for using these strains to identify ZF variants of interest from large randomized libraries (**Section 3.2**).

2. Materials

2.1. Molecular Biology Reagents

1. 10× annealing buffer (1 mL): 400 μL 1 M Tris base, pH 8.0, 200 μL 1 M $MgCl_2$, 100 μL 5 M NaCl, 300 μL H_2O.
2. Restriction enzymes (New England Biolabs): *Eco*RI, *Hind*III, *Sap*I; all 10× buffers are included with the enzymes.
3. Cloned *Pfu* polymerase and associated 10× reaction buffer (Stratagene).
4. Quick Ligation Kit (New England Biolabs).
5. Expand High-Fidelity thermostable polymerase and associated 10× Expand buffer (Roche).
6. QIAprep Spin Miniprep kit (Qiagen).
7. QIAquick PCR Purification kit (Qiagen).

2.2. Primer Sequences

1. OK.181 sequencing primer: 5′-CCAGAGCATGTATCATA TGGTCCAGAAACCC-3′
2. His3 2F primer: 5′-CGTATCACGAGGCCCTTTC-3′
3. His3 2R primer: 5′-GCAAATCCTGATCCAAACCT-3′
4. OK.61 sequencing primer: 5′-GGGTAGTACGATGA CGGAACCTGTC-3′

2.3. Bacterial Strains and Plasmids

1. CSH100 (genotype: F'*lac* *proA*$^+$*B*$^+$ *(lacI*^{q} *lacPL8)/ara*$^-$ *Δ(gpt-lac)5*)
2. KJ1C (genotype: F$^-$ *ΔhisB463 Δ(gpt-proAB-arg-lac)XIII zaj::Tn10*)
3. Reporter plasmid pKJ1712
4. Expression plasmid pAC-alphaGal4

These strains and plasmids are available by request from the Joung laboratory.

2.4. Bacterial Media

1. Ingredients for M9 minimal medium plates: Bacto Agar, M9 salts, $MgSO_4$, glucose, 100 mM $CaCl_2$.
2. Ingredients for NM medium and plates: 10× M9 salts (see below), 20% glucose, 20 mM adenine, amino acid mixture (see below), 1 M $MgSO_4$, 10 mg/mL thiamine, 10 mM $ZnSO_4$, and 100 mM $CaCl_2$, isopropyl-beta-D-thiogalactopyranoside (IPTG) and 3-aminotriazole (3-AT), Bacto Agar.
3. 200× amino acid mixture: each of the six solutions below should be made separately with ingredients mixed together in the order listed. The six solutions are then mixed together, filter sterilized, and stored at 4°C (*see* **Note 1**). This yields a 200× stock containing all amino acids except histidine, cysteine, and methionine.
   a. Solution I (100 mL): 0.99 g phenylalanine, 1.10 g lysine, and 2.50 g arginine in H_2O.
   b. Solution II (100 mL): 0.20 g glycine, 0.70 g valine, 0.84 g alanine, and 0.41 g tryptophan in H_2O.
   c. Solution III (100 mL): 0.71 g threonine, 8.40 g serine, 4.60 g proline, and 0.96 g asparagine in H_2O.
   d. Solution IV (100 mL): 9.1 mL 36.5% HCl to 80 mL H_2O, dissolve 1.04 g aspartate, and 14.60 g glutamine. Adjust final volume to 100 mL with H_2O.
   e. Solution V (100 mL): Dissolve 18.70 g potassium glutamate in 80 mL H_2O, dissolve 0.36 g tyrosine and 4 g NaOH pellets; adjust final volume to 100 mL with H_2O to 100 mL.
   f. Solution VI (100 mL): 0.79 g isoleucine and 0.79 g leucine in H_2O.
4. 10× M9 Salts (for 1 L): 67.8 g disodium phosphate (anhydrous), 30 g sodium chloride, 5 g monopotassium phosphate, 10 g ammonium chloride; fill to 1 L with ddH_2O and filter sterilize.
5. Antibiotics and other media additives:
   a. Carbenicillin (100 μg/mL in plates); stock is 50 mg/mL in ddH_2O (*see* **Note 14**).
   b. Chloramphenicol (30 μg/mL in plates); stock is 30 mg/mL in EtOH.
   c. Kanamycin (30 μg/mL in plates); stock is 30 mg/mL in ddH_2O.
   d. Tetracycline (12.5 μg/mL in plates); stock is 12.5 mg/mL in 80% EtOH.
   e. Sucrose (5%); stock is 50% in ddH_2O.
   f. IPTG (50 μM); stock is 50 mM in ddH_2O.

g. 3-AT (10–60 mM); stock is 1 M in ddH_2O (*see* **Note 2**).
h. Streptomycin (20–80 μg/mL); stock is 100 mg/mL in ddH_2O.

6. LB/TKS plates, which contain tetracycline, kanamycin, and sucrose.
7. LB/TK plates, which contain tetracycline and kanamycin.
8. LB/Kan plates, which contain kanamycin.
9. LB/CCK plates, which contain carbenicillin, chloramphenicol, and kanamycin.
10. LB/CK plates, which contain chloramphenicol and kanamycin.
11. LB/Tet plates, which contain tetracycline.

2.5. Equipment/ Consumables

1. 245 mm square plates (Corning)
2. 100 mm × 15 mm round Petri plates (Fisher)
3. 96-well flat-bottom, microtiter plates (Corning-Costar)
4. Glass beads, 3 mm (Fisher)
5. Sterile wooden sticks (Fisher)
6. 25 mm glass culture tubes (Fisher)
7. 18 mm glass culture tubes (Fisher)
8. 100 mm ×100 mm plates (VWRr)

3. Methods

3.1. Selection Strain Construction

Selection strains are constructed in two steps: initially, a target DNA site of interest is synthesized and then cloned into a multi-copy plasmid reporter vector designed in our laboratory (**Section 3.1.1**). In a second step, a portion of this reporter plasmid is recombined to a single copy F' episome in bacterial strain CSH100 and the resulting recombinant F' is then transferred by conjugation to KJ1C, an F-strain in which one can select for *HIS3* and *aadA* expression (12) (**Section 3.1.2**). We use recombination to place the reporter construct onto the F' because the F' episome is too large to be manipulated using standard subcloning techniques. Our method of selection strain construction is similar to one previously described by Whipple (13) but utilizes a counterselection step which simplifies identification of desired double recombinants (see below). Finally, the creation of the selection strain is verified both by prototrophy tests and by direct sequencing (**Section 3.1.3**).

3.1.1. Reporter Plasmid Construction

1. Digest ~1 μg of the reporter plasmid pKJ1712 with *Sap*I. pKJ1712 contains two closely positioned *Sap*I sites (*see* **Fig. 2.2**) and therefore digestion with this enzyme releases a small 45 bp fragment. Use the following quantities:

   1 μg Plasmid pKJ1712

   5 μL 10× NEB Buffer 4

   5 μL *Sap*I (2 U/μL)

   Fill to 50 μL with H_2O

   Incubate at 37°C for 2 h.

2. Isolate the 8,678 bp pKJ1712 vector backbone on either an agarose or polyacrylamide gel using standard methods. Resuspend the final purified digested vector in 20 μL of ddH_2O.

3. Treat the purified vector with *Pfu* polymerase to create extended overhangs. Cloned *Pfu* DNA Polymerase has a 3′ to 5′ exonuclease activity and by providing only one nucleotide (dCTP) to the reaction, the enzyme will

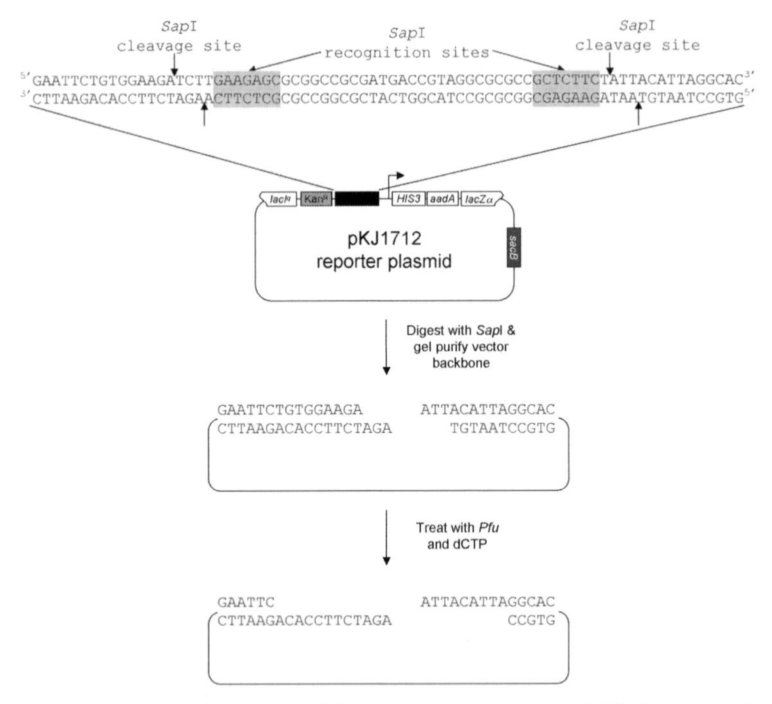

Fig. 2.2. Structure and sequence of the reporter plasmid vector pKJ1712. A schematic of the reporter plasmid pKJ1712 is shown. The region of the plasmid into which target DNA-binding sites are cloned is represented as a black rectangle with corresponding detailed sequence shown. Digestion of pKJ1712 with TypeIIS *Sap*I restriction enzyme releases a 45 bp fragment. Further treatment of the *Sap*I-digested vector backbone with *Pfu* DNA polymerase in the presence of dCTP nucleotide results in formation of the illustrated DNA overhangs.

degrade DNA until it reaches a position that can be filled in with dCTP. At this point, the forward synthesis and reverse exonuclease activities will reach equilibrium, thereby leaving extended overhangs as shown in **Fig. 2.2**. We utilize this approach to create overhangs for historical reasons related to subcloning in plasmid pKJ1712. Reaction conditions for *Pfu* treatment are as follows:

2 μL 10 mM dCTP

2 μL 10× Cloned *Pfu* Buffer

10 μL pKJ1712 *Sap*I-digested vector

1.2 μL Cloned *Pfu* (2.5 U/μL)

4.8 μL H_2O

Incubate at 72°C for 15 min, then at 4°C for >2 min.

4. As illustrated in **Fig. 2.3**, design a pair of oligonucleotides that when annealed together will form a double-stranded DNA fragment bearing the target DNA-binding site and extended overhangs compatible with the pKJ1712 vector prepared in Step 3 above.
5. Anneal the target DNA-binding site oligonucleotides together as follows:

   1 μL oligonucleotide 1 (10 pmol/μL)

   1 μL oligonucleotide 2 (10 pmol/μL)

   20 μL 10× annealing buffer

   178 μL H_2O

   Incubate at 95°C for 2 min, then shut off heat block and let tubes slow cool to 35°C. Put on ice. Alternatively, using a thermocycler, incubate at 95°C for 2 min, then cool at a rate of 1°C/min to 25°C. Change temperature to 4°C.
6. Ligate the purified pKJ1712 vector backbone to the annealed oligonucleotide binding site as follows:

   2 μL purified *Sap*I-digested, *Pfu*-treated pKJ1712 vector

   8 μL annealed binding site oligonucleotides

   10 μL 2× Quick Ligase Buffer

   1 μL T4 DNA Ligase (400 U/μL)

   Incubate at room temperature for 5 min then store on ice.
7. Transform ligations into XL-1 Blue *E. coli* competent cells using standard protocols and plate 1/3 of the transformations on LB/Kan plates. Incubate plates at 37°C overnight.
8. Isolate miniprep plasmid DNA from transformants. Typically, if there are at least 20-fold more colonies than

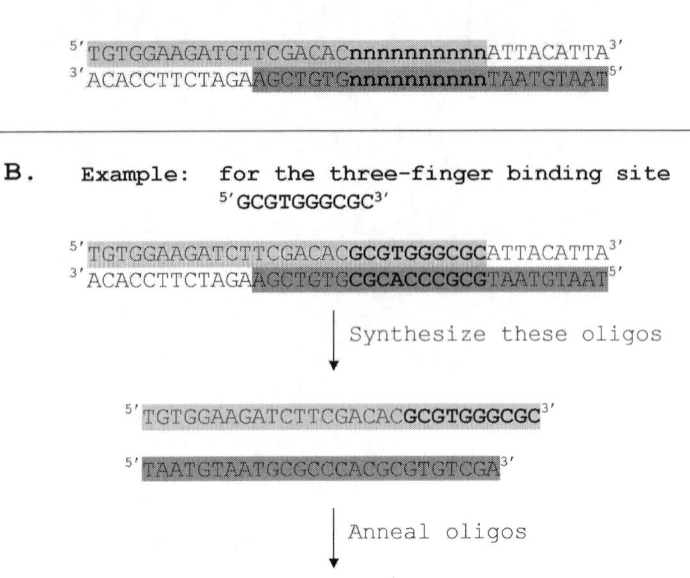

Fig. 2.3. Template for design of oligonucleotides harboring a target DNA site. (**A**) The target DNA site of interest is substituted for the "nnnnnnnnnn" sequence illustrated in the template (Note that this sequence can be longer than 10 bp if desired). Oligonucleotides corresponding to the highlighted *light and dark gray* sequences are then synthesized and annealed together to create an insert that can be ligated to reporter plasmid pKJ1712 (compare overhangs of the annealed oligonucleotide complex with the overhangs of *Sap*I-digested, *Pfu*-treated pKJ1712 shown at the *bottom* of **Fig. 2.2**). (**B**) Design of binding site oligonucleotides for the 10 bp Zif268 target DNA site. As an example, this figure illustrates the design of binding site oligonucleotides for the 10 bp Zif268 binding site 5′GCGTGGGCGC3′. In this example, the amino-terminal finger of Zif268 binds to the 3′ end of the target sequence while the carboxy-terminal finger binds to the 5′ end.

the control transformation plate (control ligation containing only the *Sap*I-digested pKJ1712 vector (i.e., without annealed oligonucleotides)), we prepare DNA from two different colonies. Since pKJ1712 is a low copy number plasmid, 10 mL overnight cultures in liquid LB/kanamycin (30 μg/mL) will yield an adequate DNA concentration when using the QIAgen's QIAprep Spin Miniprep Kit. We typically elute DNA is eluted in 50 μL 0.1× EB (a 1× EB stock is provided in the QIAgen kit).

9. To verify uptake of the annealed oligonucleotides in the pKJ1712 plasmid, we digest 5 μL of each candidate DNA with *Eco*RI and *Hind*III. As a control, we also digest plasmid pKJ1712 with *Eco*RI and *Hind*III. Recombinants that have taken up the annealed oligonucleotide insert should yield five bands of sizes 6,109, 1,006, 963, **431**, and

190 bp. By contrast, pKJ1712 should yield five bands of sizes 6,109, 1,006, 963, **456**, and 190 bp (*see* **Note 3**).

10. Plasmids that look correct by restriction digest should be sequenced to confirm the target DNA-binding site. We use primer OK.181, a primer which anneals to sense DNA strand just downstream of (and points back toward) the target binding site, to verify the sequence of the insert (*see* **Note 4**).

3.1.2. F' Episome Recombination and Transfer

3.1.2.1. Recombination of Reporter Plasmid Sequences onto the F' of Strain CSH100

As shown in **Fig. 2.4**, the reporter plasmid contains portions of the *lacI^q^* and *lacZ* genes that are also present in the F' found in strain CSH100. These regions of matching sequence can serve as point of recombination between the reporter plasmid and the F'. A double crossover event at both regions of sequence identity can lead to transfer of a portion of the reporter plasmid onto the F'.

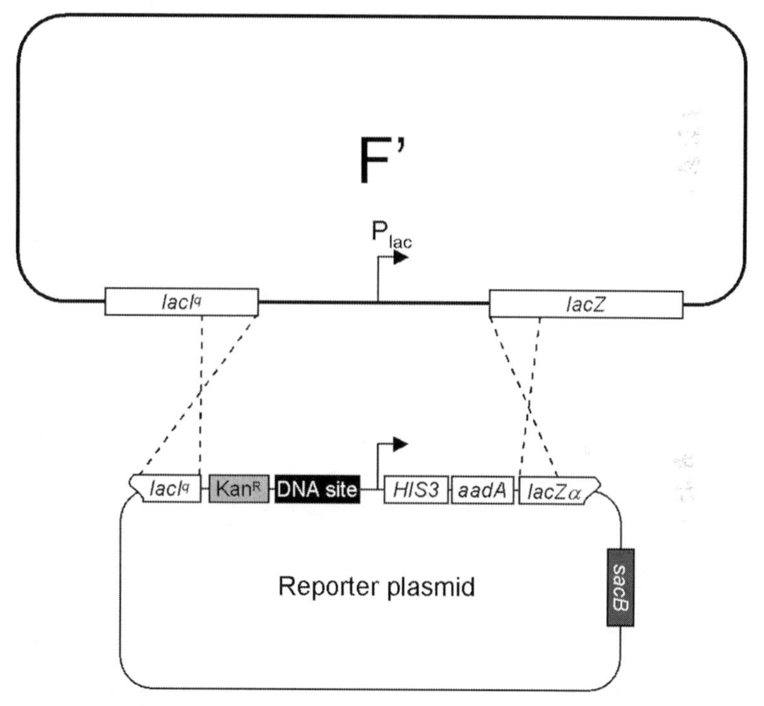

Fig. 2.4. Homologous recombination between reporter plasmids and the CSH100 strain F' mediated by regions of sequence identity. Schematic of a reporter plasmid showing its structure and its regions of sequence identity with the F' from *E. coli* strain CSH100. A double homologous recombination event between the two regions of sequence identity (from the *lacI^q^* and *lacZ* genes) leads to insertion of a fragment consisting of the kanamycin resistance (Kan^R) gene, the target DNA-binding site, and the *HIS3/aadA* reporter into the F'. Note that a double homologous recombination event does not transfer the counterselectable *sacB* gene from the reporter plasmid to the F'.

1. Streak out F-strain KJ1C on a LB/Tet plate and grow at 37°C overnight.
2. The next day, transform F+ strain CSH100 with the reporter plasmid and plate enough to obtain a confluent lawn of transformants on a LB/Kan plate. Incubate overnight at 37°C. Also inoculate an overnight culture of KJ1C in LB containing 12.5 μg/mL of tetracycline.

3.1.2.2. Transfer of F's from CSH100 to KJ1C by Bacterial Mating

The population of transformed CSH100 cells will contain a small number of cells in which a single recombination event has led to integration of the entire reporter plasmid to the F' (*see* **Fig. 2.5**). In an even smaller number of cells, a double recombination event will have exchanged only the target DNA-binding site and promoter present on the reporter plasmid with the promoter on the F' (*see* **Fig. 2.5**). As we describe in this step, all F's (recombinant and non-recombinant) in the CSH100 transformants are transferred to the F-strain KJ1C by mating. In a subsequent step, KJ1C cells harboring the desired double recombinant F' can be identified by plating on appropriate selective medium (*see* **Fig. 2.5** and below).

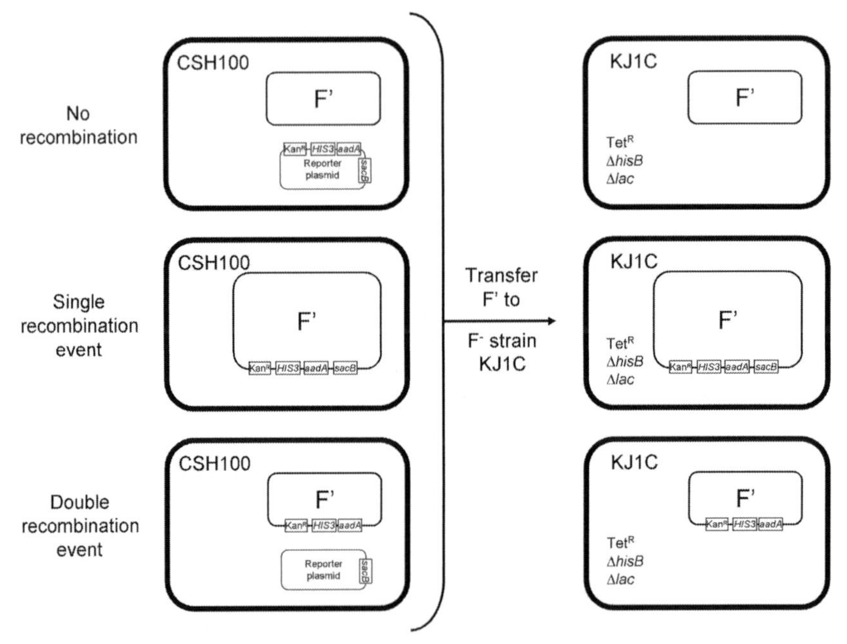

Fig. 2.5. Construction and identification of selection strains by recombination, mating, and selection. Reporter plasmids transformed into CSH100 either undergo no recombination (*top left*), single recombination (*middle left*), or double recombination (*bottom left*) with the F' present in this strain. Mating of these CSH100 cells with the tetracycline-resistant, F-strain KJ1C results in transfer of the three different F's into KJ1C cells (*top, middle, and bottom right*). When these cells are plated on medium containing tetracycline, kanamycin, and sucrose, only KJ1C cells that have received the double recombinant F' (the desired selection strain, *bottom right*) will survive. All of CSH100 cells (*left side*) will fail to grow due to sensitivity to tetracycline. KJ1C cells that do not receive an F' or that receive a non-recombinant F' (*top right*) will fail to grow due to sensitivity to kanamycin. KJ1C cells that receive a single recombinant F' (*middle right*) will fail to grow due to their sensitivity to sucrose owing to the presence of the *sacB* gene. KJ1C cells harboring a double recombinant F' will be resistant to kanamycin, tetracycline, and sucrose and thus will be the only surviving cells on the plate.

1. Scrape the confluent plate of CSH100 transformants with a sterile wooden stick and transfer cell paste to a sterile 25-mm glass tube containing 10 mL of LB (*see* **Note 5**).
2. Vortex to resuspend the CSH100 transformants (*see* **Note 6**) and transfer ~200 μL of this cell resuspension to a fresh 25-mm tube with 5 mL of LB (*see* **Note 7**).
3. Transfer ~200 μL of an overnight culture of KJ1C cells (inoculated the night before, grown in LB containing 12.5 μg/mL of tetracycline) to a sterile 25 mm tube containing 10 mL of LB.
4. Prepare a control 25 mm tube containing 10 mL of LB.
5. Incubate all tubes from Steps 2, 3, and 4 for 2 h at 37°C without agitation, thereby allowing cells to grow to log phase and for CSH100 cells to form F pili.
6. To perform matings, mix together the following combinations of the cultures from Steps 2, 3, and 4 in sterile 18 mm glass tubes. Use 1 mL of each liquid culture (i.e., each mating will consist of a total of 2 mL).
   a. CSH100 transformants + KJ1C (actual mating)
   b. CSH100 transformants + LB (negative control)
   c. KJ1C + LB (negative control)
   d. LB + LB (negative control)

   Allow matings to proceed at 37°C for 1 h *without agitation.*
7. Put tubes on a rotating wheel at 37°C for 90 min.
8. Plate 300 μL of the actual mating from Step 6a above (CSH100 transformants + KJ1C) on a LB/TK plate and on a LB/TKS plate. Spot 5 μL of each of the negative controls (Steps 6b, 6c, and 6d above) on a LB/TK plate and on a LB/TKS plate. Incubate all plates overnight at 37°C.

   As shown in **Fig. 2.5**, only KJ1C cells harboring the desired double recombinant F' should grow on LB/TKS plates. We plate all the matings on LB/TK plates as well to check that the counterselectable marker (*sacB*) is working to eliminate KJ1C cells harboring single recombinant and non-recombinant F's (*see* **Fig. 2.5**). (Note that expression of the *sacB* gene is lethal in *E. coli* cells when they are plated on medium containing sucrose.) We typically observe at least a 10-fold reduction in the number of colonies when comparing the number of colonies on LB/TK to the number of colonies on LB/TKS plates.
9. After KJ1C cells harboring the desired double recombinant F' have been identified on the basis of growth on LB/TKS plates (*see* **Note 8**), we purify clonal isolates by re-streaking candidates two times to LB/TKS plates. We typically choose two candidates (which we designate "A" and "B") for each

desired clone because a small percentage of cells that survive on LB/TKS plates will fail additional subsequent verification tests (as described below).

3.1.3. Selection Strain Verification

3.1.3.1. Confirmation of F' Transfer by Prototrophy Test (*See* **Note 9**)

1. Prepare M9 plates: For 500 mL, autoclave 439 mL H_2O with 7.5 g Bacto Agar. After agar has cooled to ~65°C, add 50 mL 10× M9 salts, 1 mL 1 M $MgSO_4$, 10 mL 20% glucose, and 0.5 mL 100 mM $CaCl_2$, and then pour plates.
2. Resuspend each colony (A & B candidates) in 100 μL of 1× M9 salts (we typically do this in wells of a sterile 96-well plate). Spot 5 μL of each cell suspension on an M9 minimal medium plate. Do not discard the resuspended colonies because they will be required for Step 1 of **Section 3.1.3.2** below.
3. After overnight growth at 37°C, verify growth of cells within the spots. Candidates that fail to grow should be discarded.

3.1.3.2. Sequencing of the Recombinant F' Reporter

For candidates that successfully demonstrate growth on M9 minimal medium plates, we amplify the portion of the F' harboring the target DNA site and perform DNA sequencing to verify the reporter.

1. Use 20 μL of the cell suspension from Step 1 of **Section 3.1.3.1** above, to inoculate a 4 mL LB/kanamycin (30 μg/mL) culture and grow overnight on a roller wheel or shaker at 37°C.
2. Transfer ~100 μL of the saturated overnight culture to a 1.5 mL microcentrifuge tube. Spin at maximum speed for 1 min in a microfuge. Remove as much of the media as possible using a pipet tip and then resuspend cells in 100 μL PCR-grade ddH_2O. Heat the cell resuspension at 95°C for 10 min and then spin at maximum speed for 1 min in a microfuge. Remove 50 μL of the supernatant to a fresh tube and use this crude preparation of F' DNA as template for a PCR reaction as follows:

PCR conditions	**Cycling**
5 μL crude F' DNA	95°C, 5 min
5 μL 10× Expand Buffer (Roche)	94°C, 30 s[a]
4 μL 10 mM dNTPs	55°C, 30 s[a]
1 μL His3 2R primer (10 pmol/μL)	72°C, 1 min[a]
1 μL His3 2F primer (10 pmol/μL)	72°C, 7 min
0.375 μL Expand enzyme (Roche)	33.625 μL H_2O

[a]Indicates repeated steps in 35 cycles.

3. Purify PCR product using a Qiagen QIAquick PCR purification kit and elute in 50 μL of 0.1× EB. Sequencing of the target DNA-binding site can be performed using primer OK.181 (we typically send 10 μL of purified PCR for sequencing).
4. We typically also make prepare glycerol stocks and competent cells of the selection strain using the overnight culture inoculated in Step 1 of **Section 3.1.3.2** above.

3.1.3.3. Transformation of Recombinant F' KJ1C Strain with Plasmid pAC-alphaGal4

The final step in preparation of the selection strain is to transform the KJ1C strain harboring a sequence verified F' reporter episome with the pAC-alphaGal4 plasmid (12) (*see* **Note 10**).

1. KJ1C cells bearing a sequence-verified recombinant F' reporter episome are transformed with plasmid pAC-alphaGal4 using standard chemical transformation protocols. Transformations are plated on LB/CK plates since the recombinant F' episome in the KJ1C cells confers resistance to kanamycin while the pAC-alphaGal4 plasmid encodes a chloramphenicol resistance gene.
2. Transformants are inoculated into overnight LB cultures containing chloramphenicol (30 μg/mL) and kanamycin (30 μg/mL) and grown overnight at 37°C.
3. Glycerol stocks of the final selection strains are prepared using the overnight culture.

3.2. Selection of C2H2 ZFs Using the Bacterial Two-Hybrid System

To perform selections, phagemid libraries encoding randomized zinc fingers are introduced into a selection strain and then plated on selective media. These zinc finger variants are expressed as fusions to a fragment of the yeast Gal11P protein that interacts with the fragment of the yeast Gal4 protein present in the alphaGal4 protein expressed in selection strains. Binding of a zinc finger domain to the target DNA sequence leads to recruitment of RNA polymerase complexes that have incorporated the alphaGal4 hybrid protein. This recruitment in turn leads to activation of reporter gene expression and survival on selective medium.

We typically perform selections in two stages. In Stage A selections, large numbers of transformants (typically $\sim 10^9$) are plated on a low stringency selection plate. Zinc finger-encoding phagemids are rescued from surviving cells (**Section 3.2.1**). In Stage B selections, this enriched population of phagemids are then re-introduced into the selection strain cells and plated on a series of higher stringency selection plates (**Section 3.2.2**). Phagemid DNA is then isolated from cells that grow on the highest stringency plate and sequenced to determine the identity of the fingers.

3.2.1. Stage A Selections

3.2.1.1. Streak Plates

Streak out the selection strain on an LB/CK plate and incubate overnight at 37°C.

3.2.1.2. Prepare NM Medium and Plates

1. Autoclave 836 mL H_2O, 15 g Bacto Agar and a magnetic stir bar together.
2. While agar is cooling, mix together the following components in a sterile flask: 100 mL of 10× M9 salts, 20 mL of 20% glucose, 10 mL of 20 mM adenine, 30 mL of 200× amino acid mixture (see below), 1 mL of 1 M $MgSO_4$, 1 mL of 10 mg/mL thiamine, 1 mL of 10 mM $ZnSO_4$, 1 mL of 100 mM $CaCl_2$, and antibiotics, IPTG and 3-AT as desired.
3. When agar has reached a temperature of ~65°C, add the above mixture to the agar, stir well and pour plates.
4. NM/CCKI plates are NM agar plates supplemented with carbenicillin (100 μg/mL), chloramphenicol (30 μg/mL), kanamycin (30 μg/mL), and IPTG (50 μM).

3.2.1.3. Culture Cells

Use a fresh, well-isolated colony to start an overnight culture of your strain in 20 mL of NM media supplemented with chloramphenicol (30 μg/mL), kanamycin (30 μg/mL), and 50 μM IPTG. Because selection strain cells are sensitive to detergents and rapid agitation, this culture should be grown in a sterile 125 mL glass flask that was rinsed thoroughly with MilliQ water before autoclaving and with shaking at 110 rpm at 37°C for ~16–24 h.

3.2.1.4. Introduction of Zinc Finger Phagemid Libraries into Selection Strain Cells

The construction of the randomized zinc finger phagemid libraries used in this step has been previously described (11, 12).

1. Thaw phagemid phage library on ice. Extreme care is required to prevent phage contamination in the laboratory as it may persist. We change our bench paper and gloves frequently and we use racks designated for phage work only. In addition, we use barrier pipet tips and expel used tips into a bucket of soapy water to inactivate the phage.
2. Transfer 5 mL of the saturated overnight culture of selection strain cells to a sterile 125 mL flask.
3. Add ~6 × 10^8 ATU (ampicillin transducing units; *see* **Note 11**) of phagemid library to the selection strain cells and gently swirl immediately. Allow the cell/phagemid mixture to sit at room temperature for 30 min without agitation.
4. Add 20 mL of pre-warmed NM medium supplemented with chloramphenicol (30 μg/mL), kanamycin (30 μg/mL), and 50 μM IPTG to the infected cells. Shake at 110 rpm, 37°C for 90 min.

5. Transfer culture to a sterile 50 mL conical tube and pellet cells by spinning at 2,500 rpm in a table top centrifuge for 25 min at room temperature.
6. Pour off medium and resuspend the cell pellet in 2.5 mL of pre-warmed NM medium supplemented with chloramphenicol (30 μg/mL), kanamycin (30 μg/mL), and 50 μM IPTG.
7. Serially dilute 100 μL aliquots of the cell resuspension in a sterile 96-well microtiter plate. Perform three independent dilution sets using NM medium supplemented with chloramphenicol (30 μg/mL), kanamycin (30 μg/mL), and 50 μM IPTG. Perform dilutions from 10^{-1} to 10^{-8} (Note that you will only plate dilutions 10^{-3}–10^{-8}).
8. Spot 5 μL of the 10^{-3}–10^{-8} dilutions each in triplicate on LB/CK, LB/CCK, and NM/CCK/50 μM IPTG plates. Incubate LB/CK and LB/CCK plates for 16–18 h at 37°C and NM/CCK/50 μM IPTG plates for 24 h at 37°C. Titers from these plates can be calculated the next day after colonies have formed (see below).
9. Pour some sterile glass beads (3 mm) onto a large (245 mm × 245 mm) NM/CCK/50 μM IPTG/10 mM 3-AT plate.
10. Measure the remaining suspension volume, record the value (this volume will be used when calculating titers), and then add the cells to the NM/CCK/50 μM IPTG/10 mM 3-AT plate.
11. Spread the cells with circular motions using the beads to distribute the cells evenly.
12. When plates have dried, turn plates over, and tap beads from the agar onto the inverted plate cover. Incubate at 37°C for 24 h and then at room temperature for 18 h.
13. The next day, count colonies on the serial dilution plates to calculate the total number of cells and total number of transformed cells plated on the selection plate. LB/CK plates are used to determine the total number of cells plated and the LB/CCK and NM/CCK/I50 plates are used to determine the total number of transformed cells plated. The following formula is used to perform these calculations:

$$(\#\ \text{colonies/volume of spots}\ (\mu\text{L})) \times \text{dilution factor} \times$$
$$\text{volume}\ (\mu\text{L})\ \text{plated on large dish}$$

Example: 65 total colonies in nine 5 μL spots of a 10^{-6} dilution together with 2,650 μL plated on the large plate would give the following equation:

$$(65 \text{ colonies}/45\mu\text{L}) \times 10^6 \times 2,650\mu\text{L} = 3.83 \times 10^9 \text{cells}$$

3.2.1.5. Recovery of Zinc Finger-Encoding Plasmids from Cells Surviving the Selection

In this step, ZF-encoding plasmids from surviving cells are rescued as phagemids by infecting these cells with M13K07 helper phage.

1. Turn the large selection plate over and tap the glass beads back onto the agar and add 15 mL of pre-warmed NM media to the plate. Agitate the plates in a circular motion using the glass beads to resuspend the cells in the media.
2. Transfer the suspension to a sterile 25 mm glass tube.
3. Remove 3 mL of cell resuspension to make glycerol stocks in case this recovery step needs to be redone.
4. Add enough of the cell suspension to 90 mL of 2×YT supplemented with carbenicillin (50 μg/mL) and kanamycin (30 μg/mL) to give it a pre-log appearance (i.e., an OD_{600} of ~0.1). Shake this culture at 120 rpm, 37°C for 1 h.
5. Infect the log phase culture with 10^{12} kanamycin-transducing units (KTU) of M13K07 helper phage. Allow the phage to adsorb to the cells, without shaking, at room temperature for 30 min.
6. Add kanamycin to a final concentration of 100 μg/mL (including the original 30 μg/mL present in the culture). Shake the culture at 125 rpm, 37°C for 6 h. During this incubation, ZF-encoding phagemids from the cells will be packaged as infectious phage particles harboring single-stranded DNA molecules and extruded into the culture medium.
7. Harvest the ZF-encoding phagemid phage by filtering the culture though a 0.2 μm PES filter membrane (no need to centrifuge the cells away first). This enriched phagemid phage library can be stored at 4°C for several weeks (for long-term storage, we freeze the phage at –80°C).

3.2.2. Stage B Selections

3.2.2.1. Introduction of Enriched Library of Zinc Finger-Encoding Phagemid Phage into the Selection Strain

1. Start a 20 mL overnight culture of the selection strain in NM medium supplemented with chloramphenicol (30 μg/mL), kanamycin (30 μg/mL), and 50 μM IPTG in a sterile 125 mL flask. Shake for 16–24 h at 110 rpm, 37°C.
2. In a 96-well plate, aliquot 50 μL of the selection strain overnight culture into five wells.
3. In another column of a 96-well plate, add 100 μL of the enriched phagemid phage library to one well. Perform serial 10-fold dilutions of the enriched library by removing 10 μL of phage and adding it to a well containing 90 μL of NM

medium supplemented with chloramphenicol (30 μg/mL), kanamycin (30 μg/mL), and IPTG (50 μM). Repeat to create 10^{-1}, 10^{-2}, 10^{-3}, and 10^{-4} dilutions.

4. Infect each of the wells containing 50 μL of selection strain overnight culture with 10 μL of each of the following: undiluted enriched phagemid library and 10^{-1}, 10^{-2}, 10^{-3}, and 10^{-4} dilutions of the phagemid library. Allow phage to adsorb by leaving them (without shaking) at room temperature for 30 min.
5. Add 190 μL of pre-warmed NM medium containing chloramphenicol (30 μg/mL), kanamycin (30 μg/mL), and IPTG (50 μM) to each well. Incubate for 2 h at 37°C (no shaking).
6. Spot 5 μL aliquots of the phagemid-infected selection strain cells on the following plates:
   a. NM/CCKI plates (six 5 μL spots on standard Petri dishes)
   b. NM/CCKI/20 mM 3-AT/20 μg/mL streptomycin (ten 5-μL spots on small square 100 mm × 100 mm plates)
   c. NM/CCKI/25 mM 3-AT/40 μg/mL streptomycin (ten 5-μL spots on small square 100 mm × 100 mm plates)
   d. NM/CCKI/40 mM 3-AT/60 μg/mL streptomycin (ten 5-μL spots on small square 100 mm ×100 mm plates)
7. Incubate plates 37°C for 48 h and inspect for colonies. Colonies may not form on NM/CCKI/40 mM 3-AT/60 μg/mL streptomycin plates until ~72–96 h of incubation at 37°C.

3.2.2.2. Isolation and Sequencing of Plasmid DNA from Selected Colonies

1. Pick 8–12 well isolated colonies from the highest stringency selection plate on which colonies appear and inoculate them into 4 mL of LB supplemented with carbenicillin (50 μg/mL). Incubate overnight at 37°C with agitation.
2. Prepare miniprep plasmid DNA from the saturated 4 mL overnight cultures using QIAgen's QIAprep Spin Miniprep Kit and their protocol but with the following differences:
   a. Perform triple washes with both PB and PE buffers (*see* **Note 12**).
   b. Elute the DNA with 60 μL of pre-warmed (60°C) 0.1× EB (*see* **Note 13**).
3. Send the plasmids for sequencing with sequencing primer OK.61, a sense strand primer which anneals just upstream of the region encoding the zinc finger domains.

4. Notes

1. It is important to dissolve the amino acids in each of the six solutions in precisely the order listed as this avoids potential solubility issues. We typically keep the amino acid mixture for no more than 2–3 months.
2. 3-AT should be prepared using gloves. In addition, we have found that the solubility and purity of 3-AT vary from lot to lot. Some preparations have the appearance of a white powder, whereas others look like brown flakes. For certain lots, we have found that heating the solution to 50°C can aid with solubility.
3. We run these digests on 5% polyacrylamide gels made with 0.5× TBE buffer to visualize the relatively small change in fragment size in clones that have taken up the annealed oligonucleotide insert.
4. We sequence verify the entire sequence between the unique *Eco*RI site (positioned just upstream of the ZF binding site) and the unique *Sal*I site (positioned at the start site of transcription). Verifying this entire span of sequence ensures that both the ZF binding site and the promoter do not have undesired mutations.
5. We have found that using a resuspension of multiple transformed CSH100 colonies rather than an overnight culture grown from a single transformed colony helps ensure that a relatively consistent percent of transformed CSH100 cells contain the desired double recombinant F'.
6. Set vortex to half-maximum speed to ensure that resuspension does not spill over the top of the glass tube.
7. The initial density of this subculture should correspond to OD_{600} of ~0.1 (i.e., pre-logarithmic phase). Depending upon the density of the resuspension culture, we add more or less culture as needed to achieve this target OD_{600}.
8. We occasionally see some small colonies on the LB/TKS plates. We avoid picking these colonies because we have found that these colonies do not yield the desired recombinants.
9. KJ1C cells will not grow unless histidine and proline are provided in their medium due to deletion of the *proAB* gene cluster and a deletion within the *hisB* gene, respectively. A double recombinant F' transferred from CSH100 cells harbors an intact *proAB* gene cluster and expresses a basal level of the yeast *HIS3* gene which is sufficient to complement the *hisB* deletion of strain KJ1C. Thus, KJ1C

cells that receive a double recombinant F' should be able to grow on M9 minimal medium lacking proline and histidine.

10. pAC-alphaGal4 encodes a fusion protein consisting of the N-terminal domain and inter-domain linker of the *E. coli* RNA polymerase alpha-subunit fused to amino acid residues 58–97 of the yeast Gal4 protein. Expression of the alphaGal4 hybrid protein from pAC-alphaGal4 is directed by a strong, IPTG-inducible semi-synthetic *lpp/lacUV5* promoter. The pAC-alphaGal4 plasmid possesses a p15A origin of replication and confers resistance to chloramphenicol.
11. We typically aim for a threefold oversampling of the size of the randomized library being interrogated and for a fivefold ratio of total cells to transformed cells. For example, for a randomized library with a complexity of $\sim 2 \times 10^8$, we would aim to plate a total of $\sim 6 \times 10^8$ transformed cells and of $>3 \times 10^9$ total cells on the selection plate.
12. We have found that these triple washes are critical for obtaining good quality sequencing reads. We believe these washes help reduce contaminating endonuclease activity from the *endA*$^+$ selection strains.
13. 0.1× EB is Buffer EB from the QIAgen miniprep kit diluted 10-fold with ddH_2O.
14. We use carbenicillin at a final concentration of 50 μg/mL in liquid media.

Acknowledgments

This work was supported by grants from the National Institutes of Health (K08 DK002883 and R01 GM069906) and start-up funds from the Massachusetts General Hospital Department of Pathology.

References

1. Jamieson, A.C., Miller, J.C., and Pabo, C.O. (2003) Drug discovery with engineered zinc-finger proteins. *Nat Rev Drug Discov.* 2, 361–368.
2. Blancafort, P., Segal, D.J., and Barbas, C.F., III (2004) Designing transcription factor architectures for drug discovery. *Mol Pharmacol.* 66, 1361–1371.
3. Durai, S., Mani, M., Kandavelou, K., Wu, J., Porteus, M.H., and Chandrasegaran, S. (2005) Zinc finger nucleases: custom-designed molecular scissors for genome engineering of plant and mammalian cells. *Nucleic Acids Res.* 33, 5978–5990.
4. Carroll, D. (2008) Progress and prospects: zinc-finger nucleases as gene therapy agents. *Gene Ther.* 15, 1463–1468.
5. Porteus, M.H. and Carroll, D. (2005) Gene targeting using zinc finger nucleases. *Nat Biotechnol.* 23, 967–973.

6. Cathomen, T. and Joung, J.K. (2008) Zinc-finger nucleases: the next generation emerges. *Mol Ther.* 16, 1200–1207.
7. Camenisch, T.D., Brilliant, M.H., and Segal, D.J. (2008) Critical parameters for genome editing using zinc finger nucleases. *Mini Rev Med Chem.* 8, 669–676.
8. Beerli, R.R. and Barbas, C.F., III (2002) Engineering polydactyl zinc-finger transcription factors. *Nat Biotechnol.* 20, 135–141.
9. Choo, Y. and Klug, A. (1995) Designing DNA-binding proteins on the surface of filamentous phage. *Curr Opin Biotechnol.* 6, 431–436.
10. Pabo, C.O., Peisach, E., and Grant, R.A. (2001) Design and selection of novel Cys2His2 zinc finger proteins. *Annu Rev Biochem.* 70, 313–340.
11. Hurt, J.A., Thibodeau, S.A., Hirsh, A.S., Pabo, C.O., and Joung, J.K. (2003) Highly specific zinc finger proteins obtained by directed domain shuffling and cell-based selection. *Proc Natl Acad Sci USA.* 100, 12271–12276.
12. Joung, J.K., Ramm, E.I., and Pabo, C.O. (2000) A bacterial two-hybrid selection system for studying protein-DNA and protein-protein interactions. *Proc Natl Acad Sci USA.* 97, 7382–7387.
13. Whipple, F.W. (1998) Genetic analysis of prokaryotic and eukaryotic DNA-binding proteins in *Escherichia coli. Nucleic Acids Res.* 26, 3700–3706.

Chapter 3

Bipartite Selection of Zinc Fingers by Phage Display for Any 9-bp DNA Target Site

Jia-Ching Shieh

Abstract

Phage display has been used to engineer DNA-binding proteins with new sequence specificities, which has allowed applications in the blockage or enhancement of gene expression as well as targeting specific sites on DNA for methylation, recombination, and cleavage. To effectively and quickly conduct selections that consider the synergistic mode of DNA binding by zinc fingers, Isalan and Choo in Aaron Klug's lab devised a bipartite phage display approach that enables selection and recombination of variants of zinc finger DNA-binding domains from a pair of premade complementary phage libraries for any given 9-bp DNA sequence. The bipartite phage display has the advantage of rapid, high-throughput selection of sequence-specific zinc finger DNA-binding domains for use in diverse applications of expression control and gene targeting.

Key words: Phage display, zinc finger domain, bipartite selection, phage ELISA.

1. Introduction

Phage display has been used to investigate protein–DNA interaction because of its ability to screen large numbers of protein variants simultaneously (1–3). In particular, we and others have used phage display to study DNA binding by Zif268 Cys_2His_2 zinc finger domains (ZFDs) (4–7), all of which have an identical structural framework but achieve chemical distinctiveness through variations in key residues (8). Such modular design offers a large number of combinatorial possibilities for DNA recognition. All the approaches involving phage display have been based on the display of ZFDs with randomized DNA-contacting residues from

J.P. Mackay, D.J. Segal (eds.), *Engineered Zinc Finger Proteins*, Methods in Molecular Biology 649,
DOI 10.1007/978-1-60761-753-2_3, © Springer Science+Business Media, LLC 2010

the surface of filamentous bacteriophage (4, 9). This approach has allowed for the selection of DNA-binding domains that can sequence-specifically interact with target genes and alter relevant cellular functions when coupled with or without effector domains (10–17).

In the phage display method, the fundamental principles for selection of DNA sequence-specific clones are to clone DNA sequences encoding peptide or protein domains into vectors derived from filamentous phage. The proteins are displayed as fusions to a coat protein encoded by *gIII* gene (18, 19). Proteins of interest can be screened from a library of variants by affinity refinement using a DNA ligand bound to a solid support. Consequently, the tightly bound clones are retained after washing and can be amplified rapidly through a bacterial host. Sequential rounds of selection and amplification enrich those clones with the highest affinity for the target DNA (20).

To date, three selection approaches using phage display, namely modular, sequential, and bipartite ones, have been developed (21) to generate proteins that bind DNA specifically and with high affinity, as compared with the prototypic zinc finger protein Zif268 and its cognate target sequence (15, 22, 23). The bipartite selection approach, developed by Isalan and colleagues (20), makes use of a pair of pre-constructed libraries, in which one-and-a-half fingers of the three-finger Zif268 are randomized. In contrast to the other two strategies, the bipartite approach requires far less time to create fingers with new specificities and exploits the possibility for interdomain cooperativity that allows targeting of any given DNA sequence. The bipartite selection approach has been used successfully in several studies to create sequence-specific DNA-binding proteins directed against targets in yeast, virus, and mitochondria (13–15, 17, 20, 24, 25).

This chapter describes the setup, selection, and testing of new zinc finger proteins (ZFPs) by the bipartite selection approach. The steps involved in cloning phage display libraries of DNA-binding proteins are described, including the preparation of the phage vector for cloning (**Section 3.1**), the preparation of cassettes encoding the zinc finger variants to be expressed on the phage, and (**Section 3.2**) construction of libraries by ligation of the vector and cassette and their subsequent transformation into a bacterial host (**Section 3.3**). Methods are also described to perform bipartite selection of ZFPs (**Section 3.4**), determine the binding affinity and specificity of selected clones by phage enzyme-linked immunosorbent assay (phage ELISA) (**Section 3.5**), and recombine the bipartite-selected phage clones (**Section 3.6**).

2. Materials

2.1. Preparation of Phage Vector

1. *Escherichia coli* strain capable of hosting a F' plasmid that generates a pilus (e.g., TG1 F'[*traD36 proAB lacI^qZ* ΔM15] *supE thi* 1-Δ*(lac-proAB)* Δ(M*crBhsd-SM)5*($r_k^-m_k^-$), Stratagene).
2. Phage vector for phage display (e.g., Fd-TET-SN (1)) (*see* **Fig. 3.1** and **Note 1**). This and noncommercial reagents are available from the author upon request, after completion of a Material Transfer Agreement from the Medical Research Council, UK.

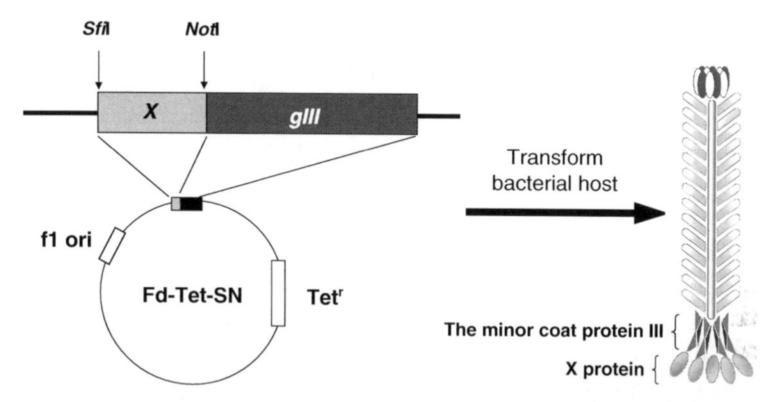

Fig. 3.1. A diagram of phage vector Fd-Tet-SN and display of the protein of interest. The *SfiI* and *NotI* cloning sites enable N-terminal fusion of a protein encoded by DNA cassette "*X*", in frame with the phage minor coat protein encoded by *gIII*. The recombinant phage plasmids can be introduced into *E. coli* to produce tetracycline-resistant transformants (due to the presence of Tet^R selection marker). The presence of an f1 ori enables single-stranded phage genomes to be generated in the *E. coli* and translocated to the cell membrane for assembly into phage particles. The N-terminal fusion of protein X is displayed on the tip of minor coat protein III.

3. Plasmid purification system (e.g., Maxiprep kit, Qiagen).
4. TE buffer: 10 mM Tris–HCl, pH 7.4, and 1 mM EDTA.
5. TYE agar: 10 g/L Bactoyeast extract, 16 g/L Bactotryptone, 5 g/L NaCl, and 15 g/L agar.
6. 2× TY medium: 10 g/L Bactoyeast extract, 16 g/L Bactotryptone, 5 g/L NaCl.
7. 10 mg/mL tetracycline hydrochloride in 50% ethanol (v/v).
8. Cesium chloride (CsCl) (Sigma).

9. 5 mL ultracentrifuge tubes (Beckman).
10. 4 L sterile conical flasks.
11. Ultracentrifuge (e.g., Beckman, with a Vti 65.2 rotor).
12. Milli-Q water-saturated 2-butanol.
13. 20,000 U/mL *Sfi*I with NEBuffer 2 and 10 mg/mL bovine serum albumin (BSA) (New England Biolabs).
14. 10,000 U/mL *Not*I with NEBuffer3 (New England Biolabs).
15. Buffer-saturated phenol (e.g., Sigma). *Note*, phenol is corrosive; handle with gloves and use appropriate protection.
16. Chloroform/isoamyl alcohol (IAA) 24:1 (e.g., Sigma).
17. 50× TAE buffer: 242 g Tris base, 57.1 mL glacial acetic acid, 37.2 g $Na_2EDTA.2H_2O$, and H_2O to 1 L.
18. 10 mg/mL ethidium bromide.
19. Agarase I (Sigma).
20. Mineral oil.
21. Low-melting point agarose.

2.2. Construction of Two Pools of DNA Cassette Coding for Two Collections of Zinc Finger Variants

1. Designed and guide DNA oligonucleotides (*see* **Fig. 3.2**, **Table 3.1**, and **Note 2**).
2. 1× TBE buffer: 100 mM Tris–HCl, pH 8.0, 100 mM boric acid, 10 mM EDTA.
3. Denaturing polyacrylamide-urea gel: Gel is made with 6–15% acrylamide from 40% stock solution (bis-acrylamide ratio, 19:1) in 1× TBE with 48% (w/v) urea.
4. Oligonucleotide elution buffer: 10 mM $MgCl_2$, 0.3 M CH_3COOK, pH 5.5.
5. 10 U/μL T4 polynucleotide kinase (New England Biolabs).
6. 400 U/μL T4 DNA ligase (New England Biolabs).
7. *Taq* DNA polymerase (New England Biolabs).
8. 10× *Taq* DNA polymerase buffer with 15 mM $MgCl_2$ (New England Biolabs).
9. Mix equal molar deoxyribonucleotide triphosphates (dNTPs), 25 mM of each G, A, T, and C.
10. 10 mM ATP.
11. A pair of PCR primers that anneal at the ends of the DNA cassette with either *Sfi*I or *Not*I restriction enzyme site

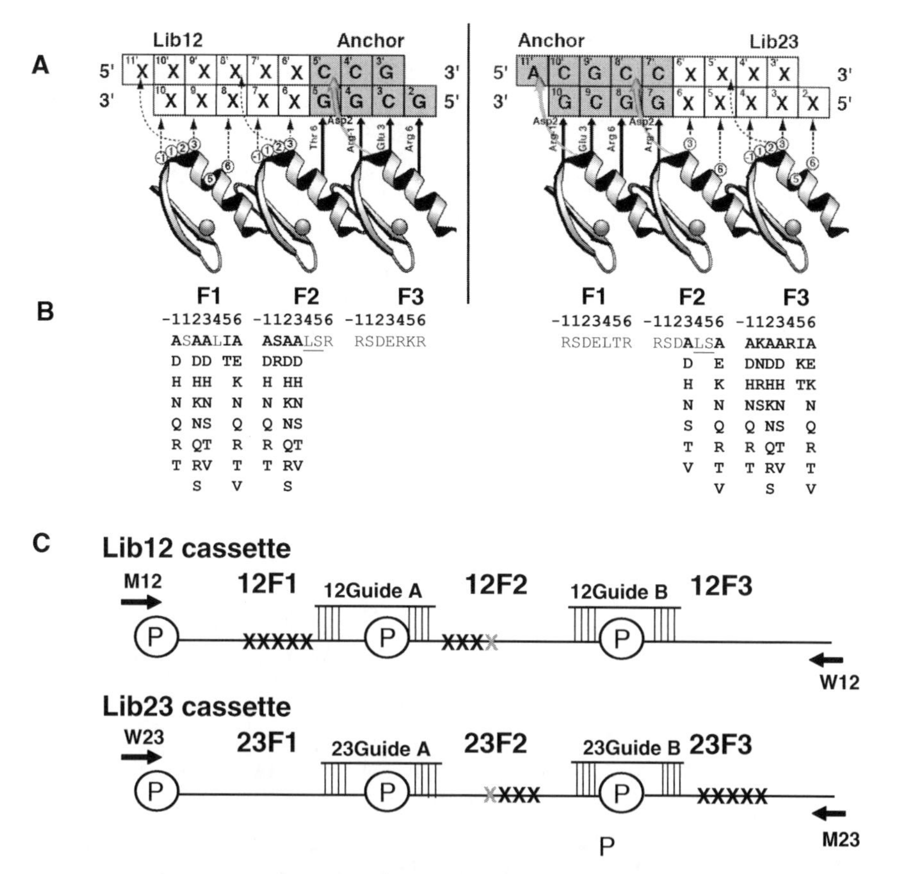

Fig. 3.2. Construction of two complementary libraries for bipartite selection approach (**A**) Mode of contact between the DNA and the ZFD based on the Zif268 framework that makes up libraries Lib12 and Lib23. The amino acids are shown as *circles*, numbered relative to their helical positions. (**B**) Potential composition of randomized amino acid residues (*in bold*) in Lib12 and Lib23. Recombination can be accomplished by the digestion and relegation of a *Dde*I restriction enzyme site (CTGAGC) that encodes Leu and Ser in positions 4 and 5 of F2 (LS) in both Lib12 and Lib23. (**C**) Generation of a DNA cassette coding for the ZFD library to be displayed on the phage. The pool of oligonucleotides is constructed by end-to-end ligation of synthetic "template" oligonucleotides (F1, F2, F3) containing randomized nucleotides (marked "×" *in black*), directed by "guide" oligonucleotides (**a**, **b**) that link the junctions by sequence-specific annealing. At the end of ligation, the full-length DNA cassette is amplified by PCR with a pair of primers that incorporate *Sfi*I (M12 and W23) and *Not*I (W12 and M23) sites for cloning into the Fd-Tet-SN phage vector. The position of break point for recombination is indicated as *light gray* "×.".

incorporated for cloning into phage vector Fd-TET-SN, each at 10 pmol/μL.

12. Steps 13–20 of **Section 2.1**.

2.3. Construction of a Pair of Phage Libraries with DNA Cassette Coding for Zinc Finger Variants

1. Electrocompetent *E. Coli* (e.g., strain TG1).
2. Electroporator and cuvettes (2 mm width) (e.g., Bio-Rad).
3. SOC medium: 0.5% (w/v) yeast extract, 2% (w/v) tryptone, 10 mM NaCl, 2.5 mM KCl, 10 mM $MgCl_2$, 10 mM

Table 3.1
Oligonucleotides used in the library construction

Oligo	Sequence
M12	GCGACGGC GGCCCAGCCGG [a]CCATGGCGGAA GAGA
12F1[b]	CCATGGCGGAAGAGAGGCCCTACGCATGCCCTGT CGAGTCCTGCGATCGCCGCTTTTCT$x^{1}x^{2}x^{3}$TCGx^{4} $x^{5}x^{6}x^{7}x^{8}x^{9}CTTx^{10}x^{11}x^{12}x^{13}x^{14}x^{15}$CATATCCGCAT CCACACCGG
12Guide A	CACTGGAAGGGCTTCTGACCGGTGTGGATGCG GATA
12F2[b]	TCAGAAGCCCTTCCAGTGTCGAATCTGCATGCGTA ACTTCAGT$x^{16}x^{17}x^{18}x^{19}x^{20}x^{21}x^{22}x^{23}x^{24}x^{25}x^{26}x^{27}$ CTGAG [c]CCGCCACATCCGCACCCACACAGG CGAG
12Guide B	GTCACAGGCAAAAGGCTTCTCGCCTGTGTGGG TGCG
12F3	AAGCCTTTTGCCTGTGACATTTGTGGGAGGAAATT TGCCAGGAGTGATGAACGCAAGAGGCATACCAAA ATCCATTTAAGACAGAAGGAC
W12	GAGTCATTCT GCGGCCGC [d]GTCCTTCTGTCTT AAA
Lib23	
Oligo	Sequence
W23	GCAACTGC GGCCCAGCCGG [a]CCATGGCAGAG GAAC
23F1	CCATGGCAGAGGAACGCCCATATGCTTGCCCTGTC GAGTCCTGCGATCGCCGCTTTTCTCGCTCGGATG AGCTTACCCGCCATATCCG
23Guide A	TTCTGACCGGTGTGGATGCGGATATGGCGGGT AAGC
23F2[b]	CATCCACACCGGTCAGAAGCCCTTCCAGTGTCGAA TCTGCATGCGTAACTTCAGTCGTAGTGAC$x^{28}x^{29}$ x^{30} CTGAG [c]$x^{31}x^{32}x^{33}$CCACATCCGCACCCACA CAGG
23Guide B	CAGGCAAAAGGCTTCTCGCCTGTGTGGGTGCG GATG

Table 3.1 (continued)

Oligo	Sequence
23F3[b]	CGAGAAGCCTTTTGCCTGTGACATTTGTGGGAGGA AATTTGCC$x^{34}x^{35}x^{36}x^{37}x^{38}x^{39}x^{40}x^{41}x^{42}x^{43}x^{44}x^{45}$C GC$x^{46}x^{47}x^{48}x^{49}x^{50}x^{51}$CATACCAAGATACACCTGCG CCAAAAAGAT
M23	CAATCACTCC GCGGCCGC [d]ATCTTTTTGGCG CAGG

[a]GGCCCAGCCGG is the recognition sequence of *Sfi*I restriction enzyme.
[b]Compositions of randomized nucleotides at specific positions – $x^1x^2x^3$:crs/rms; $x^4x^5x^6$:gmc/mrs; $x^7x^8x^9$:cac/avc/ghc; $x^{10}x^{11}x^{12}$:ayc; $x^{13}x^{14}x^{15}$:cry/gha/ams; $x^{16}x^{17}x^{18}$:crs/rmc; $x^{19}x^{20}x^{21}$:ags $x^{22}x^{23}x^{24}$:gms/mrs; $x^{25}x^{26}x^{27}$:cac/avc/ghc; $x^{28}x^{29}x^{30}$:cac/avc/ghc; $x^{31}x^{32}x^{33}$:crg/gha/ams; $x^{34}x^{35}x^{36}$: crs/rmc; $x^{37}x^{38}x^{39}$:ars; $x^{40}x^{41}x^{42}$:gmc/mrs; $x^{43}x^{44}x^{45}$:cac/avc/ghc; $x^{46}x^{47}x^{48}$:aha; $x^{49}x^{50}x^{51}$: crg/gha/ams
¶Randomized nucleotide symbols: m=A/C; s=G/C; b=C/G/T; n=A/C/G/T; r=A/C; y=T/C; v=A/C/G; w=A/T; k=T/G; h=A/C/T; d=A/G/T
[c]CTGAG is the recognition sequence of *Dde*I restriction enzyme.
[d]GCGGCCGC is the recognition sequence of *Not*I restriction enzyme.

MgSO$_4$, and 20 mM glucose. 10 mL each of 1 M MgCl$_2$ and 1 M MgSO$_4$, and 20 mL of 1 M filter-sterilized glucose are added to 1 L H$_2$O after autoclaving.

4. 1 M ZnCl$_2$.
5. Steps 5 and 6 of **Section 2.1**.

2.4. Bipartite Selection of Phage Clones Expressing Zinc Fingers Bound to DNA Targets

1. PEG/NaCl: 20% polyethylene glycol 8,000, 2.5 M NaCl.
2. Biotinylated oligonucleotide primer: 20 nt, biotin labeled at 5′ end (i.e., 5′-BiotinGACGTGTGGACTGA-CTGTGA-3′) (*see* **Fig. 3.3B**).
3. Non-biotinylated DNA target oligonucleotide, 37 nt, with 20 nt complementary to the biotinylated oligonucleotide primer (*see* **Fig. 3.3B**).
4. 2× nucleic acid annealing buffer: 40 mM Tris–HCl, pH 8.0, 200 mM NaCl.
5. TM buffer: 100 mM Tris–HCl, pH 8.0, 100 mM MgCl$_2$.
6. 2× Klenow Fill-In buffer: 8 μL of TM buffer, 4.2 μL of 20 mM dNTPs, 4.2 μL of 0.1 M DTT, and 21.6 μL of sterile Milli-Q H$_2$O.
7. 2,500 U/mL Klenow fragment of DNA polymerase.
8. Streptavidin-coated immunotubes or microtiter plates (Roche).

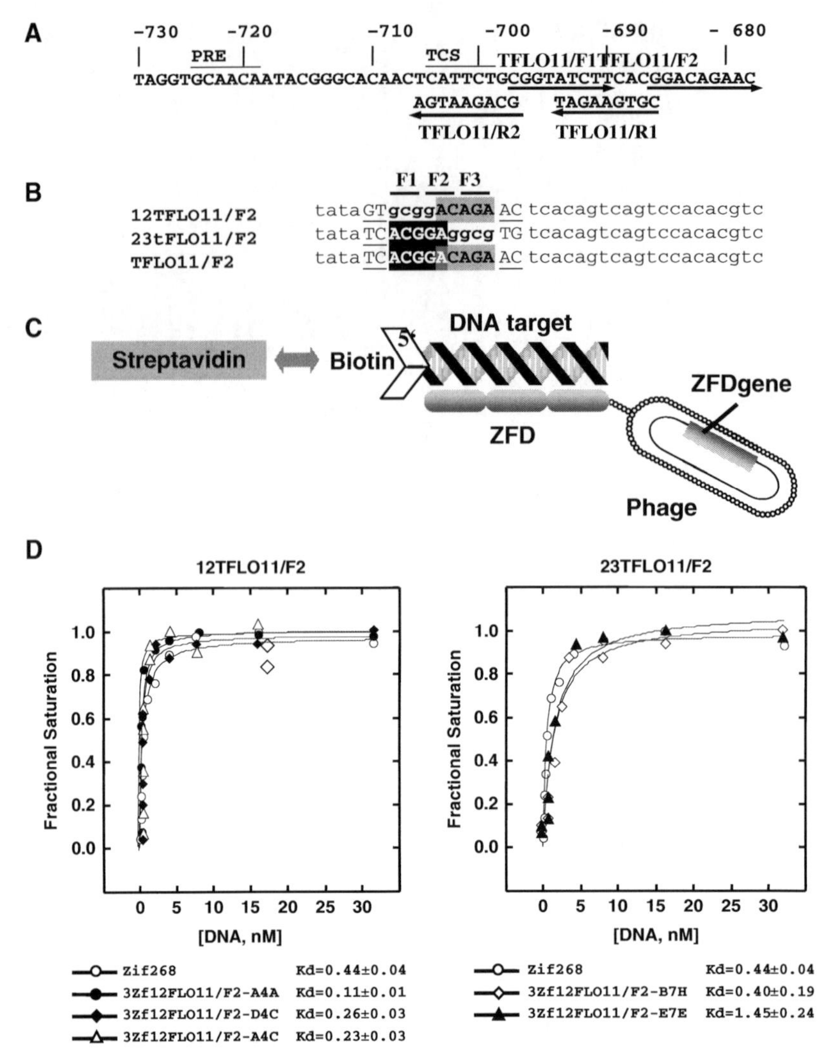

Fig. 3.3. Bipartite selection of ZFD that binds to upstream sequences of the *FLO11* gene of *Saccharomyces cerevisiae*. (**A**) The upstream sequences of *FLO11* used to select sequence-specific three-finger DNA-binding domains. Nucleotides are numbered relative to the translational start site. The 9-bp core sequences of the target sites for TFLO11/F1, TFLO11/F2, TFLO11/R1, and TFLO11/R2 selections are shown with *arrows* indicating the direction of the sequence from 5′ to 3′. Sequences of known binding sites of PRE and TCS are indicated above. (**B**) Oligonucleotide targets used in the phage selection of TFLO11/F2, including 2 bp flanking each site. The *shaded areas* represent the core selection sequences with the position of individual fingers indicated by F1, F2, and F3. The selection targets were created such that "half-site" is comprised of a gcgg anchor (*in small letters*) and a 5-bp target (*shaded light gray*) for selection of 12TFLO12/F2 from Lib12. A 5-bp target (*shaded black*) with a ggcg anchor (*in small letters*) was used for selection of 23TFLO12/F2 from Lib23. The final TFLO12/F2 selection target is shown below, with overlapping shared region *shaded gray*. The sequences flanking the core target sequences are underlined. Only one strand of the double-stranded oligonucleotide targets is shown. The complementary strand was obtained by extension with a 5′ biotinylated primer complementary to the sequence 5′-tcacagtcagtccacacgtc-3′. (**C**) Schematic representation of phage display selection. ZFD variants displayed on the surface of phage are bound to a 5′ biotinylated DNA target. ZFD–DNA complexes can be recovered by capture on a solid streptavidin-coated matrix. (**D**) The apparent K_ds of three clones bound to the 2TFLO11/F2 DNA "half-site" and two clones bound to the 23TFLO11/F2 "half-site" were determined by Phage-ELISA. The K_ds are comparable to that of Zif268 for its cognate DNA site. (**a**, **b**, and **d** are reproduced and modified from Shieh et al. (15)).

9. 10× phosphate-buffered saline (PBS): 80 g NaCl, 2 g KCl, 11.5 g $Na_2HPO_4.7H_2O$, 2 g KH_2PO_4, Milli-Q H_2O to 1 L, followed by autoclaving.
10. Tween-20.
11. PBS-Tween: PBS, 1% (v/v) Tween-20.
12. Fat-free freeze-dried milk.
13. 10 mg/L sonicated, heat-denatured salmon sperm DNA (ss-DNA), nonspecific single-stranded DNA.
14. 1 M $ZnCl_2$.
15. Selection buffer: PBS, 2% (w/v) fat-free freeze-dried milk, 1% (v/v) Tween 20, 50 μM $ZnCl_2$.
16. 0.1 M triethanolamine.
17. 1 M Tris–HCl, pH 7.4.
18. *E. coli*, recombinase-deficient (*rec*A1) F' strain (e.g., JM109 F′[*traD36 proAB lacI^qZ* ΔM15] *end*A1 *rec*A1 gyrA96 *thi hsd*R17($r_k^-m_k^+$) *Rel*A1 *supE44* λ-Δ*(lac-proAB)*).
19. M9 minimal agar: 6 g/L Na_2HPO_4, 3 g/L KH_2PO_4, 5 g/L NaCl, and 1 g/L NH_4Cl, 0.5 g/L NaCl, 0.1 mM $CaCl_2$, 1 mM $MgSO_4$, 1 mM thiamine-HCl, 2 g/L glucose, and 15 g/L agar. After autoclaving, add 0.1 mL of 1 M $CaCl_2$, 1 mL each of 1 M $MgSO_4$ and 1 M thiamine-HCl, and 10 mL of filter-sterile 20% (w/v) glucose, up to 1 L with H_2O.
20. Acid-treated sterile glass beads.
21. Steps 1, 5–7 of **Section 2.1**.

2.5. Assessment of Binding Properties of Selected Clones by Phage-ELISA

1. Round-bottom, sterile 96-well plates for cell culture (e.g., Corning).
2. Streptavidin-coated microtiter well plates (Roche).
3. Horseradish peroxidase (HRP)-conjugated anti-M13 IgG (Amersham Biosciences).
4. 3,3′,5,5′-tetramethyl-bezidine (TMB), 1 mg tablets (Sigma).
5. Dimethyl sulfoxide (DMSO).
6. 3 M sodium acetate, pH 5.5.
7. 30% (v/v) hydrogen peroxide.
8. ELISA developing solution: dissolve a 1 mg TMB tablet in 100 μL of DMSO. Add this solution to 10 mL of 0.1 M sodium acetate (pH 5.5) and 2 μL of 30% (v/v) hydrogen peroxide. Prepare freshly just before use.

9. 1 M sulfuric acid.
10. ELISA microplate absorbance reader.
11. Rotating incubator that accommodates 96-well plates.
12. Swing bucket centrifuge that accommodates 96-well plates.
13. Steps 5–7 of **Section 2.1**, and Steps 10–14 of **Section 2.4**.

2.6. Recombination of the Bipartite-Selected Phage Clones

1. Two pairs of PCR primers (M12/W12 or W23/M23) that anneal at the ends of the DNA cassette with either *Sfi*I or *Not*I restriction enzyme sites for cloning into phage vector Fd-TET-SN, each at 10 pmol/μL (*see* **Fig. 3.4a** and **Table 3.1**).
2. A pair of PCR primers (M12/M23) that anneal at the ends of the DNA cassette with either *Sfi*I or *Not*I restriction enzyme sites for cloning into phage vector Fd-TET-SN and that selectively PCR amplify the recombinant phage clones derived from bipartite selection, each at 10 pmol/μL (*see* **Fig. 3.4A** and **Table 3.1**).
3. 20,000 U/mL *Dde*I (New England Biolabs).
4. 100 μg/mL bovine serum albumin (BSA).
5. PCR purification kit (e.g., QIAquick, Qiagen).
6. Steps 2–21 of **Section 2.1**, Steps 6–11 of **Section 2.2**, Step 19 of **Section 2.4**.

3. Methods

3.1. Preparation of Phage Vector

1. Transform a phage vector for phage display such as Fd-Tet-SN (*see* **Fig. 3.1** and **Note 1**) into the TG1 strain of *E. coli* and select for tetracycline-resistant transformants on TYE agar plates supplemented with 15 μg/mL tetracycline.
2. Select a single colony from a transformed cell and use it to inoculate a 2 mL of 2× YT medium containing 15 μg/mL tetracycline. Incubate at 37°C for 4 h with shaking at 220 rpm. Transfer the culture into a 4 L flask with 1 L of 2× YT medium for an additional 15 h of incubation.
3. Prepare double-stranded replicative form of the Fd-Tet-SN phage vector from the 1 L bacterial culture using a Maxiprep kit.
4. Purify the phage vector further by cesium chloride gradient (*see* **Note 3**). Dissolve 10–100 μg of purified phage vector DNA in 4 mL TE buffer with 4 g of CsCl and 0.1 mL

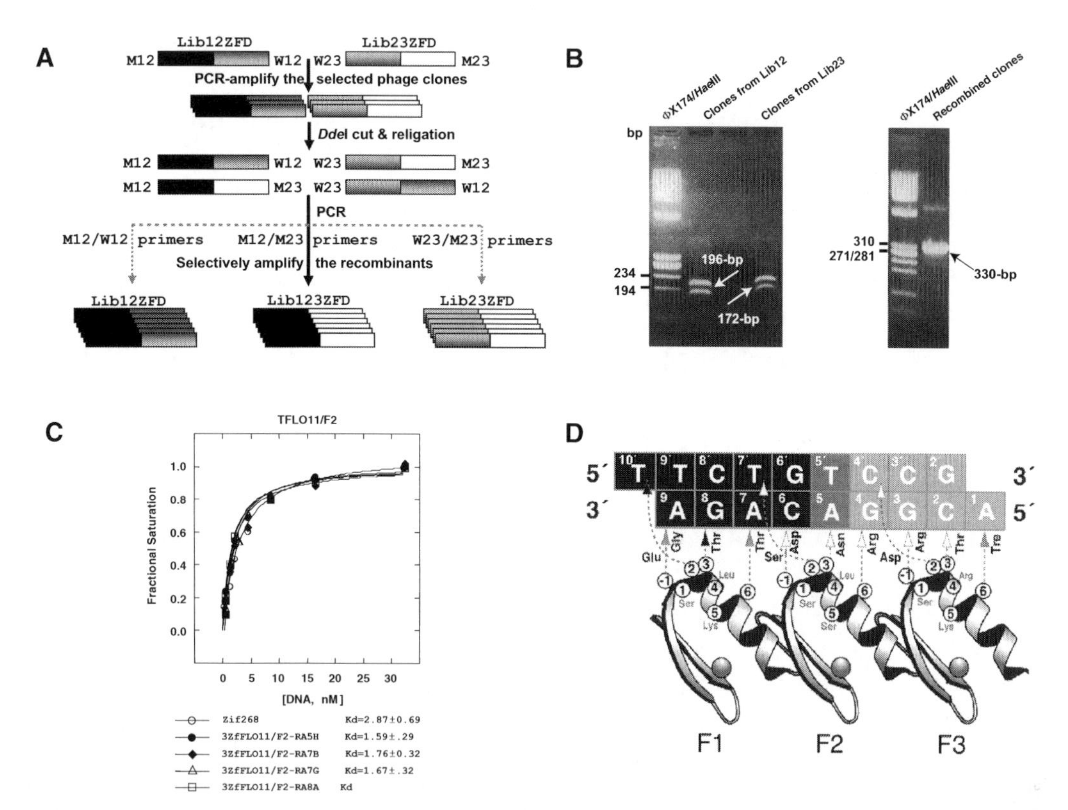

Fig. 3.4. (**A**) Schematic presentation of selective PCR to recover recombined parts of Lib12 and Lib23. After *Dde*I digestion and relegation, Lib123ZFD can be amplified with selective primers M12 and M23 from the initially selected clones of Lib12ZFD and Lib23ZFD. Other possibilities of recombination, such as initially selected clones (*light gray dotted line*), can be completely excluded. (**B**) Agarose gels (2%) of the *Dde*I-digested selected half-site clones and the recombined clones from selective PCR. The *Dde*I-digested DNA was from a pool of three clones from Lib12 that bound the 12TFLO11/F2 target (196 bp) and two from Lib23 that bound the 23TFLO11/F2 target (172 bp). The full size of the recombined clones is about 330 bp. (**C**) The apparent K_ds of four independent recombined clones derived from (**B**) bound to the TFLO11/F2 "full site" were determined by Phage-ELISA and are similar to that of Zif268 for its cognate DNA site. (**D**) Corresponding amino acids of phage-selected clones and their proposed mode of contact with the DNA target site for TFLO11/F2. The deduced amino acids represent four independent recombined clones, which appeared to be the same. The amino acids are shown as *circles*, numbered relative to their helical positions. Putative amino acid–DNA base interactions are shown by *shaded arrows* to indicate that the contact is either bad (*in black*), good (*in white*), or okay (*in gray*) based on the principles of stereochemistry. (**c** and **d** are reproduced and modified from Shieh et al. (15)).

of 10 mg/mL ethidium bromide. Dispense the solution into 5 mL ultracentrifuge tubes. Perform ultracentrifugation at 370,000×*g* using a Beckman ultracentrifuge with a Vti 65.2 rotor at 20°C for 20 h.

5. Collect the lower band (supercoiled plasmid) of the two with a syringe under long-wave UV light. Remove the ethidium bromide by addition of an equal volume of water-saturated 2-butanol, followed by mixing, centrifuging, and discarding the upper (organic) phase, for four times. Precipitate the vector DNA by standard ethanol precipitation

to remove traces of CsCl, prior to resuspending in TE at a final concentration of 1 μg/μL.

6. Precipitate 40 μL (40 μg) of the purified phage vector DNA by ethanol from Step 5 of **Section 3.1** and resuspend in 460 μL of 1× NEBuffer 2 with 100 μg/mL BSA. Add 10 μL of *Sfi*I, overlay with mineral oil to prevent evaporation of the buffer and incubate at 50°C. Boost the reaction three times with 10 μL of *Sfi*I at 2-h intervals.
7. Purify the *Sfi*I-cut phage vector DNA by extracting once with phenol and once with chloroform prior to ethanol precipitation. (*Note*, phenol is corrosive; handle with gloves and use appropriate protection.)
8. Resuspend the *Sfi*I-cut phage vector DNA in 460 μL of 1× NEBuffer 3 with 100 μg/mL BSA. Add 10 μL of *Not*I and incubate at 37°C. Boost the reaction three times with 10 μL of *Not*I at 2-h intervals.
9. Run the cut phage vector DNA on a 1% low-melting point agarose gel in 1× TAE with 0.5 μg/mL ethidium bromide.
10. Recover the phage vector DNA by excising the band with the phage vector DNA under long-wave UV light. Digest the gel slice containing phage vector DNA with Agarase I, followed by extracting once with phenol and once with chloroform before ethanol precipitation. (*Note*, phenol is corrosive; handle with gloves and use appropriate protection.)
11. Resuspend the DNA in Milli-Q sterile water and determine the concentration by spectrophotometer. The phage vector DNA can be stored at −20°C in aliquots.

3.2. Construction of Two Pools of DNA Cassette Coding for Two Collections of Zinc Finger Variants

1. Synthesize two sets of ZFD-encoding DNA cassettes (*see* **Fig. 3.2** and **Table 3.1**). By ligation end to end with two guide DNA oligonucleotide primers, each set of template oligonucleotides combine to form the three coding sequence regions of the three-finger ZFD Zif268, with randomized codons at the desired positions. One DNA cassette covers variants in the first one-and-a-half of the three-finger Zif268 and will be used to construct Lib12. The other DNA cassette covers variants in the last one-and-a-half of the three-finger ZFP Zif268 and will be used to construct Lib23 (*see* **Note 2**).
2. Run 100 μg of each of the synthesized degenerate oligonucleotides on a denaturing polyacrylamide-urea gel to purify and eliminate the secondary structure of the oligonucleotides.

3. Excise the gel with band corresponding to the full-length product of oligonucleotides that are identifiable under long-wave UV.
4. Elute the full-length oligonucleotides by crushing the excised gel in oligonucleotide elution buffer for 15 h.
5. Recover the oligonucleotides by injecting the eluate through a 0.2 μm filter disc, prior to ethanol precipitation. Resuspend the precipitated oligonucleotides in Milli-Q sterile water and adjust the concentration to 10 pmol/μL.
6. Carry out a kinase reaction by the addition of 10 μL of purified "template" oligonucleotides, 1.2 μL of 10× T_4 polynucleotide kinase buffer with 1 mM ATP, and 1 μL of T_4 polynucleotide kinase, and incubate for 1 h at 37°C. Inactivate the enzyme for 10 min at 65°C.
7. Put 10 μL of each "guide" oligonucleotide (10 pmol/μL in 1× T_4 DNA ligase buffer) into 12 μL of kinase reaction mix containing the phosphorylated "template" oligonucleotides.
8. Carry out the annealing reaction of template and guide oligonucleotides by heating the reaction tube to 94°C for 3 min, followed by cooling to 4°C at a rate of 2°C per min on a thermal cycler.
9. Perform the ligation reaction by the addition of an equal volume of 1× T_4 DNA ligase buffer to the annealing mix and 4 μL of 1× T_4 DNA ligase and incubate for 16 h at 16°C.
10. Generate the double-stranded DNA cassette by setting up a PCR reaction mix containing 391 μL of Milli-Q H_2O, 50 μL of 10× *Taq* DNA polymerase buffer, 25 μL of 25 mM dNTP, 12.5 μL each of the 10 pmol/μL flanking primers (containing either a *Sfi*I or a *Not*I restriction enzyme site), 5 μL of ligation mix, and 4 μL of *Taq* DNA polymerase. Allocate the PCR mix into 10 PCR tubes, spin the tubes briefly, overlay the solutions with mineral oil (*see* **Note 4**), and put the tubes in a thermal cycler.
11. Conduct 25 cycles of PCR reactions with the following conditions: 1 min at 94°C, 1 min at melting temperature of flanking primers, and 1 min at 72°C, before a further 10 min incubation at 72°C.
12. Purify the DNA cassette of PCR product by extracting once with phenol and once with chloroform, prior to ethanol precipitation. *Note*, phenol is corrosive.
13. Set up a *Sfi*I and *Not*I digestion and recover the DNA cassette as in Steps 6–11 of **Section 3.1**.

3.3. Construction of a Pair of Phage Libraries with DNA Cassettes Coding for Zinc Finger Variants

1. Conduct library construction for Lib12 and Lib23 separately. The two libraries are complementary to each other to allow bipartite selection (*see* **Figs. 3.2** and **3.3** and **Note 2**).
2. Perform ligation reaction in a total volume of 30 μL in 1× T_4 DNA ligase buffer with 1 mM ATP containing 1 μg of *Sfi*I/*Not*I cut phage vector DNA from Step 11 of **Section 3.1**, an appropriate amount of DNA cassette (*see* **Note 5**), and 3 μL T_4 DNA ligase. Incubate the ligation mixture at 16°C for 16 h.
3. Purify the ligation products by extracting once with phenol and once with chloroform before ethanol precipitation and air-dry the pellet. Resuspend the dried pellet in Milli-Q purified H_2O to a final concentration of 0.1 μg/μL ligation products.
4. Prepare the electrocompetent TG1 *E. coli* cells to a concentration of about 1–3 × 10^{10} cells/mL (*see* **Note 6**).
5. Set an electroporation apparatus to 2.5 kV and 25 μF, with a pulse controller setting to 200 Ω.
6. Put the cuvette in a pre-chilled chamber slide of the electroporation apparatus and pulse cells before immediately resuspending cells in 1 mL 37°C pre-warmed SOC medium and transferring to a polypropylene tube. Incubate for 1 h at 37°C with shaking at 220 rpm.
7. Plate out cells on TYE agar plates with 15 μg/mL tetracycline and grow at 30°C for 16 h (*see* **Note 7**).
8. Harvest transformed cells from the TYE agar plates by gently scraping colonies into an appropriate amount of 2× TY (~1 mL per 10 cm^2 plate area) with 15 μg/mL tetracycline and 50 μM $ZnCl_2$. Add 50% (v/v) glycerol to the cell suspension and store at –70°C. The bacterial library contains a pool of TG1 cells, each of which carries a specific Fd-Tet-SN-based plasmid capable of forming and releasing out of bacterium a phage particle that expresses a variant of zinc finger fused to minor coat protein encoded by gIII.
9. Grow 1 mL of transformed bacterial cells from the bacterial library in 200 mL 2× TY with 15 μg/mL tetracycline and 50 μM $ZnCl_2$ at 30°C for 16 h.
10. Prepare pure phage-containing supernatant by centrifuging the culture 13,000×*g* at 4°C for 10 min, followed by filtration through a 2 μm filter. Store the phage library at 4°C or –20°C or use in selection. Approximately 10^6 phage particles per μL of supernatant can be obtained.

3.4. Bipartite Selection of Phage Clones Expressing Zinc Fingers Bound to DNA Targets

1. The DNA target sites of interested can theoretically be derived from any part of a gene but are normally selected from upstream sequence, such as the promoter of the gene (*see* **Fig. 3.3a**). Bipartite selection is carried out in parallel against 9-bp DNA target sites where either the first or the last 5 bp of the 9-bp target site contains the sequence of interest (*see* **Fig. 3.3B**, **c** and **Note 8**).
2. Grow a 2 mL fresh culture of *E. coli* TG1 bacteria overnight in 2× TY with 15 μg/mL tetracycline and 50 μM $ZnCl_2$ (*see* **Note 9**) from a single colony of TG1 on a M9 minimal agar plate (*see* **Note 10**). Dilute 100× of the culture to 50 mL of 2× TY medium with 15 μg/mL tetracycline and 50 μM $ZnCl_2$ and grow the cells to log phase (OD_{600} of 0.6).
3. Perform phage infection by mixing 1 mL of phage library into the 50 mL log-phage TG1 culture at 37°C for 1 h without shaking.
4. Amplify the phage by growing the infected bacteria at 30°C overnight.
5. Prepare pure phage-containing supernatants by centrifuging the culture at 13,000×*g* for 10 min at 4°C, followed by filtration through a 2 μm filter.
6. Prepare the phage for selection by mixing 40 mL of pure phage-containing supernatant with 10 mL of PEG/NaCl at 4°C for at least 30 min before centrifuging at 13,000×*g* for 30 min at 4°C (*see* **Note 11**).
7. Resuspend the pellet in 2.2 mL of Selection Buffer supplemented with 20 μg/mL ss-DNA. Dispense 200 μL of the mixture into each of 10 wells in a non-streptavidin-coated microtiter plate (*see* **Note 12**).
8. Prepare biotinylated double-stranded DNA target sites prior to phage selection by mixing 10 μL of each of the oligonucleotide containing target site and primer (10 pmol/μL) with 20 μL of 2× nucleic acid annealing buffer (*see* **Fig. 3.3B** and **Note 13**). Heat the mixture at 94°C for 3 min before cooling to 4°C at a rate of 2°C per min in a thermal cycler to allow annealing. Perform primer extension by diluting the annealing reaction twofold with 40 μL of Klenow fill-in mix containing 38 μL of Klenow Fill-In Buffer and 2 μL Klenow and incubate at 37°C for 1 h. Add 20 μL of sterile Milli-Q H_2O to make up a total volume of 100 μL with the double-stranded DNA target site at 1 pmol/μL. Store the biotinylated DNA target site at –20°C.

9. Perform DNA-phage binding by adding 1 μL of 1 pmol/μL biotinylated DNA target site into each phage-containing well (1 pmol DNA per 200 μL or 5 nM of the target DNA) at room temperature for 1 h.
10. Attach biotinylated DNA target site bound with potential phage clones to a matrix coated with streptavidin by transferring the mixture from Step 8 to streptavidin-coated wells and binding for 5 min.
11. Wash the wells 18 times with Selection Buffer and twice with PBS containing 50 μM zinc chloride. Empty the final wash buffer from the wells.
12. Elute phage clones by adding 100 μL of 0.1 M triethylamine before immediately neutralizing with 100 μL of 1 M Tris, pH 7.4. The triethylamine is able to elute phage by disrupting protein–DNA interaction.
13. Infect 150 μL of log-phage TG1 culture with 100 μL of eluate in wells of microtiter plate at 37°C for 1 h without shaking.
14. Plate phage-infected TG1 cells with glass beads on TYE agar plates containing 15 μg/mL tetracycline and grow at 37°C overnight (*see* **Note 14**).
15. Harvest colonies from each plate in 2 mL of 2× TY with 15 μg/mL tetracycline and 50 μM $ZnCl_2$.
16. Amplify the selected phage pools by growing 1 mL of the harvested cells in 24-well plates at 30°C for 4 h.
17. Prepare pure phage-containing supernatant by centrifuging the culture at 13,000×*g* for 10 min at 4°C, followed by filtration through a 2 μm filter.
18. Set up a second round of selection by making a mix in each well of microtiter plate with 175 μL of Selection Buffer as in Step 6, 20 μL of phage-containing supernatant, 0.4 μL of 10 mg/mL ss-DNA, 4 μL of 1 pmol/μL competitor of DNA target site (4 pmol DNA per 200 μL, or 20 nM of the competitor, *see* **Note 15**), and 4 μL of 0.1 pmol/μL of DNA target site (0.4 pmol DNA per 200 μL, or 2 nM of the target DNA), followed by Steps 8–16.
19. Set up a third round of selection by making a mix in each well of a microtiter plate with 185 μL of Selection Buffer as in Step 6, 10 μL of phage-containing supernatant, 0.4 μL of 10 mg/mL ss-DNA, 4 μL of 1 pmol/μL competitor of DNA target site, and 4 μL of 0.1 pmol/μL of DNA target site, followed by Steps 8–16, except changing the bacterial strain from TG1 to JM109 (*see* **Note 16**).

20. Set up a fourth round of selection by making a mix in each well of microtiter plate with 150 μL of Selection Buffer as in Step 6, 10 μL of phage-containing supernatant, 0.4 μL of 10 mg/mL ss-DNA, 40 μL of 1 pmol/μL competitor of DNA target site (40 pmol DNA per 200 μL, or 200 nM of the competitor), and 4 μL of 0.1 pmol/μL of DNA target site, followed by Steps 8–16, except substituting the bacterial strain from TG1 to JM109.
21. Plate also a 1 in 100 dilution of phage-infected JM109 cells from Step 13 of the fourth round selection with glass beads on TYE agar plates containing 15 μg/mL tetracycline and grow at 37°C overnight to obtain single colonies for phage-ELISA.

3.5. Assessment of Binding Properties of Selected Clones by Phage-ELISA

1. Pick single bacterial colonies derived from fourth round of library selection (Step 20 of **Section 3.4**) to the wells of sterile round-bottom microtiter plates containing 2× TY with 15 μg/mL tetracycline and 50 μM $ZnCl_2$, followed by incubation at 30°C for 16 h with shaking at 220 rpm.
2. Prepare phage-containing supernatant by centrifuging the culture at 3,700×*g* for 15 min at 4°C in an appropriate swinging-bucket centrifuge. Recover the supernatants to a fresh plate and immediately use for phage-ELISA or store at 4°C.
3. Attach biotinylated DNA target sites (*see* Step 7 of **Section 3.4**) to a streptavidin matrix by addition of 0.5 μL of 1:10 diluted 1 pmol/μL DNA target site in 50 μL of PBS (0.05 pmol DNA per 50 μL, or 1 nM of the target DNA) (*see* **Note 17**) to each well of a streptavidin-coated microtiter plate and incubate at 20°C for 30 min. Use the DNA target site for Zif268 as a positive control. Use PBS without a DNA target site as a negative control.
4. Block the microtiter plate wells by addition of 150 μL of PBS containing 4% (w/v) fat-free freeze-dried milk and incubate the plate at 20°C for 1 h.
5. Prepare 50 μL of phage binding mixture for each well by mixing 5 μL of phage supernatant prepared from Step 2 with 45 μL of Selection Buffer and 1 μL of 1:10 diluted 10 mg/μL ss-DNA.
6. Perform phage binding by adding 50 μL of phage binding mixture to each well of the microtiter plate after emptying the blocking mixture. Incubate the plate at 20°C for 1 h.
7. Remove the unbound phage by discarding the binding mixture, followed by seven washes of each microtiter-plate

well with 200 μL of PBS-Tween and three washes of each microtiter-plate well with 200 μL of PBS.

8. Conduct antibody-phage binding by removing the PBS and adding to each well an antibody binding mixture containing 100 μL of PBS, 2% (w/v) fat-free freeze-dried milk, and 0.02% (v/v) HRP-conjugated anti-M13 IgG. Incubate the plate at 20°C for 1 h.
9. Remove the unbound antibody by discarding the antibody binding mixture, followed by three washes of each microtiter-plate well with 200 μL of PBS-Tween and three washes with 200 μL of PBS.
10. Prepare the ELISA developing solution. Then, empty all traces of PBS from the final wash of the microtiter plates.
11. Immediately perform the colorimetric reaction by adding 100 μL of ELISA Developing Solution to each well of microtiter plate for 5 min.
12. Stop the colorimetric reaction by addition of 100 μL of 1 M sulfuric acid to each well.
13. Quantitate the absorbance of the wells of the microtiter plate by an ELISA microplate reader at 450 nm. Clones with an absorbance value similar to that of Zif268 to its cognate DNA target site are consider as potential positive clones.
14. The apparent equilibrium dissociation constant (K_d) value is used to determine the binding affinities of the selected clones. To establish K_d values, the phage clones are assessed by phage-ELISA against a serial dilutions of DNA target sites, with a range from 0.125 to 32 nM (*see* **Fig. 3.3D** and **Note 18**). Wild-type Zif268 on its cognate DNA target site is used as a control.

3.6. Recombination of the Bipartite-Selected Phage Clones

1. To allow selective PCR, cassettes of Lib12 and Lib23 were designed with unique sequences at their 5′ and 3′ ends. The selectivity is achieved by the fact that PCR using *Taq* DNA polymerase cannot proceed efficiently if primers contain 3′ mismatches with their template DNA. As a result, only those cassettes with regions from the two randomized parts of Lib12 and Lib23 are amplified with a pair of mutant primers of M12 and M23 (*see* **Fig. 3.4a** and **Table 3.1**).
2. Amplify each of the coding regions of ZFD from the phage clones of bipartite selection by setting up a PCR

reaction mix containing 78 μL of Milli-Q H_2O, 10 μL of 10× *Taq* DNA polymerase buffer, 5 μL of 25 mM dNTP, 2.5 μL of each of 10 pmol/μL flanking primers (containing either a *Sfi*I or a *Not*I restriction enzyme site), 1 μL of phage supernatant from Step 2 of **Section 3.5**, and 1 μL of *Taq* DNA polymerase. Overlay the solutions with mineral oil and put the tubes in a thermal cycler.

3. Conduct 30 cycles of PCR with the following conditions: 1 min at 94°C, 1 min at melting temperature of flanking primers, and 1 min at 72°C, before a further 10 min at 72°C.
4. Purify the PCR products by extracting once with phenol and once with chloroform before ethanol precipitation and air-drying the pellet. Resuspend the dried pellet in 50 μL of 1× NEBuffer 3 with 100 μg/mL BSA. Add 1 μL of *Dde*I and incubate at 37°C.
5. Run the *Dde*I cut PCR products on a 1% low-melting point agarose gel in 1× TAE with 0.5 μg/mL ethidium bromide.
6. Recover the *Dde*I-digested PCR products by excising either the lower band containing the DNA encoding the ZFDs from Lib12 (Lib12ZFD) or the higher band containing ZFDs from Lib23 (Lib23ZFD) under long-wave UV light (*see* **Fig. 3.4B**). Digest the gel slices with Agarase I, purify using a Qiagen PCR purification kit, and resuspend the DNA in 50 μL of Milli-Q sterile water.
7. Perform a ligation reaction in a total volume of 100 μL in 1× T_4 DNA ligase buffer with 1 mM ATP, containing 45 μL of *Dde*I-digested DNA encoding ZFD from Lib12 and 45 μL of *Dde*I-digested DNA encoding ZFD from Lib23 and 1 μL T_4 DNA ligase. Incubate the ligation mixture at 16°C for 16 h.
8. Generate the DNA encoding recombinant ZFD (Lib123ZFD) by setting up a PCR reaction mix containing 35 μL of Milli-Q H_2O, 5 μL of 10× *Taq* DNA polymerase buffer, 2.5 μL of 25 mM dNTP, 1.25 μL of each of 10 pmol/μL flanking primers M12 and M23 (recombinant-specific and containing either *Sfi*I or *Not*I restriction enzyme sites), 4 μL of the ligation mixture from Step 7, and 1 μL of *Taq* DNA polymerase (*see* **Note 19**). Overlay the solutions with mineral oil, and put the tubes in a thermal cycler. Perform the PCR reaction and purification of the PCR product as in Steps 3 and 4, respectively.

9. Set up a *Sfi*I and *Not*I digestion and recover the DNA encoding recombinant ZFD as in Steps 6–11 of **Section 3.1**, but scale down fourfold.
10. Perform cloning as in Steps 2–7 of **Section 3.3**, except replacing TG1 with JM109 *E. coli* strain to obtain single colonies on TYE agar plates containing 15 μg/mL tetracycline.
11. Assess the K_d of recombinant phage clones by phage-ELISA (*see* **Fig. 3.4C**) as in **Section 3.5**.
12. Determine the DNA sequence, the deduced amino acid residues, and the predicted mode of DNA-ZFD contacts for the recombinant phage clones (*see* **Fig. 3.4D**).

4. Notes

1. DNA-binding domains such as zinc fingers have been displayed as fusions to the gIII-encoded minor coat on the surface of filamentous bacteriophage. Either a phagemid or phage vector is suitable for use in phage display. Phagemid vectors are advantageous for higher efficiency (up to 10^8 cfu/μg DNA) in *E. coli* transformation over phage vectors (~10^6 cfu/μg DNA) and hence can accommodate larger combinatorial libraries. However, we have found that phage vectors are more appropriate for phage-ELISA due to the polyvalent display of proteins for stronger binding to ligands on a solid support. The phage vector used for library construction is Fd-Tet-SN. This vector possesses *Sfi*I and *Not*I cloning site that allows the N-terminal fusion of a DNA-binding domain encoded by a DNA cassette in frame with phage minor coat gene *gIII*. The selection of transformed *E. coli* with phage vectors carrying a DNA cassette is facilitated by the presence of a Tet^R gene that confers tetracycline resistance. The DNA-binding ZFP encoded by DNA cassette is expressed as a fusion to the minor coat protein III, translocated through the cell membrane of *E. coli*, and assembled onto the tip of the particle that encloses the phage genome (19) (*see* **Fig. 3.1**). It should be noted that although DNA-binding ZFPs based on Zif268 can be expressed by phage as a fusion with the minor coat protein, other DNA-binding domains may prove to be more difficult to display on phage.
2. The two cassettes capable of coding for each of Lib12 and Lib23 when cloned into phage vector Fd-Tet-SN. Essen-

tially, each library contains randomizations in the α-helical DNA-contacting residues (positions –1, 2, 3, and 6) across "one-and-a-half" fingers' within a three-finger scaffold of Zif268 (*see* **Fig. 3.2A**). The remaining non-randomized DNA-contacting residues carry the nucleotide specificity of the parental Zif268 DNA-binding ZFD. The two libraries are complementary since Lib12 covers the first one-and-a-half of three fingers of ZFP Zif268, whereas Lib23 covers the last one-and-a-half fingers of ZFP Zif268. To overcome the size limitations of library construction, each randomization presents only a subset of the 20 amino acids that are known to be responsible for DNA recognition (1, 26) (*see* **Fig. 3.2B**). In addition, the bipartite system allows the recombination in vitro of the complementary portions of the two libraries, without purification steps (*see* **Fig. 3.4A**).

3. To ensure highest transformation efficiencies for construction of phage libraries, CsCl-purified vector DNA is critically essential.
4. It is unnecessary to use mineral oil if a thermal cycler with a heated lid is used.
5. Typically, 1 μg of phage vector is ligated with a molar ratio of three- or fivefold excess of DNA cassette. However, it is always necessary to conduct small-scale trial ligations, followed by transformation, to determine optimum ligation efficiency.
6. Since transformation efficiency determines the size of library, the highest possible efficiency is required during library construction. Because different *E. coli* strains exhibit diverse competency of electroporation, it is essential to establish an optimum transformation efficiency of electroporation in the initial stage of library construction. Best efficiency can be obtained using fresh (unfrozen) electrocompetent cells with a competency of no less than 10^9 cfu/μg supercoiled pUC19 plasmid. However, with such efficiency, only ~10^6–10^7 cfu/μg of phage vector can be obtained due to the large size of Fd-Tet-SN phage vector (~8 kb).
7. Plate out also a dilution series of the transformation to estimate the total number of colonies that corresponds to the library size. It is equally important to check the cloning efficiency between vector and DNA cassette by PCR from 20 randomly selected colonies with primers that amplify the cloned cassette. Lower cloning efficiency will reduce the actual library size. In addition, it is worthwhile to check the competency of phage particle assembly from the phage vector with DNA cassette. The phage titer (competency for phage particle assembly) can be estimated by reinfecting

fresh *E. coli* from a liquid culture of an individual colony to mid-log phase and plating on TYE agar containing 15 μg/mL tetracycline. The randomly selected phage clones can also be sequenced to verify if the variants within the library contain unbiased randomizations at the expected positions.

8. Selections from these two libraries are performed in parallel using DNA sequences of the form 5′-GCGG××××××-3′ for Lib12 and 5′-×××××GCGG-3′ for Lib23, where the underlined bases are bound by the wild-type portion of the ZFD of Zif268 and the letter "×" represents any given nucleotide (*see* **Fig. 3.2A**). After selections, two sets of DNA-binding ZFD variants are obtained, each recognizes novel 5-bp sequences. The selected parts of the ZFD variants from the two libraries are complementary, in that they are capable of being recombined to generate a new ZFD that recognizes a composite 9- or 10-bp sequence. Importantly, since a contact by the residue at position 2 of F1 to the cross-strand nucleotide 11′ (*see* **Fig. 3.2A**) can occur, the nucleotide at position 11 next to the end at position 10 of the 9-bp target is also subjected to the selection using Lib12. As a result, one should consider that it is the 10-bp rather than the 9-bp target using the Lib12 for selection because the cross-strand nucleotide 11′ is equally critical to determine whether good ZFD binders can be obtained. To enable recombination, a unique restriction site for the enzyme *Dde*I is designed into the two libraries at the coding sequence for the midpoint of F2 (*see* **Fig. 3.2A**). While Zif268 F2 contains Leu4 and Thr5, which do not make contact with DNA but play a structural role, the *Dde*I site (CTGAGC; *Dde*I underlined) encodes Leu4 and Ser5, which minimize disruption to the zinc finger recognition helix.

9. Certain DNA-binding domains require supplements to the growth medium to support biological activity of the domains. ZFDs, for example, are stabilized by 50 μM $ZnCl_2$ which is recommended as a supplement in all media and selection buffers.

10. Phage infection relies on the presence of pili on the *E. coli* cell, which requires the presence of an F' plasmid. To ensure generation of the pilus, TG1 and JM109 cells are derived from colonies grown freshly on M9 minimal agar plates, which selects for cells with a F' plasmid.

11. Precipitation of phage by PEG/NaCl can ensure high concentrations of phage ($\geq 10^{10}$ cfu/mL) that are particularly useful during the early rounds of selection, when phage

clones with the desired binding affinity to specific DNA target sites are rare in the library.

12. Alternatively, selections using streptavidin-coated paramagnetic beads (e.g., Dynabeads, Dyna) have been shown to be more reliable than selections using streptavidin-coated microtiter-plates or tubes. However, nearly all of our clone-obtaining selections have been performed by the methods presented here. Hence, streptavidin-coated microtiter-plates are still recommended for use.
13. The doubled-stranded biotinylated DNA target site can also be prepared without primer extension from two fully complementary oligonucleotides, with one biotinylated at 5′ end. The preparation is then reduced to the annealing Step 8 of **Section 3.4**. However, this approach requires multiple biotinylated oligonucleotides, which makes this step more expensive.
14. It is crucial to grow cells on agar plates to prevent competition between individual clones in liquid medium. The plating process can increase the number of uncommon slow-growing phages that are generated in an overnight culture. The selection bias for faster-growing clones can therefore be reduced.
15. In addition to nonspecific competitor such as sonicated salmon sperm DNA (ss-DNA) in the early rounds of selection, it is highly recommended to introduce a specific competitor in later rounds of selection (e.g., the 2nd round onward) to ensure selection of clones with specific binding affinity to the DNA target sites. Competitor sites are oligonucleotides with sequences closely related to the desired target site but containing systemic base variations. Considering the cost and efficiency of competitor sites, weighted competitor sites can be used. For each position, the synthesis mixture includes 40% of the target base and 20% of each of the other three bases. Weighting is also carried out for the other synthesis mixtures at each position in turn. The frequency of 1-base variants from target = $[(0.6)\ 9] \times 9 \times 100\% = 9\%$. Only one set of competitor oligonucleotides needs to be synthesized per DNA target site. During selection, the competitors are not biotinylated so that they remain in solution and are removed by washing after binding step of selection.
16. JM109 is a recombinase deficient (*recA*1) *E. coli* strain. We recommend to use it in later rounds of selection (e.g., the 3rd round onward) to reduce obtaining truncated phage mutants when proliferated in TG1 strain expressing a functional recombinase.

17. Since the concentration of binding site is 1 nM, phage that bind with a K_d (*see* **Note 18**) below the nanomolar range will be retained. However, the concentration of binding site can be increased if DNA-binding domains that bind their targets more weakly are to be selected.
18. The mass action equation is used to explain the concentration of protein–DNA complex (DNA.P), free DNA (DNA), and free protein (P) at equilibrium and can be transformed into a Michaelis–Menten-like equation in terms of the total protein concentration (P_{total}) as $[DNA.P] = [P_{total}] [DNA]/ [DNA]+ K_d$. Since the concentration of phage (10^{10} pfu/mL = pM) is relatively small, the concentration of free DNA approximates that of input DNA. Values for [DNA.P] at various [DNA] are fitted to the equation using the computer software KaleidaGraph (Abelbeck Software) that calculates the OD value at which binding is saturated and individual OD values are converted to a fractional saturation of binding. The data are presented as a plot of fraction of binding against the concentration of input DNA. The K_d is calculated as the concentration of DNA required to give 50% saturation of binding. Comparison of K_d values between the 9-bp selection target site and sites with only 1-bp difference can be used to determine the specificity of DNA target site bound with the selected phage clones. It is recommended to perform this analysis in order to obtain clones with highest specificity to the DNA target site.
19. During selective amplification by PCR, it is important to use the nonproofreading DNA polymerase *Taq* that lacks $5' \rightarrow 3'$ exonuclease activity, since enzymes such as *Pfu*, which has $5' \rightarrow 3'$ proofreading activity, erode the DNA ends and therefore abolish selective priming.

Acknowledgments

The author thanks Professor Aaron Klug for giving me the opportunity to work on zinc fingers with the yeast model of my choice and for his advice. The author is also indebted to colleagues in Aaron's lab at the MRC-Laboratory of Molecular Biology, particularly Yen Choo and Mark Isalan who have originally developed the method of bipartite phage display, and most of materials and methods are directly derived from their work. This work is supported by the Medical Research Council of the United Kingdom and a grant from National Science Council (NSC92-2311-B-040-007) of the Republic of China.

References

1. Choo, Y. and Klug, A. (1994) Toward a code for the interactions of zinc fingers with DNA: selection of randomized fingers displayed on phage. *Proc Natl Acad Sci USA.* **91**, 11163–11167.
2. Rebar, E.J. and Pabo, C.O. (1994) Zinc finger phage: affinity selection of fingers with new DNA-binding specificities. *Science.* **263**, 671–673.
3. Jamieson, A.C., Kim, S.H., and Wells, J.A. (1994) In vitro selection of zinc fingers with altered DNA-binding specificity. *Biochemistry.* **33**, 5689–5695.
4. Choo, Y. and Klug, A. (1995) Designing DNA-binding proteins on the surface of filamentous phage. *Curr Opin Biotechnol.* **6**, 431–436.
5. Choo, Y. and Klug, A. (1994) Selection of DNA binding sites for zinc fingers using rationally randomized DNA reveals coded interactions. *Proc Natl Acad Sci USA.* **91**, 11168–11172.
6. Rebar, E.J., Greisman, H.A., and Pabo, C.O. (1996) Phage display methods for selecting zinc finger proteins with novel DNA-binding specificities. *Methods Enzymol.* **267**, 129–149.
7. Wu, H., Yang, W.P., and Barbas, C.F., III (1995) Building zinc fingers by selection: toward a therapeutic application. *Proc Natl Acad Sci USA.* **92**, 344–348.
8. Klug, A. (1999) Zinc finger peptides for the regulation of gene expression. *J Mol Biol.* **293**, 215–218.
9. Greisman, H.A. and Pabo, C.O. (1997) A general strategy for selecting high-affinity zinc finger proteins for diverse DNA target sites. *Science.* **275**, 657–661.
10. Bartsevich, V.V., Miller, J.C., Case, C.C., and Pabo, C.O. (2003) Engineered zinc finger proteins for controlling stem cell fate. *Stem Cells.* **21**, 632–637.
11. Bibikova, M., Beumer, K., Trautman, J.K., and Carroll, D. (2003) Enhancing gene targeting with designed zinc finger nucleases. *Science.* **300**, 764.
12. Dai, Q., Huang, J., Klitzman, B., Dong, C., Goldschmidt-Clermont, P.J., March, K.L., Rokovich, J., Johnstone, B., Rebar, E.J., Spratt, S.K., Case, C.C., Kontos, C.D., and Annex, B.H. (2004) Engineered zinc finger-activating vascular endothelial growth factor transcription factor plasmid DNA induces therapeutic angiogenesis in rabbits with hindlimb ischemia. *Circulation.* **110**, 2467–2475.
13. Papworth, M., Moore, M., Isalan, M., Minczuk, M., Choo, Y., and Klug, A. (2003) Inhibition of herpes simplex virus 1 gene expression by designer zinc-finger transcription factors. *Proc Natl Acad Sci USA.* **100**, 1621–1626.
14. Reynolds, L., Ullman, C., Moore, M., Isalan, M., West, M.J., Clapham, P., Klug, A., and Choo, Y. (2003) Repression of the HIV-1 5′ LTR promoter and inhibition of HIV-1 replication by using engineered zinc-finger transcription factors. *Proc Natl Acad Sci USA.* **100**, 1615–1620.
15. Shieh, J.C., Cheng, Y.C., Su, M.C., Moore, M., Choo, Y., and Klug, A. (2007) Tailor-made zinc-finger transcription factors activate FLO11 gene expression with phenotypic consequences in the yeast Saccharomyces cerevisiae. *PLoS ONE.* *2*(8), e746.
16. Tan, W., Zhu, K., Segal, D.J., Barbas, C.F., III, and Chow, S.A. (2004) Fusion proteins consisting of human immunodeficiency virus type 1 integrase and the designed polydactyl zinc finger protein E2C direct integration of viral DNA into specific sites. *J Virol.* **78**, 1301–1313.
17. Li, F., Papworth, M., Minczuk, M., Rohde, C., Zhang, Y., Ragozin, S., and Jeltsch, A. (2007) Chimeric DNA methyltransferases target DNA methylation to specific DNA sequences and repress expression of target genes. *Nucleic Acids Res.* **35**, 100–112.
18. McCafferty, J., Griffiths, A.D., Winter, G., and Chiswell, D.J. (1990) Phage antibodies: filamentous phage displaying antibody variable domains. *Nature.* **348**, 552–554.
19. Smith, G.P. (1985) Filamentous fusion phage: novel expression vectors that display cloned antigens on the virion surface. *Science.* **228**, 1315–1317.
20. Isalan, M., Klug, A., and Choo, Y. (2001) A rapid, generally applicable method to engineer zinc fingers illustrated by targeting the HIV-1 promoter. *Nat Biotechnol.* **19**, 656–660.
21. Beerli, R.R. and Barbas, C.F., III (2002) Engineering polydactyl zinc-finger transcription factors. *Nat Biotechnol.* **20**, 135–141.
22. Segal, D.J. and Barbas, C.F., III (2000) Design of novel sequence-specific DNA-binding proteins. *Curr Opin Chem Biol.* **4**, 34–39.
23. Wolfe, S.A., Greisman, H.A., Ramm, E.I., and Pabo, C.O. (1999) Analysis of zinc fingers optimized via phage display: evaluating

the utility of a recognition code. *J Mol Biol.* **285**, 1917–1934.

24. Minczuk, M., Papworth, M.A., Kolasinska, P., Murphy, M.P., and Klug, A. (2006) Sequence-specific modification of mitochondrial DNA using a chimeric zinc finger methylase. *Proc Natl Acad Sci USA.* **103**, 19689–19694.
25. Minczuk, M., Papworth, M.A., Miller, J.C., Murphy, M.P., and Klug, A. (2008) Development of a single-chain, quasi-dimeric zinc-finger nuclease for the selective degradation of mutated human mitochondrial DNA. *Nucleic Acids Res.* **36**, 3926–3938.
26. Isalan, M., Klug, A., and Choo, Y. (1998) Comprehensive DNA recognition through concerted interactions from adjacent zinc fingers. *Biochemistry.* **37**, 12026–12033.

Chapter 4

Structure-Based DNA-Binding Prediction and Design

Andreu Alibés, Luis Serrano, and Alejandro D. Nadra

Abstract

Structure-based DNA-binding prediction is a powerful tool to infer protein-binding sites and design new specificities. It can limit experiments in scope and help focus them toward candidates with higher chances of success. The zinc finger domain is an excellent scaffold for design due to its small and robust fold and relatively simple interaction pattern. It presents some degree of modularity, and modeling can be used to guide experiments and help increase zinc finger module libraries. In this chapter we present a fast and simple but still powerful method for predicting and designing DNA-binding specificities applied to C_2H_2 zinc finger proteins, based on FoldX, a semiautomatic protein design tool. Given a template structure, this method generates candidate mutants for a given target DNA sequence selected by energetic criteria.

Key words: Structure-based prediction, design, DNA binding, zinc finger, in silico, FoldX.

1. Introduction

The zinc finger domain is one of the most abundant DNA-binding motifs. Protein–DNA interactions are key to the regulation of cellular activities such as transcriptional regulation and replication. To accomplish their function, DNA-binding proteins must find and bind infrequent and small specific binding sites and discriminate them from a huge excess of nonspecific DNA.

Several methods have been described in the literature for discovering transcription factor binding sites (reviewed in (1, 2)). On the experimental side, there are methods such as SELEX (3) and B1H (4) that challenge the protein with a library of DNA sequences and successively enrich those with high DNA binding activity. Protein binding microarrays (PBMs) (5, 6), which challenge the protein with all possible sequence combinations, is

J.P. Mackay, D.J. Segal (eds.), *Engineered Zinc Finger Proteins*, Methods in Molecular Biology 649,
DOI 10.1007/978-1-60761-753-2_4, © Springer Science+Business Media, LLC 2010

the only exhaustive technique, but is limited to relatively short binding sites (5). On the other hand, theoretical methods can be sequence based, finding conserved sequences in promoter regions (summarized in (7–9)) or inferred by homology modeling (10). In the last few years there have been also successful attempts to predict binding specificity from structure, either by using existing crystal structures (11–15) or by using a docking approach (9). These predictions were particularly evaluated in zinc fingers (12, 15, 16) where a sensitivity to docking geometry was reported (17), highlighting the importance of having multiple templates.

The C_2H_2 zinc finger domain is the most frequent motif in transcription factors. Its abundance and apparent simplicity of interaction, having just four interacting residues, combined with its modularity and with very robust folding are some of the reasons for it having been extensively studied. A "canonical" C_2H_2 zinc finger interaction with DNA has been described, where four bases interact with four side chains one to one (*see* **Fig. 4.1**). This inspired the idea of a code for C_2H_2 zinc fingers and DNA (18–21). In principle, the code could be used to engineer new multi-finger proteins with altered DNA-binding specificity. However, it was soon evident that in addition to the canonical contacts there were other bases/residues that could be involved in the interaction and that inter-finger interactions as well as linker

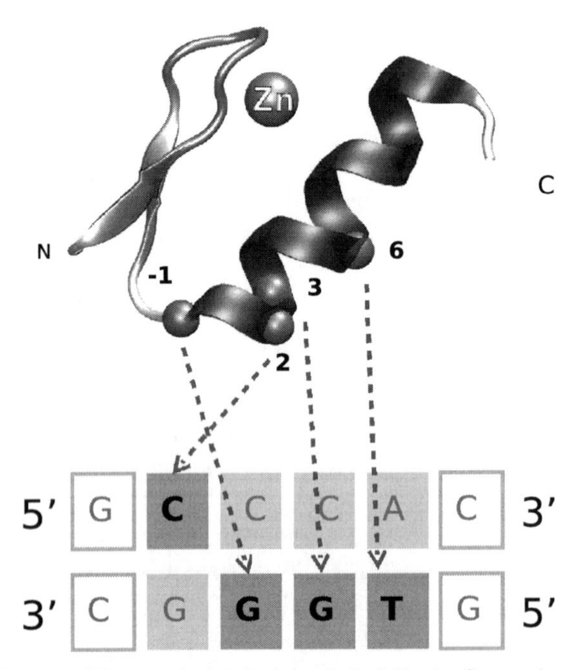

Fig. 4.1. Scheme of the canonical contacts made by C_2H_2 zinc fingers. Arrows indicate specific contacts between four residues in the alpha helix (numbered) and four bases (*dark gray*) in the DNA. Protein and DNA extremities are annotated.

sequences were also important for binding site recognition and discrimination (21–24).

Another engineering approach was to combine existing modules with known specificities to generate new multi-finger proteins targeting the desired sites, the so-called modular assembly approach (25–29). This approach had not only more success but also many drawbacks (30), probably coming from the consideration of fingers as independent modules contacting just three bases. This assumption had the obvious problem of incompatibility with the fourth omitted base (shown as the only contacting base in the top strand in **Fig. 4.1**) and generally more subtle finger–finger interface clashes.

A further improvement was the context-dependent approach, in which individual finger domains were tested at each position and in combination with other finger domains to verify their specificity while taking into account the surrounding zinc finger modules (31–36). Very recently a new protocol considering the context was reported, OPEN, that also requires the preexistence of sequence specific modules (37). The design of fingers with new specificities that could be combined would expand the available target space for these engineering approaches. Structure-based predictions may enable us to generate new modular specificities, as well as combine them in a rational way to generate longer sequence recognition by multi-finger arrays with fingers that could behave as modules.

There are different approaches for structure-based DNA-binding prediction, generally using the ROSETTA, AMBER, or FoldX force fields. In this chapter, we describe a new method to derive DNA-binding specificity from structure using FoldX (38), a fast and extensively validated software tool for the evaluation of changes in stability and interaction due to mutations. The capability to handle protein–DNA complexes has been recently added to FoldX, and its predictions have already been validated (39–41). The predictions are based on energies derived from structures, by evaluating all possible DNA bases at each position. The predictions are therefore independent of any library size effect but may be biased by a particular structure, although this concern has so far not been observed.

Structure-based prediction is the focus of this chapter. This approach has developed in the last few years, and there exist some examples of successful binding site prediction for several proteins and particularly for zinc fingers. There is, however, only one example of structure-based design of a new specificity (42), underlining the state-of-the-art nature of the approach. These techniques usually require a lot of expertise and are difficult to reproduce by nonexperts. The aim of this chapter is to establish a simple protocol for structure-based prediction of the binding sites for three-finger arrays. This protocol can be used to derive the binding site of an existing zinc finger array and also to design new specificities. Designed to be simple, this protocol may be somehow limited, as we tried to make a balance between power and simplicity in the protocol. Some hints on how to fine tune the predictions to make them more powerful are presented in the Notes section.

The general strategy is to generate mutations in a three-finger protein in order to maximize its interaction with a 10-bp target site found within the locus of interest. To do this, it is first necessary to find a template structure (**Section 3.1**) with a DNA sequence that is most similar to the 10-bp stretch in the target sequence (**Section 3.2**). Then, the DNA is mutated to match the preferred sequence, and the protein is mutated to generate a large number of mutants (**Section 3.3**) from which to select those with the desired affinity (**Section 3.4**) and specificity (**Section 3.5**; *see* **Fig. 4.2**). **Section 3.6** suggests how this approach can be applied to the construction of zinc finger nucleases.

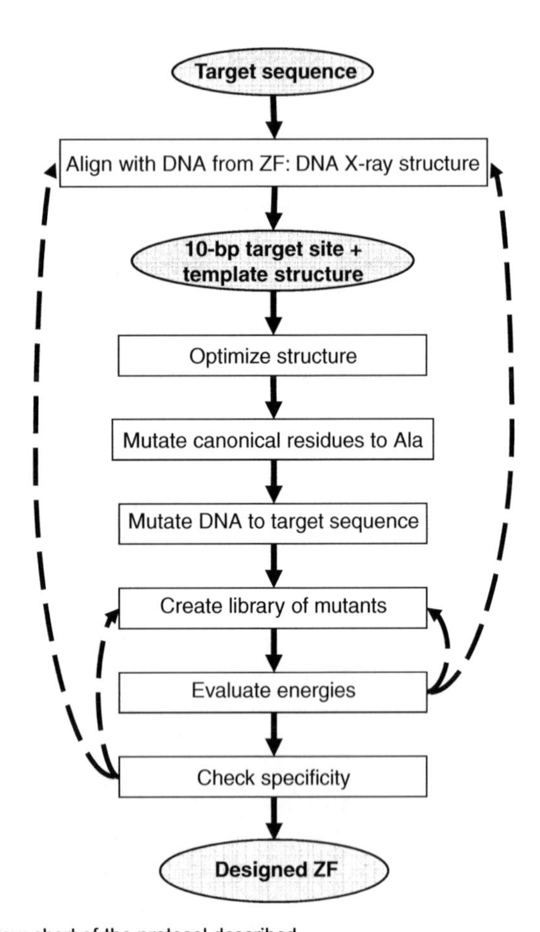

Fig. 4.2. Flow chart of the protocol described.

2. Materials

2.1. Obtaining Zinc Finger–DNA Templates from the PDB

The RCSB Protein Data Bank Web page (43) (http://www.rcsb.org) is the central repository of protein and protein–DNA complex structures. Many different criteria can be applied when searching it, using the Advanced Search feature.

2.2. Aligning the DNA

Kalign (44) is a user friendly Web application (http://www.ebi.ac.uk/Tools/kalign/) for aligning protein and DNA sequences. Its output can be directly analyzed with Jalview, a useful Java applet for coloring and sorting alignments.

2.3. Building the Library

FoldX (38–41) (http://foldx.crg.es) is an application that provides a fast and quantitative estimation of the importance of the interactions contributing to the stability of proteins, protein–protein complexes, and protein–DNA complexes. Its main feature in the framework of this chapter is its ability to mutate both protein and DNA and evaluate the effect of these mutations on the interaction energy and the stability of the complex. FoldX is freely available for academic users upon registration. Sample run-files for FoldX following the protocol described in this chapter are available at http://foldx.crg.es/zfprotocols.

2.4. Selecting and Evaluating the Best Finger

Any simple spreadsheet application can be used, as the only features used will be sorting data and adding a simple formula.

2.5. Supplementary Material

To make the procedure more straightforward to follow, a Web page (http://foldx.crg.es/zfprotocols) is available where examples and data sets are provided. All relevant crystal structures available at the time of writing (12/2008) can be directly downloaded there, as well as the DNA-binding sequences of crystallized zinc finger complexes.

3. Methods

3.1. Obtaining Zinc Finger–DNA Templates from the PDB

The first step is to look for the best zinc finger–DNA templates in a database. Look at the RCSB-PDB Web site for crystal structures of zinc fingers bound to DNA (structures solved by NMR can also be used but usually crystal structures give better results). As the prediction results can vary depending on the template, only high-resolution structures should be considered. Using structures with resolutions better than 2.8 Å is a reasonable cutoff. If the structure had poorly determined regions, they should not occur in the protein–DNA interface.

Ideally multiple templates should be used. Using the criteria defined above, there are currently 22 crystal structures of zinc finger–DNA complexes in the PDB. In this chapter we will describe how to engineer a protein using only the best model, but using multiple template structures is recommended for better results.

As the precise structure of the final protein is not known, selecting the adequate template may be challenging. Use a template that has a DNA sequence that is as close as possible to the target.

To download the complex structures, go to the RCSB Protein Data Bank Web page. In the "Advanced Search" option, select the following conditions:

1. Medical Subject Headings (MeSH): type "zinc finger"
2. Molecule/chain type: click yes in "contains DNA" and "contains protein"
3. SCOP classification: classic zinc finger, C_2H_2
4. X-ray resolution: between 0 and 2.8 Å
5. Number of chains: three or more (to allow any number, but at least a protein and double-stranded DNA)

Download all relevant structures and keep them as a source of templates (*see* **Note 1**).

3.2. Aligning the DNA

To find the 10-bp region in the locus of interest where to design a three-finger array, align the target site sequence with all the DNA sequences from the crystal co-complexes (a "library" list in fasta format from the structures currently available in the PDB can be downloaded from http://foldx.crg.es/zfprotocols/crystallizedDNAs.fa). Different applications can be used. As an example, using the Kalign Web server

1. Paste the target site sequence together with the library (*see* **Note 2**).
2. Using the standard parameters and pressing "Run," Kalign outputs an alignment that can be easily seen by clicking on the "Jalview" applet.
3. Sorting the aligned sequences by *identity* makes it easy to identify the closest template DNA, as it is next to the target site sequence (*see* **Note 3**).

Save the output and identify the closest DNA to use its structure as the template (*see* **Note 4**).

3.3. Building the Library

1. Prepare the structure to be used by FoldX. PDB structures contain a lot of information and, usually, more structures than those needed. Delete everything but the molecules you want to model (i.e., both DNA strands, protein chain, Zn^{2+} ions, and water molecules). This can be done with any text editor (*see* **Note 5**). Alternatively use any structure visualization program, such as VMD, PyMol, or Swiss-PdbViewer. Save the file in a new directory where all results will be produced and stored.
2. Optimize the structure. The next step requires preparing the reference structure for the simulations. To optimize the crystal structure, use the FoldX RepairPDB function to minimize the energy according to the FoldX force field, removing small van der Waals' clashes; generating missing side chains; and choosing the correct histidine, glutamine, and asparagine rotamers (*see* **Note 6**). After running the RepairPDB function, check the generated PDB file for the removal of relevant unrecognized atoms (*see* **Note** 7). This function creates a new file called "RepairPDB_", followed by the name of the original structure file. This file will be used in the next step.
3. Mutate the interface residues to alanine. To avoid van der Waals' clashes between the new DNA structure and the protein, it is better to first mutate the interface residues to alanine. It would take too much time to evaluate every individual interface residue that may be structurally important and/or making nonspecific contacts to DNA to increase binding affinity. Therefore, a balance is to mutate to alanine the key residues involved in specific (side chain-base) contacts. The four canonically contacting residues, at positions –1, 2, 3, and 6, will be considered for mutation here (*see* **Fig. 4.1** and **Note 8**).
4. Mutate the DNA to the target sequence. Using the structure with the alanine interface obtained in the previous step, mutate the DNA sequence in the template structure to the target sequence.
5. Mutate the protein to discriminate the target sequence. This step will define the quality of the result. Ideally this step should produce the highest number of variants to decide which the best is, but making an exhaustive library could be very computationally expensive. Options can vary from mutating each of the four positions to every amino acid and combining them (20^4 structures for each zinc finger domain) to mutating each residue to the remaining 19 but one at a time (20×4 variants per finger domain). Alternatively some previous biological knowledge can be applied to

reduce the number of mutations. For instance, some residues are never present in the interface and could be excluded from the combinations. In this case, restricting variants to include only those residues that are over-represented at each position in natural zinc fingers will reduce drastically the number of combinations, allowing for thorough combinatorial scanning.

Another variable to consider is the number of repetitions. When mutating several amino acids/bases, it is better to make several runs for the same mutations. The reason is that FoldX does not guarantee energy convergence and therefore sometimes it can be trapped far away from the global energy minimum. Every run will use a different set of rotamers and increase the likelihood of finding the energy minimum. Use five and then consider only the structure with the minimum value for the interaction energy.

3.4. Selecting the Best Fingers

Once the protein variants have been generated, a criterion to order them is needed. The main characteristics to consider are stability and interaction affinity (*see* **Note 9**). When mutating residues, some variants that give good interaction energies but introduce internal van der Waals' clashes (either with the DNA or with the protein) can be wrongly selected as a good option. To take this into account, add the intraclashes energy when the mutant has higher values than the reference (optimized structure) to the difference in interaction energy. As a rule of the thumb intraclashes smaller than 0.5 kcal/mol could be omitted. Rank the results according to $\Omega = \Delta\Delta G_{\text{interaction}} + \Delta\text{intraclashes}$. To avoid disrupting the interaction with the mutants introduced, the interaction energy should not be much higher than that of the crystal structure. A reasonable cutoff value would be to consider only those mutants below the threshold of $\Omega = 2$ kcal/mol.

Parse the results in order to filter those with interesting values. This can be done in a spreadsheet, ordering by Ω. Remember to consider the mutant with the minimum Ω of the five repetitions run per mutation. If no candidate mutant below the threshold is found, either extend the number of interfacial residues to mutate (*see* **Section 3.3.5**) and/or select another template structure (*see* **Section 3.2**).

At this point it is advisable to review the candidate structure to evaluate whether the mutations introduce specific (base to side chain) contacts. Visually inspecting a protein–DNA complex structure can be done with many freely available applications such as VMD, PyMol, or Swiss-PdbViewer.

3.5. Evaluating the Specificity

Now that a candidate protein that binds the target DNA with good affinity has been selected, check its specificity. Mutate each base pair of the 10-bp target site in the candidate PDB structure to the remaining three base pairs (3 × 10 variants plus the original sequence) and determine position by position that the target base is the best or that the difference between that variant and the best sequence is small (*see* **Notes 10** and **11**). If no candidate mutant fulfills this specificity requirement, either extend the number of residues to mutate (*see* **Section 3.3.5**) and/or select another template structure (*see* **Section 3.2**).

3.6. Designing a Zinc Finger Nuclease

The general architecture of zinc finger nucleases is described in other chapters in this volume (e.g., *see* **Chapters 1 and 14**). To design a nuclease, create two zinc finger arrays spaced by a distance compatible with FokI nuclease activity (5, 6, 7, or 16 base pairs (37)) (*see* **Note 12**).

Proceed as described above to design each three-finger array independently.

4. Notes

1. This search may yield less zinc finger structures than those available in the PDB Web site. Try removing the MeSH or the SCOP criteria to obtain the remaining structures.
2. The target sequence must contain a minimum of 10 bp. When using larger sequences, the alignment procedure will identify the closest 10 bp to any DNA in the template structures of the library. This 10-bp sequence will be used as target in Step 4 of **Section 3.3**.
3. Alternatively, identity for just those bases that are contacted in the structure could be considered.
4. If several structures share the same DNA sequence, use the one with the better resolution.
5. Take care to save it as plain text.
6. The file to be optimized, the FoldX application, the accompanying *rotabase.txt* file, and the runfiles should be in the same directory. For detailed information, refer to the FoldX manual.
7. FoldX will create a file called Unrecognized_atoms.txt, in case any unrecognized atom is found.
8. If one of these residues is a glycine, do not mutate it to alanine, as the only purpose of this mutation is to allow the DNA to mutate more freely in the next step.
9. A very important criterion is the discrimination power of the mutant. It will be considered qualitatively for the best candidate only.

10. As this mutation is done one residue at a time and not generating all possible combinations, some information regarding neighboring bases can be lost.
11. It is very difficult to find a protein that discriminate at every single base. A residue that does not prefer the target base can be selected if it does not discriminate against it.
12. Keep in mind that the FokI nuclease should be bound to the C-terminal finger when selecting the target sequences to model.

Acknowledgments

This work was supported by the EU project ZNIP (LSHB-CT-2006-037783)

References

1. Tompa, M., Li, N., Bailey, T., Church, G., De Moor, B., Eskin, E., Favorov, A., Frith, M., Fu, Y., Kent, J., Makeev, V., Mironov, A., Noble, W., Pavesi, G., Pesole, G., Régnier, M., Simonis, N., Sinha, S., Thijs, G., van Helden, J., Vandenbogaert, M., Weng, Z., Workman, C., Ye, C., and Zhu, Z. (2005) Assessing computational tools for the discovery of transcription factor binding sites. *Nat Biotechnol.* **23**, 137–144.
2. Bulyk, M.L. (2003) Computational prediction of transcription-factor binding site locations. *Genome Biol.* **5**, 201.
3. Roulet, E., Busso, S., Camargo, A.A., Simpson, A.J., Mermod, N., and Bucher, P. (2002) High-throughput SELEX SAGE method for quantitative modeling of transcription-factor binding sites. *Nat Biotechnol.* **20**, 831–835.
4. Meng, X., Brodsky, M.H., and Wolfe, S.A. (2005) A bacterial one-hybrid system for determining the DNA-binding specificity of transcription factors. *Nat Biotechnol.* **23**, 988–994.
5. Berger, M.F., Philippakis, A.A., Qureshi, A.M., He, F.S., Estep, P.W., 3rd, and Bulyk, M.L. (2006) Compact, universal DNA microarrays to comprehensively determine transcription-factor binding site specificities. *Nat Biotechnol.* **24**, 1429–1435.
6. Warren, C.L., Kratochvil, N.C., Hauschild, K.E., Foister, S., Brezinski, M.L., Dervan, P.B., Phillips, G.N., Jr., and Ansari, A.Z. (2006) Defining the sequence-recognition profile of DNA-binding molecules. *Proc Natl Acad Sci USA.* **103**, 867–872.
7. Mahony, S., Auron, P., and Benos, P. (2007) DNA familial binding profiles made easy: comparison of various Motif alignment and clustering strategies. *PLoS Comput Biol.* **3**, e61.
8. Persikov, A.V., Osada, R., and Singh, M. (2008) Predicting DNA recognition by Cys2His2 zinc finger proteins. *Bioinformatics.* **25**, 22–29.
9. Liu, Z., Guo, J.T., Li, T., and Xu, Y. (2008) Structure-based prediction of transcription factor binding sites using a protein-DNA docking approach. *Proteins.* **72**, 1114–1124.
10. Morozov, A.V. and Siggia, E.D. (2007) Connecting protein structure with predictions of regulatory sites. *Proc Natl Acad Sci USA.* **104**, 7068–7073.
11. Morozov, A., Havranek, J., Baker, D., and Siggia, E. (2005) Protein-DNA binding specificity predictions with structural models. *Nucleic Acids Res.* **33**, 5781–5798.
12. Havranek, J.J., Duarte, C.M., and Baker, D. (2004) A simple physical model for the prediction and design of protein-DNA interactions. *J Mol Biol.* **344**, 59–70.
13. Endres, R.G. and Wingreen, N.S. (2006) Weight matrices for protein-DNA binding sites from a single co-crystal structure. *Phys Rev E Stat Nonlinear Soft Matter Phys.* **73**, 061921.
14. Jamal Rahi, S., Virnau, P., Mirny, L.A., and Kardar, M. (2008) Predicting transcription

factor specificity with all-atom models. *Nucleic Acids Res.* **36**, 6209–6217.
15. Paillard, G., Deremble, C., and Lavery, R. (2004) Looking into DNA recognition: zinc finger binding specificity. *Nucleic Acids Res.* **32**, 6673–6682.
16. Benos, P.V., Lapedes, A.S., and Stormo, G.D. (2002) Probabilistic code for DNA recognition by proteins of the EGR family. *J Mol Biol.* **323**, 701–727.
17. Siggers, T.W. and Honig, B. (2007) Structure-based prediction of C2H2 zinc-finger binding specificity: sensitivity to docking geometry. *Nucleic Acids Res.* **35**, 1085–1097.
18. Choo, Y. and Klug, A. (1994) Toward a code for the interactions of zinc fingers with DNA: selection of randomized fingers displayed on phage. *Proc Natl Acad Sci USA.* **91**, 11163–11167.
19. Choo, Y. and Klug, A. (1997) Physical basis of a protein-DNA recognition code. *Curr Opin Struct Biol.* 7, 117–125.
20. Desjarlais, J.R. and Berg, J.M. (1992) Toward rules relating zinc finger protein sequences and DNA binding site preferences. *Proc Natl Acad Sci USA.* **89**, 7345–7349.
21. Pabo, C.O. and Nekludova, L. (2000) Geometric analysis and comparison of protein-DNA interfaces: why is there no simple code for recognition? *J Mol Biol.* **301**, 597–624.
22. Miller, J.C. and Pabo, C.O. (2001) Rearrangement of side-chains in a Zif268 mutant highlights the complexities of zinc finger-DNA recognition. *J Mol Biol.* **313**, 309–315.
23. Wolfe, S.A., Grant, R.A., Elrod-Erickson, M., and Pabo, C.O. (2001) Beyond the "recognition code": structures of two Cys2His2 zinc finger/TATA box complexes. *Structure.* **9**, 717–723.
24. Wuttke, D.S., Foster, M.P., Case, D.A., Gottesfeld, J.M., and Wright, P.E. (1997) Solution structure of the first three zinc fingers of TFIIIA bound to the cognate DNA sequence: determinants of affinity and sequence specificity. *J Mol Biol.* **273**, 183–206.
25. Bae, K.H., Kwon, Y.D., Shin, H.C., Hwang, M.S., Ryu, E.H., Park, K.S., Yang, H.Y., Lee, D.K., Lee, Y., Park, J., Kwon, H.S., Kim, H.W., Yeh, B.I., Lee, H.W., Sohn, S.H., Yoon, J., Seol, W., and Kim, J.S. (2003) Human zinc fingers as building blocks in the construction of artificial transcription factors. *Nat Biotechnol.* **21**, 275–280.
26. Beerli, R.R. and Barbas, C.F., III (2002) Engineering polydactyl zinc-finger transcription factors. *Nat Biotechnol.* **20**, 135–141.
27. Liu, Q., Xia, Z., Zhong, X., and Case, C.C. (2002) Validated zinc finger protein designs for all 16 GNN DNA triplet targets. *J Biol Chem.* **277**, 3850–3856.
28. Mandell, J.G. and Barbas, C.F., III (2006) Zinc finger tools: custom DNA-binding domains for transcription factors and nucleases. *Nucleic Acids Res.* **34**, W516–W523.
29. Segal, D.J., Beerli, R.R., Blancafort, P., Dreier, B., Effertz, K., Huber, A., Koksch, B., Lund, C.V., Magnenat, L., Valente, D., and Barbas, C.F., III (2003) Evaluation of a modular strategy for the construction of novel polydactyl zinc finger DNA-binding proteins. *Biochemistry.* **42**, 2137–2148.
30. Ramirez, C.L., Foley, J.E., Wright, D.A., Muller-Lerch, F., Rahman, S.H., Cornu, T.I., Winfrey, R.J., Sander, J.D., Fu, F., Townsend, J.A., Cathomen, T., Voytas, D.F., and Joung, J.K. (2008) Unexpected failure rates for modular assembly of engineered zinc fingers. *Nat Methods.* **5**, 374–375.
31. Cornu, T.I., Thibodeau-Beganny, S., Guhl, E., Alwin, S., Eichtinger, M., Joung, J.K., and Cathomen, T. (2008) DNA-binding specificity is a major determinant of the activity and toxicity of zinc-finger nucleases. *Mol Ther.* **16**, 352–358.
32. Greisman, H.A. and Pabo, C.O. (1997) A general strategy for selecting high-affinity zinc finger proteins for diverse DNA target sites. *Science.* **275**, 657–661.
33. Hurt, J.A., Thibodeau, S.A., Hirsh, A.S., Pabo, C.O., and Joung, J.K. (2003) Highly specific zinc finger proteins obtained by directed domain shuffling and cell-based selection. *Proc Natl Acad Sci USA.* **100**, 12271–12276.
34. Isalan, M., Klug, A., and Choo, Y. (2001) A rapid, generally applicable method to engineer zinc fingers illustrated by targeting the HIV-1 promoter. *Nat Biotechnol.* **19**, 656–660.
35. Pruett-Miller, S.M., Connelly, J.P., Maeder, M.L., Joung, J.K., and Porteus, M.H. (2008) Comparison of zinc finger nucleases for use in gene targeting in mammalian cells. *Mol Ther.* **16**, 707–717.
36. Segal, D.J., Dreier, B., Beerli, R.R., and Barbas, C.F., III (1999) Toward controlling gene expression at will: selection and design of zinc finger domains recognizing each of the 5′-GNN-3′ DNA target sequences. *Proc Natl Acad Sci USA.* **96**, 2758–2763.
37. Handel, E.M., Alwin, S., and Cathomen, T. (2009) Expanding or restricting the target site repertoire of zinc-finger nucleases: the inter-domain linker as a major determinant of target site selectivity. *Mol Ther.* **17**, 104–111.

38. Schymkowitz, J., Borg, J., Stricher, F., Nys, R., Rousseau, F., and Serrano, L. (2005) The FoldX web server: an online force field. *Nucleic Acids Res.* **33**, W382–W388.
39. Marcaida, M.J., Prieto, J., Redondo, P., Nadra, A.D., Alibés, A., Serrano, L., Grizot, S., Duchateau, P., Paques, F., Blanco, F.J., and Montoya, G. (2008) Crystal structure of I-DmoI in complex with its target DNA provides new insights into meganuclease engineering. *Proc Natl Acad Sci USA.* **105**, 16888–16893.
40. Redondo, P., Prieto, J., Munoz, I.G., Alibés, A., Stricher, F., Serrano, L., Cabaniols, J.P., Daboussi, F., Arnould, S., Perez, C., Duchateau, P., Paques, F., Blanco, F.J., and Montoya, G. (2008) Molecular basis of xeroderma pigmentosum group C DNA recognition by engineered meganucleases. *Nature.* **456**, 107–111.
41. Arnould, S., Chames, P., Perez, C., Lacroix, E., Duclert, A., Epinat, J.C., Stricher, F., Petit, A.S., Patin, A., Guillier, S., Rolland, S., Prieto, J., Blanco, F.J., Bravo, J., Montoya, G., Serrano, L., Duchateau, P., and Paques, F. (2006) Engineering of large numbers of highly specific homing endonucleases that induce recombination on novel DNA targets. *J Mol Biol.* **355**, 443–458.
42. Ashworth, J., Havranek, J.J., Duarte, C.M., Sussman, D., Monnat, R.J., Jr., Stoddard, B.L., and Baker, D. (2006) Computational redesign of endonuclease DNA binding and cleavage specificity. *Nature.* **441**, 656–659.
43. Berman, H.M., Westbrook, J., Feng, Z., Gilliland, G., Bhat, T.N., Weissig, H., Shindyalov, I.N., and Bourne, P.E. (2000) The protein data bank. *Nucleic Acids Res.* **28**, 235–242.
44. Lassmann, T. and Sonnhammer, E.L. (2005) Kalign–an accurate and fast multiple sequence alignment algorithm. *BMC Bioinformatics.* **6**, 298.

Section II

Artificial Transcription Factors

Chapter 5

Generation of Cell-Permeable Artificial Zinc Finger Protein Variants

Takashi Sera

Abstract

Designed or artificial zinc finger proteins (ZFPs) are one of the most promising DNA-binding proteins that target genomic sequences of interest in vitro and in vivo. Conjugation of other functional domains such as transcriptional regulatory domains and endonucleases to ZFPs provided powerful molecular tools to modulate endogenous gene expression and genetic information. These ZFP variants have been introduced into cells as DNA-encoding ZFP variants by using plasmids or viral vectors. As an alternative delivery method of ZFP variants, we developed cell-permeable ZFP variants by fusing cell-penetrating peptides to ZFP variants. We will describe how to generate cell-permeable artificial ZFP variants and how to examine the cell permeabilities by immunofluorescent staining.

Key words: Zinc finger protein, cell permeability, cell-penetrating peptide, ion-exchange chromatography, immunofluorescent staining.

1. Introduction

Designed or artificial zinc finger proteins (ZFPs) are one of the most promising classes of DNA-binding proteins that target genomic sequences of interest in vitro and in vivo (1). Conjugation of other functional domains such as transcriptional regulatory domains and endonucleases to ZFPs have provided powerful molecular tools to modulate endogenous gene expression and genetic information (2, 3). These ZFP variants have been introduced into cells as DNA-encoding ZFP variants by using plasmids or viral vectors and functioned in host cells after expression.

J.P. Mackay, D.J. Segal (eds.), *Engineered Zinc Finger Proteins*, Methods in Molecular Biology 649,
DOI 10.1007/978-1-60761-753-2_5, © Springer Science+Business Media, LLC 2010

To expand ZFP technologies, we explored direct delivery of ZFP variant proteins into eukaryotic cells. We fused cell-penetrating peptides (CPPs) (or protein-transduction domains) (4) to ZFP variants. These cell-permeable ZFP variants entered into mammalian cells and modulated endogenous gene expression (5) or inhibited DNA virus replication in living cells (6) when exogenously added to the culture medium. In this chapter, we will describe methods to generate and purify a cell-permeable artificial ZFP variant that we call a designed regulatory protein (DRP, designated TAT-ATF in (5)) **(Section 3.1)**. The TAT-ATF contains, from the N-terminus to the C-terminus, the CPP of HIV-1 TAT (residues 47–57, 7), a nuclear localization signal (NLS) from the simian virus 40 large T antigen, an artificial ZFP with six finger domains, herpes simplex virus VP16 activation domain (residues 415-490, 8), and a FLAG epitope tag, and was used to activate the endogenous VEGF-A gene (5). We will also describe methods to determine its cellular localization by immunofluorescent staining **(Section 3.2)**.

2. Materials

2.1. Purification of DRP

1. A DNA encoding the TAT-ATF. The design of other ZFPs for use as DRPs is described in other chapters in this volume (e.g., *see* **Chapters 1, 7,** and **9**) and elsewhere (e.g., 2).
2. An *Escherichia coli* expression vector (e.g., pET vector, Novagen).
3. An *E. coli* strain harboring a plasmid encoding the T7 RNA polymerase (e.g., BL21).
4. LB medium: dissolve 15 g of Luria Broth in 600 mL of deionized water. Sterilize by autoclaving.
5. Antibiotics: ampicillin, dissolve in deionized water at 100 mg/mL and filter-sterilize; chloramphenicol, dissolve in ethanol at 30 mg/mL.
6. Isopropyl β-D-thiogalactoside (IPTG): dissolve in deionized water at 1 M and filter-sterilize.
7. DTT: dissolve in deionized water at 1 M and filter-sterilize.
8. Glycerol: sterilize by autoclaving.
9. 50 mM sodium phosphate, pH 7.2.
10. Lysis buffer: 50 mM sodium phosphate, pH 7.2;1 M NaCl, 5 mM DTT, 0.1 mM $ZnCl_2$, 1 tablet of a protease inhibitor cocktail, Complete, EDTA-free (Roche Diagnostics) per 10 mL.

11. Washing buffer A: 50 mM sodium phosphate, pH 7.2; 0.2 mM DTT, 0.1 mM $ZnCl_2$.
12. Washing buffer B: 50 mM sodium phosphate, pH 7.2; 200 mM NaCl, 0.2 mM DTT, 0.1 mM $ZnCl_2$.
13. Washing buffer C: 50 mM sodium phosphate, pH 7.2, 300 mM NaCl, 0.2 mM DTT, 0.1 mM $ZnCl_2$.
14. Washing buffer D: 50 mM sodium phosphate, pH 7.2, 600 mM NaCl, 0.2 mM DTT, 0.1 mM $ZnCl_2$.
15. Sonicator (e.g., Sonic Dismembrator Model 60, Fisher).
16. DEAE Sepharose Fast Flow column (Amersham) and Bio-Rex 70 column (Bio-Rad).
17. Standard reagents and equipment for SDS-PAGE with a 14% or 10–20% Tris-Glycine gel.
18. Centrifugal filter devices for protein concentration (e.g., Amicon Ultra, 5,000 MWCO, Millipore) and sterilization (e.g., Ultrafree-MS, Millipore).
19. Protein Assay ESL kit (Roche Diagnostics).

2.2. Immunofluorescent Staining

1. HEK293 cells (American Type Culture Collection (ATCC)).
2. 8-well culture slides coated with poly(D-Lysine) (BD Biosciences).
3. Dulbecco's Modified Eagles's Medium (DMEM) (Gibco/BRL) supplemented with 0.1 mM nonessential amino acid (Gibco/BRL) and 10% fetal bovine serum (Gibco/BRL).
4. TBS: Tris-buffered saline, 50 mM Tris, 150 mM NaCl, pH 7.4.
5. Paraformaldehyde solution: suspend 0.4 g of paraformaldehyde in 9 mL of H_2O and add 5 μL of 1 M NaOH. Incubate in 60°C-water bath until dissolved completely. Finally, add 1 mL of 10× TBS to the solution.
6. 0.2% (v/v) Triton X-100 in TBS.
7. Serum solution: 10% (v/v) goat serum (Sigma), 100 mM NH_4Cl in TBS.
8. Anti-FLAG–FITC solution: 50 μg/mL anti-FLAG M2 monoclonal antibody-FITC conjugate (Sigma), 0.1% Tween-20 in TBS.
9. DAPI solution: prepare 1 mg/mL 4′,6-diamidino-2-phenylindole dihydrochloride (DAPI) in H_2O. Add 9 volumes of MeOH to the aqueous solution to the final concentration of 100 μg/mL DAPI and store it at –20°C. Just before use, mix 1 volume of the DAPI stock solution with 999 volumes of TBS.

10. Mounting solution: 70% (v/v) glycerol, 2.5% (w/v) 1,4-diazabicyclo[2.2.2]octane (Sigma) in TBS.
11. Fluorescence microscope (e.g., IX70, Olympus).

3. Methods

3.1. Purification of DRP

1. The DNA encoding the TAT-ATF is cloned into a pET vector (*see* **Note 1**).
2. *E. coli* harboring a plasmid encoding T7 RNA polymerase is transformed with the TAT-ATF expression plasmid.
3. A 600-mL culture is grown in LB medium supplemented with 30 mg of ampicillin and 18 mg of chloramphenicol at 37°C until the optical density at 600 nm reaches 0.6-0.75. The culture is induced with 1 mM IPTG for 3 h.
4. After centrifugation at 5,000×*g* for 10 min at 4°C, the harvested *E. coli* cells are resuspended in 10 mL of cold lysis buffer (*see* **Note 2**).
5. The suspension is frozen at –80°C and thawed on ice. This process is repeated three times.
6. The *E. coli* cells are then lysed by ultrasonication on ice for six cycles of 15 s each with 30 s rests between each cycle.
7. The supernatant cleared by centrifugation at 15,000×*g* for 10 min at 4°C is diluted with 40 mL of washing buffer A so that the final concentration of NaCl is 200 mM.
8. Apply the supernatant to a DEAE Sepharose Fast Flow column (bed volume: 10 mL) equilibrated with 50 mM sodium phosphate, pH 7.2 (*see* **Note 3**).
9. Collect the gravity-flow-through fraction (DEAE-unbound fraction) and wash the column with 50 mL of washing buffer B, then combine these solutions.
10. Apply the combined flow-through solution to a Bio-Rex 70 column (bed volume: 5 mL) equilibrated with 50 mM sodium phosphate, pH 7.2.
11. Wash the column by gravity flow with 100 mL of the washing buffer C.
12. Elute a TAT-ATF protein with washing buffer D.
13. Each fraction is examined by SDS-polyacrylamide gel electrophoresis with a 14% or 10–20% Tris-Glycine Gel.
14. Combine the pure fractions and concentrate them to 100 μL using an Amicon Ultra (5,000 MWCO) centrifugal filter device (4,000×*g*).

15. Filter the protein solution using an Ultrafree-MC sterile centrifugal filter unit (5,000×*g*) for filter-sterilization (*see* **Note 4**).
16. The protein concentration is determined by using the Protein Assay ESL kit.
17. Adjust the final concentration of DTT and glycerol to 5 mM and 50%, respectively, and store the final protein solution at –20C (*see* **Note 5**).

3.2. Immunofluorescent Staining

1. A total of 2 × 10^4 HEK293 cells is plated onto an 8-well culture slide coated with poly(D-Lysine). The cells, in 200 μL of DMEM, are incubated at 37°C in a humidified atmosphere of 5% CO_2.
2. After incubation for 36 h, rinse the cells once with 300 μL of DMEM to remove non-adhered cells and then add 200 μL of fresh DMEM containing 2 μM TAT-ATF.
3. Incubate the slide at 37°C in a humidified atmosphere of 5% CO_2.
4. After incubation for 4 h, rinse the cells with 300 μL of TBS three times.
5. Fix the cells with 300 μL of paraformaldehyde solution for 15 min at room temperature.
6. Rinse the cells with 300 μL of TBS three times.
7. Permeabilize the cells with 300 μL of 0.2% (v/v) Triton X-100 in TBS for 5 min at room temperature.
8. Rinse the cells with 300 μL of TBS three times.
9. Incubate the cells with serum solution for 15 min at room temperature.
10. Rinse the cells with 300 μL of TBS four times.
11. Incubate the cells with 300 μL of Anti-FLAG–FITC solution for 1 h at room temperature in the dark.
12. Rinse the cells with 300 μL of TBS twice, and then once with 300 μL of DAPI solution.
13. Cover the cells with the DAPI solution and incubate for 30 min at room temperature in the dark.
14. Wash the cells with 300 μL of TBS three times.
15. Disassemble the culture slide (*see* **Note 6**).
16. Mount the cells with mounting solution.
17. Fluorescence is analyzed under an Olympus IX70 fluorescence microscope (*see* **Note 7**).

4. Notes

1. Fusion of certain basic CPPs such as a 9-mer of arginine (9) to the N-terminus of ZFP variants may reduce DRP expression in *E. coli*. In that case, insertion of the CPP to the C-terminus of a ZFP variant may improve expression (6).
2. The TAT-ATF protein was more soluble in sodium phosphate buffer than in Tris–HCl buffer.
3. When DEAE resin from Bio-Rad was used, the eluent contained more *E.coli*-derived proteins.
4. This step should not be omitted. Mammalian cells were clearly contaminated when TAT-ATF that was not filter-sterilized was added to the culture medium.
5. Filter-sterilized DTT and autoclaved glycerol should be used.
6. The well frame can be easily and cleanly removed and no glue is left.
7. Because TAT-ATF harbored a NLS and a FLAG epitope tag, nuclei are stained by an anti-FLAG-FITC antibody (5).

References

1. Papworth, M., Kolansinska, P., and Minczuk, M. (2006) Designer zinc-finger proteins and their applications. *Gene.* **366**, 27–38.
2. Blancafort, P., Segal, D.J., and Barbas, C.F., III (2004) Designing transcription factor architectures for drug discovery. *Mol Pharmacol.* **66**, 1361–1371.
3. Porteus, M.H. and Carroll, D. (2005) Gene targeting using zinc finger nucleases. *Nat Biotechnol.* **23**, 967–973.
4. Gupta, B., Levchenko, T.S., and Torchilin, V.P. (2005) Intracellular delivery of large molecules and small particles by cell-penetrating proteins and peptides. *Adv Drug Deliv Rev.* **57**, 637–651.
5. Tachikawa, K., Schröder, O., Frey, G., Briggs, S.P., and Sera, T. (2004) Regulation of the endogenous VEGF-A gene by exogenous designed regulatory proteins. *Proc Natl Acad Sci USA.* **101**, 15225–15230.
6. Mino, T., Mori, T., Aoyama, Y., and Sera, T. (2008) Cell-permeable artificial zinc-finger proteins as potent antiviral drugs for human papillomaviruses. *Arch Virol.* **153**, 1291–1298.
7. Vivès, E., Brodin, P., and Lebleu, B. (1997) A truncated HIV-1 Tat protein basic domain rapidly translocates through the plasma membrane and accumulates in the cell nucleus. *J Biol Chem.* **272**, 16010–16017.
8. Triezenberg, S.J., Kingsbury, R.C., and McKnight, S.L. (1988) Functional dissection of VP16, the trans-activator of herpes simplex virus immediate early gene expression. *Genes Dev.* 2, 718–729.
9. Wender, P.A., Mitchell, D.J., Pattabiraman, K., Pelkey, E.T., Steinman, L., and Rothbard, J.B. (2000) The design, synthesis, and evaluation of molecules that enable or enhance cellular uptake: peptoid molecular transporters. *Proc Natl Acad Sci USA.* **97**, 13003–13008.

Chapter 6

Inhibition of Viral Transcription Using Designed Zinc Finger Proteins

Kimberley A. Hoeksema and D. Lorne J. Tyrrell

Abstract

Currently available therapeutics for hepatitis B virus (HBV) infection have limited effectiveness in patients and often do not clear HBV from the liver due to the persistence of the stable, double-stranded (ds) DNA genome of HBV. By designing zinc finger proteins (ZFPs) to bind the dsDNA genome of a model virus, duck HBV (DHBV), we were able to inhibit viral transcription, and subsequently, viral protein and progeny production. This inhibition is likely due to competition for DNA binding sites between the ZFPs and transcription factors, and interference with read-through transcription by RNA polymerase across the ZFP-binding region. Taking into account some design considerations, this method of inhibiting viral transcription can be applied to other viral infections where viral dsDNA occurs.

Key words: Hepatitis B virus (HBV), targeting enhancer, DNA binding protein, antiviral, double-stranded DNA, transcription.

1. Introduction

The highly stable hepatitis B virus (HBV) DNA genome, called covalently closed circular DNA (cccDNA), resides in the nucleus of infected hepatocytes and serves as the template for the production of viral transcripts and progeny genomes (1). Nucleoside analog therapeutics do not directly target cccDNA and often patients cannot completely clear their infection because the cccDNA reservoir is not eliminated during treatment and reinitiates virus production after cessation of therapy (2, 3). Zinc finger proteins (ZFP), which are Cys_2-His_2 DNA-binding proteins, can be designed to bind specific sequences of DNA (4–8). They

J.P. Mackay, D.J. Segal (eds.), *Engineered Zinc Finger Proteins*, Methods in Molecular Biology 649,
DOI 10.1007/978-1-60761-753-2_6, © Springer Science+Business Media, LLC 2010

can be used to target a double-stranded DNA (dsDNA) genome like that of HBV, where the sites of transcriptional regulation and the characteristics of the transcripts are known. Other viruses have been targeted with ZFPs to inhibit viral transcription and replication, including human papillomavirus type 18 (9), HIV-1 proviral DNA (10), and some agriculturally relevant viruses such as tomato yellow curl leaf virus (11) and beet severe curly top virus (12).

In this chapter, we will describe methods to target DHBV cccDNA and inhibit viral transcription with designed ZFPs (**Section 3.1**). These ZFPs have inhibited viral transcription in tissue culture as evidenced by significant decreases in viral RNA, protein, and progeny production (13). The methods described here assume prior creation of an appropriately designed ZFP, expression and purification of the ZFP from a bacterial expression vector, and cloning into a eukaryotic expression vector (for related methods, *see* **Chapters 1–5, 12** and **23**). This chapter will describe protocols to assess the binding characteristics of the designed ZFP, including apparent equilibrium dissociation constants (**Section 3.2**) and target sequence specificity (**Section 3.3**), using electrophoretic mobility shift assays (EMSA) (**Section 3.4**). Further, we will describe the transfection of mammalian cells with DNA to express the designed ZFP (**Section 3.5**), assessment of inhibition of viral transcription via SDS-polyacrylamide gel electrophoresis and Western blot (**Section 3.6**), and glyoxal-based Northern blot techniques (**Section 3.7**).

2. Materials

2.1. Design Considerations

1. Virus of interest with a dsDNA genome, or with a dsDNA form occurring during the virus life cycle.

2.2. Electrophoretic Mobility Shift Assay-1 (EMSA-1) – Determination of Apparent Equilibrium Dissociation Constant

1. A purified sample of the ZFP of interest.
2. Complementary oligonucleotides containing the ZFP target site or a non-cognate target site.
3. 1× Annealing Buffer: 10 mM Tris–HCl, pH 7.5, 10 mM $MgCl_2$, 50 mM NaCl. Store at room temperature.
4. 2× Gel Shift Buffer: 50 mM Tris–HCl, pH 8.0, 200 mM NaCl, 4 mM DTT, 200 μM $ZnCl_2$, 20% (v/v) glycerol, 100 μg/mL bovine serum albumin (BSA) fraction V (Sigma), 8 μg/mL poly (dI:dC) in water. Add five drops of saturated bromophenol blue stock solution to 10 mL of Gel Shift Buffer to ease sample loading. Aliquot into tubes and store at –20°C.

2.3. EMSA-2 – Determination of Affinity for Target Sequence

1. T4 polynucleotide kinase and 5× Forward Reaction Buffer (Invitrogen).
2. [γ-^{32}P]dATP at 10 μCi/μL (specific activity 3,000 Ci/mmol, PerkinElmer). *Note*: this material is radioactive. Use appropriate precautions.
3. QIAquick Nucleotide Removal Kit (Qiagen).
4. DE81 Ion Exchange filter paper (Whatman).
5. EcoLite Scintillation Fluid (MP Biomedicals) and scintillation vials (Fisher).
6. 2× Gel Shift Buffer, as in **Section 2.2**.
7. 5% (w/v) Na_2HPO_4, pH 9.0.
8. 95% (v/v) ethanol.

2.4. Native Polyacrylamide Gel Electrophoresis

1. 5× TBE: 0.45 M Tris base, 0.45 M boric acid, 10 mM ethylenediamine tetraacetic acid (EDTA). Store at room temperature. A stock solution of 0.5 M EDTA may be used and is produced by adding EDTA to water and dissolving it by bringing the pH to 8.0 with NaOH.
2. Ammonium persulfate (APS): Prepare 10% (w/v) solution in water. Freeze single-use aliquots at –20°C or prepare fresh solution for each use.
3. 40% acrylamide/bisacrylamide solution (29:1, Bio-Rad) and *N*,*N*,*N*,*N*′-tetramethyl-ethylenediamine (TEMED, EMD). Store both at 4°C. *Note*: acrylamide is a neurotoxin. Use appropriate precautions.
4. Native Gel Running Buffer (1× TBE): Dilute 5×TBE to 1× by adding 200 mL of 5×TBE to 800 mL of water.
5. Electrophoretic Mobility Shift Assay Kit (Molecular Probes/Invitrogen).
6. Mini-gel polyacrylamide gel electrophoresis apparatus.
7. Enzyme Kinetics analysis program (Trinity Software).
8. Phosphor-imaging plate and phosphorimager (e.g., Fujifilm).

2.5. Transfection of Mammalian Cells

1. A plasmid capable of expressing GFP in mammalian cells.
2. Mammalian cells of interest (i.e., harboring the virus of interest).
3. 6-Well tissue culture plates (e.g., Sarstedt).
4. Growth medium without antibiotics: a common medium is Dulbecco's Modified Eagle Medium (DMEM, Invitrogen) supplemented with 10% fetal bovine serum (FBS, Sigma). Use an appropriate medium for each cell line.

5. ATV: In 800 mL of water, add 8.0 g KCl, 0.4 g NaCl, 1.0 g $NaHCO_3$, 0.58 g dextrose, 0.5 g trypsin (Invitrogen), 0.2 g disodium EDTA. Adjust the pH to 7.4–7.5, bring the volume up to 1 L and filter sterilize. Store at 4°C.
6. Lipofectamine™ 2000 (Invitrogen) and Opti-MEM® I reduced serum medium (Invitrogen).

2.6. SDS-Polyacrylamide Gel Electrophoresis (SDS-PAGE) and Western Blot for Viral Gene Products

1. A plasmid capable of expressing the ZFP in mammalian cells.
2. Antibodies: Primary antibody specific for viral gene products; secondary antibody conjugated to horseradish peroxidase (HRP).
3. 1× Phosphate-buffered saline (PBS): In 800 mL of water, add 8.0 g NaCl, 0.2 g KCl, 0.92 g Na_2HPO_4, and 0.2 g KH_2PO_4. Adjust the pH to 7.15–7.4 and bring the volume up to 1 L with water. Filter sterilize and store at 4°C.
4. 1× Radioactive immunoprecipitation assay (RIPA) Lysis Buffer: Prepare 10 mM Tris–HCl, pH 8.0, 140 mM NaCl, 0.025% sodium azide, 1% (v/v) Triton X-100, 0.1% (w/v) SDS, 1% (w/v) sodium deoxycholate in 100 mL of water. Store at 4°C.
5. MicroBCA Protein Assay Kit (Pierce) and bovine serum albumin (BSA, Sigma) standards ranging from 20 μg/mL down to 2 μg/mL at intervals of 3 μg/mL.
6. 4× SDS-PAGE Loading Buffer: 125 mM Tris–HCl, pH 6.8, 5% (w/v) SDS, 10% (v/v) 2-mercaptoethanol, 15% (v/v) glycerol, 0.1% (w/v) bromophenol blue. Add 2-mercaptoethanol before use. Store at room temperature.
7. 40% acrylamide/bisacrylamide solution (29:1, Bio-Rad) and TEMED (EMD). Store both at 4°C. *Note*: acrylamide is a neurotoxin. Use appropriate precautions.
8. Ammonium persulfate (APS): Prepare 10% (w/v) solution in water. Freeze single-use aliquots at –20°C or prepare fresh solution for each use.
9. 25% (w/v) SDS: Dissolve 12.5 g SDS into a 50 mL tube. Add water to 50 mL and dissolve SDS by alternating between warming at 42°C and rocking at room temperature. Store at room temperature.
10. 2 M Tris–HCl, pH 8.8. Add 242.28 g Tris base to 700 mL of water. Bring pH to 8.8 using concentrated HCl. Bring volume up to 1 L with water. Store at room temperature.
11. 0.5 M Tris–HCl, pH 6.8. Add 30.29 g Tris base to 400 mL of water. Bring pH to 6.8 using concentrated HCl. Bring volume up to 500 mL with water. Store at room temperature.

12. 1-Butanol.
13. SDS-PAGE Running Buffer: In 800 mL of water, add 30 g Tris, 144 g glycine, and 10 g SDS. Bring volume up to 1 L with water and store at room temperature.
14. Semi-dry Transfer Buffer: In 600 mL of water, add 200 mL methanol, 14.27 g glycine, and 3.0 g Tris base. Bring volume up to 1 L with water and store at room temperature.
15. Standard semi-dry transfer reagents and other materials: Whatman filter paper, nitrocellulose membrane, semi-dry transfer apparatus, plastic bags and heat sealer, acetate cover.
16. 10× Tris-buffered saline (TBS): In 800 mL of water, add 15.76 g of Tris base, and 87.66 g of NaCl. Adjust pH to 8.0 and bring volume up to 1 L with water. Store at room temperature.
17. Skim Milk Block: Dilute 10× TBS to 1× by adding 20 mL of 10× TBS to 180 mL of water. Add Tween 20 (Fisher) to 0.1% (v/v) and skim milk powder to 2.5% (w/v).
18. TBS-T: 1× TBS with 0.1% (v/v) Tween 20.
19. Chemiluminescent substrate (e.g., Supersignal West Dura Extended Duration Substrate, Pierce).
20. X-ray film.

2.7. Glyoxal-Based Northern Blot for Viral Transcripts

1. RNase AWAY (Molecular BioProducts).
2. 10× MOPS solution: 0.4 M MOPS (3-(*N*-morpholino) propanesulfonic acid), 0.1 M sodium acetate, 10 mM EDTA. Make 1 L in water and add 1 mL of diethyl pyrocarbonate (DEPC, Sigma). Let sit overnight at room temperature to degrade RNases, then autoclave the solution to degrade remaining DEPC.
3. Agarose (e.g., Sigma).
4. Agarose gel electrophoresis apparatus.
5. DEPC-treated water: Add 3 mL of DEPC (Sigma) to 3 L of water. Let sit overnight at room temperature and then autoclave.
6. Dimethyl sulfoxide (DMSO, Sigma), tissue culture grade. Aliquot and store at –80°C.
7. Mixed Bed Resin AG 501-X8 (Bio-Rad) and 0.8 × 4 cm disposable chromatography columns (Bio-Rad).
8. 6 M glyoxal (Sigma), deionized. Glyoxal is purchased as a 6 M (40%) solution and must be deionized before use. To deionize, load 5 mL of mixed-bed resin into each of three disposable chromatography columns. Measure the pH of the starting 6 M glyoxal using pH paper; it will probably be

<1. Pour 10 mL of glyoxal over the first column, collecting the eluate in an RNase-free 15 mL centrifuge tube. Pass the eluate over the column again, collecting the eluate in a new tube. Continue passaging the solution twice over each of the two remaining columns or until the pH of the solution appears to have reached a steady state at pH 4.5–5. Measure the pH after each column using pH paper. Prepare ~50 μL aliquots of the deionized solution in RNase-free screw cap tubes and freeze at –80°C.

9. RNA Loading Buffer: 50% (v/v) glycerol, 1 mM EDTA, 0.25% (w/v) bromophenol blue, 0.25% (w/v) xylene cyanol FF. Make 5 mL in a 10 mL glass tube, then add 5 μL of DEPC. Let sit overnight at room temperature, then autoclave. Store at room temperature.
10. 50 mM NaOH and 7.5 mM NaOH, both RNase free. Dilute 10 M NaOH in DEPC-treated water to the appropriate concentration.
11. Standard Northern blot reagents and equipment: positively charged nylon membrane (e.g., Hybond-XL, GE Healthcare), hybridization tubes, and oven.
12. Wash 1: 2× SSC, 0.1% (w/v) SDS; Wash 2: 1× SSC, 0.1% (w/v) SDS; and Wash 3: 0.1× SSC, 0.1% (w/v) SDS, all RNase free. Make 20× SSC containing 3 M NaCl and 280 mM sodium citrate in 1 L of water. Treat with DEPC, let sit overnight and autoclave. Use 25% (w/v) SDS stock from **Section 2.6** and 20× SSC to make the three wash solutions, using DEPC-treated water as diluent.
13. Pre-hybridization solution: 6× SSC, 2× Denhardt's reagent, and 0.1% (w/v) SDS. Denhardt's reagent is prepared at 50× stock containing 10 g/L Ficoll (Type 400), 10 g/L polyvinylpyrrolidone, and 10 g/L bovine serum albumin (fraction V, Sigma). Filter sterilize 50× Denhardt's reagent and store aliquots at –20°C.
14. Random Primer Labeling Kit (Invitrogen) and [α-^{32}P] dCTP at 10 μCi/μL (specific activity 3,000 Ci/mmol, PerkinElmer). *Note*: this material is radioactive. Use appropriate precautions.
15. 1 M NaOH and 1 M HCl. Store at room temperature.

3. Methods

3.1. Design Considerations

In order to design ZFPs to inhibit viral transcription, one needs (1): a virus of interest with a dsDNA genome or with a dsDNA form occurring at some point during the viral life cycle;

(2) knowledge of the location of the virus in the cell; (3) information about enhancers, promoters, and transcription factor-binding sites within the viral genome; and (4) information about the transcripts produced from the viral genome.

1. The first major consideration when designing ZFPs is the form of viral genome. If the virus does not have a dsDNA genome or a dsDNA intermediate, one cannot design ZFPs to bind it with the current technology. Single-stranded DNA genomes could presumably be targeted if they were to form hairpin structures that had long enough stretches of double-stranded regions; however, this type of design will not be discussed in this chapter. The integrated proviral DNA of retroviruses can also be targeted (10).

2. The location of the genome within the cellular compartments is the second major consideration. Does the viral genome localize to the nucleus or the cytoplasm? Does it form viral factories in the cytoplasm, and if it does, would the genome be accessible to an incoming ZFP? If the genome is located in the nucleus, it becomes necessary to engineer a nuclear localization signal (NLS) onto the designed ZFP, whereas this step is not required if the genome is located in the cytoplasm (*see* **Note 1**).

3. If designing the ZFPs to bind a nuclear viral genome, further consideration must be given to inherent chromosome structure and accessibility of the DNA to DNA-binding proteins (DBP). Enhancers and promoters are generally accessible to DBPs and ZFPs should be targeted to bind in these regions. Literature searches of DNase I footprinting assays of the viral genome can reveal accessible regions (*see* **Fig. 6.1**). Keep in mind that revelations of DBP sites can direct you to the right region for design of the ZFP; however, competition with endogenous DBPs may limit the ability of a designed ZFP to bind that sequence. For viral genomes found within the cytoplasm, knowledge of protein-binding sites can also lead to regions that are accessible to ZFPs.

4. Designed ZFPs that inhibit viral transcription are presumed to work in one of the two ways. First, by binding to enhancer

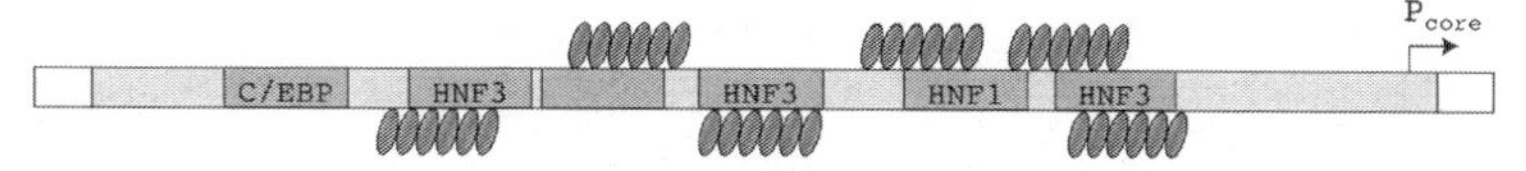

Fig. 6.1. Schematic representation of the enhancer region of DHBV. The *light grey rectangle* represents the enhancer region within the DHBV genome and the *white rectangles* represent adjacent sequences. The *dark grey rectangles* within the enhancer region represent binding sites of transcription factors, including CCAAT/enhancer binding protein-β (C/EBP) and hepatocyte nuclear factors 1 and 3 (HNF1, HNF3). The *grey ovals* represent ZFPs and demonstrate the different regions within the enhancer to which they were designed to bind (13). The promoter for the core open reading frame is shown as P_{core}.

or promoter regions, the ZFP physically interferes with either the binding of transcription factors or the formation of the transcriptional complex. In both of these cases, initiation of transcription is inhibited. Second, ZFPs binding in an open reading frame can prevent read-through transcription across that region, causing the transcript to fall off uncompleted and lacking a stabilizing polyadenylated tail. In the case of DHBV, overlapping open reading frames and long transcript length allow ZFPs binding at one location to prevent the read-through transcription of all three transcripts, allowing a global inhibition of viral transcripts. This may not be the case with other viruses, but should be considered when designing ZFPs. If the ZFPs can be designed to bind a location where the ZFP can function in both ways (inhibiting transcription initiation and transcript read-through), this may optimize the chances for successfully inhibiting viral transcription.

5. Plan ahead for cloning the designed ZFPs into bacterial expression vectors for protein purification and eukaryotic expression vectors for overexpression in a tissue culture model system. Design useful restriction enzyme sites at either end for directional cloning, as well as movement to many common vectors. Ensure that the DNA sequence of the designed ZFPs does not contain common restriction sites to avoid the complication of cutting the ZFP gene during cloning.

3.2. EMSA-1 – Determination of Apparent Equilibrium Dissociation Constant

EMSAs measure the ability of DNA-binding proteins, such as ZFPs, to bind their cognate DNA sequences by visualizing an increase in molecular weight of the small target DNA sequence to that of a large molecular weight complex, representing the ZFP bound to the DNA. Quantification of the large molecular weight complex at different ZFP concentrations allows calculation of the apparent equilibrium dissociation constant. This calculation is described in Step 6 of **Section 3.4**.

1. Prepare the native polyacrylamide gel, as described in **Section 3.4**, prior to setting up the EMSA reaction.
2. This method requires dsDNA oligonucleotides that contain the DNA sequence specific for the designed ZFP. Complementary single-stranded oligonucleotides can be purchased from many biotechnology companies and annealed to produce the requisite dsDNA for the EMSA (*see* **Note 2**). To anneal the two single-stranded oligonucleotides, mix the oligonucleotides at 250 μM concentration each in Annealing Buffer. Place the tubes in boiling water for 5 min, then allow annealing to occur by slow cooling (*see* **Note 3**).

3. In a total volume of 30 μL, add 15 μL of 2× Gel Shift Buffer, dsDNA oligonucleotides to 1.0 μM and ZFP to final concentrations ranging from 150 nM down to 9.5 nM using serial dilutions (*see* **Fig. 6.2** and **Note 4**) (14). Include a negative control (no ZFP) and perform duplicates for all samples. Incubate at room temperature for 1 h. Load the entire sample on a 7% native polyacrylamide gel as detailed in **Section 3.4**.

3.3. EMSA-2 – Determination of Target Sequence Specificity

This EMSA demonstrates the target sequence specificity of the ZFP by including competitor dsDNA oligonucleotides, in addition to the specific ^{32}P-labeled oligonucleotide probe. Designed ZFPs should bind with higher affinity to DNA containing the specific target sequences (probe) than to non-specific sequences (competitor DNA). Excess competitor DNA that contains specific target sequences will lead to a reduction in the signal from

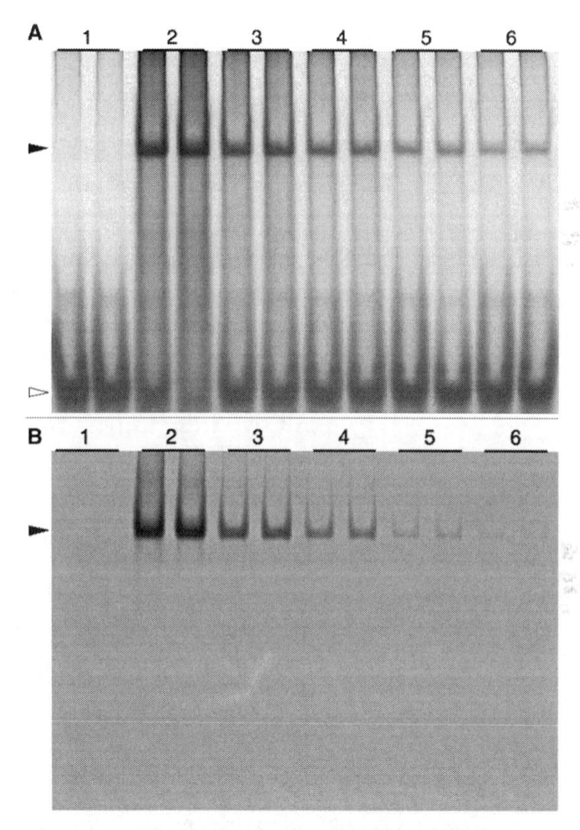

Fig. 6.2. Electrophoretic mobility shift assay-1 (EMSA-1): Determination of the apparent equilibrium dissociation constant. EMSA samples were run on native polyacrylamide gels and stained with (**A**) SYBR-Green stain or (**B**) SYPRO-Ruby stain. *Lanes 1*: Specific oligonucleotide without ZFP; *Lanes 2*: Specific oligonucleotide with ZFP at 150 nM, serially diluted 1 in 2 (*Lanes 3–5*) down to 9.5 nM (*Lanes 6*). *Black arrow*: oligonucleotide bound to ZFP. *White arrow*: unbound oligonucleotide. Each pair of lanes represents duplicate samples.

the ZFP–probe complex, as ZFP is redistributed to the unlabeled DNA. In contrast, excess non-specific competitor DNA should not result in a signal reduction, as ZFP should not redistribute to unlabeled non-target DNA. If the addition of non-specific competitor DNA leads to a reduction in the signal intensity of the ZFP–probe complex, this indicates the ZFP may have off-target binding characteristics, which is an undesirable attribute for inhibition of viral transcription (15).

1. Prepare the native polyacrylamide gel, as described in **Section 3.4**, prior to setting up the EMSA reaction.
2. Dilute dsDNA oligonucleotides produced in **Section 3.2** to 5 pmol with 5 μL of 5× Forward Reaction Buffer, 1 μL of T4 polynucleotide kinase, and 2.5 μL of [γ-^{32}P]dATP (10 μCi/μL). Bring up to 25 μL total volume with water. Mix thoroughly by pipetting up and down (do not mix vigorously). Centrifuge briefly to bring the contents to the bottom of the tube and incubate at 37°C for 10 min. *Note*: this and all subsequent steps contain radioactive materials. Use appropriate precautions.
3. Stop the reaction by heat inactivating the kinase at 65°C for 10 min. Centrifuge briefly.
4. Remove unincorporated [γ-^{32}P]dATP using the Qiagen QIAquick Nucleotide Removal Kit (16).
5. Measure the final activity of the probe by diluting 1:10 with water and spotting 1 μL onto a DE81 filter paper labeled with a pencil. Wash the filter by placing it in a glass beaker with 5 mL of 5% Na_2HPO_4 and gently mixing by rotation for 2 min. Pour off the aqueous solution and repeat the wash two more times. Dehydrate the filter by washing with 5 mL of 95% ethanol for 2 min and dry the filter by placing it face up on Whatman 3 MM filter paper. Place the filter in a scintillation vial when dry, add scintillation fluid, and count the probe activity with a scintillation counter. Dilute to 10,000 cpm/μL in water for use in the EMSA.
6. The total volume of the final reaction will be 30 μL in 1× Gel Shift Buffer. Add 1 μL of probe that has 10,000 cpm/μL. To three different samples, add 5×, 10× or 50× of unlabeled specific oligonucleotide competitor or 50× of unlabeled non-specific oligonucleotide competitor (16). Include a negative control (no ZFP) and a control with just ZFP and probe (no competitors) (*see* **Fig. 6.3**). Add the ZFP last at a final concentration near that of the dissociation constant (*see* **Note 5**). Incubate at room temperature for 1 h. Load the entire sample on a 7% native polyacrylamide gel as detailed in **Section 3.4**.

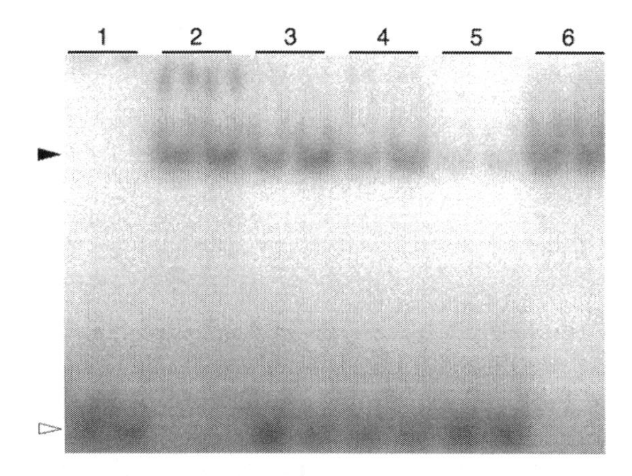

Fig. 6.3. Electrophoretic mobility shift assay-2 (EMSA-2): Determination of affinity for the target sequence. *Lanes 1*: Probe without ZFP; *Lanes 2*: Probe with 150 nM ZFP; *Lanes 3–5*: Probe with 150 nM ZFP and 5×, 10×, or 50× (respectively) of unlabeled specific oligonucleotide competitors; *Lanes 6*: Probe with 150 nM ZFP and 50× of unlabeled non-specific oligonucleotides. *Black arrow*: Probe bound to ZFP. *White arrow*: Unbound probe. Each pair of lanes represents duplicate samples.

3.4. Native Polyacrylamide Gel Electrophoresis

1. These instructions assume the use of a mini-gel system. Scrub the plates with warm soapy water and rinse with 95% ethanol.
2. Pour a 7% native polyacrylamide gel before setting up the EMSA reactions because these gels are slow to polymerize and must be pre-run before loading the samples. Prepare a 1.5-mm thick gel by mixing 2 mL of 5× TBE, 6.2 mL of water, 70 μL of APS, and 1.75 mL of 40% acrylamide/bisacrylamide. Before adding TEMED, remove 1 mL of the gel solution and keep aside (*see* **Note 6**). Add 5 μL of TEMED and pour the gel three quarters of the way, then add the comb and bring volume to the top of the gel. The gel will polymerize completely within 60 min. After polymerization, add 2 μL of TEMED to the extra aliquot of gel solution and use this to top up the wells to completeness. Wait another 15 min for polymerization.
3. Gently remove the comb from the top of the gel. Assemble the gel apparatus and fill the inner and outer chambers with Native Gel Running Buffer. Using a Pasteur pipette, rinse out the wells. Pre-run the gel for 30 min at 100 V, then rinse the wells again using a Pasteur pipette (*see* **Note 7**).
4. Slowly load the EMSA samples into the wells of the gel (*see* **Note 8**). Run the gel at 100 V until the bromophenol blue

dye front is 1 cm above the bottom of the gel (generally takes 1–1.5 h).

5. *For EMSA-1*: Remove the gel from the glass plates and place it into a plastic dish (not a glass dish, as it may absorb the dye). Stain the gel with SYBR Green (supplied in the EMSA Kit) for 20 min (*see* **Note 9**) with gentle agitation by diluting 5 μL of SYBR green in 50 mL of 1× TBE. Keep the gel protected from the light. Wash the gel twice with 150 mL of water for approximately 10 s to remove excess stain.
6. Scan the gel directly using a phosphoimager, with excitation at 495 nm and emission at 520 nm. The dissociation constant can be calculated by plotting the intensity of the shifted, high molecular weight band against the concentration of ZFP, using a program such as Enzyme Kinetics (*see* **Fig. 6.2** and **Notes 10** and **11**).
7. *For EMSA*-2: Remove the gel from the glass plates and seal in a plastic bag. Place the gel in a cassette and expose to a phosphor imaging plate overnight. Scan the plate using a phosphorimager (*see* **Note 12**).

3.5. Transfection of Mammalian Cells

1. These instructions will outline transfection using the lipid-based reagent Lipofectamine™ 2000 (LF2000™) with adherent cells (*see* **Note 13**).
2. Optimize the DNA to LF2000™ ratio in the cell line using a plasmid expressing GFP. Plate cells in 6-well tissue culture plates at a density of 4 × 10^5 cells/well in 2 mL of growth medium lacking antibiotics. The cells should be 90–95% confluent after 24 h; adjust the plating density up or down if the cells are too sparse or too dense. Test DNA (μg):LF2000™ (μL) ratios ranging from 2:1 to 1:5. Use a minimum of 4 μg of DNA for each ratio.
3. Prepare the DNA–LF2000™ complexes at the above ratios by diluting the DNA in a tube with 250 μL per well of Opti-MEM® I reduced serum medium. Dilute the LF2000™ in the same volume in a second tube. Let the tubes sit at room temperature for 5 min, then gently add the LF2000™ mixture to the DNA mixture in a drop-wise fashion. Mix gently and let the tubes sit at room temperature for 20 min. Gently add the entire solution to the plated cells. Mix by gently rotating the plates in a circular and back-and-forth motion. Return the cells to a 37°C/5% CO_2 incubator for 24 h, then assess the transfection efficiency via GFP expression using a fluorescent microscope. For subsequent experiments, use the DNA:LF2000™ ratio that provides the highest transfection efficiency, which should be at least 20–25% GFP positive cells.

3.6. SDS-PAGE and Western Blot for Viral Gene Products

The quickest way to assess the effectiveness of the designed ZFP for inhibiting viral transcription is to look for a reduction in the amount of viral gene products using Western blot. Use antibodies specific to each virus, assaying for structural and non-structural proteins if possible.

1. Transfect the cells with a plasmid encoding the ZFP and incubate for 24–48 h. To harvest the cells, wash once with 1 mL of PBS, remove, then add 400 μL of RIPA Lysis Buffer per well. Let sit for 2 min then collect the samples into microcentrifuge tubes. The samples can be stored at –20°C. Measure the protein concentration of samples using the MicroBCA Protein Assay Kit and BSA standards. Aliquot at least 30 μg of each sample into a new tube and add $^1/_4$ of the total volume of 4× SDS-PAGE Loading Buffer. Boil the samples for 5 min, allow to cool, and centrifuge briefly.
2. Using a mini-gel system, scrub the plates with warm soapy water and rinse with 95% ethanol. Prepare a 1.5 mm thick, 10% separating gel by mixing 1.88 mL of 2 M Tris–HCl, pH 8.8, 2.5 mL of 40% acrylamide/bisacrylamide, 5.5 mL of water, 40 μL of 25% SDS, 100 μL of 10% APS, and 10 μL of TEMED. Pour the gel and leave 1.5 cm at the top for the stacking gel. Layer butanol on top of the separating gel and let the gel set for 30 min. Pour off the butanol and rinse away with water. Dry the top of the gel with a piece of filter paper.
3. Prepare the stacking gel by mixing 625 μL of 0.5 M Tris–HCl, pH 6.8, 244 μL of 40% acrylamide/bisacrylamide, 1.6 mL of water, 10 μL of 25% SDS, 25 μL of 10% APS, and 2.5 μL of TEMED. Add the comb and use a pipette to pour the gel around the comb. Let gel set for 30 min.
4. Gently remove the comb from the top of the gel. Assemble the gel apparatus and fill the inner and outer chambers with SDS-PAGE Running Buffer. Load molecular weight markers into one well and the samples into the remaining wells. Run the gel at 100 V for 15 min, until the samples exit the stacking gel, then increase to 160 V until the bromphenol blue dye front has nearly reached the bottom of the gel.
5. Remove the gel from the apparatus and separate the glass plates so that the gel remains on one side. Use a scalpel to cut away the stacking gel and discard it. Transfer the gel into a dish containing Semi-dry Transfer Buffer and soak for 5 min. In another dish, soak two pieces of Whatman filter paper (cut to just larger than gel size) and a nitrocellulose membrane with Semi-dry Transfer Buffer for 5 min.

6. Pour a small amount of Semi-dry Transfer Buffer onto the bottom surface of the semi-dry transfer apparatus. Set up the transfer from bottom to top as follows: Whatman filter paper, SDS-PAGE gel, nitrocellulose membrane, Whatman filter paper. Place the lid on top and tighten evenly on all sides. Plug the apparatus into a power pack and set the limit to 500 mA for 1 h with a maximum voltage of 29 V (*see* **Note 14**).
7. After transfer, move the membrane into a small dish and add 50 mL of Skim Milk Block for 1 h at room temperature with gentle rotation. Wash the membrane three times for 5 min each with 25 mL of TBS-T.
8. Transfer the membrane to a plastic bag and seal three sides using a heat sealer. Dilute the primary antibody in 2 mL of skim milk block and add the solution to the membrane in the bag. Seal the last side of the bag after the air bubbles are pushed out. Incubate for 1 h at room temperature with gentle rotation then transfer the membrane back to a small dish. Wash the membrane three times for 5 min each with 25 mL of TBS-T.
9. Repeat Step 8 with a secondary antibody conjugated to HRP.
10. Prepare the chemiluminescent substrate in a separate tube by combining 500 μL of each supplied reagent. Pour off the TBS-T from the membrane and add the substrate on top of the membrane, shaking gently to spread it around. Wait 5 min then drain the excess fluid and place the membrane in an acetate cover. Expose the membrane to film in a dark room and develop.

3.7. Glyoxal-Based Northern Blot for Viral Transcripts

The most direct way to assess the effectiveness of the designed ZFP to inhibit viral transcription is to look for a reduction in the amount of viral transcripts using Northern blot (*see* **Note 15**).

1. Spray an agarose gel plate and comb with RNase AWAY and wipe off with paper towel (*see* **Note 16**). Allow to dry. Use tape to seal the ends of the plate and place the comb 1 cm from the top of the gel. Prepare the gel by preheating 10 mL of 10× MOPS buffer to 55°C. Add 1.0–1.5 g of agarose to 90 mL of DEPC-treated water and microwave until all of the agarose powder is dissolved. Allow solution to cool to 55°C and add the preheated 10× MOPS buffer. Pour the gel mixture into the gel plate and allow to solidify for 30 min. Remove the tape from the ends and transfer the gel to an agarose gel box, filling the reservoir with 1× MOPS as running buffer.
2. Mix 15–20 μg of RNA with 15 μL of DMSO, 5.4 μL of 6 M deionized glyoxal, and 3 μL of 10× MOPS buffer.

Bring the volume up to 30 μL with DEPC-treated water. Heat the samples at 50°C for 60 min. After denaturation, add 3 μL of RNA Loading Buffer. Mix by pipetting and load onto agarose gel. Run the gel at 2.5–5 V/cm at room temperature for 60 min.

3. Transfer the gel to a plastic dish, cut away any unused areas of the gel with a razor blade, and soak for 20 min in 50 mM NaOH to partially hydrolyze the RNA and improve the efficiency of the transfer. Place a Plexiglas support spanning a glass baking dish and put a piece of Whatman filter paper that is longer and wider than the gel over the support. Fill the dish with 7.5 mM NaOH until the liquid level is just below the top of the support. Wet the filter paper and smooth out any air bubbles.
4. Cut a piece of Hybond-XL nylon membrane slightly larger than the gel. Cut one corner off the membrane to facilitate orientation. Using forceps transfer the membrane to a container with 7.5 mM NaOH and soak for 5 min.
5. Place the gel on the center of the support, smoothing out any air bubbles between the gel and the filter paper. Using old pieces of film surround the gel on all four sides (*see* **Note 17**). This is to prevent liquid from bypassing the gel during the transfer process.
6. Place the membrane on top of the gel, using the cut corner to orient the membrane over the gel. Cut a piece of Whatman filter paper that is just larger than the membrane, wet it briefly in 7.5 mM NaOH, and place on top of the membrane. Place a stack of paper towels (5–10 cm high) on top of the filter paper. Place a glass plate on top of the paper towels and a weight of approximately 300 g. Allow the transfer to proceed overnight, as the movement of liquid up into the paper towels carries the RNA with it and deposits it on the nylon membrane.
7. Dismantle the transfer setup and place the membrane into a plastic dish. Rinse it briefly with Wash 1 and then dry it face up on paper towel for 30 min at room temperature.
8. Place the membrane in a glass hybridization tube and add 10 mL of pre-hybridization solution, pre-warmed to 68°C. Incubate the membrane for 2 h at 68°C with rotation in a hybridization oven.
9. Prepare the radioactive probe by using a Random Primer Labeling Kit (Invitrogen). Take a fragment of DNA encoding the viral target gene of interest and add 25 ng in 20 μL of water. Boil for 5 min to denature and then transfer immediately to ice. Add 2 μL each of dATP, dGTP, dTTP (0.5 mM each) and 15 μL of the supplied Random Primer

Buffer Mixture. Bring water up to 44 μL and add 5 μL of [α-^{32}P]dCTP and 1 μL of supplied Klenow fragment. Mix thoroughly by pipetting and incubate at room temperature for 3–4 h. Add 12.5 μL of 1 M NaOH for 5 min to denature, then neutralize with 12.5 μL of 1 M HCl. Store at –20°C in a lead container. *Note*: this and all subsequent steps contain radioactive materials. Use appropriate precautions.

10. Add 10 μL of the radioactive probe to the prehydridization solution in the hybridization tube. Incubate overnight at 68°C with rotation in a hybridization oven.
11. Wash the membrane twice for 15 min in Wash 2 and twice for 15 min in Wash 3. Pour off the liquid and wrap the membrane in Saran Wrap. Place the membrane in a cassette and expose to a phosphor image plate overnight. Scan the image plate using a phosphoimager (**Fig. 6.4**).

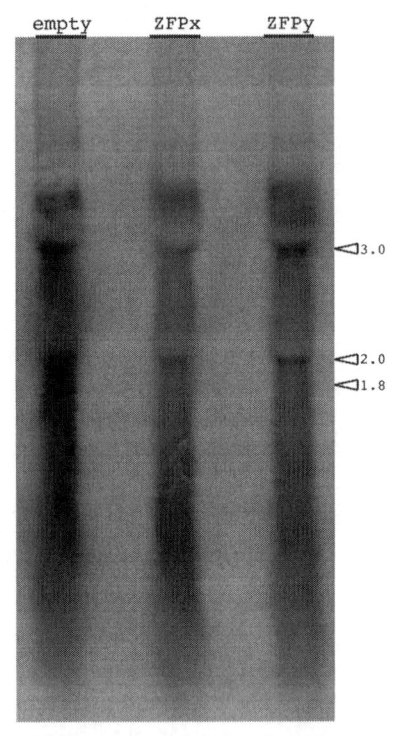

Fig. 6.4. Northern blot of total RNA from chicken hepatoma (LMH) cells transfected with designed ZFPs. Total RNA was isolated from LMH cells 24 h post-transfection with pcDNA3.1(+)–ZFPx, –ZFPy, or an empty vector. The glyoxal-based Northern blot protocol was performed as described in **Section 3.7** using 5 μg of total RNA from each sample. A DHBV-specific probe was used for hybridization. The *arrows* represent the different transcripts produced by DHBV, with sizes of 3.0, 2.0, and 1.8 kb.

4. Notes

1. The Simian Virus 40 (SV40) nuclear localization signal is composed of NH_2-Pro-Lys-Lys-Lys-Arg-Lys-Val-COOH and should be engineered in-frame at the 5′ end of the construct (17).
2. When designing single-stranded oligonucleotides for the EMSAs, we added 12 additional nucleotides flanking the specific target sequence (six on each side) to stabilize the double-stranded oligonucleotide when annealed. For example, if the target sequence was **5′-AAA TTT CCC GGG-3′**, we designed the full-length oligonucleotide to be 5′-*ACT AGT* **AAA TTT CCC GGG** *ACT AGT-3′*.
3. To ensure the oligonucleotides are double stranded after annealing, run an aliquot of the single and double-stranded DNA on a 7% native polyacrylamide gel (*see* **Section 3.4**) and stain the gel with 0.02% methylene blue (20 mg methylene blue, 2 mL 5× TBE, 98 mL water) to visualize the DNA bands. Alternatively, stain the gel with SYBR-green from the EMSA Kit (Molecular Probes/Invitrogen).
4. If the dissociation constant of the ZFP is known prior to the EMSA, use a range of ZFP concentrations around that dissociation constant. If there is no binding by the ZFP at the recommended concentrations increase the concentration of the ZFP in the reaction mixture; however, this indicates that the dissociation constant is high and the ZFP is likely not a very good binder.
5. The concentration of the ZFP should be around the dissociation constant, calculated in **Section 3.4**. Also, in the EMSA-2 procedure, it is important to add the ZFP last, otherwise there may be unequal opportunity for the ZFP to bind the probe or unlabeled competitor and the results may be skewed.
6. Native polyacrylamide gels shrink in size much more than SDS-polyacrylamide gels. By keeping aside some of the gel solution before adding TEMED, the gel can be topped up after solidifying so that the wells are complete for loading samples.
7. Native polyacrylamide gels can be run in the cold room (at 4°C) to reduce heating while running. The heat that is generated may affect the ZFP–dsDNA complexes that are formed in the EMSA reaction as they run through the gel, encouraging their dissociation. However, we have found

no difference between room temperature versus cold room EMSAs and run all of our native gels at room temperature.

8. Load EMSA samples without the addition of any loading buffer. The glycerol and trace of bromophenol blue in the Gel Shift Buffer are sufficient to allow the samples to sink into the well and to visualize them while loading. If necessary, load a separate lane with concentrated bromophenol blue, in order to better track the migration through the gel.
9. Depending on the thickness of the gel, the optimal staining time with SYBR green may vary. Thicker gels or higher percentage polyacrylamide gels will need longer staining times.
10. The EMSA Kit (Molecular Probes/Invitrogen) contains the SYPRO Ruby protein stain in addition to the SYBR green DNA stain. If desired, the gel can be stained with SYPRO Ruby (*see* **Fig. 6.2B**) after the SYBR green stain, which will simply confirm the presence of any proteins in the gel. However, this step is not necessary to calculate the dissociation constant of the ZFP.
11. We have designed a number of different ZFPs targeting both DHBV and HBV and have found large variations in dissociation constants using EMSA-1 (from pM to μM). In general, ZFPs that had dissociation constants higher than 250 nM tended to have little to no effect on the production of viral products in the tissue culture system, whereas those with dissociation constants below this value, in some cases, were found to cause decreases in viral protein and RNA production of up to 80% after 24 h of expression. In addition, some ZFPs that had good dissociation constants (lower than 250 nM) had off target-binding effects, as demonstrated by EMSA-2. These ZFPs were competed off their target sequence with non-specific competitor DNA, indicating they were able to bind DNA sequences other than their cognate sequence. We excluded these ZFPs from further consideration in the tissue culture system.
12. Alternatively, the gel can be exposed to film at room temperature overnight and developed in a film processor. One can also vacuum/heat dry radioactive EMSAs before exposing to film at –80°C; however, we have found that the images of these gels are inferior to exposing an image plate to the wet gel.
13. Depending on the cell type, electroporation-, calcium phosphate-, or lipid-based transfection systems can be used to transfect plasmid DNA. Other methods include the production of adenoviruses or lentiviruses to transduce the gene of interest, for hard-to-transfect cell lines. It is best to

test all of these methods with each cell line using a plasmid expressing GFP and to optimize the transfection efficiency for the system using GFP expression as a read-out.

14. At the start of the transfer, the voltage setting of 29 V will be above what is necessary to reach the limit of 500 mA, but as the transfer progresses, the voltage will need to be continually increased to reach that same limit. If three or four gels are being transferred simultaneously in the apparatus, the transfer time should be increased to 1.5 h.
15. Real-time quantitative PCR on cDNA made from total RNA can also be used to measure transcript levels, depending on the characteristics of the viral genome. Due to the overlapping transcripts of DHBV, each transcript cannot be measured independently from the others, thus northern blots give the most accurate description of each transcript. Assessment of progeny virus production or viral DNA using Southern blots can also be used to examine other effects of the ZFPs on the viral life cycle.
16. All of the glasswares and plasticwares for the Northern blot procedure must be sprayed with RNase AWAY, wiped with paper towel, and allowed to dry before use. Be sure to wear gloves for the entire procedure and spray the gloves with RNase AWAY as well.
17. Saran wrap or parafilm can also be used to surround the gel. We have found that old pieces of developed X-ray film are an easy method of surrounding the gel to prevent bypass fluid movement.

References

1. Ganem, D. and Schneider, R. (2007) Hepadnaviridae. In: (Knipe, D.M., and Howley, P.M., Eds.), Fields Virology, Vol. 2, 5th ed., pp. 2977–3030. Lippincott Williams & Wilkins, Philadelphia.
2. Chang, T.T., Gish, R.G., de Man, R., et al. (2006) A comparison of entecavir and lamivudine for HBeAg-positive chronic hepatitis B. *N Engl J Med.* **354**, 1001–1010.
3. Schreibman, I.R. and Schiff, E.R. (2006) Prevention and treatment of recurrent hepatitis B after liver transplantation: the current role of nucleoside and nucleotide analogues. *Ann Clin Microbiol Antimicrob.* **5**, 8–15.
4. Dreier, B., Segal, D.J., and Barbas, C.F., III (2000) Insights into the molecular recognition of the 5′-GNN-3′ family of DNA sequences by zinc finger domains. *J Mol Biol.* **303**, 489–502.
5. Dreier, B., Beerli, R.R., Segal, D.J., Flippin, J.D., and Barbas, C.F., III (2001) Development of zinc finger domains for recognition of the 5′-ANN-3′ family of DNA sequences and their use in the construction of artificial transcription factors. *J Biol Chem.* **276**, 29466–29478.
6. Dreier, B., Fuller, R.P., Segal, D.J., et al. (2005) Development of zinc finger domains for recognition of the 5′-CNN-3′ family DNA sequences and their use in the construction of artificial transcription factors. *J Biol Chem.* **280**, 35588–35597.
7. Mandell, J.G. and Barbas, C.F., III (2006) Zinc Finger Tools: custom DNA-binding domains for transcription factors and nucleases. *Nucleic Acids Res.* **34**(Web Server issue), W516–W523.
8. Segal, D.J., Dreier, B., Beerli, R.R., and Barbas, C.F., III (1999) Toward controlling

gene expression at will: selection and design of zinc finger domains recognizing each of the 5′-GNN3′ DNA target sequences. *Proc Natl Acad Sci USA.* **96**, 2758–2763.
9. Mino, T., Hatono, T., Matsumoto, N., et al. (2006) Inhibition of DNA replication of human papillomavirus by artificial zinc finger proteins. *J Virol.* **80**, 5405–5412.
10. Reynolds, L., Ullman, C., Moore, M., et al. (2003) Repression of the HIV-1 5′ LTR promoter and inhibition of HIV-1 replication by using engineered zinc-finger transcription factors. *Proc Natl Acad Sci USA.* **100**, 1615–1620.
11. Takenaka, K., Koshino-Kimura, Y., Aoyama, Y., and Sera, T. (2007) Inhibition of tomato yellow leaf curl virus replication by artificial zinc-finger proteins. *Nucleic Acid Symp Ser (Oxf).* **51**, 429–430.
12. Sera, T. (2005) Inhibition of virus DNA replication by artificial zinc finger proteins. *J Virol.* 79, 2614–2619.
13. Zimmerman, K.A., Fischer, K.P., Joyce, M.A., and Tyrrell, D.L.J. (2008) Zinc finger proteins designed to specifically target duck hepatitis B virus covalently closed circular DNA inhibit viral transcription in tissue culture. *J Virol.* **82**, 8013–8021.
14. Moore, M., Klug, A., and Choo, Y. (2001) Improved DNA binding specificity from polyzinc finger peptides by using strings of two-finger units. *Proc Natl Acad Sci USA.* **98**, 1437–1441.
15. Savery, N.J. and Busby, S.J.W. (1998) Mobility Shift Assays. In: (Rapley, R., and Walker, J.M., Eds.), Molecular Biomethods Handbook, pp. 121–129. Humana Press, Totowa.
16. Smith, J., Berg, J.M., and Chandrasegaran, S. (1999) A detailed study of the substrate specificity of a chimeric restriction enzyme. *Nucleic Acids Res.* **27**, 674–681.
17. Yoneda, Y. (1997) How proteins are transported from cytoplasm to the nucleus. *J Biochem.* **121**, 811–817.

Chapter 7

Modulation of Gene Expression Using Zinc Finger-Based Artificial Transcription Factors

Sabine Stolzenburg, Alan Bilsland, W. Nicol Keith, and Marianne G. Rots

Abstract

Artificial transcription factors (ATFs) consist of a transcriptional effector domain fused to a DNA-binding domain such as an engineered zinc finger protein (ZFP). Depending on the effector domain, ATFs can up- or downregulate gene expression and thus represent powerful tools in biomedical research and allow novel approaches in clinical practice. Here, we describe the construction of ATFs directed against the promoter of the epithelial cell adhesion molecule and against the promoter of the RNA component of telomerase. Methods to assess DNA binding of the engineered ZFP as well as to determine and improve the cellular effect of ATFs on (endogenous) promoter activity are described.

Key words: Transient gene expression modulation, VP64, Krüppel-associated box (KRAB), artificial transcription factors, valproic acid.

1. Introduction

An increasing number of diseases are described to be associated with aberrant gene expression profiles due to genetic mutations or epigenetic factors. Turning gene expression on or off at will in a gene-specific way thus represents a powerful tool in biomedical research and opens up new possibilities to regulate the activity of (as yet) undruggable targets in clinical practice. Several approaches are currently being explored to design such gene-specific artificial transcription factors (ATFs) that, at a minimum, consist of a sequence-specific DNA recognition module coupled to a transcriptional effector domain. Depending on the effector

J.P. Mackay, D.J. Segal (eds.), *Engineered Zinc Finger Proteins*, Methods in Molecular Biology 649,
DOI 10.1007/978-1-60761-753-2_7,

domain, gene expression can be transiently upregulated or downregulated (and even permanent modulation can be envisioned by targeting epigenetic enzymes to the promoter of interest). As ATF-mediated upregulation induces the expression of all splice variants in their natural ratios, ATF approaches mimic nature more closely as compared to conventional gene therapy where generally one cDNA is transferred into cells. With respect to downregulation, the ATF approach might be applied as a powerful alternative to (or in synergistic combination with) RNA-targeting approaches as generally only two copies of the DNA have to be bound per cell as compared to the numerous mRNA molecules addressed by siRNA approaches.

To date, several different DNA sequence recognition modules have been developed, including triple-helix forming oligonucleotides (TFOs), synthetic polyamides, and designer zinc finger proteins (1). The DNA-binding domain in most ATFs is based on engineered zinc finger proteins (ZFPs), which can be designed to target unique sequences of 18 bp or longer, as described extensively in this volume. The expression of a number of endogenous genes has been modulated using ZFP-based ATFs (1). In vivo animal models further demonstrated therapeutic relevance of ATFs by studying the endogenous upregulation of *utrophin* (2) and *maspin* (3), as well as the downregulation of *VEGF-A* (4). In fact, convincing therapeutic effects in several animal studies performed by Sangamo Bioscience have allowed the initiation of clinical trails (now at phase II) to test a three finger ZFP-based ATF, which targets 9 bp in the VEGF-A promotor, for its ability to induce therapeutic angiogenesis (5).

To specifically upregulate gene expression, transcriptional activation effector domains are fused to engineered ZFPs, such as the herpes simplex virus protein VP16 (or its tetramer VP64, which consist of four VP16 activation subunits). VP16 activates gene expression by facilitating the assembly of a RNA polymerase II preinitiation complex and interaction with chromatin-remodeling enzymes (e.g., components of the SWI/SNF complex). Alternatively, the NF-κB transcription factor p65 domain or β-catenin subunits can be used to recruit transcription promoting factors to genes of interest (6).

The most commonly used repressive effector domain in ZF-ATFs is the human Krüppel-associated box (KRAB) domain of the KOX1 protein. The KRAB domain consists of an A and B box and a stretch of 45 aa in the KRAB A box has been defined to function as repressor of transcription. Repression occurs by the recruitment of enzymes responsible for changing the epigenetic environment; the action of these enzymes interferes with the TATA box complex assembly and results in relocation of the gene to heterochromatin compartments in the nucleus. Other powerful tools with transcriptional repressive effects are the mSin3 interaction domain (SID, which consists of 35 aa located in the N-terminus of the transcription factor Mad), the ligand-dependent thyroid hormone receptor (TR) or its viral relative vErbA (6).

Unfortunately, at this point, no predictions can be distilled from the literature on which effector domain will work best for a given promoter. Although some studies suggest specific promoter characteristics for certain effector domains to function well, empirical data do not support any such strict rules. Similarly, no absolute rules for optimal target sequences within a promoter can be given, as one effector domain might work efficiently when targeted by a specific ZFP while another gives no effect when fused to the same ZFP (1). Apart from affecting the function of effector domains, the chromatin environment also affects the binding of the ZFP. Combining ATF approaches with chromatin opening drugs is thus likely to result in synergism, as recently shown for ATF-induced upregulation of the epigenetically silenced *maspin* gene (7).

In this chapter we present data on three- and six-finger ZFPs, constructed according to a modular design strategy (8) to target the promoter of epithelial cell adhesion molecule (*EpCAM*) (9) or of the RNA fragment of telomerase (*hTR*) (10). The *hTR* gene is an integral part of the telomerase holoenzyme, which is specifically expressed in the majority of cancer cells and directly confers immortality (11). Similar to the situation for *EpCAM*, levels of *hTR* are low in normal cells, but are substantially increased in cancer cells. Strategies to manipulate these expression levels are of substantial clinical and scientific interest in the cancer field.

Sections 3.1 and **3.2** describe cloning strategies to obtain prokaryotic as well as eukaryotic expression plasmids for ZFPs. Approaches to obtain purified ZFPs are described followed by assays to validate correct expression (western blot) and target DNA binding (electromobility shift assay). After validation, regulatory effector domains (e.g., VP64 or KRAB for up- or down-regulation, respectively) are cloned up- or downstream of the ZFP in the eukaryotic expression plasmid. The resulting ATFs can be tested for functionality after transfection of the eukaryotic expression plasmid in living cells. In **Section 3.3** we describe co-transfection experiments with reporter luciferase plasmids to assess the modulation of *hTR* promoter activity by luciferase assay. For surface proteins, the endogenous gene expression regulation can be visualized by flow cytometry analyses as described in this chapter for EpCAM.

2. Materials

2.1. Construction of the Eukaryotic Expression Plasmid for ATFs

1. TE: 10 mM Tris–HCl, pH 8.0, 1 mM EDTA.
2. Linker oligonucleotides to construct a eukaryotic ATF expression plasmid, as well as oligonucleotides to obtain the ZFPs and effector domains, as depicted in **Table 7.1** (*see* **Note 1**). Oligonucleotides are dissolved in TE.

Table 7.1
Oligonucleotide sequences. Sequences of oligonucleotides required for constructing ZFPs and their ATFs. Primers are dissolved in TE to 1 μg/μL (work solution 1:20) and stored at –20°C unless stated differently. (A) KOZAK-linker sequence to create the pcDNA3.1mnhKOZAK backbone. (B) Primers for a six-finger ZFP (F1–6) construct targeting *hTR* (nucleotides encoding amino acids –1 to 6 are shown in *bold*). (C) Oligos for primer elongation of the downregulatory KRAB effector domain. Restriction sites are shown in *capitals* and complementary sequences are *underlined*. (D) Oligos for primer elongation of the upregulatory VP64 effector domain. Restriction sites are shown in *capitals* and complementary sequences are *underlined*.

A	KOZAK upper	5′-agcttccgccatggttagatctcaaagaagaagagaaaagttaccggtggatcctc
	KOZAK lower	5′-tcgagaggatccaccggtaacttttctcttcttctttggagatctaaggatggcggga
B	i. Const F1-f	5′-gaggagggatccCCCGGGgagaagccctatgcttgtccggaatgtggtaagtccttcagc
	ii. hTR-F1-f	5′-tgtggtaagtccttcagc**cataaaaatgcactgcaaaat**caccagcgtacccatacgg
	iii. Const F2-r	5′-ctgaaagacttgccgcactctgggcatttatacggtttttcacccgtatgggtacgctgg
	iv. hTR-F2-f	5′-gtgcggcaagtctttcagt**gattgtagagatctggcaaga**catcaacgcacccacactgg
	v. Const F3-r	5′-gagaaggacttgccacattctggacatttgtatggcttctcgccagtgtgggtgcgttg
	vi. hTR-F3-f	5′-atgtggcaagtccttctct**agaagtgatgaactggtaaga**caccaacgtactcacaccg
	vii. Term F3-r	5′ctcctcaagctttcaACCGGTgtgagtacgttggtg
	viii. hTR-F4-f	5′-tgtggtaagtccttcagc**agaagtgatgaactggtaaga**caccagcgtacccatacgg
	ix. hTR-F5-f	5′-gtgcggcaagtctttcagt**agaagtgataaactggtaaga**catcaacgcacccacactgg
	x. hTR-F6-f	5′-atgtggcaagtccttctct**caaagagcacatctggaaaga**caccaacgtactcacaccg
	xi. Term F3-r LINKER	5′-ctcctcaagctttcaaccggt*gcttcctccgcttcctcc*gtgagtacgttggtg
C	KRAB: forward primer	5′-aacGGATCCagaacattagtaactttcaaagacgtattcgtagacttcacaagagaagaat ggaaattattagacacagcacaa
	KRAB: reverse primer	5′-tttGGATCCtaagcttactaagtttttgtagttttctaacattacgtttctgtatactatttgttg tgctgtgtctaataatt
D	VP64:forward primer	5′-aaaGGATCCgatgcattagatgactttgacttagatatgctaggatctgacgcgctagacg atttcgatctggacatgttgggcagcgatgctc
	VP64:reverse primer	5′-tttAGATCTcagcatgtcgaggtcgaagtcatccagggcatccgagccaagcatatctaaat cgaaatcgtccagagcatcgctgcccaacatg

3. Eukaryotic expression plasmid pcDNA3.1(+)C (Invitrogen).
4. Restriction enzymes: *Age*I, *Bam*HI, *Bgl*II, *Hind*III, *Pme*I, *Xho*I, *Xma*I (New England Biolabs).
5. Mung bean nuclease (New England Biolabs).
6. T4-DNA ligase and calf intestinal alkaline phosphatase (Fermentas).
7. *Escherichia coli* strain DH5α, Luria Broth base medium (Invitrogen), and ampicillin.
8. High-fidelity Pfu Turbo polymerase (Stratagene).
9. Plasmid preparation kits, PCR product purification, and gel extraction kits (Qiagen).

2.2. Protein Studies

2.2.1. ZFP Protein Induction and Purification (pGEX4T1-ZFP)

1. Prokaryotic expression plasmid pGEX4T1 vector, which contains a GST tag for protein purification (GE Healthcare).
2. *E. coli* strain BL21, 2YT medium (Invitrogen), and ampicillin.
3. Glucose.
4. Isopropyl-β-D-thiogalactopyranoside (IPTG, Invitrogen).
5. Binding buffer: PBS (140 mM NaCl, 2.7 mM KCl, 10 mM Na_2HPO_4, 1.8 mM KH_2PO_4, pH 7.3).
6. Filter 0.45 μm (Corning).
7. Elution Buffer: 50 mM Tris–HCl, pH 8.0, 10 mM reduced glutathione (Sigma).
8. GSTrap HP 1 mL columns (GE Healthcare). Store in 20% EtOH at 4°C.
9. FPLC system (e.g., ÄKTA Explorer, GE Healthcare).
10. French Press (Thermo).
11. 2× ZF Protein Storage Buffer: 20 mM Tris–HCl, pH 7.9, 10 mM DTT, 0.1 mM $ZnCl_2$, 2 M NaCl, 10% glycerol, 0.25× protease inhibitor cocktail.

2.2.2. Electromobility Shift Assay (EMSA)

1. Oligonucleotides: 20 bp containing the target sequence or non-target sequences. Only one strand of the target sequence needs to be labeled with IRdy681. For annealing of oligonucleotides, use 10× PCR buffer without Mg^{2+} (Fermentas).
2. Anti-c-Myc antibody (mouse immunoglobulin, Sigma-Aldrich).
3. Gel casting system (Bio-Rad).

4. Gel analysis system (e.g., Odyssey Scanner, Li-Cor Biosciences).
5. TBE: 95.4 g Tris, 2.75 g boric acid, 2 mL 0.5 M EDTA, pH 8, in 100 mL water.
6. Polyacrylamide gel materials: 40% acrylamide (acrylamide:bisacrylamide 29:1, *Warning*: toxic when soluble!) (Bio-Rad), *N, N, N, N′*-tetramethyl-ethylenediamine (TEMED, Bio-rad), ammonium persulfate (APS, Pharmacia Biotech).
7. Loading buffer: 20% Ficoll 400 (Qbiogene), bromophenol blue.
8. Binding buffer: 100 mM HEPES, pH 7.5, 500 mM KCl, 10 mM DTT.
9. Polydeoxyinosinic-deoxycytidylic acid (poly dI-dC, Sigma).

2.2.3. Western Blot

1. 1° antibody: anti-c-Myc (*see* **Sections 2.2.2**); 2° antibody: anti-mouse IgG peroxidase (DAKO Cytomation).
2. Blotting filters (Whatman paper, Whatman).
3. Nitrocellulose membrane (Whatman).
4. Luminol (SuperSignal West Dura Extended Duration Substrate, Pierce).
5. Polyacrylamide gel materials: 40% acrylamide, 0.5 M Tris–HCl, pH 6.8, 1.5 M Tris–HCl, pH 8.8, 10% sodium dodecyl sulfate (SDS), TEMED, APS.
6. 2× Electrophoresis sample buffer: 1 mL of 0.5 M Tris–HCl, pH 6.8, 1.6 mL of 10% SDS, 800 μL of glycerol, 400 μL of 1% bromophenol blue, 400 μL of β-mercaptoethanol. Add 3.8 mL Milli-Q-purified water to 8 mL final volume.
7. 10×Running buffer: 30.3 g Tris (0.25 M final), 144 g glycine (1.92 M final), and 10 g SDS (1% final) in 1 L Milli-Q-purified water.
8. 10× Blotting buffer: 30.3 g Tris (0.25 M final) and 144 g glycine (1.92 M final) in 1 L Milli-Q-purified water. pH should be 8.3 *(Do not adjust)*.
9. Blocking buffer (make fresh): skim milk (5% in PBS) and Tween 0.1%.

2.3. ATF Validation

2.3.1. Cell Culture and Transfection

1. Cells with or without the expression of the gene of interest (ATCC).
2. 96-well tissue culture treated luminometer plates (Nunc, #136101).

3. Transfection reagent (e.g., Superfect, Qiagen, or SAINT mix, Synvolux Therapeutics).

2.3.2. Luciferase Assay

1. Luciferase Stop and Glo assay kit (Promega).
2. Luciferase reporter plasmid containing the ZFP target sequence (e.g., pGL3hProm867, which contains part of the *hTR* gene) (10).
3. Renilla luciferase control plasmid pRL-SV40 (Promega).
4. Microplate luminometer (e.g., Lumicount, Packard).

2.3.3. Flow Cytometry Analysis

1. Flow cytometer (e.g., FACSCalibur, BD Bioscience).
2. 1° antibody: an EpCAM-specific mouse monoclonal antibody (IgG1) produced by a hybridoma cell line and purified by protein A column chromatography (e.g., Prosep A high capacity, Millipore).
3. 2° antibody: polyclonal rabbit-anti-mouse $F(ab)_2$-FITC (DAKO).
4. Valproic acid (VPA, Sigma-Aldrich).

3. Methods

3.1. Construction of the Eukaryotic Expression Plasmid for ATFs (see *Note 1*)

3.1.1. Cloning of the Backbone pcDNA3.1mnhKOZAK (see *Note 2*)

1. Anneal KOZAK oligonucleotides (*see* **Table 7.1**) by adding 5 nmol of each to a final volume of 50 μL of TE, incubate at 94°C for 1 min, lower the temperature by 2°C every cycle down to 20°C (cycles 2–36).
2. Digest pcDNA3.1(+)C with *Age*I and *Pme*I (removing the *Age*I site and the His-tag sequence) and blunt the overhangs using mung bean nuclease. Religate (*see* **Note 3**).
3. Digest the resulting pcDNA3.1myc-nonhis plasmid with *Xho*I and *Hind*III and ligate with the annealed KOZAK linker to obtain pcDNA3.1mnhKOZAK.

3.1.2. Obtaining Three-Finger ZFPs (Seven-Primer Method) (see *Note 4*)

1. Mix 50 ng of each primer (i, ii, iii, iv, v, vi, and vii for F1–3; and i, viii, iii, ix, v, x, vii (or xi, *see* **Note 5**) for F4–6; **Table 7.1**) with 5 U high-fidelity Pfu Turbo polymerase, 0.4 μL of dNTPs (100 mM), and 5 μL of 10× polymerase buffer; fill up to 50 μL with Milli-Q-purified water; and run a PCR reaction (*cycle 1*: 5 min at 95°C; *cycles 2–11*: 30 s 94°C, 30 s 62°C, decrease annealing temperature

1°C each cycle, 30 s 72°C; *cycles 12–30*: 30 s 94°C, 30 s 52°C, 30 s 72°C; *cycle 31*: 5 min at 72°C; samples can be on hold at 4°C).

2. Purify the PCR product of 285 bp out of a 2% agarose gel, elute in 30 μL of H_2O, and digest the product using *Xma*I and *Age*I immediately before **Section 3.1.3**.

3.1.3. Cloning of Six-Finger ZFP-ATF Plasmids

1. Digest pcDNA3.1mnhKOZAK with *Age*I, dephosphorylate, and ligate to the digested ZFP–PCR product F1–3.
2. Digest resulting pcDNA3.1mnhK-F1–3 with *Age*I, dephosphorylate, and ligate to the digested ZFP-PCR product F4–6.
3. Obtain effector domains by primer extension of described oligonucleotides (*see* **Table 7.1**) and digest the obtained product using *Bam*HI only (for KRAB = 135 bp) (*see* **Note 6**) or *Bam*HI and *Bgl*II (for VP64 = 156 bp).
4. Digest pcDNA3.1mnhKOZAK-F1–6 with *Bam*HI, dephosphorylate, and ligate to digested effector domain fragment (*see* **Notes 3 and 7**).

3.2. In Vitro ZFP Protein Studies

3.2.1. ZFP Protein Induction and Purification (pGEX4T1-ZFP)

1. Digest pcDNA3.1mnhKOZAK-ZFP with *Bgl*II (*see* **Note 6**) and *Bam*HI and ligate the ZFP-containing fragment to *Bam*HI-digested, dephosphorylated pGEX4T1. Transform into BL21 *E. coli*.
2. Use a correct clone of BL21 the transformants to inoculate a 6-mL starter culture in 2YT medium supplemented with 10% glucose and 1% ampicillin overnight at 37°C.
3. Inoculate 500 mL of 2YT medium with 0.1% glucose and 1% ampicillin using 5 mL of the starter culture. Grow at 37°C in the shaker until OD_{600} = 0.5 (about 2.5 h).
4. Add IPTG in a final concentration of 0.5 μM to induce protein production.
5. Grow for 4 h at 37°C in the shaker.
6. Harvest by centrifugation at 4,000×*g* for 30 min at 4°C.
7. Freeze the pellet at –20°C (can be stored at –20°C until further use).
8. Resuspend the pellet in 10 mL of PBS (*see* **Note 8**).
9. Use a French Press (3×at 2,500 psi) to squeeze the bacterial suspension.

10. Centrifuge the bacterial lysate for 15 min at 2,300×*g*.
11. Filter supernatant through a 0.45-μm filter.
12. Bind and elute the protein as described by the equipment manufacturer (e.g., using GSTrap HP 1-mL columns on an ÄKTA Explorer).
13. Store the protein eluates in 1×ZF protein storage buffer at –80°C.

3.2.2. Electromobility Shift Assay (EMSA)

1. Dissolve each oligonucleotide in TE, pH 8, to 100 μM (store at –20°C).
2. Anneal upper and lower oligonucleotides (keep dark) by mixing 2.5 μL of sense oligonucleotide, 2.5 μL of anti-sense oligonucleotide, 5 μL of 10× PCR buffer, and 40 μL of Milli-Q-purified water. Boil for 5 min at 100°C and slowly cool it to room temperature by allowing the boiling water to cool down. Store at –20°C.
3. Prepare a 4% polyacrylamide gel by mixing 0.75 mL of 5× TBE, 0.75 mL of 40% acrylamide, 6 mL of Milli-Q-purified water, 4.5 μL of TEMED, and 37.5 μL of APS. Pipet carefully between the glass plates until you get a little overflow.
4. Place the combs. Let the gel polymerize for 45 min. Pre-run the gel in 0.5×TBE for 30–60 min at 100 V.
5. For a supershift, first incubate 10 ng of purified ZFP (*see* **Note 9**) 1:1 (w/w) with anti-c-Myc antibody for 30 min at room temperature.
6. Preparation of the samples: Prepare a master mix for all tubes to contain 2 μL of 10× binding buffer and 1 μL of poly dI:dC. Add the appropriate amount of Milli-Q-purified water for a final volume of 20 μL. Subsequently, add the appropriate amount of cold probe for competition experiments (we generally use 50× and 100× excess (1 μL = 10 pmol and 2 μL = 20 pmol, respectively). Add 10 ng of purified ZFP to all tubes except for the first (probe only tube). Always add the hot probe last (2 μL of a 1:100 dilution = 0.2 pmol).
7. Incubate reaction tubes for 20 min at room temperature. Add 7 μL of loading buffer and load 20 μL on the polyacrylamide gel.
8. Run the gel for 30 min at 100 V and analyze on the Odyssey Scanner.

3.2.3. Western Blot

1. Prepare a 12% acrylamide solution for the separating gel by mixing 3 mL of 40% acrylamide, 2.5 mL of 1.5 M

Tris–HCl, pH 8.8, 100 μL of 10% SDS, 4.35 mL of distilled deionized water. Add 5 μL of TEMED and 50 μL of 10% APS and carefully pipette between the glass plates. Overlay with isobutanol to have a straight surface.

2. Prepare a 4% stacking gel by mixing 1 mL of 40% acrylamide, 2.5 mL of 0.5 M Tris–HCl, pH 6.8, 100 μL of 10% SDS, and 6.33 mL of distilled deionized water.
3. When the 12% separating gel is polymerized, remove the layer of isobutanol, and add 10 μL of TEMED and 50 μL of 10% APS to the 4% acrylamide solution. Pour the stacking gel carefully on the separating gel and place the comb.
4. Dilute your samples 1:2 in 2× electrophoresis sample buffer.
5. Heat samples to 95°C for 3–5 min (i.e., in boiling water).
6. Load the samples and a protein marker.
7. Electrophorese the samples in 1× running buffer. Set the generator on 80 V until the samples have penetrated the stacking gel (15–30 min), then go to 200 V until the colored front of the loading buffer has reached the bottom of the gel.
8. For blotting, prepare a sandwich by placing a soaked blotting sponge, a soaked Whatman filter, the gel, a soaked nitrocellulose membrane, a soaked Whatman filter, and a soaked blotting sponge in the cassette. All are soaked in 1× blotting buffer.
9. Insert the cassette in a blotting chamber and add prechilled 1× blotting buffer.
12. Transfer at 100 V for 1 h.
13. When finished immerse the membrane in Blocking Buffer, and incubate overnight at 4°C.
14. Add the primary anti-c-Myc antibody to the Blocking Buffer.
15. Wash the membrane three times for 10 min with PBS containing 0.1% Tween.
16. Incubate with the secondary antibody: anti-mouse IgG peroxidase in blocking buffer.
17. Wash four times 10 min with PBS containing 0.1% Tween.
18. Develop the membrane by incubating for 5 min with equal parts of the Stable Peroxide Solution and the Luminol Enhancer Solution according to manufacturer ($0.1\ mL/cm^2$). Cover the blot with a clear plastic wrap and expose to X-ray film.

3.3. Cell-Based Validation of ATFs

3.3.1. Co-transfection Using Reporter Plasmids

Co-transfection and luciferase assay are performed in a single vessel format using 96-well luminometer plates (*see* **Fig. 7.1**).

1. Seed cells in a volume of 200 μL into triplicate wells of 96-well luminometer plates 24 h prior to transfection, at a density sufficient give 70–90% confluence at the time of transfection (*see* **Note 10**). For the cell lines used in **Fig. 7.1**, seed 5×10^4 cells per well.
2. Appropriate controls for each assay include: a promoter reporter of interest, a Renilla luciferase expression vector for normalization of firefly luciferase activity, each ATF expression construct of interest, an ATF construct not specific for the promoter of interest, a vector-only control, an effector domain-only control. It may also be desirable to include mutant promoter constructs modified at the predicted ATF binding sites.

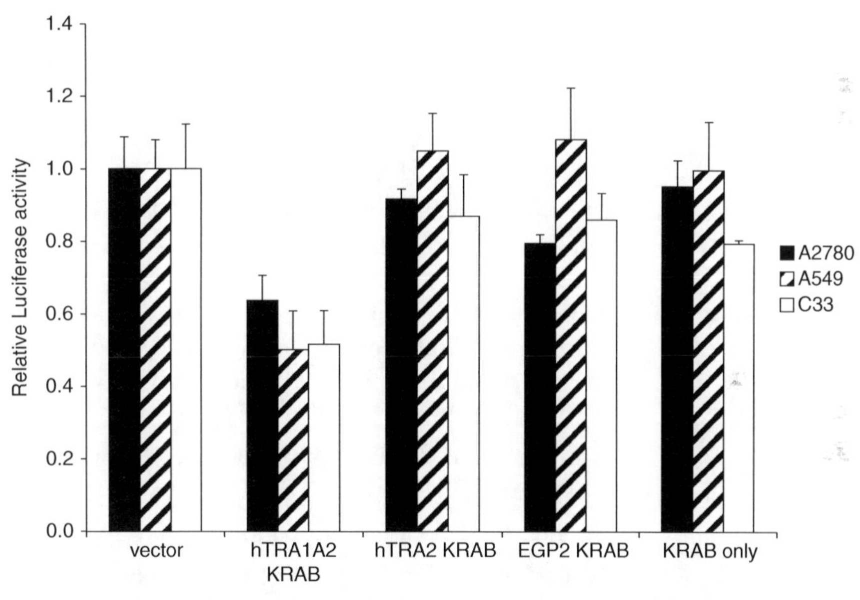

Fig. 7.1. Luciferase activity assays for ZFP-based ATFs. The graph shows luciferase activity following the co-transfection of ATF plasmids with a reporter plasmid containing the luciferase gene under the control of a fragment of the hTR promoter. The ATF plasmids contain a six-finger ZFP directed against *hTR* (A1A2), a three-finger ZFP against *hTR* (A2), or a three-finger ZFP against *EpCAM* (EGP2), all fused to the KRAB repressor domain. The pcDNA3.1mnhKOZAK KRAB plasmid containing an untargeted KRAB domain only (KRAB) was also included. Co-transfections were performed in three different *hTR* positive cell lines (ovarian cancer A2780, lung cancer A549 and cervical cancer C33). Results are the mean of the mean of three independent experiments performed in triplicate. Inhibition of *hTR* promoter activity was up to 50% compared to the vector-only control; $p = 0.02$ for all three cell lines. No effect was seen on three irrelevant promoters (data not shown).

3. For each well to be transfected, dilute 250 ng of firefly luciferase expression vector containing promoter of interest along with 125 ng of ATF expression construct and 30 ng of Renilla luciferase expression construct in 30 μL of serum-free growth medium (*see* **Note 11**). For triplicates, make a 4× DNA Master Mix for each transfection condition.
4. For each well to be transfected, dilute 0.625 μL of Superfect transfection reagent in 20 μL of serum-free growth medium. Make a Superfect Master Mix sufficient for all wells (*see* **Note 12**).
5. Add 80 μL of Superfect Master Mix to each 120 μL of 4× DNA Master Mix and incubate at room temperature for 15 min.
6. To each tube containing the complex of Superfect and DNA, add 600 μL of complete growth medium.
7. Remove the cells from incubation. Aspirate the growth medium from each well using, for example, a 16-gauge syringe needle. Rinse each well once with PBS. Aspirate the PBS.
8. Add 200 μL of the medium containing Superfect and DNA to triplicate wells and return the cells to incubation for 2.5 h.
9. Aspirate the transfection complexes from each well and rinse once with PBS. Remove the PBS and add back 200 μL of complete growth medium to each well.
10. Incubate the cells for 48 h prior to the luciferase assay.

3.3.2. Luciferase Assay (see *Note 10*)

1. Aspirate the medium from all wells using, for example, a 16-gauge syringe needle attached to a vacuum line. Aspirate the medium from the edge of the well to avoid damage to the cell layer.
2. Using a multi-channel pipette, rinse the wells once with 200 μL of PBS.
3. Aspirate the PBS as described for medium in **Section 3.3.1**.
4. To each well, add 50 μL of 1× passive lysis buffer, diluted with distilled water from the 5× solution provided in the Stop and Glo assay kit.
5. Incubate at room temperature for 20 min with gentle shaking using, for example, a rocking table or vortexer with microplate attachment on low speed.
6. Read the luciferase activities of the control and ATF-transfected wells using a microplate luminometer. Fifty microliters of luciferase assay substrate dissolved in luciferase

assay buffer II is added to each well, followed by 50 μL of Stop and Glo substrate dissolved in Stop and Glo buffer. Measure each luciferase activity over 5 s with a 2-s delay following injection of each reagent.

7. Normalize the firefly luciferase activity to the Renilla activity of each well using the formula $L_N = L_W \times (R_M/R_W)$, which preserves the magnitude of the firefly luciferase activity (L_N = normalized firefly luciferase activity, L_W = firefly luciferase activity of a well, R_M = mean Renilla activity of the plate, and R_W = Renilla activity of the well).
8. Calculate the mean and standard errors of triplicate wells using appropriate software (e.g., Microsoft Excel).
9. The relative effect of each ATF on a promoter of interest, such as promoter repression, is estimated by determining the relative luciferase activity: divide the mean luciferase activity of triplicate ATF transfectants by that of vector-only transfectants (*see* **Fig. 7.1**). The transfection/luciferase assay should be repeated at least three times.

3.3.3. Flow Cytometry Analysis of Targeting the Endogenous Promotor (EpCAM)

1. Plate the cells at a density of, e.g., 250,000 cells/well in a 6-well plate (*see* **Note 10**) (*see* **Fig. 7.2**).
2. Transfect the cells at a confluency of about 70% (*see* **Note 12**).
3. Culture the cells for 72 h at 5% CO_2 and 37°C, with or without VPA (*see* **Note 13**).
4. Harvest the cells using TEP (Trypsin/EDTA/PBS) (*see* **Note 14**).
5. Wash cells two times with 2 mL of PBS, and centrifuge 5 min at 500×*g* at 4°C.
6. Add 100 μL of MOC-31 supernatant, resuspend the cells, and incubate for 1 h at 4°C.
7. Wash twice with 2 mL of PBS, and centrifuge 5 min at 500×*g* at 4°C.
8. Aspirate the supernatant and add 100 μL of polyclonal rabbit-anti-mouse Ig-FITC antibody diluted 1:40 in 5% pool serum. Resuspend.
9. Incubate for 30 min at 4°C in the dark.
10. Wash two times with 2 mL of PBS, and centrifuge 5 min at 500×*g* at 4°C.
11. Aspirate the supernatant and resuspend the pellet in 200 μL of PBS.
12. Measure on the FACSCalibur 100,000 cells. Analyze the results using WinList software.

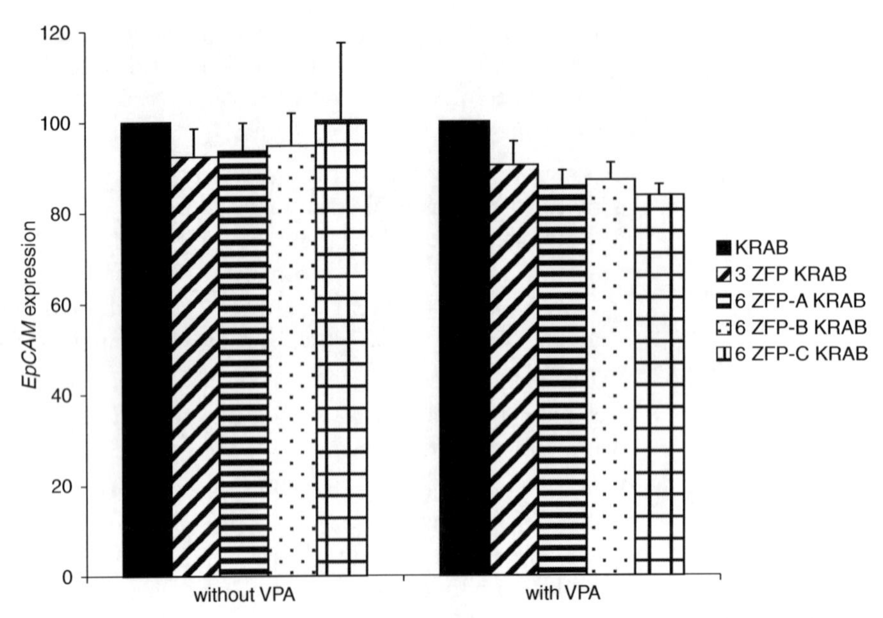

Fig. 7.2. Downregulation of the endogenous *EpCAM* promoter. SKOV-3 cells were transfected with ATF expression plasmids (described in (9)) in the presence or absence of the chromatin opener VPA. No effect on endogenous promoter activity could be observed in the absence of VPA. However, the combinatorial treatment of ATF constructs and VPA resulted in a slight downregulation of up to 15% of the endogenous *EpCAM* expression as measured with flow cytometry. Results are the mean of the mean of three independent experiments performed in duplicate ($^{*}p < 0.05$; $^{**}p < 0.01$; $^{***}p < 0.001$).

4. Notes

1. All plasmids described in this chapter will be provided by the authors upon request.
2. Linkers can be designed to contain any choice of restriction sites. The described KOZAK linker contains the Kozak sequence (underlined in **Table 7.1**), followed by a *Bgl*II restriction site (*see* **Note 6**), a nuclear localization signal, *Age*I and *Bam*HI sites, and one additional nucleotide to keep the reading frame for Myc tag expression. After annealing, the linker is flanked with overhanging *Hind*III and *Xho*I sites.
3. To monitor unwanted effects of mung bean nuclease on double stranded DNA, always sequence your plasmid before continuation. Obviously, all further described primer elongation products and similar materials should also be verified by sequencing analysis.
4. Please refer to reference (*8*) for further information on the 7-primer method, as well as other relevant chapters in this volume regarding ZFP design (*see* **Chapters 1–4**). Check

for the absence of *Bam*HI restriction sites when designing the ZFPs. Always construct several ZFPs for targeting a specific promoter: For *EpCAM*, we constructed one three-finger ZFP and two six-finger ZFPs, all of which worked nicely in reporter studies (9). In contrast, for hTR we constructed five three-finger and two six-finger ZFPs, for which only one of the six-finger ZFP worked consistently.

5. To improve flexibility within the ATF, include a GGSGGS linker in the reverse primer of the last finger (F3 in three-finger ZFPs or F6 in six-finger ZFPs) (*see* **Table 7.1**, primer xi).
6. We created a *Hind*III site in the KRAB construct to verify orientation. VP64 can be checked using *Bam*HI. Obviously, other restriction enzyme recognition sites can be easily included in the linker as well as in the effector domains to use your favorite restriction enzyme. You might also want to clone the effector domain upstream of the ZFP by using *Bgl*II partial digestion (*Bgl*II is also present in the pcDNA3.1 backbone 904 bp 5′ of the *Bgl*II site in the linker).
7. Also, construct plasmids containing the effector domain only as controls.
8. DTT (1–10 mM) can be added to the binding and elution buffer to increase binding of the protein to the column.
9. Always test several amounts of your purified ZFP first for binding to the probe as affinities differ for different ZFPs.
10. Appropriate cell seeding densities should be determined for the culture area and cell type prior to performing the transfection. In the 96-well white luminometer plate format, confluence cannot be examined microscopically prior to transfection. Therefore, determine appropriate seeding densities by seeding a range of concentrations in clear 96-well plates before performing the transfection in luminometer plates. It is also common to perform transfections and harvesting in a clear vessel format, then transfer cell lysates to luminometer plates for luciferase assay. In the single vessel format, inclusion of Renilla luciferase provides a control for transfection efficiency and minor variations in well to well densities.
11. Optimization of plasmid vector amounts may be required in different cells. Thirty nanogram pRL-SV40 provides sufficient signal for normalization in the cells used here. To determine optimal amounts of vectors, titration of the ATF and Renilla constructs should be performed, using empty vector to keep the total DNA amount constant.

12. The optimal ratio of Superfect to plasmid DNA varies according to the cell line. The amount of Superfect stated in the method has been used to successfully transfect several cell lines, including the C33A, A2780 and A549 cells shown in **Fig. 7.1**. However, optimization of Superfect volume may be necessary for other cell lines. Also other transfection agents can be used (SAINT transfections have been performed for the FACS experiments).
13. Other epigenetic drugs can be used to open up the chromatin. Always determine the optimal, non-toxic dose for each cell line.
14. Include positive and negative cells as control, also incubate a control with conjugated 2° antibody only and include an unstained sample to exclude autofluorescence.

References

1. Visser, A.E., Verschure, P.J., Gommans, W.M., Haisma, H.J., and Rots, M.G. (2006) Step into the groove: engineered transcription factors as modulators of gene expression. *Adv Genet.* **56**, 131–161.
2. Lu, Y., Tian, C., Danialou, G., Gilbert, R., Petrof, B.J., Karpati, G., and Nalbantoglu, J. (2008) Targeting artificial transcription factors to the utrophin A promoter: effects on dystrophic pathology and muscle function. *J Biol Chem.* **283**, 34720–34727.
3. Beltran, A., Parikh, S., Liu, Y., Cuevas, B.D., Johnson, G.L., Futscher, B.W., and Blancafort, P. (2007) Re-activation of a dormant tumor suppressor gene maspin by designed transcription factors. *Oncogene.* **26**, 2791–2798.
4. Kang, Y.A., Shin, H.C., Yoo, J.Y., Kim, J.H., Kim, J.S., and Yun, C.O. (2008) Novel cancer antiangiotherapy using the VEGF promoter-targeted artificial zinc-finger protein and oncolytic adenovirus. *Mol Ther.* **16**, 1033–1040.
5. Sera, T. (2009) Zinc-finger-based artificial transcription factors and their applications. *Adv Drug Deliv Rev.* **61**, 513–526.
6. Verschure, P.J., Visser, A.E., and Rots, M.G. (2006) Step out of the groove: epigenetic gene control systems and engineered transcription factors. *Adv Genet.* **56**, 163–204.
7. Beltran, A.S., Sun, X., Lizardi, P.M., and Blancafort, P. (2008) Reprogramming epigenetic silencing: artificial transcription factors synergize with chromatin remodeling drugs to reactivate the tumor suppressor mammary serine protease inhibitor. *Mol Cancer Ther.* **7**, 1080–1090.
8. Carroll, D., Morton, J.J., Beumer, K.J., and Segal, D.J. (2006) Design, construction and in vitro testing of zinc finger nucleases. *Nat Protoc.* **1**, 1329–1341.
9. Gommans, W.M., McLaughlin, P.M., Lindhout, B.I., Segal, D.J., Wiegman, D.J., Haisma, H.J., van der Zaal, B.J., and Rots, M.G. (2007) Engineering zinc finger protein transcription factors to downregulate the epithelial glycoprotein-2 promoter as a novel anti-cancer treatment. *Mol Carcinog.* **46**, 391–401.
10. Zhao, J.Q., Hoare, S.F., McFarlane, R., Muir, S., Parkinson, E.K., Black, D.M., and Keith, W.N. (1998) Cloning and characterization of human and mouse telomerase RNA gene promoter sequences. *Oncogene.* **16**, 1345–1350.
11. Shay, J.W. and Keith, W.N. (2008) Targeting telomerase for cancer therapeutics. *Br J Cancer.* **98**, 677–683.

Chapter 8

Construction of Combinatorial Libraries that Encode Zinc Finger-Based Transcription Factors

Seokjoong Kim, Eun Ji Kim, and Jin-Soo Kim

Abstract

Combinatorial retroviral libraries of zinc finger transcription factors (ZF-TFs) can be constructed that encode tens of thousands of different, multi-finger zinc finger proteins (ZFPs) with distinct DNA-binding specificities, each of which is fused to a transcriptional activation or repression domain. Individual zinc fingers (ZFs) recognize their target DNA subsites and retain their binding specificities in the context of artificially constructed multi-finger ZFPs. Because of this modular nature, expression libraries that specify diverse multi-finger ZF-TFs can be created by the combinatorial stitching together of individual modules in a pool of single-ZF-encoding DNA segments. When these libraries are introduced into cells, the encoded ZF-TFs are expressed and can then activate or repress the transcription of endogenous target genes. Ideally, the ZF-TF-encoding retroviral vectors in the library enter cells randomly at a ratio of ~1 per cell. As a result, individual cells express different ZF-TFs and thus display distinct phenotypical changes. Using an appropriate screening or selection method, one can isolate clonal cells that display phenotypes of interest. One can then identify the ZF-TFs responsible for the phenotypes and, ultimately, the genes that are targeted by the selected ZF-TFs. Here, we provide protocols for the preparation of retroviral libraries that encode ZF-TFs for use in mammalian cells.

Key words: Cellular reprogramming, artificial transcription factor, zinc finger, metabolic engineering, stem cell, strain improvement.

1. Introduction

Transcription factors induce diverse phenotypical changes in cells and organisms by regulating the expression of endogenous genes in the course of differentiation, development, and adaptation to environmental challenges. These processes can be mimicked by artificial transcription factors composed of DNA-binding ZF

J.P. Mackay, D.J. Segal (eds.), *Engineered Zinc Finger Proteins*, Methods in Molecular Biology 649,
DOI 10.1007/978-1-60761-753-2_8, © Springer Science+Business Media, LLC 2010

modules and transcriptional effector (activation or repression) domains. Individual ZFs retain their DNA-binding specificities in the context of artificially constructed multi-finger ZF proteins (ZFPs). Because of this modular nature, combinatorial expression libraries can be created that specify diverse, artificial, multi-finger ZF-TFs composed of subsets of the tens of thousands of possible different single-ZF modules (1). We and others have used ZF-TF libraries to induce diverse phenotypical changes in various cell lines and organisms from bacteria to plants to mammalian cells (**Table 8.1**) (2–11).

ZF-TF libraries are highly efficient tools for strain improvement of both prokaryotic and eukaryotic microorganisms. Various desirable traits, such as drug resistance, enhanced efficiency of protein production, and tolerance under stressful conditions, have been induced in yeast and *E. coli* using ZF-TFs. Complex phenotypes such as these are unlikely to be controlled by the down-regulation or up-regulation of a single gene. Therefore, ZF-TFs may exert their effects by regulating the expression of multiple target genes or by inducing a global perturbation of gene expression (12).

ZF-TF libraries are also useful for the reprogramming of mammalian cells. After screening thousands of artificial ZF-TFs, we identified one that induced the trans-differentiation of mouse myoblasts to osteoblasts and another that induced neurogenesis of a mouse neuroblastoma cell line in the absence of retinoic acid, a natural signaling molecule (6). We and others also demonstrated that ZF-TFs could induce Taxol resistance in human cervical cancer cells (7, 8).

Recent advances in stem cell research have focused on the controlled differentiation of stem cells and the de-differentiation or trans-differentiation of somatic cells. Of particular note is the recent derivation of induced pluripotent stem cells (iPS cells) by introducing, into somatic cells, genes that encode four naturally occurring transcription factors, OCT4, SOX2, KLF4, and c-MYC (13). The Yamanaka group screened dozens of transcription factors in murine somatic cells before they identified these four factors that, together, induce the de-differentiation of somatic cells and their subsequent reprogramming into pluripotent states (14). In the same vein, we envision that one could screen ZF-TF libraries for artificial transcription factors that induce or enhance the differentiation of stem cells and de-differentiation or trans-differentiation of somatic cells.

In principle, three different sets of ZF-TF libraries can be employed in a variety of organisms; these include a ZFP–transcriptional activator domain fusion library, a ZFP–transcriptional repressor domain fusion library, and an isolated ZFP library. The multi-finger ZFPs encoded in an isolated ZFP library neither contain transcriptional activator nor repressor

Table 8.1
Summary of results from previous studies that employed ZF-TF libraries and phenotype-based screening in various model systems

Organism	No. of fingers	Library complexity	Effector domain	Phenotype	Related gene(s)	References
Arabidopsis	3	3.8×10^3	VP16	Increasing somatic HR	–	Lindhout et al. (2)
E. coli	3/4	$6.4 \times 10^4/2.6 \times 10^6$	None	Thermotolerance	ubiX	Park et al. (3)
	3	6.4×10^4	CRP D1	Thermotolerance	cpxP, ompW, marRAB	Lee et al. (4)
				Osmotic tolerance		
				Cold shock resistance		
S. cerevisiae	3/4	$6.4 \times 10^4/4 \times 10^5$	Gal4 /Ume6	Thermotolerance	PDR5, YLL053C	Park et al. (6)
				Osmotic tolerance		
				Drug resistance		
	4	4×10^5	Gal4/Ume6	Enhancing protein productivity	–	Park (5)
Mammalian cells	3/4	$6.4 \times 10^4/4 \times 10^5$	p65/Kid	Enhancing protein productivity	–	Park (5)
	4	4×10^5	p65	Taxol resistance	R92425, AA421265	Lee (7)
	6	8.42×10^7	VP64	Taxol resistance	E48, AGT, IL-13Rα1	Blancafort et al. (8)
	3/6	$1 \times 10^4/8.4 \times 10^7$	VP64	Drug resistance	hDMP1α/p21$^{WAF1/CIP1}$	Blancafort et al. (9)
	3/6	$1 \times 10^4/8.4 \times 10^7$	VP64	Induction of endothelial marker	CDH5	Blancafort et al. (10)
	4	4×10^5	KRAB/VP16	Improving antibody production	–	Kwon et al. (11)
	4	4×10^5	p65/KRAB	Neurogenesis	–	Park et al. (6)
				Osteogenesis		

domains, but ZFPs alone can function as transcriptional repressors by blocking the association of RNA polymerase and endogenous transcription factors with DNA. Transcriptional activation and repression domains usually are species specific. Thus, we use the Gal4 activation domain and the Ume6 repression domain in *Saccharomyces cerevisiae* and other yeasts and the NFkB p65 activation domain and the Krüppel-associated box (KRAB) repression domain in mammalian cells.

In this chapter, we provide protocols for the preparation of retroviral ZF-TF libraries for use in mammalian cells. Phenotype-based screening in a mammalian cell culture system using a randomized ZF-TF library requires the following procedures: (1) construction of a ZF-TF library with randomized DNA-recognition specificity, (2) an efficient method for transfer of the ZF-TF library into cultured mammalian cells, and (3) a highly specific selection system that permits the isolation or enrichment of cells that have acquired the desired phenotype (**Fig. 8.1**). Because various selection systems for many different cellular phenotypes can be considered, depending on the interest of the researchers, in this chapter, we focus on providing detailed protocols for the construction of a combinatorial ZF-TF library by random shuffling of ZF modules in a retroviral vector system (**Sections 3.1** and **3.2**) and retrovirus-mediated delivery of the ZF-TF library into mammalian cells (**Sections 3.3** and **3.4**).

2. Materials

2.1. ZF Array Libraries

1. Library of single-ZF modules. Previously, we described the identification and characterization of 50 ZFs from the human genome that display diverse DNA-binding specificities (15). Of these, 35 ZFs are available from Addgene, a non-profit plasmid distribution service (http://www.addgene.org/pgvec1), as a part of the Zinc Finger Consortium collection.

To establish a single-ZF library, single-finger encoding modules were each amplified by PCR and subcloned individually into the p3 vector, a derivative of pcDNA3 (Invitrogen) that contains a multiple cloning site (MCS) designed to provide a framework for the construction of ZF tandem arrays using standard molecular biology techniques (**Fig. 8.2a, b**). In addition, p3 contains sequences that encode a hemagglutinin (HA) epitope and a nuclear localization signal (NLS) to facilitate, respectively, the detection and appropriate localization of expressed ZFs. Equal amounts of each single-finger vector were combined to generate

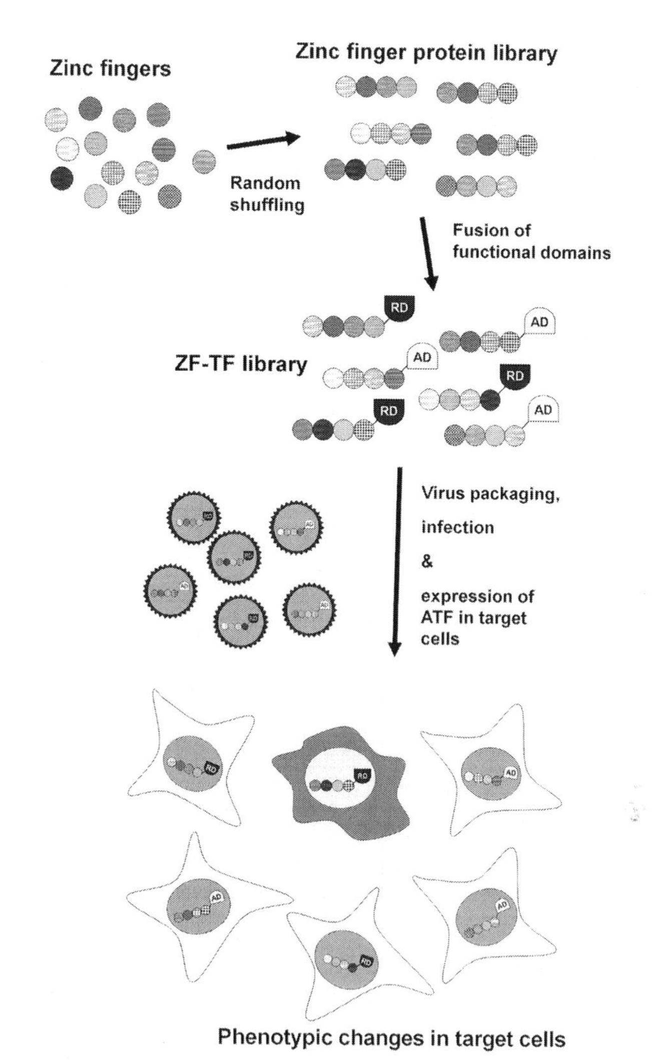

Fig. 8.1. Scheme of screening procedure for specific phenotypes using a retroviral ZF-TF library. RD, transcriptional repression domain; AD, transcriptional activation domain.

a single-finger library. The p3 backbone vector and single finger library in p3 vector system are available upon request.

2. *Eco*RI, *Xho*I, *Xma*I, *Not*I, *Age*I, T4 DNA ligase, and calf intestine phosphatase.
3. Luria–Bertani (LB) medium: 1% (w/v) Bacto-tryptone, 1% (w/v) sodium chloride, 0.5% yeast extract.
4. LB–ampicillin agar plates were prepared by pouring an autoclaved LB medium/0.2% (w/v) bacto-agar mixture supplemented with 100 μg/mL ampicillin onto conventional disposable Petri plates (100 mm diameter) or square Petri plates (245 mm × 245 mm, Corning).

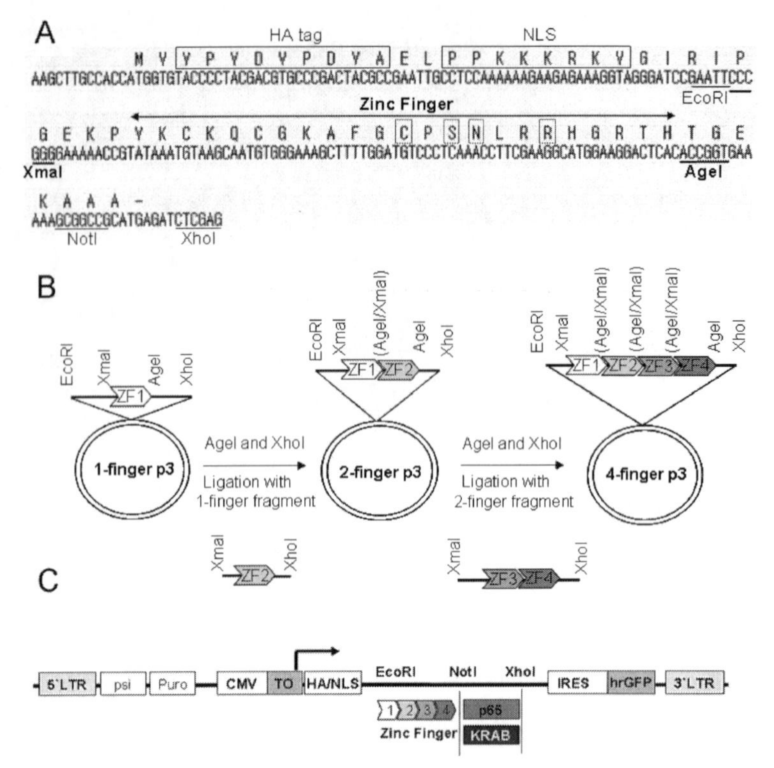

Fig. 8.2. Schematic map of vector systems. (**A**) Nucleotide sequence of the MCS of the p3 expression plasmid housing ZF107 (Zinc Finger Consortium Collection, Addgene). Amino acid residues important for DNA recognition specificity are indicated by *squares*. (**B**) Construction of a randomized four-finger ZFP library, from a pool of one-finger modules, using the p3 vector system. (**C**) Schematic map of pLPCX/TO-IG with either the p65 transcriptional activation domain or the KRAB repression domain.

5. Standard materials and reagents for agarose gel electrophoresis.
6. QIAquick Gel Extraction Kit, QIAquick PCR Purification Kit, Plasmid Prep Kit (Qiagen).
7. Chemically competent (cloning grade) and Electro Ten-Blue electroporation-competent *E. coli* cells (Stratagene).
8. Standard materials and reagents for bacterial electroporation.
9. SOC medium: 0.5% yeast extract, 2% tryptone, 10 mM sodium chloride, 2.5 mM potassium chloride, 10 mM magnesium chloride, 10 mM magnesium sulfate, and 20 mM glycerol.

2.2. Transcriptional Regulatory Domains and Retroviral Vectors

Retroviral vector. We constructed the retroviral vector pLPCX/TO-IG using a multi-step subcloning procedure as follows. A tetracycline-regulated CMV promoter (CMV/TO) from pcDNA5/TO/FRT (Invitrogen), the MCS from p3, and the

IRES–hrGFP element from pShuttle-IRES-hrGFP (Stratagene) were inserted into the pLPCX retroviral vector (Clontech) to produce pLPCX/TO-IG. Subsequently, the KRAB transcriptional repression domain of ZFN10 or the transcriptional activation domain of NFkB p65 was cloned into pLPCX/TO-IG to construct pLPCX/TO-KRAB-IG or pLPCX/TO-p65-IG, respectively (*see* **Fig. 8.2c**). These vectors are available upon request.

2.3. Production of Retroviruses

1. 293T/17 cells (CRL-11268, American Type Culture Collection (ATCC)), cultured in Dulbecco's Modified Eagle's Medium (DMEM) supplemented with 10% fetal bovine serum (FBS), 100 unit/mL penicillin, and 0.1 mg/mL streptomycin.
2. TransIT-293 transfection reagent (Mirus Bio) and Opti-MEM reduced serum medium (Invitrogen), used for transient transfection of 293T/17 cells.
3. pGP (Takara) – a retroviral Gag-Pol expression vector.
4. pVPack-VSV-G (Stratagene) – an expression vector for the vesicular stomatitis virus G (VSV-G) envelope protein.
5. pcDNA6/TR (Invitrogen) – a mammalian expression vector for the tetracycline repressor protein.

2.4. Retroviral Infection

1. IMR-90 cells (CCL-186, ATCC).
2. Fibroblast medium: Minimum Essential Medium Eagle (MEM) supplemented with 10% FBS, 1 mM sodium pyruvate, and antibiotics (Pen/Strep).
3. Polybrene (Sigma), dissolved in phosphate-buffered saline (PBS) at a concentration of 4 mg/mL, filter sterilized, and stored in aliquots at −20°C.
4. Flow cytometer.

3. Methods

3.1. Construction of Randomized ZF Array Library

1. First, determine the desired complexity of the randomized ZFP library to be constructed. To this end, one should consider the number of different single-ZF modules to be used and the number of ZFs required in the final multi-finger ZFPs. For example, we routinely construct randomized four-finger ZFP libraries using 30 different well-characterized single-finger modules. To do so, we first build a randomized two-finger ZFP library and, subsequently, use it to make a four-finger ZFP library. The theoretical complexity of a randomized two-finger array of 30 different single-ZF modules is 900 (30 fingers ×

30 fingers) and that of a four-finger array of 30 different single-ZF modules is 810,000 (30 fingers × 30 fingers × 30 fingers × 30 fingers). To ensure that greater than 90% of all possible combinations are included in a resulting library, it is recommended to obtain a number of transformed colonies that are at least three times the theoretical complexity.

ZFs in our p3 vector system are flanked by upstream *Xma*I site and downstream *Age*I site. These two restriction enzymes make compatible cohesive ends on DNA which can be efficiently ligated but with destroying the recognition sequence of either restriction enzyme. This enables the construction of long ZF array by repeating simple molecular cloning procedure of combining two ZF moieties separately digested by *Xma*I/*Xho*I or *Age*I/*Xho*I (**Fig. 8.2B**). While this chapter describes the construction of ZF-TF library with four-ZF array, because the construction of a two-finger library from a single-finger pool and the construction of a four-finger library from a two-finger library follow the same basic protocol and because it is relatively easy to construct a two-finger library with a theoretical complexity of 900, we focus here on the construction of a four-finger library by random shuffling of a two-finger library.

2. Prepare the p3–two-finger vector fragment by digesting 30 μg of a two-finger library by incubation with 90 U of *Age*I and *Xho*I overnight at 37°C. Incubate the digested DNA with 6 μL of calf intestine alkaline phosphatase for 1 h at 37°C. Separate the digested DNA fragments by agarose gel electrophoresis [0.8% agarose gel in Tris/borate/EDTA buffer (TBE)] and then purify the p3–two-finger fragment, which is ~5.5 kb in size, using a QIAquick Gel Extraction Kit.
3. Prepare the two-finger inserts by digesting 100 μg of a two-finger library by incubation with 300 U of *Xma*I and *Xho*I overnight at 37°C. Do NOT to treat this DNA with calf alkaline phosphatase. Separate the digested DNA fragments by agarose gel electrophoresis (2.5% agarose gel in TBE) and purify the two-finger fragments of ~150 bp in size using a QIAquick Gel Extraction Kit.
4. Determine the background (colonies from self-ligation) associated with the library construction cloning process by performing a small-scale ligation of the p3–ZF vector fragments and the two-finger insert fragments and then transforming *E. coli* with the ligation mixture. Briefly, ligate 100 ng of the purified p3–ZF vector fragment with 10 ng of the two-finger insert fragment using 0.5 μL of T4 DNA

ligase in a 10 μL total reaction volume. This amount of vector and insert will yield a vector:insert molar ratio of 1:3. As a negative control, ligate 100 ng of the purified p3–ZF vector fragment with no insert fragment added. Incubate the ligation reaction mixtures overnight at 16°C and then heat inactivate the reactions by incubation at 65°C for 10 min. Use the inactivated reaction mixtures to transform 200 μL each of chemically competent *E. coli* cells. When plated onto LB–amp plates, the 'vector + insert' ligation reaction should produce at least 20 times more colonies than the 'vector only' ligation reaction (*see* **Note 1**).

5. Ligate 5 μg of purified p3–ZF vector fragment with 500 ng of two-finger insert fragment using 25 μL of T4 DNA ligase in a 200 μL reaction volume. After overnight ligation at 16°C, incubate the ligation mixture at 65°C for 10 min for heat inactivation. To concentrate and desalt the ligation mixture, purify the DNA using a QIAquick PCR purification kit. For elution use 20 μL of EB buffer (*see* **Note 2**).
6. Prepare 30 LB–amp square plates. While this can be done with conventional Petri dishes, it would require several hundred, making it inconvenient to perform subsequent experiments.
7. Transform the purified ligation mixture into *E. coli* cells by electroporation. Mix 10 μL of the purified ligation mixture with 200 μL of Electro Ten-Blue Electrocompetent *E. coli* cells in a pre-chilled microfuge tube. Transfer 40 μL of the DNA–cell mixture to each of five pre-chilled 0.2-cm-gap electroporation cuvettes. Insert the cuvettes into an electroporator and treat with 2.25 kV of electricity. After electroporation, collect the cells in 10 mL of SOC medium and incubate the mixture at 37°C on a shaking incubator for 1 h. Next, add 50 mL of SOC medium (a final volume of 60 mL) and spread 2-mL aliquots of the electroporated cell suspensions onto the LB–amp square plates. Additionally, to estimate the number of total colonies, plate 20, 2, and 0.2 μL of cell suspension to 100 mm LB–amp plates (*see* **Note 3**). Incubate the plates at 37°C overnight.
8. Count the number of colonies in the 100-mm plates and use this value to estimate the total number of colonies. Ideally, a 100-mm plate with 2 μL of the cell suspension plated on it will yield between 100 and 300 colonies (*see* **Note 3**). This means that the electroporated cell suspensions will generate $1 \times 10^5 - 3 \times 10^5$ colonies per square plate and a total of $3 \times 10^6 - 9 \times 10^6$ colonies, which cover more that three times the theoretical complexity of a four-finger library composed of 30 different single fingers ($30 \times 30 \times 30 \times 30 = 8.1 \times 10^5$).

9. Collect the colonies into a 250-mL centrifuge bottle by adding 8 mL of LB medium to each square plate and carefully scraping with a spatula. Make sure to weigh the empty centrifuge bottle before adding the scraped cell suspension, because it is important to estimate the pellet weight of the collected colonies. After harvesting all of the colonies from the 30 square plates, centrifuge the cell suspension for 30 min at $5,000 \times g$ in a refrigerated centrifuge. Discard the supernatant very carefully (because of agar debris, the cell pellet is not well settled at this stage) and drain from the pellet as much of the residual liquid as possible.
10. Weigh the centrifuge bottle with the cell pellet. Subtract from this value the weight of an empty centrifuge bottle to calculate the cell pellet weight. Purify DNA from the cells using an appropriately sized Qiagen Plasmid Prep Kit (*see* **Note 4**).

3.2. Construction of the ZF-TF Library in a Retroviral Vector System

The next step is to combine the randomized four-finger arrays in the ZFP library from **Section 3.1** with transcriptional regulatory domains to form a combinatorial ZF-TF library. We use the NFkB p65 transcriptional activation domain or the KRAB transcriptional repression domain to produce an activating or repressing ZF-TF, respectively. For stable expression of ZF-TFs in target cells, we designed and constructed two retroviral vectors: pLPCX/TO-KRAB-IG and pLPCX/TO-p65-IG (*see* Step 1 of **Section 2.2**), which allow us to insert multi-finger-encoding DNA fragments from the p3-based ZFP library using a simple cloning procedure. To this end, the MCS of the p3 vector, which contains an HA epitope tag and a NLS signal, was transferred to the pLPCX vector and then DNA fragments encoding either the KRAB or the p65 transcriptional regulatory domain were cloned between the *Not*I and the *Xho*I sites. In addition, to facilitate the monitoring of retroviral infection and ZF-TF expression in mammalian cells, we engineered these retroviral vectors to contain an IRES–hrGFP element downstream of the ZF-TF coding sequence. Finally, to enable tight control of ZF-TF expression when required, we also replaced the CMV promoter of the parental pLPCX retroviral vector with the CMR/TO promoter, a tetracycline-regulated version of the CMV promoter.

Because the method for ZF-TF library construction is similar to the previously described procedure for four-finger library construction (*see* **Section 3.1**), we highlight here only the important differences between the two procedures. First, *Eco*RI and *Not*I are used to prepare both the retroviral vectors and the four-finger ZFP arrays. Second, retroviral vectors are ~10 kb and four-finger arrays are ~350 bp in size. Third, the efficiency of retroviral ZF-TF library construction appears to be less than that of four-finger library construction, perhaps due to the larger size of the retroviral vectors, relative to the p3 vector. Thus, to achieve comparable complexities in the ZFP and ZF-TF libraries, we recommend that researchers use at least 50% more DNA in ligation and electroporation for the construction of ZF-TF libraries in retroviral vectors.

3.3. Production of Retroviruses

1. 293T/17 cells should be passaged every other day to prevent them from becoming over-confluent. The day before transfection plate 3.5×10^6 cells in a 100-mm tissue culture dish.
2. The next day prepare the transfection mixtures for the 293T/17 cells. For each 100-mm tissue culture dish, add 36 μL of TransIT-293 transfection reagent to 1 mL of Opti-MEM, mix gently, and incubate at room temperature for 15 min.
3. Add 3 μg of pGP, 3 μg of pVPack-VSV-G, 2 μg of pcDNA6/TR, and 4 μg of ZF-TF library DNA to the transfection mixture, mix gently, and incubate at room temperature for 20 min (*see* **Note 5**).
4. Add the transfection mixture to the 293T/17 cells and mix gently.
5. Incubate overnight at 37°C (*see* **Note 6**).
6. At 24-h post-transfection, remove the medium and add 10 mL of fresh fibroblast medium (MEM + 10% FBS) to each tissue culture dish (*see* **Notes 7 and 8**).
7. Twenty-four hours later (48-h post-transfection), harvest the retroviral supernatants from the culture dishes and transfer to a 1-mL centrifuge tube. Centrifuge the retroviral supernatants at $100 \times g$ for 5 min in a tabletop centrifuge to remove any floating cells or cell debris.
8. After centrifugation, transfer the resulting retroviral supernatant to a fresh 15-mL conical tube. Store at –80°C (*see* **Note 9**).

3.4. Retroviral Infection of IMR-90 Cells and Analysis of Transduction Efficiency

1. IMR-90 cells should be passaged two or three times each week in fresh fibroblast medium. It is important to passage IMR-90 cells before they reach 100% confluency. The day before retroviral infection seed 3×10^4 IMR-90 cells in each well of a 12-well plate in a final volume of 1 mL of fibroblast medium.
2. For each retrovirus supernatant make three infection mixtures with 2, 20, and 200 μL of virus in a final volume of 1 mL of fibroblast medium. Supplement the infection mixture with 8 μg/mL polybrene.

3. Replace the cell culture medium with the infection mixture.
4. Twenty-four hours post-infection, remove the infection mixture and add 1 mL of fresh fibroblast medium.
5. Forty-eight hours post-infection, harvest the cells and measure green fluorescent protein (GFP) expression using flow cytometry (*see* **Notes 10 and 11**).
6. The titer of the retroviral supernatant used to infect the IMR-90 cells (*see* **Note 12**) can be calculated using the following formula: Active retroviral titer = 30,000 (number of cells seeded for infection) × (percentage of GFP positive cells/100)/(volume of virus supernatant used for infection) × 1,000 (*see* **Table 8.2** and **Note 13**).

Table 8.2
Tetracycline repressor improves the active titer of ZF-TF library retroviruses. ZF-TF library retroviruses were produced in presence or absence of tetracycline repressor. Indicated amount of virus supernatant was used to infect 30,000 IMR-90 cells. Forty-eight hours post-infection, the percentage of GFP-positive cells was determined by flow cytometry analysis

	Virus (μL)	GFP (%)	Active titer
pLPCX/TO-4ZF-KRAB-IG Library	2	1.06	1.59E+05
	20	9.38	1.41E+05
	200	47.08	7.06E+04
pLPCX/TO-4ZF-KRAB-IG Library + tetR	2	1.62	2.43E+05
	20	14.96	2.24E+05
	200	64.82	9.72E+04
pLPCX/TO-4ZF-p65-IG Library	2	0.06	9.00E+03
	20	0.28	4.20E+03
	200	2.54	3.81E+03
pLPCX/TO-4ZF-p65-IG Library + tetR	2	0.98	1.47E+05
	20	6.80	1.02E+05
	200	42.46	6.37E+04

tetR = tetracycline repressor; GFP (%) = percentage of GFP-positive cells measured by low cytometry analysis; Virus = amount of virus added to infect IMR-90 cells.

4. Notes

1. It is important to confirm the formation of a four-finger array by performing colony polymerase chain reaction (PCR) analysis with the F-NLS (5′-ctccaaaaaagaagagaaagg-3′) and R-SP6 (5′-tagaaggcacagtcgagg-3′) primers. The

expected sizes of the PCR products for 1-, 2-, and four-finger constructs in the p3 MCS cassette are about 230, 320, and 480 bp, respectively.

2. While the purification step is not absolutely required, we find that purification greatly improves the efficiency of electroporation. Thus, in the construction of a ZFP library of smaller complexity (e.g., a two-finger ZFP library construction), the purification step may be omitted.
3. In our experience, it is best to have between 1×10^5 and 3×10^5 colonies in a square Petri plate. With less than 1×10^5 colonies per square plate, it is hard to obtain the desired number of total colonies needed to cover the required complexity. On the other hand, with more than 3×10^5 colonies per square plate, the colonies come in contact with each other, which could have a negative effect on the complexity of the library. Therefore, in order to determine how much of electroporated *E. coli* cell suspension should be spread onto each square Petri plate, it is best to perform a small-scale electroporation with a purified ligation mixture and calculate the size of the aliquot needed to obtain the desired number of colonies per plate.
4. Because agar debris obtained from scraping the LB plates contaminates the cell pellet, we usually consider the actual cell pellet weight to be half of what is measured. We then use that number to determine the scale of the plasmid preparation procedure, performed according to the manufacturer's manual.
5. We have observed that the suppression of ZF-TF expression (especially one with the p65 activation domain) during retrovirus packaging by co-expression of tetracycline repressor protein can vastly increase retrovirus production (*see* **Table 8.2**).
6. Addition of sodium butyrate (NaBu) to a final concentration of 5–10 mM generally improves retroviral titers. We add NaBu to cell cultures at 6-h post-transfection.
7. One's choice of cell culture medium will depend on the type of cells to be infected with the retrovirus. In addition, be sure to treat cells gently at this point because we have found that transfected 293T/17 cells are very easily detached from culture dishes during the medium-replacing process.
8. Incubation of cells at 32°C is known to improve virus titers, because retroviruses are much more stable at this temperature than at 32°C.
9. While we describe a single harvest at the 48-h post-transfection time point, supernatant with a similar retroviral

titer can be obtained by adding fresh medium and harvesting again after additional 24-h incubation (the 72-h post-transfection time point).

10. In principle, because retrovirus-mediated gene delivery results in stable integration into the cellular genome, it will make no difference to assess the infection efficiency at a time point later than 48 h after retroviral infection. Indeed, analyses of infection efficiency at 48, 72, and 96 h post-infection yielded similar results in our hands.
11. When looking for GFP-positive cells using flow cytometry analysis, we obtained more accurate results with the FL1 plot versus the FL2 plot, which allowed us to eliminate the autofluorescence signal (16).
12. The active retroviral titer of a given supernatant can vary depending on the type of target cells used for infection. For example, we have observed that 293T cells are more easily infected by retroviruses than are human fibroblast cell lines, such as IMR-90. As a result, when 293T cells are used to determine the titer of a viral supernatant, a higher active titer of retroviruses will be estimated, relative to those obtained with human fibroblast cell lines.
13. The typical titer of ZF-TF library retroviruses we produce following this protocol ranges between 5×10^4/mL and 2×10^5/mL for IMR-90 cells (*see* **Table 8.2**) and between 2×10^5/mL and 8×10^5/mL for 293T cells.

Acknowledgments

This work was supported by a grant from the Korea Science and Engineering Foundation (R17-2007-019-01001-0).

References

1. Lee, D.K., Seol, W., and Kim, J.S. (2003) Custom DNA-binding proteins and artificial transcription factors. *Curr Topics Med Chem.* **3**, 645–657.
2. Lindhout, B.I., Pinas, J.E., Hooykaas, P.J., and van der Zaal, B.J. (2006) Employing libraries of zinc finger artificial transcription factors to screen for homologous recombination mutants in Arabidopsis. *Plant J.* **48**, 475–483.
3. Park, K.S., Jang, Y.S., Lee, H., and Kim, J.S. (2005) Phenotypic alteration and target gene identification using combinatorial libraries of zinc finger proteins in prokaryotic cells. *J Bacteriol.* **187**, 5496–5499.
4. Lee, J.Y., Sung, B.H., Yu, B.J., Lee, J.H., Lee, S.H., Kim, M.S., Koob, M.D., and Kim, S.C. (2008) Phenotypic engineering by reprogramming gene transcription using novel artificial transcription factors in Escherichia coli. *Nucleic Acids Res.* **36**, e102.
5. Park, K.S. (2005) Identification and use of zinc finger transcription factors that increase production of recombinant proteins in yeast

and mammalian cells. *Biotechnol Prog.* **21**, 664–670.

6. Park, K.S., Lee, D.K., Lee, H., Lee, Y., Jang, Y.S., Kim, Y.H., Yang, H.Y., Lee, S.I., Seol, W., and Kim, J.S. (2003) Phenotypic alteration of eukaryotic cells using randomized libraries of artificial transcription factors. *Nat Biotechnol.* **21**, 1208–1214.
7. Lee, D.K., Kim, Y.H., Kim, J.S., and Seol, W. (2004) Induction and characterization of taxol-resistance phenotypes with a transiently expressed artificial transcriptional activator library. *Nucleic Acids Res.* **32**, e116.
8. Blancafort, P., Chen, E.I., Gonzalez, B., Bergquist, S., Zijlstra, A., Guthy, D., Brachat, A., Brakenhoff, R.H., Quigley, J.P., Erdmann, D., and Barbas, C.F., 3rd (2005) Genetic reprogramming of tumor cells by zinc finger transcription factors. *Proc Natl Acad Sci USA.* **102**, 11716–11721.
9. Blancafort, P., Tschan, M.P., Bergquist, S., Guthy, D., Brachat, A., Sheeter, D.A., Torbett, B.E., Erdmann, D., and Barbas, C.F., 3rd (2008) Modulation of drug resistance by artificial transcription factors. *Mol Cancer Ther.* 7, 688–697.
10. Blancafort, P., Magnenat, L., and Barbas, C.F., 3rd (2003) Scanning the human genome with combinatorial transcription factor libraries. *Nat Biotechnol.* **21**, 269–274.
11. Kwon, R.J., Kim, S.K., Lee, S.I., Hwang, S.J., Lee, G.M., Kim, J.S., and Seol, W. (2006) Artificial transcription factors increase production of recombinant antibodies in Chinese hamster ovary cells. *Biotechnol Lett.* **28**, 9–15.
12. Santos, C.N. and Stephanopoulos, G. (2008) Combinatorial engineering of microbes for optimizing cellular phenotype. *Curr Opin Chem Biol.* **12**, 168–176.
13. Yamanaka, S. (2008) Pluripotency and nuclear reprogramming. *Philos Trans R Soc Lond B Biol Sci.* **363**, 2079–2087.
14. Takahashi, K. and Yamanaka, S. (2006) Induction of pluripotent stem cells from mouse embryonic and adult fibroblast cultures by defined factors. *Cell.* **126**, 663–676.
15. Bae, K.H., Kwon, Y.D., Shin, H.C., Hwang, M.S., Ryu, E.H., Park, K.S., Yang, H.Y., Lee, D.K., Lee, Y., Park, J., Kwon, H.S., Kim, H.W., Yeh, B.I., Lee, H.W., Sohn, S.H., Yoon, J., Seol, W., and Kim, J.S. (2003) Human zinc fingers as building blocks in the construction of artificial transcription factors. *Nat Biotechnol.* **21**, 275–280.
16. Roederer, M. and Murphy, R.F. (1986) Cell-by-cell autofluorescence correction for low signal-to-noise systems: application to epidermal growth factor endocytosis by 3T3 fibroblasts. *Cytometry.* 7, 558–565.

Chapter 9

Silencing of Gene Expression by Targeted DNA Methylation: Concepts and Approaches

Renata Z. Jurkowska and Albert Jeltsch

Abstract

Targeted DNA methylation is a novel and attractive approach for stable silencing of gene expression by epigenetic mechanisms. The potential applications of this concept include cancer treatment, treatment of viral infections and, in general, treatment of any disease that could be attenuated by the stable repression of known target genes. We review the literature on targeted DNA methylation and gene silencing, summarize the achievements and the challenges that remain, and discuss technical issues critical for this approach.

Key words: DNA methylation, gene regulation, epigenetics, protein design.

1. Introduction

All human cells carry the same genetic information, yet they follow different developmental pathways, which need to be realized with precision. Epigenetic regulatory mechanisms exist to ensure the coordinated expression of genes during cellular differentiation. These include posttranslational modification of histone tails, expression of non-coding RNA, and addition of methyl groups to DNA; all act in concert to regulate gene expression through modification of the chromatin state. Epigenetic signals are faithfully maintained through cell divisions, leading to heritable changes of gene expression (1).

In this chapter, we focus on the resetting of one type of epigenetic signal, DNA methylation, which is an essential covalent DNA modification. In mammals, DNA methylation occurs at the C5 position of cytosine residues, predominantly within CG

J.P. Mackay, D.J. Segal (eds.), *Engineered Zinc Finger Proteins*, Methods in Molecular Biology 649,
DOI 10.1007/978-1-60761-753-2_9,

dinucleotides. However, only certain CG sites are methylated, resulting in the establishment of a tissue and cell type-specific pattern of methylation. It has been estimated that approximately 70–80% of all CGs in the human genome are modified (3–8% of all cytosines) (2–4). DNA methylation is localized predominantly in repetitive sequences, whereas CG islands, which are regions with high CG content often accompanying gene promoters, remain usually unmethylated.

1.1. DNA Methylation in Mammals

In mammals, three different DNA methyltransferases (MTases) – Dnmt1, Dnmt3a, and Dnmt3b – have been identified and characterized (2–4). All are multi-domain proteins comprising a large N-terminal part and a smaller C-terminal domain. The N-terminal part consists of several domains of variable sizes that have regulatory functions; this region guides the nuclear localization of the enzymes and mediates their interactions with other proteins. The C-terminal domain is conserved among all cytosine C5 DNA MTases and harbors the catalytic center of the enzyme. Dnmt1 shows a high preference for hemimethylated DNA over unmethylated substrates and it is responsible for the preservation of the methylation patterns during cell division by methylation of the hemimethylated CG dinucleotides produced during DNA replication (5). This process makes DNA methylation a stable modification that is inherited through cell divisions. Dnmt3a and Dnmt3b do not display a significant preference between hemimethylated and unmethylated DNA (6, 7). They are involved in the establishment of the methylation patterns at specific sequences early in development, creating an initial methylation profile (8) that is inherited through the cell divisions and gets modified only during developmental changes. In contrast to Dnmt1 (9), the isolated catalytic domains of Dnmt3a and Dnmt3b are enzymatically active (6), which make them applicable for targeted DNA methylation. These enzymes do not display significant sequence specificity, except for CG sites and certain flanking sequences (10).

A third member of the Dnmt3 family, Dnmt3L (Dnmt3-like) has been identified; however, despite clear homology to Dnmt3a and Dnmt3b, Dnmt3L does not show any cytosine methyltransferase activity, due to the mutation of key catalytic residues. Dnmt3L interacts with both Dnmt3a and Dnmt3b (11) and acts as a regulatory factor in germ cells (12, 13). The Dnmt3a/3L complex forms a heterotetramer that positions the active centers of the two Dnmt3a molecules in a distance of 8–10 bp on the DNA. The heterotetramer further multimerizes on DNA, leading to the formation of large nucleoprotein filaments. Filament formation, together with the intrinsic distance of the two active centers in the heterotetramer, leads to a preference of Dnmt3a for methylation of CG sites localized in a distance of 8–10 bp (14, 15).

In mammals, the fundamental role of DNA methylation is in the regulation of gene expression. Usually, DNA methylation in the promoter region of genes is associated with transcriptional repression and heavily methylated genes are silenced, whereas active genes show low levels of gene promoter methylation (16). Two mechanisms by which DNA methylation inhibits transcription have been documented. On one hand the methyl groups directly interfere with the binding of transcription factors to their specific promoter sequences. The best characterized examples are c-Myc (17) and CTCF (18); both proteins bind only to unmethylated DNA. On the other hand, DNA methylation indirectly prevents transcription, because the methylated CG sites serve as a signal for recruitment of repressor proteins, like MeCP2, Mbd1, or Kaiso, which all preferentially bind methylated CGs (19). These repressing factors read the methylation signal, and target co-repressor complexes containing histone deacetylases, histone 3 lysine 9 (H3K9) methyltransferases and chromatin remodeling factors to specific genomic loci, leading to the formation of repressed chromatin structure and inhibition of transcription (4). Thus, an interplay between different epigenetic mechanisms controls gene activity.

The crucial role of DNA methylation in the regulation of gene expression is illustrated by the findings that aberrant methylation patterns contribute to the development of many human diseases, including cancer (20–24). Loss of genomic methylation in repetitive sequences is observed regularly in cancer cells and leads to genomic instability, a hallmark of tumor cells. The global hypomethylation of the genome is accompanied in almost every tumor type by an aberrant de novo methylation of promoter regions of genes involved in cell-cycle regulation, cell signaling, apoptosis, cell adhesion, chromatin remodeling, and DNA repair. The hypermethylation of the promoter regions of these key genes, leading to their inappropriate silencing, provides a selective advantage to the tumor cells.

1.2. The Concept of Targeted DNA Methylation

The fact that a growing number of human diseases, including cancer, have an epigenetic etiology resulted in the development of a new concept, so-called "epigenetic therapy," and new strategies that target or make use of the components of the epigenetic machinery have been designed. One of them is targeted DNA methylation. In this context, one long-term goal could be the reversion of incorrect epigenetic profiles, which represents a true causative approach to therapy. However, this aim is clearly overambitious at the current stage of technology. An alternative and more pragmatic approach aims to down-regulate genes that are crucial for disease. This approach does not attempt to reinstall the original, "healthy" epigenetic state, but rather to prevent or slow down the progression of disease by interfering with the expression of disease-promoting genes.

In the targeted methylation strategy, a DNA methyltransferase is specifically delivered to the promoter region of aberrantly up-regulated genes in order to reduce their activity. The methyltransferase then modifies the promoter of the target gene, leading to the repression of gene expression. This concept implies that an enzymatically active module – a DNA methyltransferase – is targeted to specific, pre-selected genomic regions through a fused, sequence-specific targeting module. The most popular approach in this direction involves fusing a DNA methyltransferase to a DNA-binding domain (DBD) that would specifically interact with a recognition sequence in the promoter of the target gene and therefore target the fusion protein harboring the methyltransferase activity to the neighboring region. Once targeted, the DNA methyltransferase would incorporate the methyl groups into the DNA adjacent to the recognition site of the targeting module, leading to the silencing of the target gene (**Fig. 9.1a**).

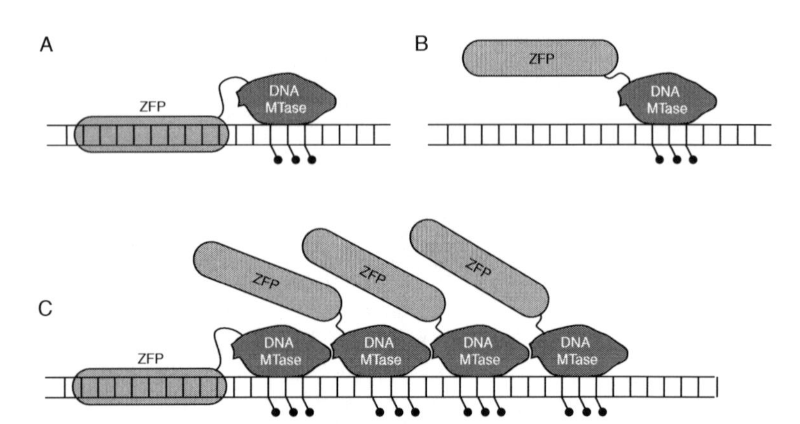

Fig. 9.1. Principles of targeted DNA methylation. (**A**) Construction of artificial DNA MTases for specific methylation (*lollypops*) of target sequences by fusing a targeting device (which can be a ZFP) with an active DNA MTase. (**B**) Non-specific methylation is observed if the MTase interacts with the DNA productively, even when the targeting domain does not bind its specific target. (**C**) Polymerization of mammalian DNA MTase Dnmt3a enzyme on the DNA leading to the methylation of a larger part of the DNA.

Since DNA methylation represents one of the most stable and heritable epigenetic signals, targeted DNA methylation is a very promising concept for epigenetic therapy. The limiting step of this approach is in the transfection efficiency of the target cells, which can usually be optimized or overcome by selection of transfected cells. A further drawback of this approach is that only the sequences for which specific DBDs are available can be targeted. However, as described throughout this volume, techniques for designing artificial zinc finger proteins (ZFPs) have recently been greatly developed and ZFP:DNA recognition codes have been established, together with powerful selection methods for creating ZFPs that are specific for a particular DNA sequence. One could, therefore, imagine that in the near-future ZFPs recognizing virtually any DNA sequence could be available. Potential applications of targeted methylation include the suppression of viral genes, oncogenes, or genes causing protein aggregation diseases. However, in particular when combined with a cell type-selective delivery, in principle any disease could be combated by silencing genes that contribute to the disease phenotype.

Due to the epigenetic inheritance of the silencing signal, targeted DNA methylation has greater potential for the regulation of endogenous genes than alternative strategies like RNA interference. Chimeric DNA MTases can be constructed at the genetic level and DNA methylation is induced after transient expression of the enzymes. After its establishment, the newly generated DNA methylation pattern is copied by the cellular maintenance methylation machinery, such that permanent expression of the chimeric MTase is not required. Thereby, many of the risks associated with stable delivery of genes in mammalian cells are avoided. Thus, targeted DNA methylation combines many prospects of gene therapy, but excludes most of the important risk factors.

The process of targeted gene silencing by chimeric DNA MTases resembles endogenous regulation, which often includes the piggyback transportation of DNA MTases to target genes by transcription factors and other chromatin-interacting proteins. For example, Dnmt3a has been reported to interact with the transcription factors p53 and RP58 (25, 26), the retinoic acid receptor (27), EZH2 and SetDB1 (both histone methyltransferases) (28, 29), Kaposi's sarcoma-associated herpes virus protein LANA (30), and both Mbd3 and Brg1 (31). Therefore, the application of chimeric MTases for external gene regulation is an example of a successful imitation of Nature in biotechnology.

2. Examples of Targeted DNA Methylation in Literature

The applicability of targeted DNA methylation was initially demonstrated by Xu and Bestor, who achieved targeted methylation of a single CG site in vitro using a chimeric protein consisting of the DBD from Zif268 and the prokaryotic DNA methyltransferase M.SssI, which is specific for CG sites (32).

Using bacterial cytosine C5 MTases, Carvin et al. confirmed that DNA methylation can be effectively targeted in living cells (33, 34). They introduced the DBD–MTase constructs into yeast *Saccharomyces cerevisiae*, which has no endogenous DNA methylation system and observed a specific delivery of cytosine methylation to the target promoters. First, they fused the native transactivator PHO4 to a viral cytosine C5 MTase M.CviPI that is specific for GC sites and observed targeting of DNA methylation to the promoters containing PHO-binding sites. They demonstrated that, in the context of chromatin, the efficiency of targeted cytosine methylation at a given site is determined by its accessibility, its rotational orientation relative to the DBD-binding site, and the higher order chromosome structure (33).

Additionally, they demonstrated that DNA methylation can be targeted in vivo by ZFPs to a potentially broad range of sequences. After fusing the zinc finger DBDs from Zif268 and Zip53 to M.CviPI and M.SssI DNA MTases, targeting of the MTases to specific sites neighboring the ZFP target sites was observed by bisulfite sequencing analysis (34). However, no gene silencing was detected, which could be due to the lack of a functional DNA methylation readout system in yeast. Moreover, in both cases untargeted methylation was observed, which could be due to the high catalytic activity or untargeted binding of the methyltransferases to their endogenous recognition sequences (**Fig. 9.1B**).

One approach to reduce non targeted methylation is to use fusion constructs with DNA MTase variants of reduced activity instead of the wild-type enzyme. Smith et al. showed that by introducing mutations into the *H*paII and *H*haI MTases, which resulted either in reduction of the catalytic activity or in reduced DNA-binding affinity of the mutants, enhanced specificity of the targeted DNA methylation could be achieved (35) (**Fig. 9.2a**).

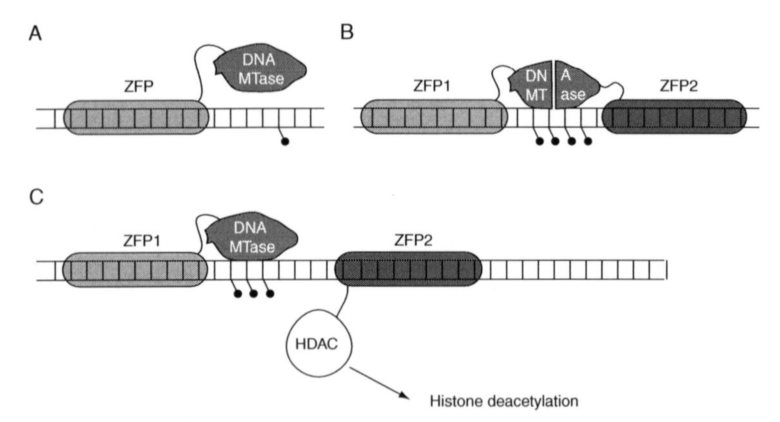

Fig. 9.2. Strategies to reduce non-specific DNA methylation. (**A**) DNA MTase domains that do not bind tightly to DNA will cause lower level of non-targeted methylation. (**B**) In the split enzyme approach, the MTase is split into two subdomains which are targeted by independent ZFP and reconstitute only after both ZFPs have bound to their specific site. (**C**) Co-targeting of different silencing activities (like DNA MTase, histone H3K9 MTase, or histone lysine deacetylase) by different ZFP to the same target region might improve efficiency and specificity of gene silencing.

Nomura and Barbas (36) described another approach to increase the specificity of targeting and reduce the methylation at off-target sites (**Fig. 9.2B**). They fused an N-terminal segment of the DNA MTase M.HhaI to a ZFP, which recognizes a specific 9-bp sequence. A C-terminal segment of M.HhaI was fused to another ZFP targeting a different 9-bp sequence localized next to the first one. In vivo, both parts of M.HhaI can reassemble and reconstitute an active DNA methyltransferase, as had been shown previously for this enzyme (37). When the two DNA MTase–ZFP fusion proteins were simultaneously expressed in *Escherichia coli*, an M.HhaI recognition site flanked by the two ZFP-binding sites became methylated, whereas other M.HhaI recognition sites remained unmethylated. The 5′-GCGC-3′ methylation specificity of M.HhaI overlaps with CG methylation (the methylated cytosine is underlined), which makes this approach directly applicable to targeted methylation of CG sites flanked by a 5′-G and a 3′-C residue. This strategy might allow the specific methylation of a single CG site in a genome in future (38).

The next step in the proof of the concept of targeted gene silencing by DNA methylation was achieved by Li et al., who demonstrated for the first time the repressive effect of targeted DNA methylation in cells (39). They fused either the Gal4 ZFP or an engineered ZFP directed toward the promoter of the human simplex virus type 1 (HSV-1) IE175k gene to the catalytic domains of mouse Dnmt3a or 3b (40). The activity of the Gal4 fusion protein was assessed in luciferase assays using reporters containing either the human TK or the ras promoters (both fused to the UAS Gal4 target sequence), whereas the engineered ZFP fusion was assayed using the IE175 promoter upstream of a luciferase reporter. With both ZFPs, dense targeted methylation could be achieved in transient co-transfection experiments that was dependent on the catalytic activity of the MTase domain, specific DNA binding of the ZFP, and the presence of the target site. No increase in DNA methylation could be found at all other genomic loci tested. In contrast to data from other laboratories, the authors observed a massive methylation of up to 57 CG sites at both target regions. This result was unexpected, but it can be explained by the ability of Dnmt3a to polymerize on DNA (**Fig. 9.1C**) (14). With all systems, they also observed efficient and specific gene silencing. Additionally, they showed that methylation-mediated gene silencing effectively represses infection with HSV-1 virus, suggesting that similar strategies could be used as anti-viral treatment.

The applicability of targeted methylation in modulating gene expression in the context of genomic DNA was recently demonstrated by Smith et al. (41). They transiently expressed a fusion protein consisting of a four zinc finger-targeting module and bacterial *Hpa*II DNA methyltransferase in mouse NIH3T3 cells and showed that targeted DNA methylation led to the repression of the integrated reporter gene and heterochromatin formation. Importantly, the repressive state was maintained epigenetically after the removal of the fusion methyltransferase from the cells.

As a further extension of the concept, Minczuk et al. showed that ZFP-targeted methylation can also be used for the modification of mitochondrial DNA (42). After addition of a mitochondrial targeting sequence to a fusion protein consisting of an engineered ZFP and the mammalian Dnmt3a catalytic domain, DNA methylation could be specifically targeted to mitochondrial DNA.

In addition, different DNA MTases have been fused to natural DNA-binding proteins to map the DNA-binding sites of the fusion partner. After expression of the fusion proteins in the cell, the DBD binds to the DNA and the neighboring methylation sites get modified. Therefore, the methylation pattern of the DNA will reflect the specificity of the DBD and the accessibility of the chromatin for binding. Among the different MTases employed for methylation probing, the *E. coli* Dam enzyme has been proven to be the most successful (43). The advantage of this approach (called DamID) is that the Dam enzyme methylates adenine residues, creating a modification that does not naturally occur in metazoans and does not cause a detectable phenotype in flies (44). DamID has been employed to map in vivo binding sites for various transcription factors including HP1, cyclin D3, Myc, Max, and Mad/Mnt (45–48), illustrating that DNA MTases can be successfully targeted to different sequences in the genome.

3. Discussion and Conclusions

Although an increasing amount of experimental data supports the successful application of targeted methylation for gene silencing, the final proof of concept, which is the heritable repression of an aberrantly expressed endogenous gene through targeted DNA methylation, has not yet been achieved.

While considering the use of targeted methylation as a tool in epigenetic therapy, one faces a potential problem of associated non-targeted methylation due to the endogenous DNA interaction of the DNA MTase part (**Fig. 9.1B**), which could have detrimental effects. The level of such undesired methylation could be reduced by using DNA methyltransferases with low catalytic activity and low-intrinsic DNA-binding specificity (**Fig. 9.2A**), examples of which are mammalian Dnmt3a and Dnmt3b. Otherwise, mutant enzymes with significantly reduced methyltransferase activity could be used instead of the wild-type enzyme or split enzymes could be targeted by two ZFPs (**Fig. 9.2B**). An additional advantage of using mammalian enzymes is that they are natural components of the endogenous epigenetic machinery; therefore, the signal they set might be properly recognized and interpreted by the co-repressor complexes, leading to the propagation of the silent state. Dnmt3a, in addition, has the ability to polymerize on DNA and cause the methylation of a whole stretch of DNA next to the target site (**Fig. 9.1C**) (14, 15), an effect that could lead to an efficient silencing of the target gene.

An interesting question regards the stability of the introduced DNA methylation. On one hand the biological process of imprinting illustrates that lifelong silencing of the expression of single alleles can be achieved by DNA methylation in the germ line (49). On the other hand, recent data obtained for the pS2/TFF1 gene promoter suggest that DNA methylation is not stable, at least at some promoters, where it has even been shown to cycle (50). This discrepancy might be explained by the fact that in imprinting (where DNA methylation is stable) a larger region of the DNA is methylated (the so-called differentially methylated region), while the differential methylation observed in the pS2 promoter (which is not stable) was restricted to individual CG sites.

The observations discussed in the last paragraph lead to the important question of how much methylation is needed to change gene expression. Genome methylation studies indicate that DNA methylation levels have a bimodal distribution with enrichment of highly methylated and unmethylated sequences (51–53). This result suggests that the methylation of individual CG sites will not be sufficient to change the expression of a gene. Therefore, approaches leading to a dense methylation of target regions might be favored. In addition, the combination of several silencing activities at one site (like DNA methylation, histone deacetylation, and H3K9 methylation) might cause stronger and more stable gene repression than just DNA methylation alone (**Fig. 9.2C**).

Another critical question concerns the specificity of the epigenetic changes introduced. This refers both to the potential silencing of the target gene in healthy (i.e., non-target) cells and the potential silencing of non-target genes. With respect to the silencing of the target gene in healthy cells, there are several scenarios in which this potential problem can be circumvented or tolerated:

(i) Viral target genes are only present in infected cells, so aberrant silencing in healthy cells cannot occur.

(ii) The target cells might be preferentially addressed by the vectors used to express the chimeric DNA MTases, for example, by using viral application in aerosols to address lung cells or exploiting the natural or designed specificity of viruses for particular target cell types.

(iii) For some genes, which play a prominent role in a disease process, some degree of silencing in healthy cells might be tolerable.

With respect to the aberrant methylation and potential silencing of non-target genes, the specificity of targeted DNA methylation has to be confirmed for each construct, prior to even considering potential therapeutic applications. For this step, modern techniques involving whole-genome bisulfite conversion and analysis are available, such as enrichment of methylated or unmethylated fractions of genomic DNA with antibodies against methylated cytosine, MBD domains, or other protein domains, followed by readout using microarray or DNA sequencing (MEDIP and MCIP) (53–57) or conversion of the genomic DNA with bisulfite and genome-wide analysis of DNA methylation by deep sequencing (52, 58, 59). Future work will reveal whether or not targeted gene silencing by chimeric DNA MTases will deliver on its promises.

Acknowledgments

Work in the authors' laboratory has been supported by the DFG and the Wilhelm Sander Foundation.

References

1. Allis, C.D., Jenuwein, T., and Reinberg, D. (2007) Epigenetics. Cold Spring Harbor Laboratory Press, Cold Spring Harbor, New York.
2. Goll, M.G. and Bestor, T.H. (2005) Eukaryotic cytosine methyltransferases. *Annu Rev Biochem.* **74**, 481–514.
3. Hermann, A., Gowher, H., and Jeltsch, A. (2004) Biochemistry and biology of mammalian DNA methyltransferases. *Cell Mol Life Sci.* **61**, 2571–2587.
4. Klose, R.J. and Bird, A.P. (2006) Genomic DNA methylation: the mark and its mediators. *Trends Biochem Sci.* **31**, 89–97.
5. Jeltsch, A. (2006) On the enzymatic properties of Dnmt1: specificity, processivity, mechanism of linear diffusion and allosteric regulation of the enzyme. *Epigenetics.* **1**, 63–66.
6. Gowher, H. and Jeltsch, A. (2001) Enzymatic properties of recombinant Dnmt3a DNA methyltransferase from mouse: the enzyme modifies DNA in a non-processive manner and also methylates non-CpG sites. *J Mol Biol.* **309**, 1201–1208.
7. Okano, M., Xie, S., and Li, E. (1998) Cloning and characterization of a family of novel mammalian DNA (cytosine-5) methyltransferases. *Nat Genet.* **19**, 219–220.
8. Okano, M., Bell, D.W., Haber, D.A., and Li, E. (1999) DNA methyltransferases Dnmt3a and Dnmt3b are essential for de novo methylation and mammalian development. *Cell.* **99**, 247–257.
9. Fatemi, M., Hermann, A., Pradhan, S., and Jeltsch, A. (2001) The activity of the murine DNA methyltransferase Dnmt1 is controlled by interaction of the catalytic domain with the N-terminal part of the enzyme leading to an allosteric activation of the enzyme after binding to methylated DNA. *J Mol Biol.* **309**, 1189–1199.
10. Handa, V. and Jeltsch, A. (2005) Profound flanking sequence preference of Dnmt3a and Dnmt3b mammalian DNA methyltrans-

ferases shape the human epigenome. *J Mol Biol.* **348**, 1103–1112.
11. Gowher, H., Liebert, K., Hermann, A., Xu, G., and Jeltsch, A. (2005) Mechanism of stimulation of catalytic activity of Dnmt3A and Dnmt3B DNA-(cytosine-C5)-methyltransferases by Dnmt3L. *J. Biol. Chem.* **280**, 13341–13348.
12. Bourc'his, D., Xu, G.L., Lin, C.S., Bollman, B., and Bestor, T.H. (2001) Dnmt3L and the establishment of maternal genomic imprints. *Science.* **294**, 2536–2539.
13. Hata, K., Okano, M., Lei, H., and Li, E. (2002) Dnmt3L cooperates with the Dnmt3 family of de novo DNA methyltransferases to establish maternal imprints in mice. *Development.* **129**, 1983–1993.
14. Jia, D., Jurkowska, R.Z., Zhang, X., Jeltsch, A., and Cheng, X. (2007) Structure of Dnmt3a bound to Dnmt3L suggests a model for de novo DNA methylation. *Nature.* **449**, 248–251.
15. Jurkowska, R.Z., Anspach, N., Urbanke, C., Jia, D., Reinhardt, R., Nellen, W., Cheng, X., and Jeltsch, A. (2008) Formation of nucleoprotein filaments by mammalian DNA methyltransferase Dnmt3a in complex with regulator Dnmt3L. *Nucleic Acids Res.* **36**, 6656–6663.
16. Kass, S.U., Pruss, D., and Wolffe, A.P. (1997) How does DNA methylation repress transcription? *Trends Genet.* **13**, 444–449.
17. Prendergast, G.C. and Ziff, E.B. (1991) Methylation-sensitive sequence-specific DNA binding by the c-Myc basic region. *Science.* **251**, 186–189.
18. Bell, A.C. and Felsenfeld, G. (2000) Methylation of a CTCF-dependent boundary controls imprinted expression of the Igf2 gene. *Nature.* **405**, 482–485.
19. Sansom, O.J., Maddison, K., and Clarke, A.R. (2007) Mechanisms of disease: methyl-binding domain proteins as potential therapeutic targets in cancer. *Nat Clin Pract Oncol.* **4**, 305–315.
20. Esteller, M. (2007) Cancer epigenomics: DNA methylomes and histone-modification maps. *Nat Rev Genet.* **8**, 286–298.
21. Feinberg, A.P. (2007) Phenotypic plasticity and the epigenetics of human disease. *Nature.* **447**, 433–440.
22. Feinberg, A.P. and Tycko, B. (2004) The history of cancer epigenetics. *Nat Rev Cancer.* **4**, 143–153.
23. Jones, P.A. and Baylin, S.B. (2007) The epigenomics of cancer. *Cell.* **128**, 683–692.
24. Robertson, K.D. (2005) DNA methylation and human disease. *Nat Rev Genet.* **6**, 597–610.
25. Fuks, F., Burgers, W.A., Godin, N., Kasai, M., and Kouzarides, T. (2001) Dnmt3a binds deacetylases and is recruited by a sequence-specific repressor to silence transcription. *EMBO J.* **20**, 2536–2544.
26. Wang, Y.A., Kamarova, Y., Shen, K.C., Jiang, Z., Hahn, M.J., Wang, Y., and Brooks, S.C. (2005) DNA methyltransferase-3a interacts with p53 and represses p53-mediated gene expression. *Cancer Biol Ther.* **4**, 1138–1143.
27. Di Croce, L., Raker, V.A., Corsaro, M., Fazi, F., Fanelli, M., Faretta, M., Fuks, F., Lo Coco, F., Kouzarides, T., Nervi, C., Minucci, S., and Pelicci, P.G. (2002) Methyltransferase recruitment and DNA hypermethylation of target promoters by an oncogenic transcription factor. *Science.* **295**, 1079–1082.
28. Li, H., Rauch, T., Chen, Z.X., Szabo, P.E., Riggs, A.D., and Pfeifer, G.P. (2006) The histone methyltransferase SETDB1 and the DNA methyltransferase DNMT3A interact directly and localize to promoters silenced in cancer cells. *J Biol Chem.* **281**, 19489–19500.
29. Vire, E., Brenner, C., Deplus, R., Blanchon, L., Fraga, M., Didelot, C., Morey, L., Van Eynde, A., Bernard, D., Vanderwinden, J.M., Bollen, M., Esteller, M., Di Croce, L., de Launoit, Y., and Fuks, F. (2006) The Polycomb group protein EZH2 directly controls DNA methylation. *Nature.* **439**, 871–874.
30. Shamay, M., Krithivas, A., Zhang, J., and Hayward, S.D. (2006) Recruitment of the de novo DNA methyltransferase Dnmt3a by Kaposi's sarcoma-associated herpesvirus LANA. *Proc Natl Acad Sci USA.* **103**, 14554–14559.
31. Datta, J., Majumder, S., Bai, S., Ghoshal, K., Kutay, H., Smith, D.S., Crabb, J.W., and Jacob, S.T. (2005) Physical and functional interaction of DNA methyltransferase 3A with Mbd3 and Brg1 in mouse lymphosarcoma cells. *Cancer Res.* **65**, 10891–10900.
32. Xu, G.L. and Bestor, T.H. (1997) Cytosine methylation targeted to pre-determined sequences. *Nat Genet.* **17**, 376–378.
33. Carvin, C.D., Dhasarathy, A., Friesenhahn, L.B., Jessen, W.J., and Kladde, M.P. (2003) Targeted cytosine methylation for in vivo detection of protein-DNA interactions. *Proc Natl Acad Sci USA.* **100**, 7743–7748.
34. Carvin, C.D., Parr, R.D., and Kladde, M.P. (2003) Site-selective in vivo targeting of cytosine-5 DNA methylation by zinc-finger proteins. *Nucleic Acids Res.* **31**, 6493–6501.

35. Smith, A.E. and Ford, K.G. (2007) Specific targeting of cytosine methylation to DNA sequences in vivo. *Nucleic Acids Res.* **35**, 740–754.
36. Nomura, W. and Barbas, C.F., 3rd. (2007) In vivo site-specific DNA methylation with a designed sequence-enabled DNA methylase. *J Am Chem Soc* **129**, 8676–8677.
37. Posfai, G., Kim, S.C., Szilak, L., Kovacs, A., and Venetianer, P. (1991) Complementation by detached parts of GGCC-specific DNA methyltransferases. *Nucleic Acids Res.* **19**, 4843–4847.
38. Kiss, A. and Weinhold, E. (2008) Functional reassembly of split enzymes on-site: a novel approach for highly sequence-specific targeted DNA methylation. *Chembiochem.* **9**, 351–353.
39. Li, F., Papworth, M., Minczuk, M., Rohde, C., Zhang, Y., Ragozin, S., and Jeltsch, A. (2007) Chimeric DNA methyltransferases target DNA methylation to specific DNA sequences and repress expression of target genes. *Nucleic Acids Res.* **35**, 100–112.
40. Gowher, H. and Jeltsch, A. (2002) Molecular enzymology of the catalytic domains of the Dnmt3a and Dnmt3b DNA methyltransferases. *J Biol Chem.* **277**, 20409–20414.
41. Smith, A.E., Hurd, P.J., Bannister, A.J., Kouzarides, T., and Ford, K.G. (2008) Heritable gene repression through the action of a directed DNA methyltransferase at a chromosomal locus. *J Biol Chem.* **283**, 9878–9885.
42. Minczuk, M., Papworth, M.A., Kolasinska, P., Murphy, M.P., and Klug, A. (2006) Sequence-specific modification of mitochondrial DNA using a chimeric zinc finger methylase. *Proc Natl Acad Sci USA.* **103**, 19689–19694.
43. Greil, F., Moorman, C., and van Steensel, B. (2006) DamID: mapping of in vivo protein-genome interactions using tethered DNA adenine methyltransferase. *Methods Enzymol.* **410**, 342–359.
44. Wines, D.R., Talbert, P.B., Clark, D.V., and Henikoff, S. (1996) Introduction of a DNA methyltransferase into Drosophila to probe chromatin structure in vivo. *Chromosoma.* **104**, 332–340.
45. de Wit, E., Greil, F., and van Steensel, B. (2005) Genome-wide HP1 binding in Drosophila: developmental plasticity and genomic targeting signals. *Genome Res.* **15**, 1265–1273.
46. Orian, A., van Steensel, B., Delrow, J., Bussemaker, H.J., Li, L., Sawado, T., Williams, E., Loo, L.W., Cowley, S.M., Yost, C., Pierce, S., Edgar, B.A., Parkhurst, S.M., and Eisenman, R.N. (2003) Genomic binding by the Drosophila Myc, Max, Mad/Mnt transcription factor network. *Genes Dev.* **17**, 1101–1114.
47. Song, S., Cooperman, J., Letting, D.L., Blobel, G.A., and Choi, J.K. (2004) Identification of cyclin D3 as a direct target of E2A using DamID. *Mol Cell Biol.* **24**, 8790–8802.
48. van Steensel, B., Delrow, J., and Henikoff, S. (2001) Chromatin profiling using targeted DNA adenine methyltransferase. *Nat Genet.* **27**, 304–308.
49. Reik, W. (2007) Stability and flexibility of epigenetic gene regulation in mammalian development. *Nature.* **447**, 425–432.
50. Metivier, R., Gallais, R., Tiffoche, C., Le Peron, C., Jurkowska, R.Z., Carmouche, R.P., Ibberson, D., Barath, P., Demay, F., Reid, G., Benes, V., Jeltsch, A., Gannon, F., and Salbert, G. (2008) Cyclical DNA methylation of a transcriptionally active promoter. *Nature.* **452**, 45–50.
51. Eckhardt, F., Lewin, J., Cortese, R., Rakyan, V.K., Attwood, J., Burger, M., Burton, J., Cox, T.V., Davies, R., Down, T.A., Haefliger, C., Horton, R., Howe, K., Jackson, D.K., Kunde, J., Koenig, C., Liddle, J., Niblett, D., Otto, T., Pettett, R., Seemann, S., Thompson, C., West, T., Rogers, J., Olek, A., Berlin, K., and Beck, S. (2006) DNA methylation profiling of human chromosomes 6, 20 and 22. *Nat Genet.* **38**, 1378–1385.
52. Meissner, A., Mikkelsen, T.S., Gu, H., Wernig, M., Hanna, J., Sivachenko, A., Zhang, X., Bernstein, B.E., Nusbaum, C., Jaffe, D.B., Gnirke, A., Jaenisch, R., and Lander, E.S. (2008) Genome-scale DNA methylation maps of pluripotent and differentiated cells. *Nature.* **454**, 766–770.
53. Weber, M., Hellmann, I., Stadler, M.B., Ramos, L., Paabo, S., Rebhan, M., and Schubeler, D. (2007) Distribution, silencing potential and evolutionary impact of promoter DNA methylation in the human genome. *Nat Genet.* **39**, 457–466.
54. Illingworth, R., Kerr, A., Desousa, D., Jorgensen, H., Ellis, P., Stalker, J., Jackson, D., Clee, C., Plumb, R., Rogers, J., Humphray, S., Cox, T., Langford, C., and Bird, A. (2008) A novel CpG island set identifies tissue-specific methylation at developmental gene loci. *PLoS Biol.* **6**, e22.
55. Keshet, I., Schlesinger, Y., Farkash, S., Rand, E., Hecht, M., Segal, E., Pikarski, E., Young, R.A., Niveleau, A., Cedar, H., and Simon, I. (2006) Evidence for an instructive mechanism of de novo methylation in cancer cells. *Nat Genet.* **38**, 149–153.

56. Rakyan, V.K., Down, T.A., Thorne, N.P., Flicek, P., Kulesha, E., Graf, S., Tomazou, E.M., Backdahl, L., Johnson, N., Herberth, M., Howe, K.L., Jackson, D.K., Miretti, M.M., Fiegler, H., Marioni, J.C., Birney, E., Hubbard, T.J., Carter, N.P., Tavare, S., and Beck, S. (2008) An integrated resource for genome-wide identification and analysis of human tissue-specific differentially methylated regions (tDMRs). *Genome Res.* **18**, 1518–1529.
57. Weber, M., Davies, J.J., Wittig, D., Oakeley, E.J., Haase, M., Lam, W.L., and Schubeler, D. (2005) Chromosome-wide and promoter-specific analyses identify sites of differential DNA methylation in normal and transformed human cells. *Nat Genet.* **37**, 853–862.
58. Cokus, S.J., Feng, S., Zhang, X., Chen, Z., Merriman, B., Haudenschild, C.D., Pradhan, S., Nelson, S.F., Pellegrini, M., and Jacobsen, S.E. (2008) Shotgun bisulphite sequencing of the Arabidopsis genome reveals DNA methylation patterning. *Nature.* **452**, 215–219.
59. Lister, R., O'Malley, R.C., Tonti-Filippini, J., Gregory, B.D., Berry, C.C., Millar, A.H., and Ecker, J.R. (2008) Highly integrated single-base resolution maps of the epigenome in Arabidopsis. *Cell.* **133**, 523–536.

Chapter 10

Remodeling Genomes with Artificial Transcription Factors (ATFs)

Adriana S. Beltran and Pilar Blancafort

Abstract

Chromatin structure plays a pivotal role in defining which regions of the genome are accessible for effective transcription. Chromatin-remodeling agents are able to relax this structure, facilitating the access of transcription factors into the DNA. Herein, we describe a new method, which combines artificial transcription factors (ATFs) and chromatin-remodeling agents to specifically reactivate silenced regions of the genome and reprogram cellular phenotypes.

Key words: Zinc fingers, artificial transcription factors (ATFs), epigenetics, SAHA, 5-aza-2′-dC.

1. Introduction

In complex genomes, most of the genetic loci are not expressed at any given time. Instead, these loci are present in a "locked" state and are transcriptionally inactive. The chromatin structure in these sites is found in a compact topology, which impairs access of transcription factors and polymerase II complex to effectively engage transcription (1–3). The chromatin structure is therefore a key determinant in defining which regions of the genome are inaccessible for gene expression and which genes are primed for effective transcription (4, 5). It is generally established that the structure of the chromatin at a given promoter site is defined by specific combination of epigenetic marks or "epigenetic grammar." These marks are chemical modifications both at DNA and at specific histone tails (6, 7). Our laboratory is interested in

J.P. Mackay, D.J. Segal (eds.), *Engineered Zinc Finger Proteins*, Methods in Molecular Biology 649,
DOI 10.1007/978-1-60761-753-2_10, © Springer Science+Business Media, LLC 2010

regulating tumor suppressor genes, which are aberrantly silenced in cancer cells. Silencing of tumor suppressor gene expression occurs by methylation of DNA and histones, as well as histone deacetylation (8). These mechanisms result in chromatin condensation, which impairs transcriptional activity. It is not surprising that designed artificial transcription factors (ATFs) targeting methylated regions have limited gene regulation capabilities (9–12).

In this chapter, we describe a strategy that combines the retroviral delivery of ATFs designed against tumor suppressor gene promoters and chromatin-remodeling drugs. Chromatin-remodeling agents help to remove the repressive epigenetic marks associated with tumor suppressor gene silencing, which results in a relaxed promoter topology and facilitates the access of the ATFs into the DNA. This combination of ATFs and chromatin-remodeling agents results in a synergistic interaction in reactivating tumor suppressor gene expression (*see* **Fig. 10.1**, (10)). Although in the following sections we shall focus in the model system tumor suppressor gene *maspin*, investigators can use the same protocol to assess changes in gene expression of any silenced gene and in any given cell type. Furthermore, the same methodology can be applied to many other methyltransferase and HDAC inhibitors for targeting any tumor suppressor gene (17–19).

Reactivation of tumor suppressor genes, such as *maspin*, results in an enhancement of apoptosis (14–17). Herein, we describe a procedure to address the synergy between ATFs and chromatin-remodeling agents in inducing tumor suppressor gene re-expression and reduction of tumor cell proliferation (10). A general overview is described in **Fig. 10.2**. In the following sections we will describe the targeting of the tumor suppressor gene mammary serine protease inhibitor (*maspin*) and a specific ATF designed for this gene, which binds an 18-base pair site in the proximal promoter region (ATF-126, *see* **Fig. 10.3**) (9). We will include the methyltransferase inhibitor 5-aza-2′-dC and the histone deacetylase inhibitor SAHA as chromatin-remodeling agents. We also describe the combinatorial index (CI) method to evaluate the pharmacological synergisms between ATFs and chromatin-remodeling drugs. The CI method will be described to address synergism, in which $CI < 1$ defines a synergistic interaction, $CI > 1$ antagonistic drug interaction, and the straight line at $CI = 1$ represents additive effects. As shown in the particular example illustrated in **Fig. 10.4**, ATF-126 (which was designed to up-regulate *maspin*) synergizes with 5-aza-2′-dC, SAHA, and with both inhibitors (5-aza-2′-dC + SAHA) as the calculated CI value is lower than 1.

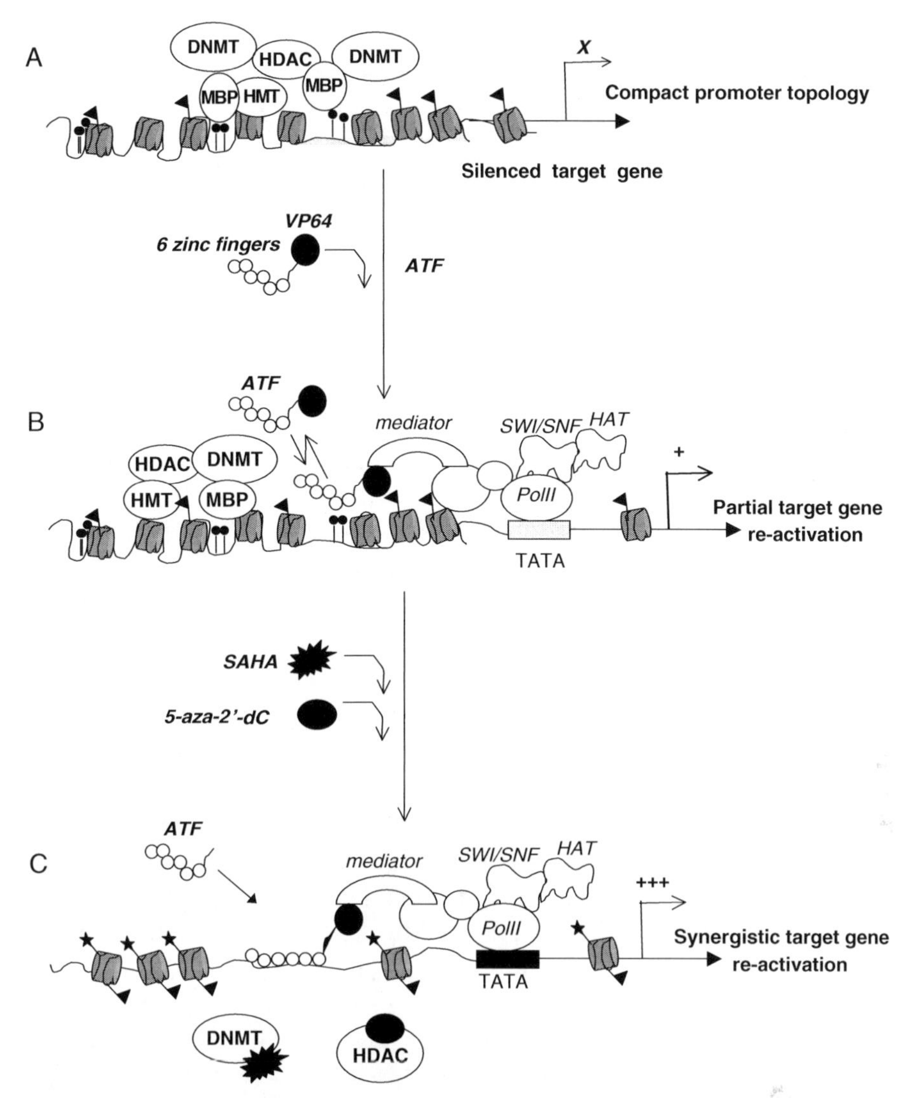

Fig. 10.1. Model explaining the synergism between ATFs and the chromatin-remodeling drugs. (**A**) Compact chromatin structure, which affects the promoter topology and silences the targeted gene. (**B**) ATF binding into the silenced promoter triggers a partial re-activation of the target gene. (**C**) Synergistic interaction between the ATF and the chromatin-remodeling drugs in re-activating a silenced gene. Upon treatment with chromatin-remodeling drugs, changes in the chromatin structure facilitate the ATF's landing on the target gene promoter, which enhance target gene re-activation. *Lollypops* represent methylated cytosines; DNMT, DNA methyltransferases; HDAC, histone deacetylase; MBP, methyl-binding proteins; HMT, histone methyltransferase; HAT, histone acetyltransferase. *Stars* and *triangles* represent activating and repressive histone modifications, respectively.

2. Materials

2.1. Retroviral Delivery of Artificial Transcription Factors (ATFs)

1. Retroviral packaging cells, denoted as 90.74 (CRL-11654, American Tissue Culture Collection).
2. MDA-MB-231 breast cancer cells (American Tissue Culture Collection).

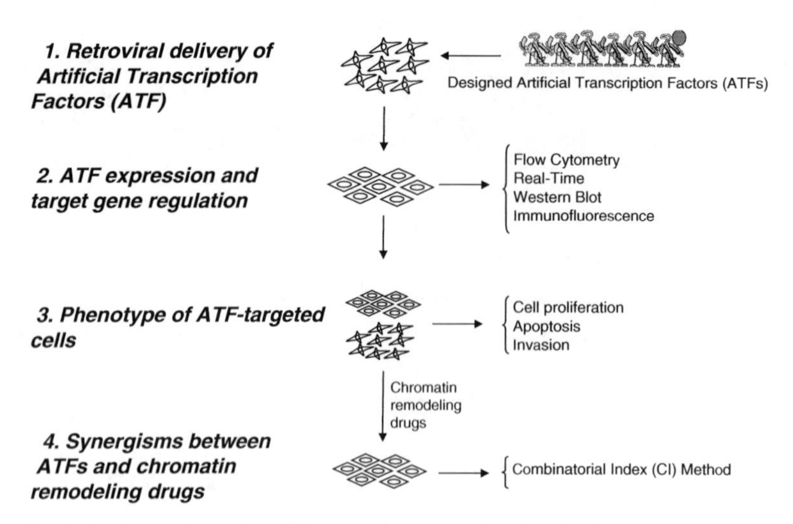

Fig. 10.2. General overview of the procedures described in this chapter.

3. Dulbecco's Modified Eagle's Medium (DMEM) (Gibco/BRL).
4. DMEM-Supplemented: DMEM, 10% fetal bovine serum (FBS), 1× penicillin–streptomycin (100× stock) (Gibco/BRL).
5. 0.05% Trypsin/EDTA solution: 0.05% trypsin, 1 mM ethylenediamine tetraacetic acid (EDTA) (Gibco/BRL).
6. Poly-D-Lysine dissolved in tissue culture water at 5 mg/mL and stored in aliquots at –20°C. Working solutions are prepared by dilution to 100 μg/mL in water.
7. Phosphate-buffered saline (PBS): 8 g NaCl, 0.2 g KCl, 1.44 g Na_2HPO_4, 0.24 g KH_2PO_4 in 800 mL of distilled H_2O. Adjust the pH to 7.4 with HCl. Add enough water to make volume up to 1 L. Sterilize by autoclaving.
8. pMDG.1 (VSV-G envelope-expressing plasmid (13)).
9. pMX-Ires-VP64-GFP control vector (13) and *maspin*-specific ATF constructs (9).
10. Plus Reagent (Invitrogen).
11. Lipofectamine (Invitrogen).
12. Polybrene (SH9268, Sigma) dissolved at 80 mg/mL in tissue culture water and store in aliquots at –20°C. The working solution of 800 μg/mL is stored at 4°C. The final concentration in the virus preparation is 8 μg/mL.

2.2. Assessment of ATF Expression and Target Gene Regulation

1. Primer design software (e.g., Primer Express, Applied Biosystems).
2. Primer and probes for *maspin* and the ATFs: *maspin* forward (5′-CGACCAGACCAAAATCCTTG-3′), reverse (5′-GA A

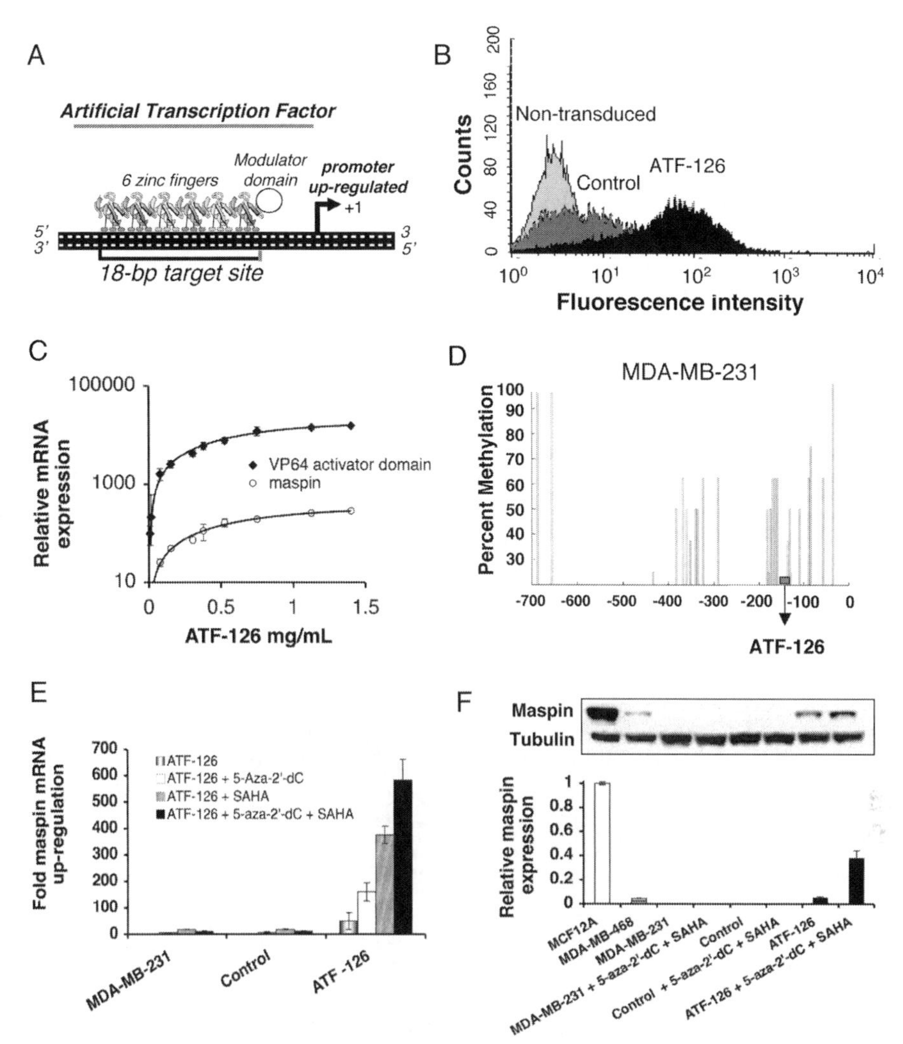

Fig. 10.3. The ATFs synergize with chromatin-remodeling drugs to reactivate *maspin*. (**A**) Schematic representation of a six zinc finger artificial transcription factor. (**B**) MDA-MB-231cells are efficiently transduced with ATF-126 as assessed by flow cytometry GFP levels. (**C**) Levels of ATF mRNA directly correlate with *maspin* mRNA levels, as evaluated by real-time PCR using primers specific for *maspin* and the ATF (the primers bind to the VP64 activator domain which is linked to the 6ZF). Before the real-time PCR, cells were transduced with different amounts of ATF-encoded DNA. (**D**) The *maspin* promoter is methylated in the MDA-MB-231 cells as assessed by sodium bisulfite sequencing. The percentage of methylation (*y*-axis) of each nucleotide position (*x*-axis) of the *maspin* promoter is relative to the first methionine and includes the ATF-binding sites (9). (**E**) Real-time quantification of *maspin* in MDA-MB-231 cells. Cells were transduced with ATF-126 and a control vector (empty retroviral vector) and treated with 5-aza-2′-dC (0.5 μg/mL) and/or SAHA (1 μg/mL). Real-time PCR was expressed as "fold change" in *maspin* mRNA expression normalized to GAPDH and relative to the vehicle-treated control. (**F**) ATF-126 in combination with chromatin-remodeling drugs restores *maspin* expression in MDA-MB-231 cell line. Western blot showing *maspin* protein in the breast cancer cell lines MCF-12A, MDA-MB-468, and MDA-MB-231. MDA-MB-231 cells were transduced with a control empty retroviral vector and ATF-126. ATF-126-transduced cells were treated with 5-aza-2′-dC (1.0 μg/mL) and SAHA (0.5 μg/mL). MCF-12A was used as a normalizing control.

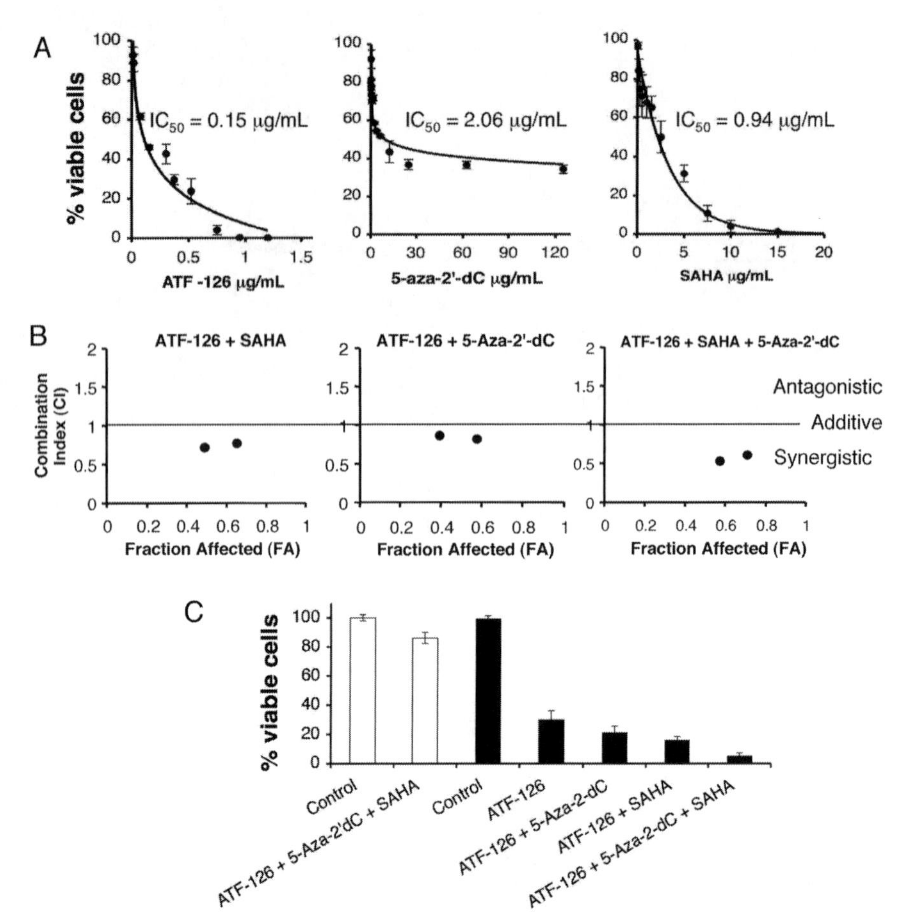

Fig. 10.4. ATF-126 synergizes with 5-aza-2′dC and SAHA in inhibiting tumor cell proliferation. (**A**) Dose–effect curve for cells transduced with ATF-126 (0–2 μg/mL) or treated with 5-aza-2′-dC (0–125 μg/mL) and SAHA (0–10 μg/mL). (**B**) Combination index (CI) plots of ATF-126 transduced cells treated with 5-aza-2′-dC [ATF-126 + 5-aza-2′-dC], SAHA [ATF-126 + SAHA], and both inhibitors [ATF-126 + 5-aza-2′-dC + SAHA]. Combination index (CI) was calculated to measure the synergistic interaction between ATF-126, 5-aza-2′-dC, and SAHA in MDA-MB-231 breast cancer cell line. The concentrations used in each combination are described in 10. CI < 1 defines a synergistic interaction, CI > 1 antagonistic drug interaction, and the straight line at CI = 1 represents additive effects. (**C**) ATF-126 inhibits tumor cell viability in combination with chromatin-remodeling drugs. MCF-12A and MDA-MB-231 cells were transduced with control empty vector and ATF-126 (0.525 μg/mL). ATF-126 transduced cells were treated with a combination of 5-aza-2′-dC (1 μg/mL) and SAHA (0.5 μg/mL) and both inhibitors together for 48 h at 37°C and 5% CO_2. Cell viability was measured using the XTT assay.

CGTGGCCTCCATGT TC-3′), probe (5′-6FAM-GACAG ACACCAAACCAGTGCAG-MBG-3′); *VP64* forward (5′-AAGCGACGCATTGGATGAC-3′), reverse (5′-GGAACG TCGTACGGGTAGTTAATT-3′), probe (5′-6FAM-TCGG CTCCGATGCT-MGB-3′).

3. Human GAPDH qRT-PCR endogenous control, 20× primers, and probe mix (Applied Biosystems #4352934E, FAM Dye/MGB Probe, Non-Primer Limited).

4. Real-time PCR machine (e.g., 7,500 Fast System, Applied Biosystems).
5. RNeasy Plus Mini Kit (Qiagen).
6. 10 U/μL RNase-free DNase I (Qiagen).
7. High-Capacity cDNA Archive Kit (Applied Biosystems) for cDNA synthesis.
8. RNase inhibitor 40 U/μL (Applied Biosystems).
9. TaqMan Fast Real-Time PCR Universal Master Mix (2×), no AmpErase UNG (Applied Biosystems).

2.3. Assessing the Phenotype of ATF-Transduced Cells

1. Apoptosis Detection Kit (e.g., Annexin V:PE Apoptosis Detection Kit I, BD Biosciences).
2. Cell Proliferation Kit II (XTT, Roche Applied Sciences).
3. 5-Aza-2′-dC (5-Aza-2′-deoxycytidine, Sigma) dissolved in DMSO. Stock concentration: 1 mg/mL. Aliquot and store at –80°C. Use a fresh working dilution from 0 to 125 μg/mL in growth medium.
4. SAHA (suberoylanilide hydroxamic acid, BioVision) dissolved in DMSO at 1 mg/mL. Aliquot and store at –80°C. Use a fresh working dilution from 0 to 10 μg/mL in growth medium.
5. Flow cytometer (e.g., FACSCalibur and CellQuest software, BD Biosciences).

2.4. Assessing Synergy Between ATFs and Chromatin-Remodeling Drugs

1. PharmToolsPro (McCary Group).
2. CalcuSyn (BIOSOFT).

3. Methods

3.1. Retroviral Delivery of Artificial Transcription Factors (ATFs)

MDA-MB-231 breast cancer cells harboring a methylated *maspin* promoter (*see* **Fig. 10.3D**) are used as model system. ATFs specifically targeting the *maspin* promoter are expressed using retroviral vectors and transduction efficiency is measured by flow cytometry (*see* **Notes 1** and **2**).

3.1.1. Seeding Retroviral Packing 90.74 Cells

1. Thaw a vial of retroviral packing 90.74 cells by swirling in a 37°C water bath until most of, but not all, the contents are thawed.
2. Spray the vial with 70% ethanol and wipe it dry before placing it in the tissue culture hood.

3. Gently add 1 mL of DMEM-Supplemented to the contents of the cryovial, mix, and transfer into a 15-mL conical tube containing 9 mL of DMEM-Supplemented.
4. Centrifuge the cells at 200×*g* at room temperature for 4 min.
5. Discard the supernatant and resuspend the cells with 10 mL of DMEM-Supplemented.
6. Determine the number of retroviral packaging 90.74 cells and adjust the concentration to 2×10^5 cells/mL.
7. Transfer 10 mL of cell suspension into a 100-mm plate. Incubate the cells at 37°C, 5% CO_2 overnight.
8. The next day, prepare a 100 μg/mL poly-D-Lysine working solution from the stock (5 mg/mL). Add 5 mL to a 100-mm plate and incubate for 45 min at room temperature.
9. Wash the plates twice with 5 mL of water and add 5 mL of DMEM medium. The plates are ready for seeding the cells.
10. Wash the retroviral packaging 90.74 plates with 5 mL of PBS.
11. Add 1 mL of 0.05% Trypsin/EDTA and allow cells to detach by incubating for 2 min at 37°C.
12. Collect cells by adding 4 mL of DMEM-Supplemented and transfer to a 15-mL conical tube.
13. Centrifuge the cells at 200×*g* at room temperature for 4 min.
14. Discard the supernatant and resuspend the cells with 5 mL of DMEM-Supplemented medium in a 50-mL conical tube.
15. Determine the number of retroviral packing 90.74 cells and adjust the concentration to 3×10^5 cells/mL.
16. Transfer 10 mL of cell suspension into the poly-D-Lysine-coated plates. Incubate the cells at 37°C, 5% CO_2 overnight. Cells should be 70–80% confluent after 24 h.

3.1.2. Transient Transfection of Retroviral Packing 90.74 Cells

1. Prepare the transfection medium, which contains the pMDG.1 (VSV-G envelope-expressing plasmid (17)), the retroviral vector (expressing the *maspin*-specific ATFs) (9), Plus Reagent, Lipofectamine, and DMEM.
   a. Prepare the following mix: 1.25 μg of pMDG.1 plasmid, 30 μL of Plus Reagent, and enough DMEM for 470 μL per 100-mm plate.
   b. Add 3.75 μg of DNA control (empty vector) to a 15-mL conical tube label as control. Repeat the procedure for ATF-126.

c. Add to each tube 500 μL of the mix from Step (a).
d. Incubate for 15 min at room temperature.
e. Prepare Lipomix: 20 μL of Lipofectamine and 480 μL of DMEM per 100-mm plate.
f. Add 500 μL of Lipomix to each retroviral vector tube from Step (c).
g. Incubate 15 min at room temperature.
h. Add 4 mL of DMEM to each tube from Step (f).

2. Aspirate the medium from the retroviral packing 90.74 plates and wash the cells with 5 mL of DMEM.
3. Label the retroviral packaging 90.74 plates with the corresponding retroviral construct.
4. Add 5 mL of the transfection medium from each tube to the corresponding retroviral packing 90.74 plate.
5. Incubate the retroviral packaging 90.74 plates for 3 h at 37°C, 5% CO_2.
6. Aspirate the transfection medium and add 6 mL of DMEM-Supplemented.
7. Incubate the plates for 48 h at 37°C, 5% CO_2.

3.1.3. Seeding Host Cells (MDA-MB-231)

The host cells need to be seeded 24 h before harvesting the retrovirus. A seeding density of 1×10^5 cells/plate in a 100-mm plate is recommended (*see* **Note 3**).

3.1.4. Harvesting the Retrovirus and Transduction of Host Cells

1. Days 1–2:
   a. Collect the retrovirus-containing medium. Filter the retroviral packaging 90.74 supernatants with a 0.22-μm filter unit to eliminate any cell debris (residual packaging cells).
   b. Add 6 mL of fresh DMEM-Supplemented to each plate and return it to the incubator.
   c. Add 10 μL/mL of polybrene (working solution 800 μg/mL) to the virus-containing medium for a final concentration of 8 μg/mL.
   d. Add 5 mL of retrovirus-containing medium to each host cell plate (e.g., MDA-MB-231) and incubate the plates for at least 8 h at 37°C, 5% CO_2.
   e. Harvest retrovirus every 8 h for a total of four times. These supernatants are used immediately to transduce the host cells.
2. Day 3:
   a. Aspirate the retrovirus-containing medium from the host cells.

b. Add fresh host cell medium (e.g., MDA-MB-231 are grown in DMEM-supplemented medium).
c. Incubate the plates for 48–72 h at 37°C, 5% CO_2.
d. 48-h post-transduction, determine the transduction efficiency by flow cytometry.

3.1.5. Calculating Transduction Efficiency by Flow Cytometry

1. Since every retroviral transduction is different, it is important to assure that cells are transduced at least 70–90% before proceeding with any experimental procedure.
2. Wash cell plates with 5 mL of PBS.
3. Add 1 mL of trypsin/EDTA and allow cells to detach by incubating for 5 min at 37°C.
4. Collect cells by adding 4 mL of host cell medium and transfer to a new 15-mL conical tube.
5. Centrifuge the cells at 200×*g* at room temperature for 4 min.
6. Discard the supernatant and wash the pellet twice with 5 mL of PBS.
7. Resuspend the cell pellet in 500 μL of PBS.
8. Process the samples by flow cytometry. A FACSCalibur and CellQuest software are used for the analysis in our laboratory. Instructions for this software are given here.
   a. Collect forward scatter (FSC) and side scatter (SSC) readings, using non-transduced cells.
   b. Draw a light scatter gate in the SSC versus FS plot to exclude cell debris.
   c. Display cells in the gate in a single-parameter histogram for the GFP and collect the GFP signal using the FL1 channel (*see* **Fig. 10.3B**).
   d. Define your positive population by using the Quadrant Marker tool from the tool palette. Choose the Quadrant Stats from the Stats menu and display the percentage (%) of GFP-positive cells for each clone.

3.2. Assessment of ATF Expression and Target Gene Regulation

3.2.1. Design Primers and Probe for Target Gene Amplification

1. There are numerous software packages available for real-time PCR primer design, some freely available and some commercial.
2. We use Primer Express from Applied Biosystems to design both primers and probes. Primer pairs are designed to

amplify a 70–110 bp product, are 20–25 bp long, and have a T_m of 58–63°C. The probe is a hydrolytic probe known as TaqMan assays (Applied Biosystems). Specific examples of primers and probes used for the *maspin* and ATF genes are listed in **Section 2**.

3.2.2. RNA Isolation

1. Isolate the RNA from ATF-transduced and control cells using a commercial RNA extraction system (e.g., RNeasy Plus Mini Kit), following the manufacturer's instructions precisely.
2. Dissolve the RNA in 35 μL of molecular biology grade water and place the tubes on ice (*see* **Note 4**).
3. Incubate the RNA with DNase I to remove residual genomic DNA that may be present in the RNA prep.
   a. In a PCR tube add 4 μL of 10 U/μL RNase-free DNase I, 5 μL of 25 mM $MgCl_2$, 2.8 μg of RNA, and enough water for 30 μL total.
   b. Mix gently and spin briefly. Using a PCR thermal cycler, incubate at 37°C for 10 min, then 90°C for 5 min to inactivate the DNase.

3.2.3. cDNA Preparation

1. Synthesize the cDNA using the High-Capacity cDNA Archive Kit, following the manufacturer's instructions. Briefly, prepare the reverse transcription master mix: 5 μL of 10× Reverse Transcription Buffer, 2 μL of 25× dNTPs, 5 μL of 10× random primers, 2.5 μL of 50 U/μL Multi-Scribe Reverse Transcriptase, 1 μL of RNase inhibitor, and 11.5 μL nuclease-free water. For more than one reaction, adjust volumes accordingly.
2. Add 25 μL of the reverse transcription master mix to a 200-μL PCR sample tube labeled with the date and sample number.
3. Dilute 2.5 μg of the DNase-treated RNA into 25 μL water and add it to the sample tube.
4. Incubate at the following temperatures using a PCR thermal cycler: 25°C for 10 min (to allow the random hexamers to anneal), 37°C for 120 min, 85°C for 5 min (to inactivate the reverse transcriptase), and 4°C for storage.
5. The resulting cDNA may be analyzed immediately by real-time PCR or stored at –20 or –80°C.

3.2.4. Real-Time Quantitative PCR (qRT-PCR)

1. Perform three PCR replicates for each cDNA sample. Place 100 ng of template DNA into a 0.2-mL PCR tube.
2. Dilute the cDNA to a final concentration of 100 ng in 5 μL of nuclease-free water. Prepare enough for the three replicates of each gene.

4. Prepare a 20× primers/probe mix for each gene which will contain 4 μM of forward primer, 4 μM of reverse primer, and 2 μM of probe.
5. Prepare a master mix for each target gene (ATF, *maspin*, and the endogenous control (e.g., GAPDH)). This mix should contain 10 μL of 2× Fast TaqMan, 1 μL of 20× primers/probe mix and 4 μL of nuclease-free water.
6. Place 5 μL of diluted cDNA into a 96-well plate.
7. Add 15 μL of master mix into each tube.
8. Place the plate into a real-time PCR machine. Run the PCR for 10 min at 95°C, and 40 cycles of 1 s at 95°C and 30 s at 65°C. Select acquisition of the data during the 65°C.
9. Calculate the fold change in gene expression using the comparative C_T method ($2\text{–}\Delta\Delta^{CT}$ method) as described in **Notes 5** and **6**.

3.2.5. Statistical Analysis

1. Perform a Student's *t* test or ANOVA test on the output data to determine whether the expression of the ATFs and the target gene vary under the experimental conditions.

3.3. Assessing the Phenotype of ATF-Transduced Cells

3.3.1. Annexin V Staining to Quantify Apoptosis

1. The Annexin V protein conjugated with phycoerythrin (PE), a component of the Annexin V:PE Apoptosis Detection Kit I, is used to quantify the early apoptotic effect of *maspin*-specific ATFs in cancer cells (*see* **Notes 7, 8, 9,** and **10**).
2. Wash the cells twice with cold PBS and resuspend them in 1× Binding Buffer (Kit supplied) at a concentration of 1 × 10^6 cells/mL.
3. Transfer 100 μL of the cell suspension (1 × 10^5 cells) into a 5-mL culture tube.
4. Add 5 μL of PE-Annexin V and 5 μL of 7-AAD.
5. Gently vortex the cells and incubate for 15 min at room temperature (25 °C) in the dark.
6. Add 400 μL of 1× Binding Buffer into each tube.
7. Analyze by flow cytometry within 1 h. The percentage of apoptosis is measured by flow cytometry using a FACSCalibur and CellQuest software.

3.3.2. Assay for Cell Proliferation (Viability Assay, XTT Assay)

The effect of the ATFs and the chromatin-remodeling drugs in inhibiting tumor cell viability is measured using the Cell Proliferation Kit II (*see* **Note 11**).

1. Collect cells by adding 4 mL of growth medium (specific to the cell line about to be plated) and transfer into a 15-mL conical tube.
2. Centrifuge the cells at 200×*g* at room temperature for 5 min.
3. Discard the supernatant and resuspend the cells in 10 mL of growth medium.
4. Determine the number of cells and adjust the concentration to 3,000 cells/well in 100 μL of medium (*see* **Note 12**).
5. Label 96-well plates and add to each well 100 μL of counted cells from Step 4.
6. Place the 96-well plates at 37°C, 5% CO_2.
7. Prepare the XTT reagent for the normalizing plate (*see* **Note 13**).
8. Thaw XTT labeling reagent and electron-coupling reagent, respectively, in a water bath at 37°C. Mix each vial thoroughly to obtain a clear solution. To perform a cell proliferation assay (XTT) with one 96-well plate, mix 5 mL of XTT labeling reagent with 0.1 mL of electron coupling reagent (1:50 dilution).
9. Immediately, add 50 μL of XTT labeling mixture to each well (final XTT concentration 0.3 mg/mL). Place it back in the incubator for 4–6 h.
10. Read the absorbance at two wavelengths: 490 nm for XTT activity and 650 nm as a reference.
11. Collect the output results for all samples.
12. Place the output results for all data points into a Microsoft Excel worksheet. Obtain the mean (*M*) and standard deviation (SD) for each clone. To calculate the percentage of viable cells, normalize using the vehicle-treated cells. Present the data in a plot as percentage of viable cells versus sample.
13. For each time point chosen (24, 48, 72 h, etc.), repeat the procedure from Steps 8–12.
14. Modifications to the proliferation assay for ATF-transduced cells that have been treated with drugs (*see* **Section 3.3.3**) are described in **Sections 3.3.2.1, 3.3.2.2**, and **3.3.2.3**.

3.3.2.1. Cell Viability Curves for Transduced Cells

1. Plate the cells in a 96-well plate at 24-h post-transduction.
2. Plate non-transduced cells as a negative control, control cells (empty vector vehicle treated), and ATF-126-transduced cells (*see* **Section 3.3.3** for the range of concentrations for control vector/ATF-126).

3. Incubate the cells for a 48-h period.
4. Follow the proliferation protocol as described previously (*see* **Fig. 10.4A**).

3.3.2.2. Cell Viability Curve for Drug-Treated Cells

1. Make serial dilutions of the drugs/vehicle in 50 μL of growth medium.
2. Collect and determine the concentration of cells. Adjust the concentration of cells to 3,000 cells/50 μL medium per well.
3. Place 50 μL of cell suspension into each well of a 96-well plate.
4. Add 50 μL of drug/vehicle dilution to each well (*see* **Section 3.3.3** for the range of concentrations for 5-aza-2′-dC and SAHA).
5. Incubate the cells for a 48-h period and follow the proliferation protocol described previously (*see* **Fig. 10.4B**).

3.3.2.3. Cell Viability Curve for Drug Combinations

1. Twenty-four hours post-transduction, collect and determine the concentration of cells. Adjust the concentration of cells to 3,000 cells/50 μL of DMEM-Supplemented.
2. Place 50 μL of cells suspension into each well of a 96-well plate.
3. Make combinations of the drugs/vehicle according to the concentrations chosen in 50 μL of growth medium.
4. Add 50 μL of drug/vehicle dilution to each well (*see* **Section 3.3.3** for the range of concentrations for 5-aza-2′-dC and SAHA).
5. Incubate the cells for a 48-h period and follow the proliferation protocol described previously (*see* **Fig. 10.4C**).

3.3.3. Drug Treatments

1. Transduce MDA-MB-231 cells with several concentrations of plasmid-encoded ATFs (from 0 to 2 μg/mL) for a period of 72 h at 37°C, 5% CO_2 incubator. Repeat the same process for the control empty vector.
2. Treat transduced cells and non-transduced control cells with a range of concentrations of drug, depending on the specific drug to be tested (e.g., for 5-aza-2′-dC, 0–125 μg/mL and for SAHA, 0–10 μg/mL). Treatments are allowed for a period of 48 h in a 37°C, 5% CO_2 incubator.
3. A total of 0.25×10^6 cells are seeded into 100-mm plates and treated with the drug dilutions. After the incubation period, collect the cells and analyze them by qRT-PCR of target gene regulation (e.g., *maspin*) as described in **Section 3.2**.
4. To perform a drug dose–response curves, use the range of concentrations described in Step 2 of **Section 3.3.3**.

3.4. Assessing Synergy Between ATFs and Chromatin-Remodeling Drugs

3.4.1. Combinatorial Index (CI) Method to Determine Synergism Between ATFs and Chromatin-Remodeling Drugs

1. The CI is a measurement of the pharmacological interaction between drugs and is based on the isobologram equation (CI-isobologram) (21–25). The CI-isobologram equation determines the additive effect of drug combinations, such as synergisms (defined as a greater-than-expected additive effect) and antagonisms (defined as a less-than-expected additive effect) (*see* **Notes 6** and 7).
2. The CI-isobologram equation for a three drug analysis is represented by $[CI=(D)_1/(Dx)_1 + (D)_2/(Dx)_2 + (D)_3/(Dx)_3]$, where CI is the combinatorial index, D is the dose of drug that gives a specific effect, and Dx is the dose of drug used in the combination. CI<1, CI=1, and CI>1, indicate synergy, additive effect, and antagonism, respectively.
3. The CI-isobologram method requires the generation of the dose–effect curves and median–effect plots to calculate the IC_{50}, the slope (m), and the correlation coefficient (r). The IC_{50} is calculated by the median effect equation: $D_x = D_m [fa/(1\text{–}fa)]^{1/m}$, where Dx is the dose, fa is the fraction of cells affected by the dose, and Dm is the median–effective dose. The Dm is equivalent to the IC_{50}.
4. There are user-friendly software tools available to calculate the combinatorial index, such as CalcuSing (26, 27) and PharmToolsPro (21).
5. Here, we will perform a manual calculation in which the readout values from the viability assay for all replicates for each data point are placed in a Microsoft Excel worksheet (*see* **Notes 14** and **15**).
6. Calculate the mean and standard deviation for each point. If a significant difference is detected (run a Student's t test or ANOVA), proceed to normalizing each point with the zero concentration.
7. Calculate the percentage of viable cells for each agent as

$$\text{Percentage of viable cells} = (\text{ODnon} - \text{treated cells}/\text{OD of treated cell}) \times 100$$

8. Plot the percentage of viable cells versus the concentration. Make the median–effect plots for each agent and calculate the IC_{50} using the formula $IC_{50} = 10^{-(-y\ \text{intercept})/m}$,

where y-intercept is the point where the curve intercepts the y-axis and m is the slope of the curve.

9. To address synergisms between ATF-126 and either 5-aza-2′-dC or SAHA, we fix the concentrations used in combination (26, 27). For *maspin*, we chose concentrations that reduced cell viability to 30–80%. The combination of drugs was made by combining the concentration of ATF-126 with 5-aza-2′-dC and SAHA that gave the desired effect (*see* **Notes 16** and **17**).
10. The following steps illustrate an example in which the CI is calculated for a specific drug combination. In this example, we use the cell viability data generated after the triple combination of ATF-126/5-aza-2′-dC/SAHA as well as data from the single treatments.
    a. Calculate the percentage of viable cells after treatment with each combination. The OD for vehicle-treated cells is 0.556 and the OD of drug-treated cell is 0.239.

$$\%\text{ viable cells} = (\text{O.D. vehicle-treated cells}/\text{O.D. of drug-treated cell}) \times 100$$

$$\%\text{ viable cells} = (0.239/0.556) \times 100 = 43.06\%$$

$$\%\text{ non-viable cells} = 100 - 43.06\% = 56.93\%$$

    b. Calculate the fraction affected (fa) for the drug combination:

$$\text{fa} = \text{non-viable cells}/100 = 56.93/100 = 0.569$$

    c. For each combination tested, calculate Dx. For ATF-126, IC_{50} $(Dm) = 0.15$ and $m = 1.05$ (*see* **Fig. 10.4A**):

$$Dx = Dm[\text{fa}/(1 - \text{fa})]1/m$$

$$\text{ATF-126} - Dx = 0.15[0.569/1 - 0.569]1/1.05 = 0.19$$

The same procedure is applied to calculate [5-aza-2′-dC–Dx] and [SAHA–Dx].

    d. Finally, calculate the CI using the CI-isobologram equation:

$$\text{CI} = (D)_1/(Dx)_1 + (D)_2/(Dx)_2 + (D)_3/(Dx)_3$$

$$\text{CI} = [0.075/0.19] + [0.132/1.0] + [0.25/14.36] = 0.54$$

Since 0.54 is less than 1, the combination is considered *synergistic*.

4. Notes

1. To follow the transduction protocol, it is important to make a calendar with major steps involved. For example, *Monday*, prepare and seed retroviral packaging 90.74 plates; *Tuesday*, transiently transfect retroviral packaging 90.74 cells; *Wednesday*, seed the host cells; *Thursday*, transduce every 8–12 h; *Friday*, transduce every 8–12 h; *Saturday*, replace the virus-contained medium with fresh growth medium; *Monday*, assess transduction efficiency by flow cytometry.
2. After transduction, cells are collected and can be processed for real-time PCR, Western blot, immunofluorescence, and other biological assays, depending on the nature of the specific target gene.
3. The seeding density for MDA-MB-231 is 1×10^5 cells/plate in a 100-mm plate. An adjustment on the seeding density is recommended for other cell lines depending on the growth ratio.
4. RNA may be stored at –80°C. It is recommended that RNA be converted to cDNA the same day of isolation.
5. The qRT-PCR is a powerful tool for quantifying the expression of the ATFs and assessing changes in the expression of the target gene. The $2^{-\Delta\Delta CT}$ method is used to calculate these changes. In qRT-PCR the quantitative endpoint is called the threshold cycle (C_T). The C_T is defined as the PCR cycle at which the fluorescent signal of the reporter dye crosses an arbitrarily placed threshold. The data represented as a C_T are acquired from the exponential phase of the amplification curve. The C_T numerical value is inversely related to the amount of amplicon in the reaction (20). This method assumes that the efficiency of the PCR is close to 1 and that the amplification efficiency is similar for both the target gene and the internal control gene.
6. The fold change in gene expression normalized by the internal control is calculated using the following formula:

$$\text{Fold change} = 2^{-\Delta\Delta CT} = (C_T \text{ target gene} - C_T \text{ internal control}) \text{ treated sample} - (C_T \text{ target gene} - C_T \text{ internal control}) \text{ untreated sample.}$$

7. The loss of the plasma membrane is among the earliest events in the apoptosis process. The phospholipid phosphatidylserine (PS) is translocated from the inner to the outer leaflet of the plasma membrane and is exposed to the

external cellular environment. Annexin V is a phospholipid-binding protein with a high affinity for PS, which is used to bind extracellular PS. In addition, 7-amino-actinomycin D (7-ADD) is used to discriminate between live and dead cells. 7-AAD intercalates into double-stranded nucleic acids of dead or dying cells, but is excluded by viable cells.

8. Controls are also needed to set up the flow cytometry experiment. These controls include unstained cells, cells stained with PE-Annexin V alone (no 7-AAD), and cells stained with 7-AAD alone (no PE-Annexin V).
9. A positive control for apoptosis is induced with 10 μM camptotecin for a period of 24 h prior to the analysis.
10. We measure apoptosis at 120 h post-transduction in MDA-MB-231 cells.
11. The effect of the ATFs and the chromatin-remodeling drugs in inhibiting tumor cell viability is measured by the ability of metabolically active cells to reduce the tetrazolium salt XTT to orange-colored compounds of formazan. Transduced/treated cells are incubated with a tetrazolium salt (XTT), a component of the Cell Proliferation Kit II, and incubated for 4–6 h. The appearance of color is measured at 490 nm. The color directly correlates with the number of viable cells in the sample.
12. We recommend seeding at least five replicates per clone.
13. We recommend seeding a normalizing plate, which should be read the same day of the seeding in order to reduce counting/seeding errors.
14. For the combination experiments, we recommend fixing the concentrations in the IC_{50} range, which results in tumor cell growth inhibition.
15. It is important to add as controls vehicle-treated cells and non-transformed cells to verify that the treatment is specific for cancer cells and not normal cells (which express high levels of tumor suppressor).
16. When targeting epigenetically silenced genes, a null or small regulation may be detected. In these cases, an appropriate chromatin-remodeling drug must be chosen. To study if these drugs synergize with the ATFs, use a combination with a previously published *Dm* (IC_{50}) if this is available or calculate the specific *Dm* for each drug (*see* **Fig. 10.4A**).
17. The combination with the Dm_{50} would give a target gene regulation that may appear as a greater-than-expected-additive effect. In this case, it is recommended to calcu-

late synergism by proposing combinations based on each drug's dose–response curve. If no regulation is detected, an adjustment on the drug dose may be necessary. Alternatively, use a different type of chromatin-remodeling drug.

References

1. Collingwood, T.N., Urnov, F.D., and Wolffe, A.P. (1999) Nuclear receptors: coactivators, corepressors and chromatin remodeling in the control of transcription. *J Mol Endocrinol.* **3**, 255–275.
2. Ohm, J.E., McGarvey, K.M., Yu, X., Cheng, L., Schuebel, K.E., Cope, L., Mohammad, H.P., Chen, W., Daniel, V.C., Yu, W., Berman, D.M., Jenuwein, T., Pruitt, K., Sharkis, S.J., Watkins, D.N., Herman, J.G., and Baylin, S.B. (2007) A stem cell-like chromatin pattern may predispose tumor suppressor genes to DNA hypermethylation and heritable silencing. *Nat Genet.* **39**, 237–242.
3. Baylin, S.B. (2005) DNA methylation and gene silencing in cancer. *Nat Clin Pract Oncol.* **1**, S4.
4. Ting, A.H., McGarvey, K.M., and Baylin, S.B. (2006) The cancer epigenome—components and functional correlates. *Genes Dev.* **20**, 3215–3231.
5. Jones, P.A. and Baylin, S.B. (2007) The epigenomics of cancer. *Cell.* **128**, 683–692.
6. Jones, P.A. and Laird, P.W. (1999) Cancer epigenetics comes of age. *Nat Genet.* **2**, 163–168.
7. Sims, R.J., III and Reinberg, D. (2008) Is there a code embedded in proteins that is based on post-translational modifications? *Nat Rev Mol Cell Biol.* **10**, 815–820.
8. Palii, S.S. and Robertson, K.D. (2007) Epigenetic control of tumor suppression. *Crit Rev Eukaryot Gene Expr.* **4**, 295–316.
9. Beltran, A., Parikh, S., Liu, Y., Cuevas, B.D., Johnson, G.L., Futscher, B.W., and Blancafort, P. (2007) Re-activation of a dormant tumor suppressor gene maspin by designed transcription factors. *Oncogene.* **19**, 2791–2798.
10. Beltran, A.S., Sun, X., Lizardi, P.M., and Blancafort, P. (2008) Reprogramming epigenetic silencing: artificial transcription factors synergize with chromatin remodeling drugs to reactivate the tumor suppressor mammary serine protease inhibitor. *Mol Cancer Ther.* **5**, 1080–1090.
11. Blancafort, P. and Beltran, A.S. (2008) Rational design, selection and specificity of artificial transcription factors (ATFs): the influence of chromatin in target gene regulation. *Comb Chem High Throughput Screen.* **11**, 146–158.
12. Beltran, A., Liu, Y., Parikh, S., Temple, B., and Blancafort, P. (2006) Interrogating genomes with combinatorial artificial transcription factor libraries: asking zinc finger questions. *Assay Drug Dev Technol.* **3**, 317–331.
13. Blancafort, P., Magnenat, L., and Barbas, I.I.I.C.F. (2003) Scanning the human genome with combinatorial transcription factor libraries. *Nat Biotechnol.* **21**, 269–274.
14. Schaefer, J.S. and Zhang, M. (2006) Targeting maspin in endothelial cells to induce cell apoptosis. *Expert Opin Ther Targets.* **3**, 401–408.
15. Zhang, W., Shi, H.Y., and Zhang, M. (2005) Maspin overexpression modulates tumor cell apoptosis through the regulation of Bcl-2 family proteins. *BMC Cancer.* **5**, 50.
16. Li, X., Chen, D., Yin, S., Meng, Y., Yang, H., Landis-Piwowar, K.R., Li, Y., Sarkar, F.H., Reddy, G.P., Dou, Q.P., and Sheng, S. (2007) Maspin augments proteasome inhibitor-induced apoptosis in prostate cancer cells. *J Cell Physiol.* **2**, 298–306.
17. Wozniak, R.J., Klimecki, W.T., Lau, S.S., Feinstein, Y., and Futscher, B.W. 5-Aza-2′-deoxycytidine-mediated reductions in G9A histone methyltransferase and histone H3 di-methylation levels are linked to tumor suppressor gene reactivation. *Oncogene.* **1**, 77–90.
18. Primeau, M., Gagnon, J., and Momparler, R.L. (2003) Synergistic antineoplastic action of DNA methylation inhibitor 5-AZA-2′-deoxycytidine and histone deacetylase inhibitor depsipeptide on human breast carcinoma cells. *Int J Cancer.* **2**, 177–184.
19. Oshiro, M.M., Watts, G.S., Wozniak, R.J., Junk, D.J., Munoz-Rodriguez, J.L., Domann, F.E., and Futscher, B.W. (2003) Mutant p53 and aberrant cytosine methylation cooperate to silence gene expression. *Oncogene.* **22**, 3624–3634.
20. Livak, K.J. and Schmittgens, T.D. (2001) Analysis of relative gene expression data using real-time quantitative PCR and the $2^{(-\Delta\Delta CT)}$ method. *Methods.* **4**, 402–408.

21. Tallarida, R.J. (2000) Drug synergism and dose-effect data analysis, pp. 15–71. CRC Press, Boca Raton
22. Berenbaum, M.C. (1989) What is synergy? *Pharmacol Rev.* **2**, 93–141.
23. Chou, T.C. and Talalay, P. (1984) Quantitative analysis of dose-effect relationships: the combined effects of multiple drugs or enzyme inhibitors. *Adv Enzyme Regul.* **22**, 27–55.
24. Chou, T.C. (1994) Assessment of synergistic and antagonistic effects of chemotherapeutic agents in vitro. *Contrib Gynecol Obstet.* **19**, 91–107.
25. Chou, T.C. (2006) Theoretical basis, experimental design, and computerized simulation of synergism and antagonism in drug combination studies. *Pharmacol Rev.* **3**, 621–681.
26. Reynolds, C.P. and Maurer, B.J. (2005) Evaluating response to antineoplastic drug combinations in tissue culture models. *Methods Mol Med.* **110**, 173–183.
27. Meczes, E.L., Pearson, A.D., Austin, C.A., and Tilby, M.J. (2002) Schedule-dependent response of neuroblastoma cell lines to combinations of etoposide and cisplatin. *Br J Cancer.* **3**, 485–489.

Chapter 11

Transgenic Mice Expressing an Artificial Zinc Finger Regulator Targeting an Endogenous Gene

Claudio Passananti, Nicoletta Corbi, Annalisa Onori, Maria Grazia Di Certo, and Elisabetta Mattei

Abstract

Zinc finger (ZF) proteins belonging to the Cys2–His2 class provide a simple and versatile framework to design novel artificial transcription factors (ATFs) targeted to the desired genes. Our work is based on ZF ATFs engineered to up-regulate the expression level of the dystrophin-related gene utrophin in Duchenne muscular dystrophy (DMD). In particular, on the basis of the "recognition code" that defines specific rules between zinc finger primary structure and potential DNA-binding sites we engineered and selected a new family of artificial transcription factors, whose DNA-binding domain consists in a three zinc finger peptide called "Jazz." Jazz protein binds specifically the 9 bp DNA sequence (5′-GCT-GCT-GCG-3′) present in the promoter region of both the human and mouse utrophin gene. We generated a transgenic mouse expressing Jazz protein fused to the strong transcriptional activation domain VP16 and under the control of the muscle specific promoter of the myosin light chain gene. Vp16-Jazz mice display a strong up-regulation of the utrophin at both mRNA and protein levels. To our knowledge, this represents the first example of a transgenic mouse expressing an artificial gene coding for a zinc finger-based transcription factor.

Key words: Zinc finger, synthetic gene, artificial transcription factor, utrophin, dystrophin, muscular dystrophy, transgenic mice, gene therapy.

1. Introduction

In the last two decades, advances in design and engineering of DNA-binding domains led to the promising field of artificial transcription factors (ATFs). Zinc finger (ZF) domains have been shown to be optimal building blocks for generating ATFs, thanks to their versatility and modularity (1, 2). Importantly, several

J.P. Mackay, D.J. Segal (eds.), *Engineered Zinc Finger Proteins*, Methods in Molecular Biology 649,
DOI 10.1007/978-1-60761-753-2_11, © Springer Science+Business Media, LLC 2010

research groups, elucidating the interaction between zinc finger proteins and DNA at the molecular level, contributed to the development of a "recognition code" that relates amino acids in specific positions of the finger domain to its DNA target sequence (3–7). Multiple synthetic ZFs fused to the proper effector domain can act as transcription modulators that are able to control the expression of virtually any desired gene in the genome.

The possibility to reprogram gene expression will offer many applications in the fields of functional genomics and gene therapy (8–10). Genes can be turned on or off with the intent to find a cure to pathologies such as viral infections, cancer, and genetic diseases (11–15). We have contributed by demonstrating the feasibility of regulating an endogenous gene in vivo using designed ATFs. In particular, our research group has been focused on the generation of ZF ATFs that are able to reprogram the expression of the dystrophin-related gene, utrophin, in Duchenne muscular dystrophy. Utrophin expression complements the lack of dystrophin function observed in this disease (16–19). It is now well established that overexpression and relocalization of utrophin protein along the sarcolemma ameliorate the dystrophic phenotype in the mdx (dystrophin-deficient) mouse model (15, 20–22). The approach we chose was to engineer ZF ATFs to target the utrophin promoter A, in a sequence close to the conserved N and E box elements, and therefore in a favorable conformation of the chromatin in terms of transcription factor accessibility (23).

Here, we report in detail the procedures to obtain transgenic mice expressing at the muscular level, the artificial three zinc finger protein Vp16-Jazz, which is able to specifically up-regulate the utrophin gene. The Vp16-Jazz transgenic mice represent the first animal model expressing an artificial zinc finger-based transcription factor (18). Significantly, the increase of utrophin mRNA ranged from threefold to fourfold in mice from different transgenic families compared with their non-transgenic littermates, and the changes in the expression of utrophin mRNA were consistent with the increase and relocalization along the sarcolemma of utrophin protein (18). The Vp16-Jazz animal model validates the strategy of transcriptional targeting of endogenous genes and provides an exciting tool for use in drug discovery and therapeutics. Compared to other approaches that increase utrophin levels in muscle, such as the use of truncated utrophin genes, the possibility of up-regulating the expression of the endogenous utrophin gene by a synthetic transcription factor offers some specific advantages. The ATF is very small, is active at a very low concentration, and mimics the body's natural regulatory mechanism, resulting in the production of all the different isoforms of the targeted protein. We crossbred Vp16-Jazz transgenic mice with mdx mice. The resulting mdx/Vp16-Jazz mice display a strong amelioration of the dystrophic phenotype (24).

These promising results make Jazz-based ZF ATFs good candidates as novel therapeutic molecules for DMD treatment.

The methods reported here describe the experimental strategies employed to make novel ATFs, as well as to obtain transgenic mice expressing these artificial molecules. **Section 3.1** describes how the artificial gene Jazz was designed using the zinc finger recognition code and selected for its high efficiency in targeting transcriptional function. Jazz was fused to the Vp16 transcriptional activation domain, derived from herpes simplex virus, to up-regulate transcription from the utrophin gene promoter A. To obtain muscle-specific expression the myosin light chain 1 (MLC1) promoter/enhancer regions was used. **Section 3.2** describes how to ensure a high yield of transgenic mouse production, with particular attention on the quality of the DNA to be injected. This step is particularly critical for embryo survival and correct DNA integration. To get high quality DNA the whole synthetic Jazz gene unit was amplified by PCR. **Sections 3.3, 3.4, 3.5,** and **3.6** describe the production of transgenic mice. However, it is important to underline that the generation of transgenic mice needs specific instruments and trained and skilled personnel (*see* **Note 1**). Once transgenic mice were obtained, the expression of the synthetic protein in skeletal muscle was revealed by western blotting, and the ability of the zinc finger polypeptide to effectively bind its target sequence on the utrophin promoter was proven by chromatin immunoprecipitation (ChIP) assay. Both real-time PCR and western blotting demonstrated the resulting utrophin up-regulation. These analyses are described in **Sections 3.7, 3.8,** and **3.9**.

2. Materials

2.1. Synthetic Gene Design

1. Two long oligonucleotides named Jazz-5′ (153 nt) and Jazz-3′ (160 nt) (*see* **Section 3.1**). The GenBank accession number of the Jazz gene is AJ243577; the GenBank accession number of the Jazz peptide is CAB52142 (16).
2. Primer TM24: 5′-GCT AGC GTG ACA GAC CCT AT-3′, primer TM25: 5′-TCC GCT CGA GAT CAT TTT GC-3′.
3. Annealing buffer: 10 mM Tris–HCl, pH 7.5, 5 mM $MgCl_2$, and 7.5 mM DTT.
4. Klenow DNA polymerase and 10 mM dNTP mix PCR Grade (Qiagen).
5. Standard cloning vector (e.g., pGEM Easy Vector, Promega).

6. The pMEX mammalian vector containing the myosin light chain (MLC1) promoter/enhancer regions (25).
7. The transcriptional activation domain "Vp16" derived from herpes simplex virus (NCBI accession number of the amino acidic sequence of Vp16 protein is P04486) from amino acid 413–490: APPTDVSLGDELHLDGEDVA MAHADALDDFDLDMLGDGDSPGPGFTPHDSAPYG ALDMADFEFEQMFTDALGIDEYGG.
8. Intron taken from the precocious region of SV40 virus (from nucleotide 276–621 of the GenBank sequence AJ012749).
9. PolyA+ addition sequence taken from SV40 large T gene: 5′-AACTTGTTTATTGCAGCTTATAATGGTTACAAAT AAAGCAATAGCATCACAAATTTCACAAATAAAGCAT TTTTTTCACTGCATTCTAGTTGTGGTTTGTCCAAA CTCATCAATGTATCTTATCATGTCTGGAT-3′.
10. A nuclear localization signal (NLS) taken from the large T precocious protein of the SV40 virus: MAPKKKRKV.
11. Myc epitope 13 amino acid tag: MEQKLISEEDLNE.

2.2. Purification and Injection of the Transgenic DNA Construct

1. Expand PCR-Plus system (Roche).
2. Jazz forward: 5′-GAAAGGGGATGTGCTGCAAGGCG-3′; Jazz reverse: 5′-ACCACATTAATGTGAAAGAAGAGC-3′.
3. 10× Tris-acetate-EDTA (TAE) buffer: 0.4 M Tris-acetate, 10 mM EDTA, pH 8.
4. Ultrapure agarose SeaKem Genetic Technology Grade (Lonza).
5. SYBR Safe DNA Gel Stain (Invitrogen).
6. TE buffer: 10 M Tris–HCl, 1 mM EDTA, pH 8.
7. 10× DNA loading buffer: 0.1% bromophenol blue and 0.1% xylene cyanole in 30% glycerol and double distilled water (to enhance density and facilitate sample loading).
8. GENECLEAN kit (Q-BIOgene-).
9. A spectrophotometer such as a Qubit (Invitrogen) or NanoDrop1000 (Thermo Scientific).
10. Injection buffer: 10 mM Tris–HCl, pH 7.5, 0.1 mM EDTA, 30 mM NaCl. Filter through a 0.22 μm filter before use, discarding the first 1–2 mL of flow through.

2.3. Preparation of Embryo Donor Mice

1. BDF1 (*see* **Note 2**) female mice aged 5–8 weeks (Charles River Laboratories) are used as embryo donors (five mice for each microinjection experiment).

2. BDF1 male mice aged 2–12 months are used for breeding with donor females.
3. Lyophilized pregnant mare's serum gonadotropin (PMSG) and human chorionic gonadotropin (hCG) for veterinary use (Intervet International). Dissolve in sterile 0.9% NaCl aqueous solution at a concentration of 50 IU (International Units)/mL. Store as single-use aliquots at –20°C. A volume of 0.1 mL is injected into each animal at a dose of 5 IU per mouse.

2.4. Embryo Collection, Culture, and Microinjection

1. 35 mm Petri dishes.
2. Glass capillary tubing.
3. M2 medium (Sigma-Aldrich) for manipulating embryos outside incubator.
4. Hyaluronidase from bovine testes (Sigma-Aldrich). Dissolve in M2 medium to a final concentration of 10 mg/mL. Store at –20°C in 1 mL aliquots. Can be freeze–thawed several times without loss of enzymatic activity.
5. M16 medium (Sigma-Aldrich) for embryo culture in CO_2 incubator (37°C, 5% CO_2 in air).
6. 200× penicillin–streptomycin solution: penicillin G-potassium salt 100 units/mL and streptomycin sulphate 50 mg/mL (both from Invitrogen) dissolved in sterile water. To be added to embryo culture media before use.
7. Mineral oil, sterile-filtered and mouse embryo tested (Sigma-Aldrich), for covering embryo culture media drops to avoid evaporation and concentration.
8. Inverted microscope equipped with two micromanipulators and pressure control devices.
9. Standard depression slides.

2.5. Preparation of Embryo Recipient Mice

1. BDF1 female mice aged 2–8 months and tested for pregnancy and delivery (Charles River Laboratories) to be used as embryo recipients.
2. BDF1 vasectomized male mice aged 3–12 months (Charles River Laboratories) to be coupled with embryo recipient females to induce pseudo-pregnancy (26).
3. Anaesthetic solution: 50 mg/mL xilazine (Rompun 2% solution, Bayer), 20 mg/mL ketamine (Ketavet 50, Intervet International) in sterile 0.9% NaCl aqueous solution. Store at 4°C.

2.6. Mice Genotyping

1. Inhalation anaesthesia: isofluorane (Forane, Abbot).
2. Tail solution: 100 mM Tris–HCl, pH 8.5, 5 mM EDTA, 0.2% (v/v) SDS, 200 mM NaCl. Prepare freshly.

3. Proteinase K, PCR Grade (Roche). Prepare stock solution at 20 mg/mL in nuclease-free water.
4. 10× PCR buffer (Roche), 50 mM $MgCl_2$, 10 mM × 4 dNTPs.
5. Primer genotyping 1: 5′-GTC GCC CCC CCG ACC GAT GTC AGC-3′ and primer genotyping 2: 5′-GGG CGA TCC AGG ATC CCC GGG AAT-3′

2.7. Western Blot Analysis of Zinc Finger and Target Gene Protein Expression Levels

2.7.1. Protein Extraction

1. Rotary-blade homogenizer (e.g., General Lab Homogenizer (GLH), OMNI International).
2. SDS-lysis solution: 2% (v/v) SDS, 5 mM EDTA.
3. DC Protein Assay Kit (Bio-Rad).
4. 2× Laemmli sample buffer (27): 125 mM Tris–HCl, pH 6.8, 4% SDS, 20% (w/v) glycerol, 10% 2-mercaptoethanol. Store at room temperature.

2.7.2. SDS-Polyacrylamide Gel Electrophoresis (SDS-PAGE)

1. 10× running buffer: 30.25 g Tris, 144 g glycine, 100 mL 10% SDS. Dissolve in water up to 1000 mL (adjust to pH 8.3). Store at room temperature.

2.7.3. Semi-dry Electroblotting

1. Semi-dry electrotransfer system (e.g., Semi-Phor transfer apparatus, Hoeffer).
2. 10× transfer buffer: 58 g Tris, 29 g glycine, 37 mL 10% SDS. Dissolve in water up to 1000 mL. Store at room temperature.
3. 3 MM paper (Whatman).
4. Protran nitrocellulose transfer membrane (0.45 μm; Whatman).
5. Ponceau S solution (Sigma).

2.7.4. Immunoblotting

1. 10× phosphate-buffered saline (PBS): 1.37 M NaCl, 27 mM KCl, 100 mM Na_2HPO_4, 18 mM KH_2PO_4 (adjust to pH 7.4). Autoclave before storage at room temperature. Prepare a 1× working solution by dilution of one part with nine parts water.
2. Blocking buffer: 4% (w/v) non-fat dry milk in PBS.
3. Primary antibodies: mouse anti-myc monoclonal (9E10 clone, hybridoma-conditioned medium), mouse

anti-utrophin monoclonal (NCL-DRP2, Novocastra), and anti-α-tubulin monoclonal antibody (Sigma).

4. Secondary antibody: anti-mouse HRP-linked antibody (cell signaling).
5. ECL Plus Western Blotting Detection System (Amersham).
6. Biomax XAR film (Kodak).

2.7.5. Stripping

1. Stripping buffer: Restore Western Blot Stripping buffer (Pierce).
2. Blocking buffer: 4% (w/v) non-fat dry milk in PBS.

2.8. Real-Time PCR Analysis of Target Gene mRNA Expression Levels

1. Trizol reagent (Invitrogen).
2. Rotary-blade homogenizer (e.g., GLH, OMNI International).
3. Oligo $(dT)_{12-18}$ primers 0.5 μg/μL (Invitrogen).
4. 10 mM dNTP mix PCR Grade (Invitrogen).
5. 5× first-strand buffer: 250 mM Tris–HCl, pH 8.3, 375 mM KCl, 15 mM $MgCl_2$.
6. RNAsi OUT Recombinant Ribonuclease Inibitor, 40 U/μL (Invitrogen).
7. SUPERSCRIPT II RNase H Reverse Transcriptase, 200 U/μL (Invitrogen).
8. RNase-free DEPC water for diluting RNA samples.
9. Real-time PCR system (e.g., ABI Prism 7000 Sequence Detection System, Applied Biosystems).
10. 2× TaqMan Universal PCR Master Mix (Applied Biosystems).
11. 20× TaqMan Gene Expression Assay Primer/Probe Mix, TaqMan MGB Probe with Fam dye (Applied Byosystems) for mouse utrophin cDNA (UTRN).
12. 20× TaqMan Endogenous Control Primer/Probe Mix, TaqMan MGB Probe with Fam dye (Applied Byosystems) for mouse β-2-microglobulin cDNA (B2M).
13. 20× TaqMan Endogenous Control Primer/Probe Mix, TaqMan MGB Probe with Fam dye (Applied Byosystems) for mouse b-glucoronidase cDNA (GUS).
14. Optical adhesive cover for 96-well assay plates (Applied Biosystems).

2.9. Chromatin Immunoprecipitation (ChIP) Assay

1. 37% formaldehyde solution (Sigma).

2. 1% formaldehyde in 1× PBS. Prepare freshly. Formaldehyde should be used in a certified fume hood.
3. Protease inhibitor cocktail tablets, EDTA-free (Roche).
4. Rotary-blade homogenizer (e.g., GLH, OMNI International).
5. Sonicator (e.g., Ultrasonic Processor W-385 equipped with a 2 mm tip, Heat Systems-Ultrasonics, Inc.).
6. ChIP lysis buffer: 1% (v/v) SDS, 1% Triton X-100, 10 mM EDTA, 50 mM Tris–HCl, pH 8.1. Complete with Protease Inhibitor cocktail. Prepare freshly.
7. ChIP dilution buffer: 0.01% (v/v) SDS, 1.1% Triton X-100, 1.2 mM EDTA, 16.7 mM Tris–HCl, pH 8.1, 167 mM NaCl. Complete with proteinase inhibitors cocktail.
8. Protein G agarose/Salmon sperm DNA (50% slurry) (Upstate).
9. Low-salt wash buffer: 0.1% (v/v) SDS, 1% Triton X-100, 2 mM EDTA, 20 mM Tris–HCl, pH 8.1, 150 mM NaCl.
10. High-salt wash buffer: 0.1% (v/v) SDS, 1% Triton X-100, 2 mM EDTA, 20 mM Tris–HCl, pH 8.1, 500 mM NaCl.
11. LiCl buffer: 0.25 M LiCl, 1% NP-40, 1 mM EDTA, 10 mM Tris–HCl, pH 8.1.
12. TE buffer: 10 M Tris–HCl, 1 mM EDTA, pH 8.
13. Elution buffer: 1% (v/v) SDS, 0.1 M $NaHCO_3$.
14. Proteinase K, PCR Grade (Roche). Prepare a stock solution at 20 mg/mL in nuclease-free water.
15. PCR primers: Utroprom forward: 5′-GCACGCACGACT GGTTCCGGGATTC-3′; Utroprom reverse: 5′-CTTTGT TCTCCCGGGGAGACCAGTC-3′
16. PCR primers: Dystroprom forward: 5′-CTGTGGCAGC TTAAGGCTTGTTCCCA-3′, Dystrprom reverse: 5′-CAT GTCGCCCGAGTGTTATTCTGTGC-3′

3. Methods

3.1. Synthetic Gene Design

1. *DNA-binding domain*: The ZF backbone used to construct Jazz is related to the Zif268 gene product. The DNA-binding specificity of Jazz has been assigned on the basis of the available list of “code” signatures that relate ZF primary structure to potential 9 bp DNA-binding targets **(Fig. 11.1)**. Each finger domain appears to behave as an

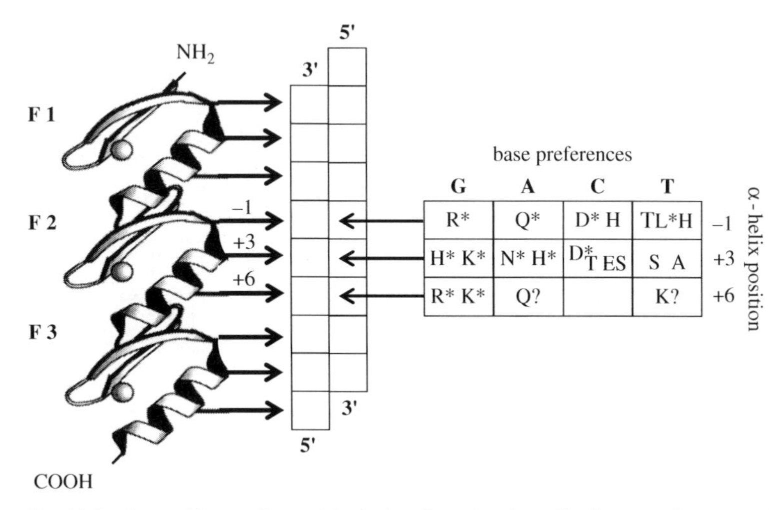

Fig. 11.1. Recognition code used to design Jazz. A schematic diagram of contacts is shown between the middle finger of a prototypic three ZF protein and its DNA-binding site. Amino acid residues at the crucial positions in the α-helix (–1, +3, +6) are listed in a matrix, relating to the four bases at each position of the DNA subsite. Amino acid residues that arise recurrently from phage display selection are in bold and asterisks indicate interactions observed in structural studies. Some uncertain interactions are indicated by a question mark. Others poorly defined are left blank.

independent DNA-binding module, specifying 3 bp in adjacent but discrete subsites, essentially on one DNA strand. In particular, the ZF section of the Zif268 gene has been modified at positions –1, +3, and +6 of the α-helix, described to be crucial for DNA-binding specificity (16). The resulting Jazz protein structure is the following: the second and third ZF domains are designed to each recognize the DNA triplet 5′-GCT-3′. The Jazz gene consists of a 282 bp DNA fragment that encodes a 95 residue peptide containing three ZF domains (**Fig. 11.2a**). The first finger recognizes the DNA triplet, 5′-GCG-3′, the same sequence that it recognizes in wild-type Zif268. ZF protein recognition of DNA involves an antiparallel arrangement of the protein: the amino-terminal ZF domain is involved in 3′-contacts with the target sequence, whereas the carboxyl-terminal domain is involved in 5′-contacts.

2. In order to produce the Jazz ZF gene portion, we synthesized two overlapping oligonucleotides, namely Jazz-5′ (153 nt) and Jazz-3′ (160 nt) (16) (*see* **Note 3** and **Fig. 11.2A**). Using a 20 nt overlapping central region, 0.5 μg of each oligonucleotide is annealed, giving a 282 bp DNA fragment in a total volume of 50 μL of annealing buffer. The reaction is heated to 75°C and slowly cooled to 37°C. Four microliters of 10 mM dNTP and 2 μL (10 U) of Klenow are added and the mixture is incubated at 37°C for 20 min. A volume

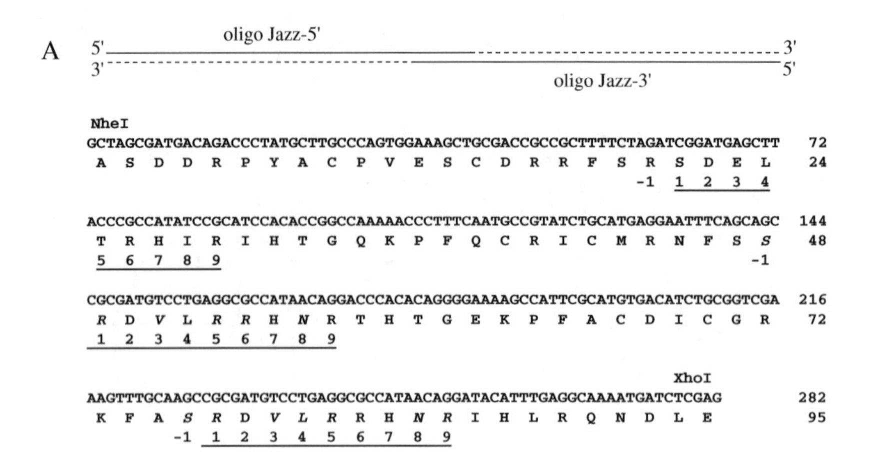

Jazz "Code programmed" target sequence: 5'-GCT GCT GCG-3'

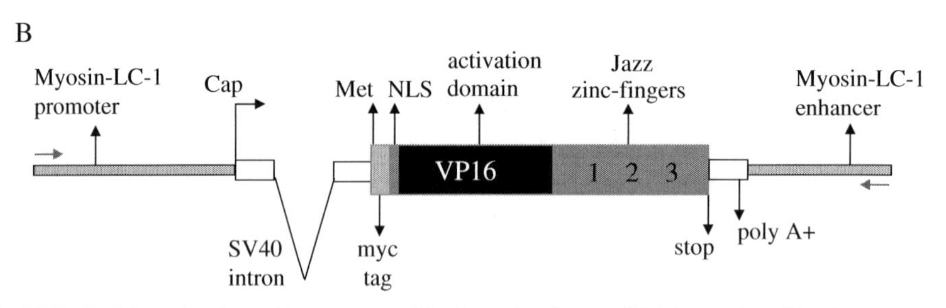

Fig. 11.2. (**A**) Nucleotide and amino acid sequences of the three zinc finger artificial gene Jazz. *Top*: Jazz was synthesized following the Zif268 finger region backbone. Two overlapping oligonucleotides were annealed and filled-in using Klenow polymerase, as shown in the scheme. The amino acid residues present in the α-helical portions of each zinc finger domain are numbered and *underlined*. Within these regions, amino acid residues modified from the original Zif268 protein are indicated in *italics*. The restriction enzyme sites used to engineer and subclone Jazz gene are indicated. The Jazz "code programmed" 9 nt DNA target sequence is indicated in the *bottom* of the panel. (**B**) Schematic representation of the Vp16-Jazz gene unit, used to generate transgenic mice. The myosin light chain 1 (MLC1) promoter/enhancer regions are indicated at the extremity of the unit. The intron region, the cap site and the poly A+ addition consensus all derived from the Simian virus SV40 are shown. The strong transcriptional activation domain Vp16 from herpes simplex virus, the epitope "myc tag" recognized by the anti-myc monoclonal 9E10 antibody, the Jazz zinc finger region and a nuclear localization signal (NLS) are indicated. The whole synthetic Jazz transgene unit was amplified using PCR.

of 0.1 μL of the resulting mixture is used as the template for the following amplification reaction: 94°C for 30 s, 60°C for 30 s, and 72°C for 1 min, for a total of 30 cycles plus 72°C for 10 min, using as primers two 20 nt oligonucleotides with sequences complementary to the ends of the 282 bp DNA fragment (primer TM24 and primer TM25). The amplification reaction is checked on a 1.5% agarose gel. A volume of 1 μL of the reaction is ligated into the pGEM Easy vector.

3. *Additional protein domains* (*see* **Section 2.1** and **Fig. 11.2B**). To efficiently activate transcription, the Jazz ZF portion was fused to the strong activation domain from the Vp16 gene. To compartmentalize the Jazz protein

in the nucleus, we added the nuclear localization signal from SV40 large T protein. To recognize Jazz protein in western blot analysis, a myc epitope tag (repeated five times) that is recognized by 9E10 anti-myc monoclonal antibody was added.

4. *Synthetic gene regulatory region.* To optimize Jazz gene expression in transgenic mouse skeletal muscle we used the human myosin light chain 1 (MLC1) promoter/enhancer region (**Fig. 11.2B**).
5. *Additional DNA sequences.* To ensure Jazz gene mRNA stability inside cells, an intron and the polyA+ addition sequence, both from SV40 large T gene, were added (**Fig. 11.2B**).

3.2. Purification and Injection of the Transgenic DNA Construct

1. To achieve efficient mouse transgenesis, the Jazz transgene unit contained in pMEX-Vp16-Jazz was obtained by PCR using the proof-reading Expand PCR-Plus enzyme. The amplified fragment was checked by electrophoresis on an agarose gel and by DNA sequencing of Vp16-Jazz coding regions. The amplification reaction was performed as follows: 94°C for 30 s, 60°C for 30 s, and 72°C for 6 min, for a total of 30 cycles plus 72°C for 20 min. The 27 nt oligonucleotide primers (Jazz forward and Jazz reverse) were located at the ends of the 4160 bp Jazz transgene unit:
2. Agarose gel electophoresis is performed using a Mini L Horizontal 1214 gel apparatus but is easily adaptable to maxigel chambers. It is recommended to use ultrapure agarose specifically designed for preparative DNA electrophoresis.
3. Prepare an 0.8% agarose gel by adding 0.8 g of SeaKem GTG agarose to 100 mL of 1× TAE and 10 μL of 10,000× SYBR Safe DNA gel stain solution (*see* **Note 4**).
4. Load 30–70 μg of the pMEX-myc-Vp16-Jazz PCR amplicon dissolved in 50–100 μL of TE and a 1× final concentration of DNA loading buffer.
5. Run the gel at 80 V for the first 10 min, then at 30 V for 4 h.
6. Then place the gel on a UV trans-illuminator at a wavelength of 365 nm to visualize the DNA. Identify the band corresponding to the pMEXmyc-Vp16-Jazz construct, which has an expected size of 4300 bp, and excise the band from the gel with the aid of a clean surgical scalpel. Minimize the time of exposure of the DNA to the UV light during this step to prevent DNA damage.
7. Expected DNA contaminants such as organic solvents, salts, agarose, and dyes must be removed before injection. Many excellent commercial kits for the extraction

and purification of DNA from agarose gels are available. We used GENECLEAN kit, following the manufacturer's instructions.

8. The volume and concentration of the DNA eluted from the column after purification are normally too low to be read by a conventional spectrophotometer. A spectrophotometer such as a Qubit or a NanoDrop 1000 can be used to read the DNA concentration in the range of μg–ng/μL. With the NanoDrop, it is possible to simultaneously obtain information on the purity of the DNA, because the ratios between the UV absorbance at 260/280 and 260/230 nm are also provided.
9. Dilute the transgenic DNA in injection buffer to a final concentration of 2–2.5 ng/μL and store at 4°C in 20–30 μL aliquots.

3.3. Preparation of Embryo Donor Mice

1. All procedures involving animals must adhere to national and institutional policies on laboratory animal care. Treat embryo donor BDF1 female mice with hormones to induce super-ovulation. Pregnant mare's serum gonadotropin (PMSG) is administered to mimic the endogenous effects of follicle-stimulating hormone. After 47 h, human chorionic gonadotropin (hCG) is injected to induce the rupture of the mature follicle and oocytes release. Both hormones are injected intraperitoneally at a dose of 5 IU per mouse.
2. Immediately after the injection of hCG, mate the females with BDF1 male mice and the day after check for the vaginal plug to identify fertilized dams.
3. Kill the fertilized female mice by CO_2 asphyxiation or cervical dislocation (*see* **Note 5**), open the abdominal cavity and cut away the oviducts.

3.4. Embryo Collection, Culture, and Microinjection

1. Put the oviducts in a 35 mm Petri dish and grasp them to release the follicle in M2 medium containing hyaluronidase (0.5–1 mg/mL final concentration), which allows the enzymatic separation of the zygotes from cumulus cells.
2. After few minutes, when the cumulus cells are shed, transfer the zygotes using a glass capillary through several drops of M2 medium to remove residual cumulus cells and debris. Use a stereomicroscope to visualize and transfer embryos.
3. Finally transfer the zygotes into a 35 mm Petri dish in a large drop of M16 medium covered with mineral oil to avoid evaporation and concentration and keep in an incubator (37°C, 5% CO_2) for a few hours until they are in the best condition for the microinjection procedure. Take into

account that the pronuclei are easier to inject when they reach their maximum size, and this happens a few hours after collection.

4. A good inverted microscope equipped with two micromanipulators and pressure control devices is needed to perform microinjection. One micromanipulator is connected to a holding glass capillary to keep the zygote fixed in the right position during injection, whereas the other controls the movements of the injection capillary. Holding and injection glass capillaries can be purchased from commercial sources or made in the lab using a capillary glass puller and a microforge.
5. Put a depression slide containing a small drop of M2 medium covered with mineral oil under the microscope and, operating by the micromanipulators, place under your visual field the holding capillary, filled with M2 medium, and the injection capillary, filled with the transgenic DNA construct, dissolved in injection buffer at a final concentration of 2–2.5 μg/mL.
6. Transfer the zygotes in the drop of M2 medium and adjust the microscope magnification until the pronuclei are clearly visible (normally at 200× magnification).
7. By a manual or automatic pressure control system, apply a negative pressure to the holding capillary to immobilize one embryo and lead the injection capillary to make a hole through the zona pellucida and the plasma membrane of the zygote, to reach one of the two pronuclei. At this time, apply a positive pressure to the injection capillary until you are able to see the pronucleus swell.
8. Release the injected embryo, applying a positive pressure to the holding capillary and proceed in the same way until all the zygotes are injected.
9. Transfer the injected embryos in a 35 mm Petri dish in a large drop of M16 medium covered with mineral oil and keep in the incubator (37°C, 5% CO_2) until you are ready to transfer the embryos into the oviduct of recipient female mice. Perform oviduct transfer in surrogate mothers immediately after the microinjection or the day after, when injected embryos have reached the two cell stage.

3.5. Preparation of Embryo Recipient Mice

1. For the successful implantation of embryos it is necessary to induce a favorable hormonal condition in the recipient females (pseudopregnancy), which is achieved by mating them with vasectomized male mice. Select recipient females in estrus and mate them with vasectomized males the day before the surgical procedure of embryo transfer.

2. The next morning check for the vaginal plug to identify females successfully mated to males.
3. Anesthetize the female to be subjected to the surgical procedure of embryo transfer by intraperitoneal injection of anesthetic solution (0.1 mL/10 g of body weight).
4. Wipe the skin with 70% ethanol and make a small incision of both skin and body wall (0.5 cm) in the left or right middle back of the recipient female. Pull out ovary, oviduct, and the upper part of the uterus, by picking with fine forceps the associated fat pad. Localize the opening of the oviduct (infundibulum) and insert into 15–25 injected embryos by the use of a glass capillary. Sew up the incision with wound clips (*see* **Note 6**).
5. Keep the mouse in a warm environment (37–38°C) until it has completely recovered from anesthesia. Delivery of progeny mice developed from injected embryos takes place 19–20 days later.

3.6. Mouse Genotyping

1. Successful transgene integration is expected in 10–30% of mice derived from injected embryos (the F0 generation). They can be revealed by several procedures such as southern blot, dot blot, or PCR, using probes or templates able to identify the transgenic sequence within the whole genomic DNA. To check the transgene integration in F0 mice and its transmission to the following F1/F2 generations we typically use PCR analysis. The procedure described here uses conditions optimized to amplify the specific transgene sequence without genomic DNA extraction. However, depending on the templates, it may be more suitable to purify the DNA after lysis of the tail using a standard phenol–chloroform extraction method or a commercial kit.
2. At the time of weaning (21–28 days after birth), cut a fragment of tail (no more than 0.5 cm) from each mouse. This operation should be performed under light gaseous anesthesia, keeping the mice in isofluorane 1.5–3% in air until they lose consciousness. Recovery is very rapid and occurs within 30–40 s.
3. Add 500 μL of Tail solution supplemented with 50 μg of proteinase K.
4. Incubate overnight at 53°C with agitation.
5. Add 1 μL of lysate to be tested to 49 μL of PCR mix containing: 36 μL of MilliQ H_2O, 5 μL of 10× PCR buffer, 0.5 μL Taq polymerase (5 U/μL), and 1 μM of the following primers: primer genotyping 1 and primer genotyping 2.

6. Carry out PCR at 94°C for 1 min, followed by 94°C for 1 min, 62°C for 2 min, and 72°C for 3 min, for a total of 30 cycles, and 72°C for 10 min.
7. Check the template amplification on 1% agarose gel.

3.7. Western Blot Analysis of Zinc Finger and Target Gene Protein Expression Levels

1. The procedure described here works generally well for frozen muscle tissue. However, we have found this protocol to be readily applicable to other tissues (brain, liver, lung, and kidney). Some changes may be required to optimize lysis and protein extraction.
2. The homogenization of the frozen tissues is based on a rotary-blade homogenizer.
3. Add 1 mL/100 mg of ice-cold SDS-lysis buffer to frozen tissue into an appropriate tube (*see* **Note 7**).
4. Homogenize for a minimum of 30 s (*see* **Note 8**).
5. Immerge the tube in a beaker with boiling water and boil the sample for 10 min.
6. Centrifuge at 13,000×*g* for 10 min at 4°C and transfer the supernatant to a new tube (*see* **Note 9**).
7. Quantify the protein concentration by the DC Protein Assay (Bio-Rad) (*see* **Note 10**).
8. Mix 25–50 μg of the sample in 2× Laemmli sample buffer.
9. Heat samples at 95°C for 5 min. The sample is ready for separation by SDS-PAGE.
10. The SDS-PAGE assumes the use of a Mini Protean II mini gel system, but it is easily adaptable to other formats.
11. Select and cast the appropriate resolving gel. The amounts of reagents required for one 1.5 mm thick gel with dimensions 8.5 × 6 cm and containing 6 or 10% acrylamide are given in **Table 11.1** (*see* **Note 11**).

Table 11.1
Formulations for polyacrylamide gels

mL	6%	10%	Stacking gel
H_2O	5.3	4.0	2.1
30% Acrylamide solution	2.0	3.3	0.5
1.5 M Tris, pH 8.8	2.5	2.5	–
0.5 M Tris, pH 6.8	–	–	0.38
10% SDS	0.1	0.1	0.03
10% APS	0.1	0.1	0.03
TEMED	0.008	0.004	0.003

12. Load the gel (approximately 30 μL of each sample). Start electrophoresis with an initial current of 20 mA and maintain at this current until the sample has completely entered the stacking gel (*see* **Note 12**).
13. The proteins that have been separated by SDS-PAGE are transferred electrophoretically to a nitrocellulose membrane by a semi-dry electrotransfer system.
14. Cut the nitrocellulose membrane (45 μm) to the size of the gel, float it on deionized water until completely wet, then soak it in 1× transfer buffer until use.
15. Cut six sheets of 3 MM blot papers to the size of the gel and soak with 1× transfer buffer (*see* **Note 13**).
16. Place three pre-wet blot papers on top of the base of the apparatus (anode). Place the pre-wet nitrocellulose membrane on top of blot papers.
17. Put the gel on top of the nitrocellulose membrane and cover the gel with the remaining three sheets of 3 MM blot papers soaked with 1× transfer buffer.
18. Place the lid (cathode) on the base holding the blot sandwich.
19. Transfer for 45 min to 2 h at a constant current of 3.0 mA/cm^2 gel at room temperature (*see* **Note 14**).
20. To check for transfer of the samples onto the nitrocellulose, immerse the membrane in Ponceau S solution for 2 min. The red staining will wash away in water followed by PBS. The membrane is now ready for the immunoblotting assay.
21. All incubation and washing steps of this procedure are performed on an orbital shaker at room temperature (*see* **Note 15**).
22. Block nonspecific-binding sites by immersing the membrane in blocking buffer for at least 1 h, or overnight in a refrigerator at 2–8°C, if more convenient.
23. Briefly wash the membrane with two changes of 1× PBS.
24. For myc-tag detection, incubate the membrane with a 1:15 dilution of the anti-myc monoclonal antibody in blocking buffer for at least 1 h (*see* **Note 16**). Examples of the signals for Vp16-Jazz protein are shown in **Fig. 11.3a**. For utrophin detection, incubate the membrane with a 1:10 dilution of the anti-utrophin monoclonal antibody in blocking buffer overnight at 4°C. For α-tubulin detection, incubate the membrane with a 1:2000 dilution of the anti-α-tubulin monoclonal antibody in blocking buffer for 1 h. Example of the signals for α-tubulin protein are shown in **Fig. 11.3B**.

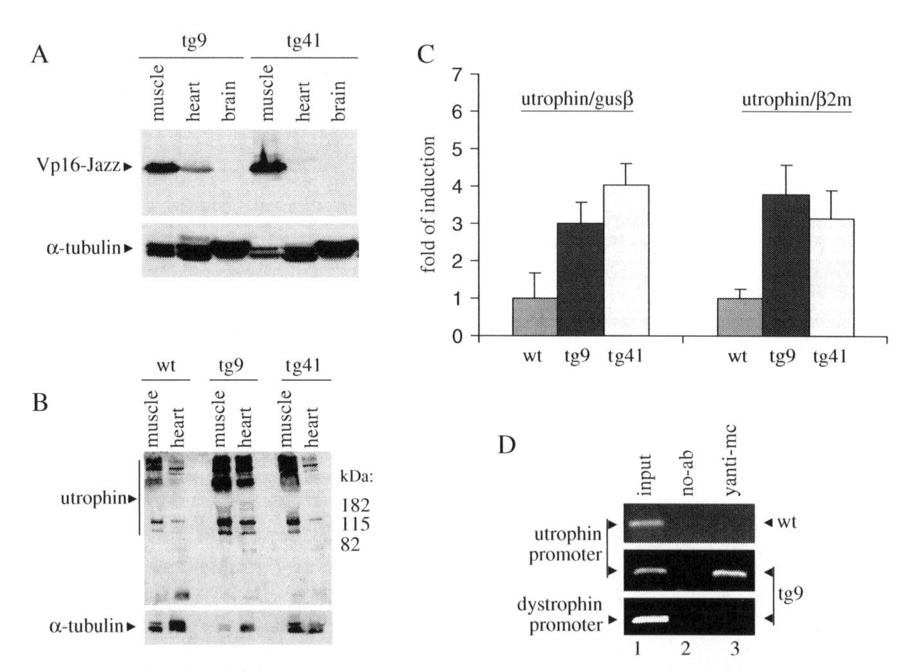

Fig. 11.3. (**A**) Western blot analysis of total proteins extracted from the skeletal muscle, heart and brain of Vp16-Jazz transgenic mice derived from two different founders (tg9 and tg41). The expression of the Vp16-Jazz transgene was monitored by the anti-myc monoclonal 9E10 antibody. Detection of α-tubulin was used to normalize the amount of proteins. (**B**) Western blot of total protein extracts derived from skeletal muscle and heart from wild-type (wt) mice and Vp16-Jazz transgenic mice (tg9 and tg41) incubated with monoclonal antibody against utrophin. The same membrane was stripped and reblotted for anti-α-tubulin monoclonal antibody for loading normalization. (**C**) Real-time PCR analysis of the utrophin gene expression level in Vp16-Jazz transgenic mice (tg9 and tg41) and control wt mice. The gene expression ratio between utrophin and β-glucoronidase (GUS) and β2-microglobulin (β2M) is shown as mean ± standard deviation (*error bars*) from three independent experiments performed in triplicate. (**D**) Vp16-Jazz chromatin immunoprecipitation, performed in skeletal muscle derived from wt mice and transgenic mice (family tg9) using myc monoclonal antibody/protein G agarose beads or protein G agarose beads as a control (no-ab). Immunoprecipitates from each sample were analyzed for the presence of utrophin promoter by PCR. A sample representing linear amplification of the total input chromatin (input) was included (lane 1). As control, samples from transgenic mice were also tested for the presence of the dystrophin promoter sequence.

25. Wash the membrane in 1× PBS for three times 5 min each.
26. Incubate the membrane with a 1:1000 dilution of the anti-mouse HRP-linked antibody in blocking buffer for 1 h.
27. Wash the membrane six times with 1× PBS, 5 min each.
28. The membrane is now ready for the ECL detection, following the manufacturer's protocol.
29. After each immunodetection the membranes may be stripped of bound antibodies and reprobed several times. This step is suitable to provide a loading control of the protein content in each gel. The complete removal of primary and secondary antibodies from the membrane is possible following several protocols (*see* **Note 17**).

3.8. Real-Time PCR Analysis of Target Gene mRNA Expression Levels

1. Add 4 mL of Trizol reagent to 100–150 mg of frozen muscle tissue.
2. Homogenize the tissue by the use of a rotary-blade homogenizer.
3. Purify the total RNA by Trizol reagent according to the manufacturer's instructions and dissolve in 100–200 μL of DEPC water.
4. Quantify the sample by a conventional spectrophotometer.
5. Perform reverse transcription in a total volume of 20 μL using 2 μg of total RNA for the sample and 8 μg of total RNA for the sample selected as the calibrator, used to make the standard curve.
6. Prepare on ice MIX 1 containing 1 μL Oligo dT primer, 1 μL dNTP mix, H_2O DEPC to 12 μL. Add this to the RNA sample and incubate at 70°C for 10 min.
7. Prepare on ice MIX 2 containing 4 μL 5× first-strand buffer, 2 μL DTT, and 1 μL RNAsi OUT. Add this to the sample and incubate at RT for 10 min and 42°C for 2 min.
8. Add to each sample 1 μL of SUPERSCRIPT.
9. Incubate at 42°C for 50 min for first-strand synthesis, then place at 70°C for 15 min for final enzyme inactivation.
10. Amplify RNA samples converted to cDNA in real-time PCR. We perform the reaction in a 96-well format in an ABI Prism 7000 Sequence Detection System instrument using TaqMan detection chemistry. This system employs fluorescently labeled probes and the 5′-exonuclease activity of the Taq DNA polymerase. The TaqMan probe anneals specifically to a complementary sequence between the forward and reverse primer sites. During amplification, the 5′-exonuclease activity of the polymerase separates the reporter from the quencher dye located at opposite ends of interrogation probe. Signal is generated as the reporter dye is released away from the quencher dye.
11. For PCR amplification use between 10 and 40 ng reverse-transcribed cDNAs (the efficiency of retrotranscription is supposed to be 100%) in a 25 μL PCR. To correct for minor variations due to differences in input RNA amount or in efficiencies of reverse transcription, housekeeping genes are also amplified, in our case β2-microglobulin (β2M) and β-glucoronidase (GUS).
12. To quantify the results prepare a standard curve; use as a calibrator serial dilutions (200, 40, 8, 1.6, and 0.32 ng) of cDNA from wild-type mouse (negative control) and

determine the relative level of expression of the gene of interest and the housekeeping genes for all experimental samples.

13. Thaw any frozen cDNA samples and 20× TaqMan primer/probe assays by placing them on ice.
14. When thawed, resuspend the samples by vortexing and then centrifuge the tubes briefly.
15. Put the cDNA template into the 96-well optical reaction plate (samples and standard curve dilutions are in triplicate or quadruplicate).
16. Prepare the PCR mix for each sample adding in sequence: H_2O to 25 μL, 1.25 μL of 20× gene expression target assay or TaqMan endogenous control assay and 12.5 μL of 2× TaqMan Universal PCR Master Mix.
17. Mix and then add to the plate containing the cDNA template.
18. Cover the plate with an optical adhesive cover.
19. Configure the sequence detector plate document and run the plate in the sequence detector.
20. Use the default thermal cycling conditions: incubation at 50°C for 2 min, denaturing at 95°C for 10 min, and then 40 cycles of the amplification step of 95°C for 15 s and 60°C for 1 min.
21. Start the run. The results are analyzed using Applied Biosystems analysis software. The data are expressed as the ratio between UTRN and β2 M or GUS mRNA expression, respectively. An example of the resulting histogram is shown in **Fig. 11.3C**.

3.9. Chromatin Immunoprecipitation (ChIP) on Muscle Tissue of Transgenic Mice

1. We have optimized this protocol for skeletal muscle tissue, but do not know if this procedure can be successfully used with other tissue types. Before starting, adjust the sonication conditions on formaldehyde cross-linked tissues in order to determine how many sonication pulses you need to get fragments of the desired size.
2. Kill the animal (*see* **Note 5**).
3. Quickly remove the muscle tissue of interest and trim away any adherent tissue that is not desired. For small samples, this can be performed under a dissecting microscope.
4. Place the tissue (100–150 mg) in a Petri dish on ice.
5. Wet the tissue with a drop of 1% formaldehyde and make several incisions with a razor blade.
6. Transfer the tissue into a tube with a screw cap lid and add 5 mL of 1% formaldehyde.

7. Rotate tube at 37°C for 15 min.
8. Centrifuge the sample at low speed, decant the supernatant.
9. Wash three times by adding 10 mL of ice-cold PBS with protease inhibitors. After the last washing, completely aspirate the supernatant.
10. Add 1 mL of the ChIP lysis buffer containing the protease inhibitors.
11. Homogenize sample by using a rotary-blade homogenizer until no intact tissue is apparent.
12. Incubate for 30 min at 2–8°C.
13. Sonicate the sample to shear the DNA to lengths between 200 and 1000 bp. Be sure to keep the sample on ice at all times (*see* **Note 18**).
14. Dilute the sonicated sample 10-fold with ChIP dilution buffer complete with protease inhibitors.
15. Centrifuge samples to remove debris at 13,000×*g* for 10 min at 8°C.
16. Store a portion of the diluted sample (1%) at –80°C. This is considered to be your input material and can be kept to control the amount of DNA present in the sample at the PCR protocol.
17. To reduce nonspecific background, pre-clear the sample with 300 μL of protein G agarose/Salmon sperm DNA (50% slurry) for 15 min at 4°C on a rotating platform.
18. Pellet the agarose by brief centrifugation and transfer the sample to a new tube.
19. Repeat the pre-clearing step three times.
20. Split sample in two tubes.
21. Add 200 μL of the immunoprecipitating antibody (anti-myc monoclonal antibody, 9E10 clone, hybridoma-conditioned medium) to one tube (Ab tube). Always use at least 2–4 μg of antibody.
22. With the second tube perform a no-antibody immunoprecipitation for a negative control (no-Ab tube).
23. Incubate at 4°C overnight on a rotating platform.
24. Collect the immunocomplexes by adding 120 μL of protein G agarose/Salmon sperm DNA (50% slurry) and incubate at 4°C for 20 min on a rotating platform.
25. Pellet immunocomplexes by centrifugation at 200×*g* for 2 min at 4°C.
26. Wash the immunocomplexes for 3 min at room temperature on a rotating platform with 4 mL of the buffers listed

in the order as indicated: low-salt wash buffer, two washes; high-salt wash buffer, two washes; LiCl buffer, two washes; TE buffer, three washes.

27. After the last wash, elute the immunocomplexes by adding 300 μL of the freshly prepared elution buffer. Shake on a vortex for at least 15 min. Pellet and transfer the supernatants to clean tubes. Repeat and combine both elutions (600 μL) in the same tube.
28. Add 5 M NaCl to a final concentration of 0.2 M to the combined eluates and reverse crosslink by heating at 65°C overnight or for a minimum of 4 h. From now on, include also the input material previously stored. At this step the samples can be stored at –20°C and the protocol continued the next day.
29. Add 10 mM EDTA, 40 mM Tris–HCl, pH 6.8, and 40 μg of proteinase K. Complete the input material only with the proteinase K.
30. Incubate for 1 h at 45°C.
31. Recover DNA by phenol/chloroform extraction and ethanol precipitation (2 h to overnight at –20°C).
32. Centrifuge DNA, let air dry, and resuspend in 20 μL of TE buffer.
33. Subject the final DNA to PCR analysis. For utrophin promoter/enhancer amplification, perform PCR in 25 μL of reaction buffer containing 0.8 mM dNTPs, 2.5 units of Taq DNA polymerase, and 1 μM of the primers specific for the mouse utrophin promoter: Utroprom forward and Utroprom reverse: 5′-CTTT GTTCTCCCGGGGAGACCAGTC-3′. Carry out the PCR at 94°C for 45 s, 66°C for 30 s, and 72°C for 30 s, for a total of 33 cycles. An example of PCR results is shown in **Fig. 11.3D**.
34. As a control, it is possible to analyze the same samples for the presence of a different promoter sequence. For dystrophin promoter/enhancer, PCRs were carried out at 94°C for 45 s, 62°C for 30 s, and 72°C for 30 s, for a total of 33 cycles by using the following primers: Dystrprom forward and Dystrprom reverse. An example of PCR results is shown in **Fig. 11.3D**.

4. Notes

1. It is advisable to contact the nearest or most convenient transgenic facility, choosing between institutional or "for profit" services. The following link to the International

Society for Transgenic Technology contains a long, if not exhaustive, list of transgenic facilities existing all over the world: http://www.transtechsociety.org/linkstg.html

2. BDF1 (first generation hybrid from C57BL/6 N and DBA2N strains) mice are widely used in transgenesis due to high fertility and good parental care, but also other strains (including inbred strains) could be used depending on individual specific needs.
3. More recently, the synthesis of constructs expressing artificial ZF proteins such as Jazz and its derivatives is performed commercially (e.g., GenScript). The company provides the genes cloned in pUC57 and flanked by *Eco*RI and *Xho*I restriction-binding sites, ready for use or subcloning.
4. This product is a cyanine dye that recently replaced the mutagen ethidium bromide for nucleic acid staining.
5. Consider the method of killing carefully. Cervical dislocation should be performed only by skilled personnel. In all cases, you should adhere to your national and institutional policy on laboratory animal care procedures.
6. The procedure is complex and requires skill and experience in microsurgery. For additional procedural details, *see* reference (25).
7. The volume of the lysis buffer may need to be optimized for a particular tissue or organ.
8. During the homogenization step, keep tissue ice cold to prevent protein degradation.
9. It is possible to freeze the extracts at –80°C until use.
10. It is important to quantify protein concentration with assays compatible with detergent solublization of the sample.
11. The freshly prepared ammonium persulfate (APS) solution and TEMED should be added last, immediately before pouring the gels, because these polymerize the gels.
12. The specific settings for electrophoresis depend considerably on the apparatus used, the length and thickness of the gel, and the acrylamide concentration of the gel. Fast runs give better results than overnight runs, especially with 10% acrylamide gels. Alternatively, a constant voltage of 10 V per gel might be set to ensure an even distribution of heat.
13. Transfer membrane and gel blotting paper must not be cut larger than the gel to ensure optimal flow of current.
14. Higher amperages and/or longer transfer times will heat up the electrodes and the blot sandwich. The longer time (2 h) is recommended for the optimal transfer efficiency of large proteins (>200 kD).

15. The washing steps are extremely important. In order to obtain the optimal signal-to-noise ratio, wash the membrane in >1 mL/cm^2.
16. Dilution, incubation times, and temperature may vary and should be optimized for each antibody batch.
17. We used the Restore Western Blot Stripping buffer (Pierce) according to the manufacture's instructions following the manufacturer's protocol.
18. The time and number of pulses will vary depending on the sonicator, tissue type, and extent of crosslinking. In our experiments, the DNA is sheared to the appropriate length with nine sets of 12 s pulses using a Sonicator Ultrasonic Processor W-385 equipped with a 2 mm tip and set to 40% of maximum power.

Acknowledgments

This work was supported by: Telethon-Italy (Grant # GGP07177), Health Ministery RF-INP-2007-653996, and Paul Blümel Stiftung für medizinische Forschung. M.G. Di Certo is a recipient of a fellowship supported by Regione Lazio/Filas fundings for "Sviluppo della Ricerca sul Cervello.".

References

1. Corbi, N., Libri, V., Onori, A., and Passananti, C. (2004) Synthetic zinc finger peptides: old and novel applications. *Biochem Cell Biol.* **82**, 428–436.
2. Blancafort, P., Segal, D.J., and Barbas, C.F., III. (2004) Designing transcription factor architectures for drug discovery. *Mol Pharmacol* **66**, 1361–1371.
3. Choo, Y. and Klug, A. (1997) Physical basis of a protein-DNA recognition code. *Curr Opin Struct Biol.* 7, 117–125.
4. Pabo, C.O., Peisach, E., and Grant, R.A. (2001) Design and selection of novel Cys2His2 zinc finger proteins. *Annu Rev Biochem.* **70**, 313–340.
5. Dreier, B., Segal, D.J., and Barbas, C.F., III. (2000) Insights into the molecular recognition of the 5′-GNN-3′ family of DNA sequences by zinc finger domains. *J Mol Biol.* **303**, 489–502.
6. Dreier, B., Beerli, R.R., Segal, D.J., Flippin, J.D., and Barbas, C.F., III. (2001) Development of zinc finger domains for recognition of the 5′-ANN-3′ family of DNA sequences and their use in the construction of artificial transcription factors. *J Biol Chem.* **276**, 29466–29478.
7. Dreier, B., Fuller, R.P., Segal, D.J., Lund, C.V., Blancafort, P., Huber, A., Koksch, B., and Barbas, C.F., III. (2005) Development of zinc finger domains for recognition of the 5′-CNN-3′ family DNA sequences and their use in the construction of artificial transcription factors. *J Biol Chem.* **280**, 35588–35597.
8. Papworth, M., Kolasinska, P., and Minczuk, M. (2006) Designer zinc-finger proteins and their applications. *Gene.* **366**, 27–38.
9. Beltran, A., Liu, Y., Parikh, S., Temple, B., and Blancafort, P. (2006) Interrogating genomes with combinatorial artificial transcription factor libraries: asking zinc finger questions. *Assay Drug Dev Technol.* **4**, 317–331.
10. Blancafort, P. and Beltran, A.S. (2008) Rational design, selection and specificity of artificial transcription factors (ATFs): the influence of chromatin in target gene regulation. *Comb Chem High Throughput Screen.* **11**(2), 146–158.

11. Papworth, M., Moore, M., Isalan, M., Minczuk, M., Choo, Y., and Klug, A. (2003) Inhibition of herpes simplex virus 1 gene expression by designer zinc-finger transcription factors. *Proc Natl Acad Sci USA.* **100**, 1621–1626.
12. Reynolds, L., Ullman, C., Moore, M., Isalan, M., West, M.J., Clapham, P., Klug, A., and Choo, Y. (2003) Repression of the HIV-1 5′ LTR promoter and inhibition of HIV-1 replication by using engineered zinc-finger transcription factors. *Proc Natl Acad Sci USA.* **100**, 1615–1620.
13. Beltran, A., Parikh, S., Liu, Y., Cuevas, B.D., Johnson, G.L., Futscher, B.W., and Blancafort, P. (2007) Re-activation of a dormant tumor suppressor gene maspin by designed transcription factors. *Oncogene.* **26**(19), 2791–2798.
14. Rebar, E.J., Huang, Y., Hickey, R., Nath, A.K., Meoli, D., Nath, S., Chen, B., Xu, L., Liang, Y., Jamieson, A.C., Zhang, L., Spratt, S.K., Case, C.C., Wolffe, A., and Giordano, F.J. (2002) Induction of angiogenesis in a mouse model using engineered transcription factors. *Nat Med* **8**, 1427–1432.
15. Lu, Y., Tian, C., Danialou, G., Gilbert, R., Petrof, B.J., Karpati, G., and Nalbantoglu, J. (2008) Targeting artificial transcription factors to the Utrophin A promoter: effects on dystrophic pathology and muscle function. *J Biol Chem.* **283**(50), 34720–34727.
16. Corbi, N., Libri, V., Fanciulli, M., Tinsley, J.M., Davies, K.E., and Passananti, C. (2000) The artificial zinc finger coding gene "Jazz" binds the utrophin promoter and activates transcription. *Gene Ther.* **7**, 1076–1083.
17. Onori, A., Desantis, A., Buontempo, S., Di Certo, M.G., Fanciulli, M., Salvatori, L., Passananti, C., and Corbi, N. (2007) The artificial 4-zinc-finger Bagly binds human utrophin promoter A at the endogenous chromosomal site and activates transcription. *Biochem Cell Biol.* **85**(3), 358–365.
18. Mattei, E., Corbi, N., Di Certo, M.G., Strimpakos, G., Severini, C., Onori, A., Desantis, A., Libri, V., Buontempo, S., Floridi, A., Fanciulli, M., Baban, D., Davies, K.E., and Passananti, C. (2007) Utrophin upregulation by an artificial transcription factor in transgenic mice. *PLoS ONE.* **22**(1), e774.
19. Desantis, A., Onori, A., Di Certo, M.G., Mattei, E., Fanciulli, M., Passananti, C., and Corbi, N. (2009) Novel activation domain derived from Che-1 cofactor coupled with the artificial protein Jazz drives utrophin upregulation. *Neuromuscular Disorders.* **19**(2), 158–162.
20. Tinsley, J.M., Potter, A.C., Phelps, S.R., Fisher, R., Trickett, J.I., and Davies, K.E. (1996) Amelioration of the dystrophic phenotype of mdx mice using a truncated utrophin transgene. *Nature.* **384**, 349–353.
21. Tinsley, J., Deconinck, N., Fisher, R., Kahn, D., Phelps, S., Gillis, J.M., and Davies, K.E. (1998) Expression of full-length utrophin prevents muscular dystrophy in mdx mice. *Nat Med.* **4**, 1441–1444.
22. Miura, P. and Jasmin, B.J. (2006) Utrophin upregulation for treating Duchenne or Becker muscular dystrophy: how close are we? *Trends Mol Med.* **3**, 122–129.
23. Dennis, C.L., Tinsley, J.M., Deconinck, A.E., and Davies, K.E. (1996) Molecular and functional analysis of the utrophin promoter. *Nucleic Acids Res.* **24**, 1646–1652.
24. Di Certo, M.G., Corbi, N., Strimpakos, G., Onori, A., Luvisetto, S., Severini, C., Guglielmotti, A., Matassa, E.M., Pisani, C., Floridi, A., Benassi, B., Fanciulli, M., Magrelli, A., Mattei, E., and Passananti, C. (2010) The artificial gene Jazz, a transcriptional regulator of utrophin, corrects the dystrophic pathology in mdx mice. *Hum Mol Genet.* **19**, 752-760.
25. Musarò, A., McCullagh, K., Paul, A., Houghton, L., Dobrowolny, G., Molinaro, M., Barton, E.R., Sweeney, H.L., and Rosenthal, N. (2001) Localized Igf-1 transgene expression sustains hypertrophy and regeneration in senescent skeletal muscle. *Nat Genet.* **27**(2), 195–200.
26. Nagy, A., Gertsenstein, M., Vintersten, K., and Behringer, R. (2003) Manipulating the Mouse Embryo: A Laboratory Manual. 3rd ed. Cold Spring Harbor Laboratory Press, New York.
27. Laemmli, U.K. (1970) Cleavege of structural proteins during the assembly of the head of bacteriophage T4. *Nature.* **227**, 680–685.

Section III

Zinc Finger Nucleases

Chapter 12

Artificial Zinc Finger Nucleases for DNA Cloning

Vardit Zeevi, Andriy Tovkach, and Tzvi Tzfira

Abstract

DNA cloning is fundamental for modern cell research and biotechnology. Various restriction enzymes have been isolated, characterized, and purified to facilitate the digestion and ligation of DNA molecules of different origins. Nevertheless, the very small numbers of enzymes capable of digesting novel and long DNA sequences and the tedious and nearly impossible task of re-engineering existing enzymes with novel specificities greatly limit the use of restriction enzymes for the construction of complex and long DNA molecules. Zinc finger nucleases (ZFNs) – hybrid restriction enzymes that can be tailor made for the digestion of both native and artificial DNA sequences – offer a unique opportunity for expanding the repertoire of restriction enzymes useful for various DNA cloning tasks. Here we present protocols for the assembly, expression, and purification of cloning-grade ZFNs and their use for DNA cloning. We focus our discussion on the assembly of a dual-cassette plant transformation vector, as an example of a task that is nearly impossible to perform using the current collection of naturally occurring and recombinant 6–8 bp long restriction enzymes.

Key words: Cloning, binary vector, plant transformation, multigene.

1. Introduction

Many 4- and 6-bp long restriction enzymes are extremely useful for recombinant DNA applications (1). However, their high occurrence within native and artificial DNA sequences hinders their use for the construction of complex DNA structures and for the cloning of long DNA molecules. Thus, for example, cloning of many of the human genes (>50% of which are longer than 5 kb)

The authors Vardit Zeevi and Andriy Tovkach contributed equally to this work.

J.P. Mackay, D.J. Segal (eds.), *Engineered Zinc Finger Proteins*, Methods in Molecular Biology 649,
DOI 10.1007/978-1-60761-753-2_12, © Springer Science+Business Media, LLC 2010

would benefit greatly from a collection of restriction enzymes capable of recognizing and cleaving target sequences longer than 6 bp (2). As another example, rare-cutting restriction enzymes (i.e., those capable of targeting extremely long sequences) have proven instrumental for the construction of multigene transformation vectors (3–6). Such vectors, which are typically composed of several independent expression cassettes, are extremely useful for plant genetic engineering as they allow the simple, one-step delivery and regeneration of transgenic plants with several novel traits (3). Unfortunately, the very small number of naturally occurring and recombinant enzymes greatly limits their use for native DNA cloning, for the assembly of multigene transformation vectors and for other recombinant DNA applications. The ability to custom engineer zinc finger nucleases (ZFNs) to target both artificial and native sequences (7) may provide a novel alternative that can overcome these limitations.

The design of ZFNs has been discussed in several excellent papers (e.g., 7–10) and book chapters (elsewhere in this volume) and is beyond the scope of our current report. Presented below are general protocols for the assembly and purification of ZFNs for the purposes of DNA cloning. More specifically, in **Section 3.1**, we describe a procedure for the assembly of novel ZFNs from overlapping oligonucleotides while **Section 3.2** is dedicated to the purification of ZFNs from *Escherichia coli* cells. We then exemplify the use of ZFNs for cloning purposes via assembly of a multigene transformation vector, a task that can rarely (if at all) be achieved using 4- and 6-bp restriction enzymes. We demonstrate how plant transformation binary vectors (**Section 3.3.1**) and donor vectors (**Section 3.3.2**) can be designed to carry semi-palindromic recognition sites for novel ZFNs and how several independent plant expression cassettes can be mounted onto a single vector by ZFNs (**Section 3.3.3**).

2. Materials

2.1. Equipment and Consumables

1. Environmentally controlled shaker (37°C) for culturing *E. coli*.
2. Environmentally controlled incubator (37°C) for growing *E. coli*.
3. Environmentally controlled shaker (22°C) for protein expression.
4. Spectrophotometer for measuring optical density of bacterial cultures.
5. Polymerase chain reaction (PCR) thermocycler.

6. French press with a large (35 mL) chamber for breaking bacterial cells.
7. Rotary shaker.
8. Dry/wet heat block.
9. Ice bucket.
10. High-volume (10–50 mL) refrigerated centrifuge for pelleting bacterial cells from cell culture.
11. Small-volume (1.5–2 mL) refrigerated centrifuge.
12. Disposable 90-mm bacterial culture dishes.
13. Disposable plastic columns for trapping Ni-NTA resin (e.g., Bio-Rad Poly-Prep chromatography columns).
14. Ni-NTA agarose (e.g., cat. no. 30210, Qiagen).
15. Plasmid DNA extraction kit (e.g., cat. no. K3030, Bioneer).
16. PCR-fragment purification kit (e.g., Illustra GFX PCR DNA and Gel Band Purification Kit, cat. no. 28-9034-70, GE Healthcare).
17. Thin-walled PCR tubes.
18. Standard materials and reagents for agarose and polyacrylamide gel electrophoresis.
19. Power supply.

2.2. Media, Antibiotics, and Chemicals

1. 0.7 M isopropyl-β-D-thiogalactopyranoside (IPTG) in double-distilled water (ddH_2O). Filter sterilize, aliquot, and store at –20°C for up to 30 days.
2. 1,000× stock solutions of antibiotics: 50 mg/mL kanamycin, 100 mg/mL ampicillin, 50 mg/mL spectinomycin, 50 mg/mL streptomycin in ddH_2O. Aliquot and store at –20°C for 30 days.
3. 0.5 M ethylenediaminetetraacetic acid (EDTA), pH 8.0. Autoclave and store at room temperature.
4. 1 M Tris(hydroxymethyl)aminomethane hydrochloride (Tris–HCl), pH 8.0. Store at room temperature.
5. 1 M imidazole, pH 8.0. Store at room temperature.
6. 50× Tris–acetate–EDTA (TAE) buffer: dissolve 241 g Tris base and 14.6 g EDTA in 700 mL ddH_2O, add 57.1 mL of glacial acetic acid and bring the final volume to 1 L using ddH_2O.
7. 1 M dithiothreitol (DTT) in ddH_2O. Aliquot and store at –20°C.
8. Lysis Buffer: 20 mM Tris–HCl, pH 8.0, 5 mM imidazole, 100 μM $ZnCl_2$, 5% (v/v) glycerol, 500 mM NaCl.

9. Wash Buffer: Lysis Buffer with 20 mM imidazole.
10. Elution Buffer: Lysis Buffer with 400 mM imidazole.
11. Protein Dilution Buffer: 50 mM Tris–HCl, pH 7.5, 50 mM NaCl, 10 μM $ZnCl_2$
12. 75% (v/v) glycerol.
13. Luria broth (LB) liquid medium: 5 g yeast extract, 10 g tryptone, and 10 g NaCl in 1 L of ddH_2O. Autoclave and store at room temperature.
14. LB solid medium: same as LB liquid medium, with the addition of 15 g/L agar before autoclaving. Store at room temperature.
15. Molecular biology-grade agarose.
16. TaKaRa EX *Taq* DNA polymerase (Pan Vera Corp.) at a concentration of 5 units/μL and EX Taq 10× reaction buffer.
17. 2.5 mM (each) dNTP mix for PCR.
18. T4 DNA ligase at a concentration of 3 U/μL (Promega).
19. Restriction endonucleases: *Spe*I, *Xho*I, *Nco*I, *Age*I, *Not*I, and *Bam*HI (New England Biolabs).
20. rAPid alkaline phosphatase (Roche).
21. ZFN Digestion Buffer: 10 mM Tris–HCl, pH 8.5, 50 mM NaCl, 1 mM DTT, 100 μM $ZnCl_2$, 50 μg/mL BSA, and 100 μg/mL tRNA.
22. 100 mM $MgCl_2$ in ddH_2O. Filter and store at room temperature.
23. 10× NEBuffer 4 (New England Biolabs).

2.3. Plasmids, Bacterial Strains, and Primers

1. pSAT6-*Fok*I, a high-copy cloning plasmid for ZFN assembly. This and other non-commercial genetic material can be obtained from the author upon request.
2. pET28-XH, a modified bacteriophage T7 RNA polymerase bacterial expression vector for cloning and expression of recombinant ZFNs.
3. Chemically competent cells of *E. coli* strain DH5α.
4. Chemically competent *E. coli* strain BL21 GOLD (DE3) pLysS (Novagen).
5. pPZP-RCS2, binary plant transformation vector.
6. pSAT6-MCS, satellite plant expression vector.
7. ZFN backbone assembly primers: BBO1: (5′GAAAAACC TTACAAGTGTCCTGAATGTGGAAAGTCTTTTTCT), BBO2:(5′CAGCGAACACACACAGGTGAGAAGCCATA TAAATGCCCAGAATGTGGTAAATCATTCAG) and

BBO3: (5′CAACGGACCCACACCGGGGAGAAGCCAT TTAAATGCCCTGAGTGCGGGAAGAGTTTTT).

8. ZFN finger-specific primers SDO1, SDO2, and SDO3 (*see* Step 1 of **Section 3.1**).
9. Zinc finger DNA-binding domain amplification forward primer: BBO1-*Xho*I-F (5′CCGCTCGAGCTGAAAAACC TTACAAGTGTCC) and zinc finger DNA-binding domain amplification reverse primer: SDO3-*Spe*I-R (5′GGACTA GTCCTCCAGTATGAGTACGTTGATG), each primer diluted to 10 μM.
10. 35S-F, pSAT6-*Fok*I sequencing primer (5′CTATCCTTC GCAAGACCCTTCC).
11. PZP-F, pPZP-RCS2 sequencing primer (5′GCTACCCTC CGCGAGATCATCC).
12. SAT-F and SAT-R, pSAT backbone sequencing primers (5′TCTTCGCTATTACGCCAGC and 5′ACGACAGGTTTCCCGACTGG, respectively).
13. pSAT6-MCS donor vector modification primers (*see* examples below).
14. pPZP-RCS2 acceptor vector modification primers (*see* examples below).

3. Methods

The rules and protocols for the design and assembly of ZFNs have been discussed in various excellent papers (e.g., 7, 10) and elsewhere in this issue. The readers are referred to these reports as well as to Web-based programs (e.g., http://zincfingertools.org and http://bindr.gdcb.iastate.edu/ZiFiT) for the design of their ZFN of choice. Here we provide a slightly modified protocol from Mani et al. (7) for the assembly of zinc finger proteins (ZFPs) into a pre-engineered ZFN assembly construct (*see* **Fig. 12.1a**). The pSAT6-*Fok*I vector allows fusing pre-assembled ZFPs between a nuclear localization signal (NLS) and the *Fok*I endonuclease domain (*see* **Fig. 12.1A**). The pET28-XH plasmid was engineered as an acceptor vector for ZFN-coding sequences from pSAT6-*Fok*I-based plasmids, allowing the expression of a 6×His-tagged ZFN fusion proteins (*see* **Figs. 12.1B** and **12.2**) in vitro and in *E. coli* cells (*see* **Note 1**). We found that the cloning efficiency of short ZFPs into the high-copy plasmid pSAT6-*Fok*I is several-fold higher than the efficiency of cloning them directly into pET28-based vectors. We thus recommend a two-step assembly procedure for the construction of newly assembled ZFNs.

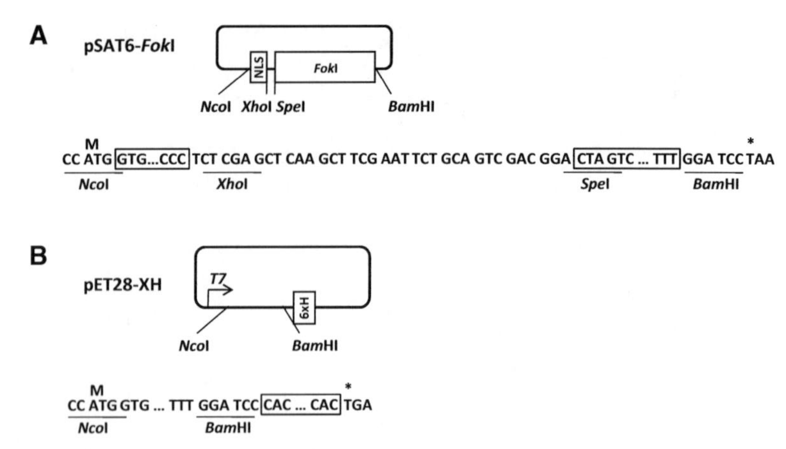

Fig. 12.1. Structure and key features of the ZFN assembly and expression vectors. (**A**) We used pSAT6-*Fok*I, a high-copy cloning vector, for the construction of custom-designed ZFNs. PCR-assembled ZFPs are cloned in frame between a nuclear localization signal (NLS) and the *Fok*I endonuclease domains using *Spe*I and *Xho*I. Also indicated are the locations of the *Nco*I and *Bam*HI sites useful for transferring the entire ZFN-coding sequence without its TAA stop codon. (**B**) We used pET28-XH for in vitro and in vivo expression of His-tagged ZFN proteins. Transferring ZFNs as *Nco*I–*Bam*HI fragments from recombinant pSAT6-*Fok*I vectors will produce a 6×His fusion coding sequence controlled by the T7 promoter. Also indicated is the TGA stop codon (*asterisk*) which is adjacent to the 6×His coding sequence in pET28-XH.

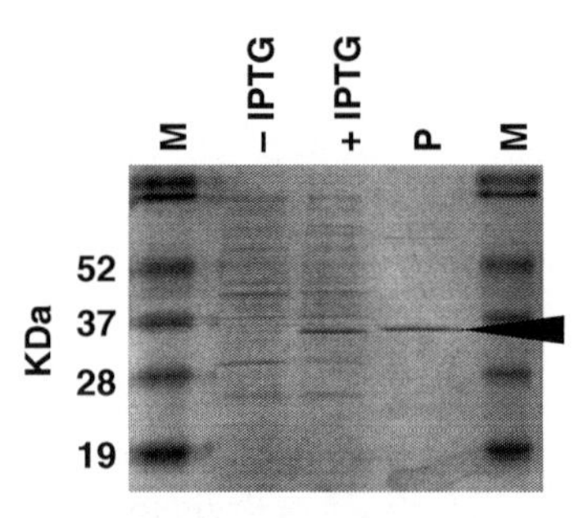

Fig. 12.2. SDS-PAGE analysis of ZFN10. Image shows the total crude extract from induced (+IPTG) and non-induced (-IPTG) ZFN10 protein-expressing *E. coli* cells and ZFN10 protein purified by Ni-NTA affinity chromatography (P). (Reproduced with permission from Tovkach et al. (14)).

3.1. Assembly of ZFPs from Overlapping Oligonucleotides

1. Dilute oligonucleotides BBO1, BBO2, BBO3, SDO1-10, SDO2-10, and SDO3-10 to a final concentration of 100 μM (*see* **Note 2**).
2. Prepare a BBO cocktail containing 1 μL of each of the 100 μM BBO primers, bringing the volume to 100 μL with ddH_2O.
3. Prepare an SDO cocktail containing 1 μL of each 100 μM SDO primers, bringing the volume to 100 μL with ddH_2O.

4. Prepare a PCR mixture with a total volume of 50 μL containing 0.5 μL BBO cocktail, 0.5 μL SDO cocktail, 1 μL each of BBO1-*Xho*I-F and SDO3-*Spe*I-R primers, 4 μL dNTPs (2.5 mM each), 2 units of EX *Taq* DNA polymerase, and 5 μL of EX Taq 10× reaction buffer.
5. Perform PCR with the following program: denaturation at 95°C for 5 min followed by 35 cycles of 45 s at 95°C, 30 s at 55°C and 30 s at 72°C.
6. Run the PCR product on a 1.5% agarose gel.
7. Excise the ~270-bp ZFP-coding sequence band from the gel and purify it using the GFX PCR purification kit. Elute the sample in 50 μL ddH_2O.
8. Digest the purified ZFP-coding sequence band with *Spe*I and *Xho*I and repurify the digested fragment with the GFX PCR purification kit.
9. Clone the ZFP-coding sequence band into the *Spe*I and *Xho*I sites of pSAT6-*Fok*I, producing pSAT6-ZFN-X (*see* **Note 3**). In our sample, we constructed pSAT-ZFN10 which codes for the ZFN10 enzyme (*see* **Note 2**).
10. Analyze the integrity and accuracy of the assembled ZFP domain and its proper fusion with the *Fok*I domain by DNA sequencing (*see* **Note 4**) using the 35S-F sequencing primer.
11. Transfer the assembled ZFN-coding sequence from pSAT6-ZFN-X as a *Nco*I–*Bam*HI fragment into pET28-XH, producing pET28-ZFN-X. In our sample, we constructed pET28-ZFN10 which codes for the ZFN10 enzyme (*see* **Note 2**).
12. Repeat this procedure using different SDO primers to create additional vectors coding for other ZFNs (we also constructed pSAT-ZFN12 and pET28-ZFN12, *see* **Note 5**).

3.2. ZFN Expression and Purification

Crude protein extracts from ZFN-expressing *E. coli* cells are usually suitable for in vitro digestion analysis of ZFN activity (e.g., 7, 11). However, the presence of nucleases and other DNA-damaging enzymes in crude bacterial extracts may hinder the use of these recombinant ZFNs for cloning purposes. The fastest and perhaps simplest strategy for obtaining cloning-grade ZFNs is to use in vitro expression kits in combination with prepacked Ni-purification columns. However, while useful for the expression and purification of various ZFNs for cloning purposes (12), in vitro expression kits are costly and yield relatively small amounts of protein. We thus recommend using *E. coli*-mediated protein expression which is expected to yield relatively high quantities of recombinant protein. Herein we present a detailed protocol for the expression and purification of cloning-grade ZFNs from *E. coli* cells.

3.2.1. Expression of ZFN in E. coli

1. Transform the ZFN-expressing construct into BL21 GOLD (DE3) pLysS cells, plate on LB solid medium supplemented with 50 μg/mL kanamycin, and grow overnight at 37°C.
2. The following day, prepare a starter culture by inoculating 5 mL of LB liquid medium with several colonies. Supplement the medium with 50 μg/mL kanamycin and grow overnight at 37°C with shaking at 250 rpm in a 100-mL flask.
3. The following day, transfer 2 mL of the starter culture into 100 mL of LB liquid medium in a 500-mL flask. Supplement the medium with 50 μg/mL kanamycin and 100 μM $ZnCl_2$. Grow for about 4 h at 22°C with shaking at 250 rpm to an A_{600} of 0.4–0.7 (*see* **Note 6**).
4. Add 100 μL of 0.7 M IPTG (to a final concentration of 0.7 mM) and let grow for another 3 h.
5. Centrifuge (3,000×*g*, 10 min, 4°C), discard the supernatant, and resuspend the cell pellet in 30 mL of Lysis Buffer cooled to 4°C. Keep on ice (*see* **Note 7**). The resuspended cells can be processed immediately or stored at –70°C for several months.
6. Break cells by passing twice through a French press at 18,000 psi. Transfer to 50-mL polypropylene centrifuge tubes and centrifuge (12,000×*g*, 10 min, 4°C). Transfer the supernatant to a fresh 50-mL tube (*see* **Note 8**).
7. Sediment 0.5 mL of Ni-NTA agarose resin by centrifuging for 2 min at 150×*g*. Discard the supernatant (containing 20% ethanol used for resin storage) and resuspend resin in 1 mL of Lysis Buffer.
8. Add the Ni-NTA resin from Step 7 to the ZFN supernatant from Step 6. Swirl gently for 45 min at 4°C on a rotating shaker.
9. Sediment the Ni-NTA resin (which should now be carrying bound ZFN protein) by centrifuging for 2 min at 150×*g*. Discard the supernatant. Resuspend the resin in 40 mL of Wash Buffer.
10. Repeat the resin wash procedure from Step 9.
11. Centrifuge Ni-NTA resin again and resuspend it in 10 mL of Wash Buffer. Transfer to a 10-mL disposable chromatography column and let the resin settle for 5 min. Open the column and let it drain.

12. Elute the ZFNs with 0.5 mL of Elution Buffer into 1.5-mL chilled (to 4°C) Eppendorf tube.
13. Add 1 mL of chilled (to 4°C) 75% glycerol to the elaute (final glycerol concentration 50%). Split into 100-μL aliquots and store at –20°C (*see* **Note 9**).
14. Repeat this procedure to produce additional enzymes (*see* **Note 10**).

3.3. Assembly of Multigene Transformation Vector by ZFNs

Various cloning tasks can potentially be achieved using ZFNs. Custom-made ZFNs can be designed to digest and cleave native DNA molecules which can then be cloned into target vectors (12). ZFNs can also be designed to target artificial DNA sequences and assist with the assembly of complex DNA structures (12), similar to the use of naturally occurring rare cutters (3). Here we demonstrate the use of ZFNs for the assembly of multigene plant transformation vectors, a task that can rarely be achieved using traditional 4- and 6-bp restriction enzymes (reviewed in Ref. 3). Our experimental approach is based on conversion of the pPZP-RCS2 (3) (**Section 3.3.1**) plant transformation vector's multicloning site's (MCS) to carry several unique ZFN recognition sites, construction of donor plasmids (based on the structure of pSAT6-MCS, 13) (**Section 3.3.2**) each carrying a functional expression cassette flanked with ZFN recognition sites, and use of purified ZFNs (**Section 3.3.3**) to clone plant expression cassettes onto the acceptor plasmid (*see* **Fig. 12.3**).

3.3.1. Modifying pPZP-RCS2 to Carry ZFN Recognition Sites

1. Design pairs of 24-nucleotide long primers that can anneal to each other and form an artificial semi-palindromic ZFN recognition site (illustrated as primers P1 and P2, and P3 and P4 in **Fig. 12.3**). Here we used 5′GGGGTACCTGCATCCATACTTACATGGATGCAGG TACCCC (P1) and 5′GGGGTACCTGCATCCATACTT ACATGGATGCAGGTACCCC (P2) and the plant binary vector pPZP-RCS2 as an example of the modification of a plant transformation vector to carry a new ZFN10 semi-palindromic recognition site (*see* **Note 11**).
2. Prepare primer-annealing cocktail with a total volume of 50 μL containing 0.25 μL of each primer, and 5 μL of 10× *Kpn*I digestion buffer in a 200-μL thin-walled PCR tube.
3. Heat the sample to 95°C in a PCR thermocycler for 30 s and cool quickly to 45°C for an additional 30 s. Lower the temperature and maintain the sample at 4°C until use.
4. Add 0.5 μL of *Kpn*I (at a concentration of 10 U/μL) to the primer–annealing mixture and digest for 2 h at 37°C.

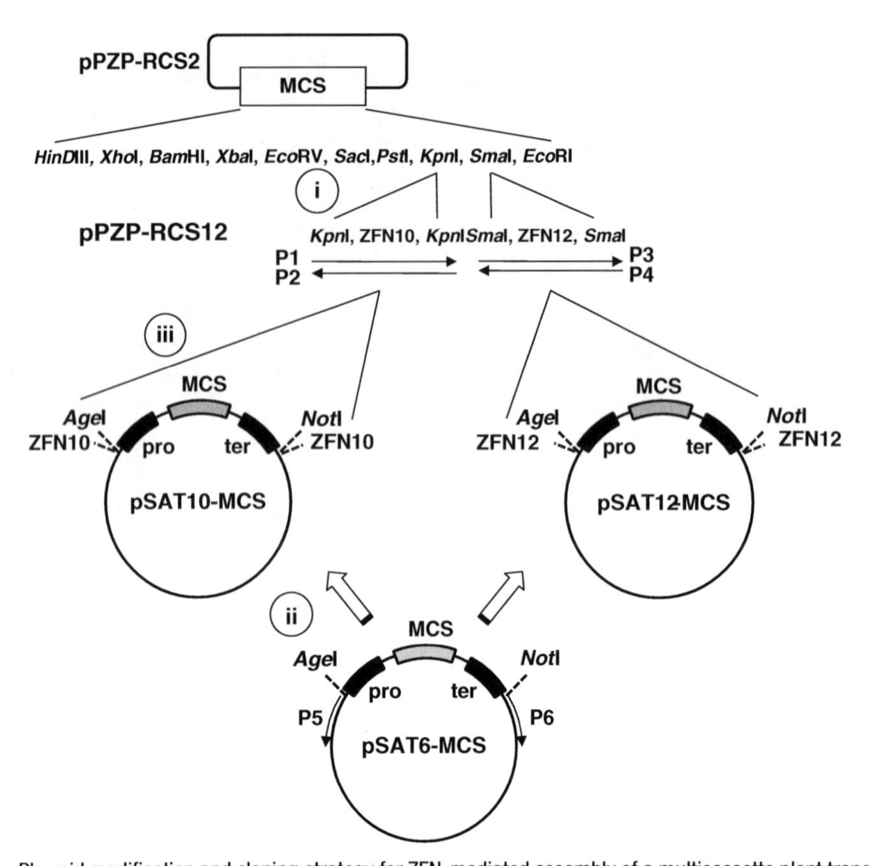

Fig. 12.3. Plasmid modification and cloning strategy for ZFN-mediated assembly of a multicassette plant transformation vector. Cloning begins with modification of the acceptor plasmid (pPZP-RCS2) to carry novel ZFN recognition sites by ligating annealed pairs of primers into the plasmid's MCS (e.g., primers P1 and P2 for ZFN10 and P3 and P4 for ZFN12) (Step i) and reconstructing a satellite vector (e.g., pSAT6-MCS) so that the plant expression cassettes on the new satellite vectors (e.g., pSAT10-MCS and pSAT12-MCS) will be flanked by novel ZFN recognition sites (Step ii). The latter can be achieved by PCR amplification of pSAT6-MCS backbone using a pair of primers which are engineered to carry the ZFN recognition sites (indicated here by P5 and P6). Plant expression cassettes, carrying different genes of interest, can now be mounted onto the modified acceptor plasmid using purified ZFNs (Step iii). Abbreviations: pro, promoter region; ter terminator region; MCS, multicloning site.

5. Purify the annealed fragments by ethanol-based precipitation or by running them through a GFX PCR purification column (*see* **Note 12**).
6. Digest pPZP-RCS2 with *Kpn*I, separate the linearized vector in a 0.7% agarose gel and purify it using the GFX PCR purification kit.
7. Treat the linearized vector with rAPid alkaline phosphatase for 7–10 min at 37°C and deactivate the enzyme by heat treatment for 3 min at 75°C.
8. Prepare a ligation cocktail with a total volume of 15 μL containing approximately 50 ng of linearized pPZP-RCS2, 10 μL of annealing mixture, 0.5 μL of T4-DNA ligase

(at a concentration of 3 U/μL), and 1.5 μL of 10× ligation buffer.

9. Perform the ligation for 2 h at room temperature and use the entire ligation mixture for transformation into chemically competent *E. coli* cells (*see* **Note 13**).
10. Pick selected colonies, extract DNA and verify the integrity and orientation of the inserted ZFN recognition sites by DNA sequencing using PZP-F sequencing primer.
11. Repeat this procedure several times to create a modified target vector with the correct type and number of ZFN recognition sites on its MCS (*see* **Note 14**). We used this strategy and produced pPZP-RCS12, a binary plant transformation vector with artificial semi-palindromic ZFN recognition site for ZFN10 and ZFN12.

3.3.2. Construction of Donor Vectors with ZFN Recognition Sites

1. Design pairs of primers capable of amplifying the entire plasmid backbone of pSAT6-MCS. These primers should be designed to allow the reconstruction of a new plasmid backbone carrying semi-palindromic ZFN recognition sites which will flank the plant expression cassette from pSAT6-MCS (illustrated as primers P5 and P6 in **Fig. 12.3**). As an example, we used 5′ATAAGAATGCGGCCGCTGCATCCATGTAAGTATGG ATGCAGTAATCATGGTCATAGCTGTTTCC (P5) and 5′GACGCACCGGTTGCATCCATACTTACATGGATGCA GGCACTGGCCGTCGTTTTACAACG (P6) primer pairs for the construction of pSAT10-MCS (*see* **Fig. 12.3** and **Note 15**).
2. Prepare a PCR mixture with a total volume of 50 μL containing 2 ng template DNA (i.e., pSAT6-MCS), 0.2 mM of each dNTP, 0.2 μM of each primer, 2 units of EX *Taq* DNA polymerase, and 5 μL of EX Taq 10× reaction buffer (*see* **Note 16**).
3. Perform PCR with the following program: denaturation at 95°C for 5 min followed by 36 cycles of 45 s at 95°C, 30 s at 55°C, and 3 min at 72°C (*see* **Note 17**).
4. Separate the PCR product on a 0.7% agarose gel and purify it using the GFX PCR purification kit.
5. Digest the purified PCR product with *Age*I and *Not*I and use it as an acceptor plasmid for an *Age*I–*Not*I expression cassette fragment from pSAT6-MCS (*see* **Note 18**).
6. Pick selected colonies, extract DNA, and verify the integrity and orientation of the new ZFN recognition sites by DNA sequencing using SAT-F and SAT-R sequencing primers.

7. Repeat this procedure to produce additional vectors carrying different types of ZFN recognition sites on their MCS (*see* **Note 19**).
8. Clone your selected genes into the MCS of the modified donor vectors (here, pSAT10-MCS) using traditional restriction enzymes (*see* **Note 20**).

3.3.3. Mounting Expression Cassettes onto a Single Acceptor Vector by ZFN

1. Prepare 1:1, 1:10, 1:100, and 1:1,000 dilutions of ZFN (ZFN10 in our example) in Protein Dilution Buffer.
2. Set up test digestion cocktails containing 200 ng of pSAT10-MCS, 1 μL of diluted ZFN10 in a total volume of 25 μL in ZFN Digestion Buffer.
3. Incubate at room temperature for 20–30 min, allowing the ZFNs to bind to their target sequences.
4. Add 2.5 μL of 100 mM $MgCl_2$ and incubate for 5–45 min at room temperature at 10-min intervals (*see* **Note 21**).
5. Analyze the quality and quantity of ZFN10-mediated digestion of pSAT10 on an agarose gel and select several combinations of enzyme dilutions and incubation periods that give the most efficient digestion activity and lowest non-specific digestion and/or DNA degradation (*see* **Note 22**).
6. Extract several digested pSAT10 backbone fragments (corresponding to the selected combinations from Step 5) from the gel and purify them by GFX PCR purification kit.
7. For each fragment, prepare a ligation cocktail with a total volume of 10 μL containing approximately 50 ng of linearized pSAT10 backbone fragment, 0.5 μL T4-DNA ligase (at a concentration of 3 U/μL), and 1 μL of 10× ligation buffer.
8. Perform the ligations for 2 h at room temperature and use the entire ligation reaction mixture for transformation into chemically competent *E. coli* cells.
9. Compare the ligation efficiency of various fragments obtained in Step 6 and verify and narrow down the number of optimal combinations of enzyme dilution and incubation period selected in Step 5.
10. Pick selected colonies, extract DNA and verify the quality of the ligation by linearization of the resultant plasmid by ZFN10 and/or by DNA sequencing using SAT-F sequencing primer (*see* **Note 23**). Use these data to verify and determine the best combination of enzyme dilution and incubation period, as selected in Steps 5 and 9.

11. Use the optimal ZFN dilution and incubation period, as determined in Steps 5 and 9–10, to digest the acceptor (pPZP-RCS12 in our example, *see* **Note 14**) and donor (pSAT10-YFP-CHS in our example, *see* **Note 20**) plasmids.
12. Separate the linearized acceptor plasmid (pPZP-RCS12) and the expression cassette (YFP-CHS) from the donor plasmid on a 0.7% agarose gel and purify the DNA fragments using the GFX PCR purification kit.
13. Prepare a ligation cocktail with a total volume of 15 μL containing approximately 50 ng of linearized pPZP-RCS12, approximately 150 ng of YFP-CHS fragments, 0.5 μL T4-DNA ligase (3 U/μL), and 1.5 μL of 10× ligation buffer.
14. Perform the ligations for 2 h at room temperature and use the entire ligation reaction mixture for transformation into chemically competent *E. coli* cells.
15. Pick selected colonies, extract DNA, and verify the integrity of the new expression cassette by ZFN10 digestion of the recombinant constructs.
16. Repeat this procedure to add additional expression cassettes onto the resultant plasmid from Step 15 (e.g., pPZP-RCS12[YFP-CHS]) using other ZFNs (e.g., use ZFN12 to add the CHRD-RFP expression cassette from pSAT12-CHRD-RFP into pPZP-RCS12[YFP-CHS] and produce pPZP-RCS12[YFP-CHS][CHRD-RFP]).
17. Verify the functionality of the resultant multigene transformation vector in plant cells (*see* **Fig. 12.4**).

4. Notes

1. While originally intended for expression in eukaryotic cells, the presence of a NLS does not hinder the in vitro activity of the ZFN.
2. We used SDO1-10 (5′ACCTGTGTGTGTTCGCTGGTG ACGACGCAAATCTCCAGACTGAGAAAAGACTTTC CACA), SDO2-10 (5′CCCGGTGTGGGTCCGTTGGTG ACGAACCAAATTTCCAGAAGTACTGAATGATTTAC CACA), and SDO3-10 (5′TCCAGTATGAGTACGTTGAT GAACATTCAATTCATCACGACGTGAAAAACTCTTC CCGCAC), as an example set of primers used for the construction of ZFN10, a ZFN monomer capable of binding to a TGCATCCAT-$(N)_x$-ATGGATGCA ($4 \leq x \leq 6$) target site. Other primers can be designed for the construction of ZFNs with different target sites.

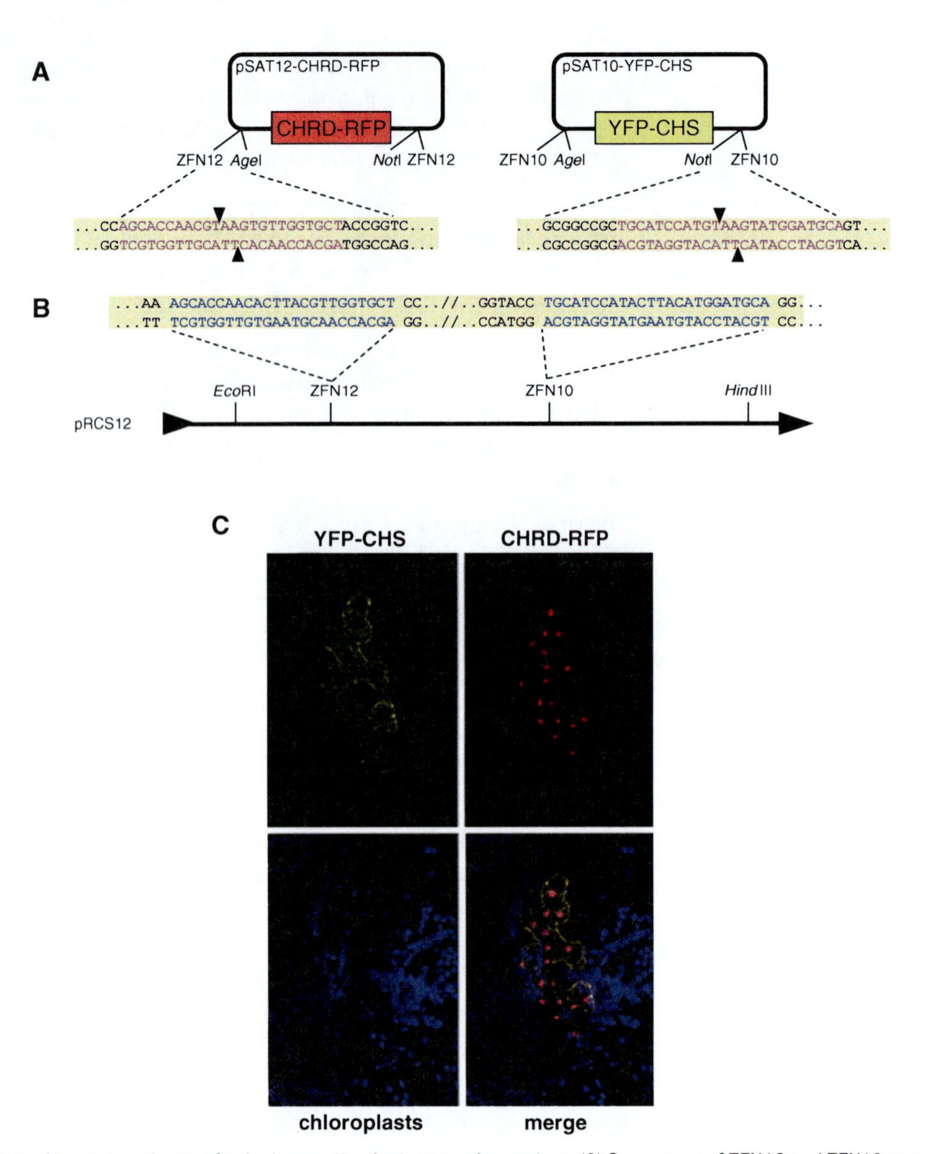

Fig. 12.4. Structure and uses of a dual-cassette plant expression system. (**A**) Sequences of ZFN12 and ZFN10 recognition sites on pSAT12-CHRD-RFP and pSAT10-YFP-CHS (carrying the RFP-tagged CHRD protein and YFP-tagged CHS protein, respectively) are shown in *purple*. (**B**) Sequences of ZFN10 and ZFN11 recognition sites on pPZP-RCS12 are shown in *blue*. (**C**) Confocal microscopy analysis of YFP-CHS (in *yellow*) and CHRD-RFP (in *red*) expression following bombardment of the dual plant-expression cassette vector (pPZP-RCS12 containing both CHRD-RFP and YFP-CHS) into plant cells. Chloroplast autofluorescence is shown in *blue*.

3. The close proximity of the *Spe*I and *Xho*I sites on pSAT6-*Fok*I may lead to partial digestion of the vector. rAPid alkaline phosphatase treatment for 7–10 min at 37°C and deactivation of the enzyme by heat treatment for 3 min at 75°C are advised.
4. The nature of the ZFP assembly process may result in a variety of DNA products and it is extremely important to

verify the final product sequence prior to using it for protein expression and DNA cloning.

5. We also constructed ZFN12 (also known as ZFN-H2a, Ref. 12), a ZFN monomer capable of binding to an AGCACCAAC-$(N)_x$-GTTGGTGCT ($4 \leq x \leq 6$) target site using the following SDO primers: SDO1-ZFN12 (5′ACCTGTGTGTGTTCGCTGGTGACGAGTAAGATCAGAAGACTGAGAAAAAGACTTTCCACA), SDO2-ZFN12 (5′CCCGGTGTGGGTCCGTTGGTGACGAACAAGATGTCCAGAAGTACTGAATGATTTACCACA), and SDO3-ZFN12 (5′TCCAGTATGAGTACGTTGATGACGAACAAGAGATCCAGAAGTTGAAAAACTCTTCCCGCAC).
6. Save 1 mL of non-induced cell culture as a control for SDS-PAGE analysis of the expressed ZFNs.
7. Save 1 mL of induced cell culture as a control for SDS-PAGE analysis of the expressed ZFNs.
8. The supernatant obtained after centrifuging cell lysates can be used directly for assaying enzymatic activity of newly designed ZFNs. However, to obtain cloning-grade ZFNs, it is important to process it through Ni-affinity purification.
9. We recommend confirming ZFN expression by running non-induced *E. coli* cell culture (**Note 6**), induced *E. coli* cell culture (**Note 7**), and Ni-affinity-purified ZFNs on SDS-PAGE, as shown in **Fig. 12.2**.
10. In addition to ZFN10, we used in vitro expression methods to produce ZFN12 and other ZFNs (12, 14).
11. Primers should actually be longer than 24 nucleotides and include recognition sites for cloning the annealed fragments into the target transformation vector. For example, the primers we used also carry a *Pst*I recognition site which allows cloning the resultant DNA fragment into the same site of the acceptor pPZP-RCS2 plasmid.
12. Alternatively, use heat deactivation instead of the extra purification step to deactivate the restriction enzyme.
13. Using electroporation may result in higher ligation efficiency, but will require purification by ethanol precipitation of the ligation reaction mixture. Do not forget to set up control ligations composed of a linear vector without the annealed primers. This control ligation will be useful for estimating the number of putative recombinant molecules, which will be nearly identical in size to their ancestors and cannot typically be assayed by restriction analysis.
14. We further modified the pPZP-RCS2 MCS by adding a semi-palindromic ZFN12 recognition site,

illustrated as primers P3 (5′TCCCCCGGGAGCA CCAACGTAAGTGTTGGTGCTCCCGGGGGA) and P4 (5′TCCCCCGGGAGCACCAACACTTACGTTGGTGCT CCCGGGGGA) in **Fig. 12.3**.

15. Primers should also carry the appropriate restriction enzyme recognition sites needed for the construction of the donor plasmid. In our system, we used *Age*I and *Not*I to clone a 1.2-kb-long empty plant expression cassette from pSAT6-MCS into pSAT10-MCS and other pSAT-based vectors.
16. Other types of *Taq* DNA polymerase can potentially be used, but the use of proofreading enzymes (i.e., *Pfu/Taq* DNA polymerase) is not essential since the plasmid backbone may be more tolerant to single nucleotide deletions and/or substitutions which may arise during the amplification process.
17. We found that a 3-min long extension step is sufficient for amplification of the 2.7-kb pSAT-based backbone when using TaKaRa EX Taq. Shorter or longer extension times should be considered when amplifying longer DNA sequences and/or using other *Taq* DNA polymerases.
18. The product is a modified linear plasmid backbone without the MCS and without the rare-cutting recognition sites on pSAT6. Ligating it with an *Age*I–*Not*I fragment from the parental pSAT6-MCS plasmid will produce a modified plasmid, in which the MCS is flanked by ZFN recognition sites.
19. We also constructed pSAT12-MCS, a variant of pSAT6-MCS with only ZFN12 recognition sites (*see* **Fig. 12.3**).
20. We used the newly constructed pSAT10 and pSAT12 satellite plasmids for the assembly of pSAT10-YFP-CHS and pSAT12-DsRed2-P, carrying the yellow fluorescent protein-tagged endoplasmic reticulum protein chalcone synthase (CHS) and the DsRed2-tagged protein P of Sonchus Yellow Net Virus (SYNV) (15), respectively.
21. It is also possible to use 10× NEBuffer 4 as the restriction cocktail buffer and perform the digestion reaction at 37°C. Digesting at 37°C may require shorter incubation periods, but may also lead to non-specific activity and possible degradation of the DNA's sticky ends.
22. The goal is to achieve high-quality digestion, as evidenced by the absence of non-specific degradation (and not necessarily complete digestion). Additional dilutions and/or various incubation periods may be required to calibrate the specific activities of different enzymes.

23. Accurate reconstruction of the ZFN restriction site following self-ligation is the most reliable criterion for the specific activity of the engineered ZFN.

Acknowledgments

We thank Dr. G.N. Drews for the gift of pHS::QQR-QEQ/2300. We also thank Dan Weinthal for his instrumental advice and support. This work in our lab is supported by grants from the Human Frontiers Science Program, the Biotechnology Research and Development Corporation (BRDC) and University of Michigan startup funds.

References

1. Roberts, R.J., Vincze, T., Posfai, J., and Macelis, D. (2007) REBASE–enzymes and genes for DNA restriction and modification. *Nucleic Acids Res.* **35**, D269–D270.
2. Veselkov, A.G., Demidov, V.V., Nielson, P.E., and Frank-Kamenetskii, M.D. (1996) A new class of genome rare cutters. *Nucleic Acids Res.* **24**, 2483–2487.
3. Dafny-Yelin, M. and Tzfira, T. (2007) Delivery of multiple transgenes to plant cells. *Plant Physiol.* **145**, 1118–1128.
4. Goderis, I.J., De Bolle, M.F., Francois, I.E., Wouters, P.F., Broekaert, W.F., and Cammue, B.P. (2002) A set of modular plant transformation vectors allowing flexible insertion of up to six expression units. *Plant Mol Biol.* **50**, 17–27.
5. Lin, L., Liu, Y.G., Xu, X., and Li, B. (2003) Efficient linking and transfer of multiple genes by a multigene assembly and transformation vector system. *Proc Natl Acad Sci USA.* **100**, 5962–5967.
6. Lu, C., Mansoorabadi, K., and Jeffries, T. (2007) Comparison of multiple gene assembly methods for metabolic engineering. *Appl Biochem Biotechnol.* **137–140**, 703–710.
7. Mani, M., Kandavelou, K., Dy, F.J., Durai, S., and Chandrasegaran, S. (2005) Design, engineering, and characterization of zinc finger nucleases. *Biochem Biophys Res Commun.* **335**, 447–457.
8. Mandell, J.G. and Barbas, C.F., 3rd (2006) Zinc finger tools: custom DNA-binding domains for transcription factors and nucleases. *Nucleic Acids Res.* **34**, W516–W523.
9. Porteus, M. (2008) Design and testing of zinc finger nucleases for use in mammalian cells. *Methods Mol Biol.* **435**, 47–61.
10. Wright, D.A., Thibodeau-Beganny, S., Sander, J.D., Winfrey, R.J., Hirsh, A.S., Eichtinger, M., Fu, F., Porteus, M.H., Dobbs, D., Voytas, D.F., and Joung, J.K. (2006) Standardized reagents and protocols for engineering zinc finger nucleases by modular assembly. *Nat Protoc.* **1**, 1637–1652.
11. Carroll, D., Morton, J.J., Beumer, K.J., and Segal, D.J. (2006) Design, construction and in vitro testing of zinc finger nucleases. *Nat Protoc.* **1**, 1329–1341.
12. Zeevi, V., Tovkach, A., and Tzfira, T. (2008) Increasing cloning possibilities using artificial zinc finger nucleases. *Proc Natl Acad Sci USA.* **105**, 12785–12790.
13. Tzfira, T., Tian, G.-W., Lacroix, B., Vyas, S., Li, J., Leitner-Dagan, Y., Krichevsky, A., Taylor, T., Vainstein, A., and Citovsky, V. (2005) pSAT vectors: a modular series of plasmids for autofluorescent protein tagging and expression of multiple genes in plants. *Plant Mol Biol.* **57**, 503–516.
14. Tovkach, A., Zeevi, V., and Tzfira, T. (2009) A toolbox and procedural notes for characterizing novel zinc finger nucleases for genome editing in plant cells. *Plant J.* **57**, 747–757.
15. Goodin, M.M., Dietzgen, R.G., Schichnes, D., Ruzin, S., and Jackson, A.O. (2002) pGD vectors: versatile tools for the expression of green and red fluorescent protein fusions in agroinfiltrated plant leaves. *Plant J.* **31**, 375–383.

Chapter 13

In Vitro Assessment of Zinc Finger Nuclease Activity

Toni Cathomen and Cem Şöllü

Abstract

The technical advances in developing artificial endonucleases, such as zinc finger nucleases (ZFNs), have opened a wide field of applications in the genome engineering arena, including the therapeutic correction of mutated genes in the human genome. Gene editing frequencies of up to 50% in human cells under non-selective conditions reveal the power of the ZFN technology. Activity and toxicity of ZFNs are determined by a number of parameters, including the specificity of DNA binding, the kinetics of dimerization of the two ZFN subunits, and the catalytic activity. In order to investigate these parameters individually, a cell-free system that models these reactions is essential. Here, we present a simple and fast method for the functional testing of ZFNs in vitro.

Key words: Zinc finger nuclease, artificial nuclease, custom nuclease, DNA double-strand break, DSB, DNA cleavage, in vitro translation.

1. Introduction

The targeted manipulation of a complex genome by zinc finger nucleases (ZFNs) is a promising technology to generate both gene knockouts by site-specific mutagenesis and precise editing of the genome by gene targeting (1, 2). Gene targeting is based on homologous recombination (HR) between the chromosomal target locus and an exogenous donor DNA. The frequency of HR in mammalian cells is relatively low, with only about one spontaneous event per 10^6 cells (3). However, the HR frequency can be significantly increased by introducing a targeted DNA double-strand break (DSB) at the site of the desired recombination event. Some studies have reported HR frequencies of up to 50% in mammalian cells when the targeted DSB was introduced by ZFNs (4, 5). In the absence of a donor DNA, ZFN-induced DSBs

J.P. Mackay, D.J. Segal (eds.), *Engineered Zinc Finger Proteins*, Methods in Molecular Biology 649,
DOI 10.1007/978-1-60761-753-2_13,

are sealed using the error-prone non-homologous end-joining (NHEJ) pathway and the ensuing insertions and deletions result in disruption of the open reading frame. Recent studies have shown that a functional gene knockout can be achieved in up to 50% of treated primary human cells (6). For these reasons, ZFNs have developed into a valuable tool to modify complex genomes ad libitum, for both genetic studies in animal models (7–11) and therapeutic applications in human patients (6).

ZFNs are made up of an engineered zinc finger (ZF) DNA-binding domain, which confers specific binding to the desired target site and the catalytic domain of the restriction enzyme *Fok*I (12). A single ZF motif consists of about 30 amino acids and includes an α-helix that contacts three bases in the major groove of DNA (13, 14). Through variation of three to six amino acids in the α-helix, a new binding specificity can be created (15). Moreover, by combining three to six ZF motifs in a tandem fashion, extended binding sites of 9–18 bp can be targeted (1, 2, 16). Because the nuclease domain needs to dimerize in order to become catalytically active, two ZFN subunits are required to bind to opposite strands of the DNA target site in a tail-to-tail orientation (*see* **Fig. 13.1A**) (17, 18). Hence, three ZF motifs in each one of the two ZFN subunits will confer binding to an 18-bp sequence, which is statistically sufficient to define a unique target site in the human genome. Due to symmetry at the *Fok*I dimerization interface, homodimers of two ZFN subunits also can form, which theoretically can lead to undesirable cleavage events at off-target sites. The targeted exchange of critical residues in the *Fok*I dimerization domain can largely prevent homodimerization

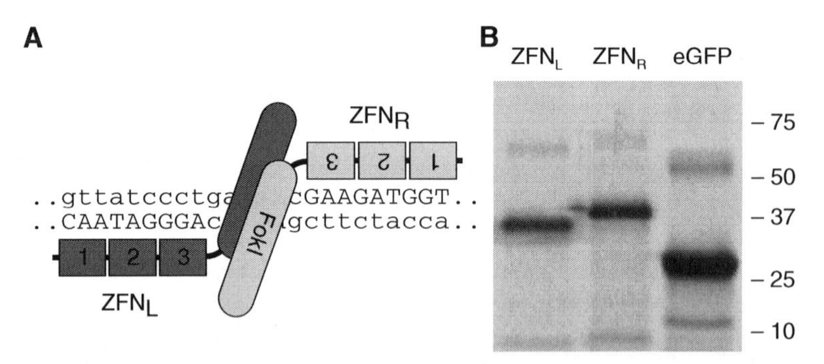

Fig. 13.1. In vitro translation of ZFNs. (**A**) Schematic of a ZFN bound to DNA. A ZFN consists of two subunits, ZFN_L and ZFN_R, which form a heterodimer. Each ZFN subunit contains three individual ZFs (1–3), which are fused to the catalytic domain of the *FokI* endonuclease (FokI). Since each ZF contacts three nucleotides of the target sequence, the full target site is made up of two 9-bp half-sites, which are separated by a short 6-bp spacer. (**B**) Autoradiography. To ascertain correct expression of ZFNs, an initial in vitro translation was performed in the presence of L-[^{35}S]-methionine. Subsequently, proteins were separated by SDS-PAGE and the dried gel exposed to X-ray film (*see* **Note 1**). Size markers (in kDa) are indicated on the *right*.

of two identical ZFN subunits and, as a result, significantly reduce ZFN-associated toxicity (19, 20).

In this chapter, we describe a method to assess ZFN activity in vitro. In vitro assays will be crucial to gain additional insights into the individual factors that affect the activity and toxicity of ZFNs, including parameters that influence DNA binding, the kinetics of dimerization, or the catalytic activity.

2. Materials

2.1. In Vitro Expression of ZFNs with Radioactive Label

1. TNT SP6 Coupled Reticulocytes Lysate System (Promega).
2. RNase inhibitor: RNaseOUT recombinant ribonuclease inhibitor (Invitrogen).
3. ZFN expression plasmids, and the EGFP negative control, containing an SP6 promoter for in vitro transcription have been described before (21–23) and can be obtained from the authors upon request.
4. L-[^{35}S]-methionine (10 mCi/mL) for radioactive labeling of TNT-expressed ZFNs. Note, this material is radioactive. Use appropriate precautions.
5. 6× SDS gel loading buffer: 350 mM Tris–HCl, pH 6.8, 36% glycerol, 12% SDS, 5% ß-mercaptoethanol, 0.01% bromophenol blue.
6. Apparatus to perform polyacrylamide gel electrophoresis: e.g., Mini-PROTEAN 3 Cell (Biorad) and power supply.
7. 40% polyacrylamide:bisacrylamide (19:1) stock solution: Rotiphorese Gel 40 (Carl Roth GmbH).
8. Fixing solution: mix 9 mL of methanol, 9 mL of H_2O, and 2 mL of glacial acetic acid.
9. 5% glycerol solution: mix 5 mL of glycerol with 95 mL of H_2O.
10. Gel dryer: e.g., model 543 (Biorad).
11. Whatman 3 MM paper.
12. Medical X-ray film.

2.2. In Vitro Expression of ZFNs Without Radioactive Label

1. TNT SP6 Coupled Reticulocytes Lysate System (Promega).
2. RNase inhibitor: RNaseOUT recombinant ribonuclease inhibitor (Invitrogen).
3. ZFN expression plasmids, and the EGFP negative control, containing an SP6 promoter for in vitro transcription have

been described before (21–23) and can be obtained from the authors upon request.

2.3. Generation of Target DNA

1. PCR kit: Pfu Ultra High Fidelity DNA Polymerase AD (Stratagene).
2. Oligonucleotides (e.g., "High Purity Salt Free" from Eurofins MWG Operon) for PCR amplification of the DNA cleavage substrates: Forward primer (pLF) 5′-tgcacgctggataacgacat, reverse primer (pER) 5′-aagtcgtgctgcttcatgtggt.
3. The template plasmid containing the ZFN-binding sites to generate the cleavage substrate by PCR has been described earlier (21) and can be obtained from the authors upon request.
4. Thermal cycler for PCR.
5. Apparatus to perform agarose gel electrophoresis (e.g., Mini-Sub Gel from Biorad) and power supply.
6. Tris–acetate–EDTA (TAE) running buffer: For 1 L of 50× TAE buffer, dissolve 242 g of Tris base (2-amino-2-hydroxymethyl-propane-1,3-diol) in 800 mL of H_2O. Add 57.1 mL of glacial acetic acid and 100 mL of 0.5 M Na_2EDTA, pH 8.0 (ethylenediamine tetraacetic acid). Add water to 1,000 mL.
7. 6× DNA loading dye: 20% (w/v) Ficoll 400, 0.25% (w/v) bromophenol blue, 0.25% (w/v) xylene cyanol FF, dissolved in water.
8. DNA gel extraction kit: e.g., Invisorb Spin DNA Extraction Kit (Invitek).
9. Spectrophotometer: e.g., DU 530 (Beckman Coulter).
10. Gel documentation system (e.g., Intas).

2.4. In Vitro Cleavage Assay

1. Water bath.
2. Apparatus to perform agarose gel electrophoresis: Mini-Sub Gel (Biorad) and power supply.
3. Tris–acetate–EDTA (TAE) running buffer: For 1 L of 50× TAE buffer dissolve 242 g of Tris base (2-amino-2-hydroxymethyl-propane-1,3-diol) in 800 mL of H_2O. Add 57.1 mL of glacial acetic acid and 100 mL of 0.5 M Na_2EDTA, pH 8.0 (ethylenediamine tetraacetic acid). Add water to 1,000 mL.
4. 6× DNA loading dye: 20% (w/v) Ficoll 400, 0.25% (w/v) bromophenol blue, 0.25% (w/v) xylene cyanol FF, dissolved in water.
5. Gel documentation system (e.g., Intas).

3. Methods

3.1. In Vitro Expression of ZFNs with Radioactive Label

Before employing the nucleases in the in vitro cleavage assay, correct expression of the ZFNs should be verified. This can be done by performing the in vitro translation in the presence of L-[^{35}S]-methionine.

1. Remove the TNT reaction components from –80°C and store the TNT SP6 RNA polymerase on ice. Thaw the TNT reticulocyte lysate by hand warming and immediately place on ice. The other reagents can be thawed at room temperature and then stored on ice.
2. Mix the reaction components in a 1.5-mL microcentrifuge tube as follows: 5.0 μL of TNT lysate, 0.4 μL of TNT reaction buffer, 0.2 μL of 1 mM amino acid mixture (minus methionine), 0.5 μL of L-[^{35}S]-methionine; 0.2 μL of TNT SP6 RNA polymerase, 0.2 μL of 40 U/μL RNA inhibitor RNaseOUT, 1.0 μL of 200 ng/μL template DNA with SP6 promoter, and water to 10 μL.
3. Incubate the reaction at 30°C for 90 min.
4. In the meantime pour a 10% polyacrylamide mini gel (thickness 0.75 mm).
5. Mix 1 μL of the TNT lysate (Step 3) with 9 μL of water and 2 μL of 6× SDS gel loading buffer and boil the samples for 3 min at 100°C.
6. Load samples on gel and separate the proteins by applying 180 V for 60 min.
7. After electrophoresis is complete, incubate the gel in 20 mL of fixing solution for 30 min.
8. Transfer the gel to a 5% glycerol solution and incubate for at least 2 h (or better overnight).
9. Vacuum dry the gel onto a 3 MM Whatman paper and expose to X-ray film for 24 h to visualize in vitro translated ZFNs (*see* **Fig. 13.1B** and **Note 1**).

3.2. In Vitro Expression Without Radioactive Label

After correct expression of ZFNs has been verified, the proteins are synthesized in the absence of a radioactive marker. This greatly simplifies the ensuing in vitro cleavage reactions as they can be performed in the regular lab.

1. Remove the TNT reaction components from –80°C and store the TNT SP6 RNA polymerase on ice. Thaw the TNT reticulocyte lysate by hand warming and immediately place on ice. The other reagents can be thawed at room temperature and then stored on ice.

2. Mix the reaction components in a 1.5-mL microcentrifuge tube as follows: 5.0 μL of TNT lysate, 0.4 μL of TNT reaction buffer, 0.1 μL of 1 mM amino acid mixture (minus methionine), 0.1 μL of 1 mM amino acid mixture (minus leucine), 0.2 μL of TNT SP6 RNA polymerase, 0.2 μL of 40 U/μL RNA inhibitor RNaseOUT, 1.0 μL of 200 ng/μL template DNA with SP6 promoter, and water to 10 μL.
3. Incubate the reaction at 30°C for 90 min.
4. Store samples on ice until further use (*see* **Note 2**).

3.3. Generation of Target DNA

The DNA target fragment is generated by PCR. For easier analysis, the ZFN target site should be located acentrically, such that two fragments of unequal lengths are generated after cleavage (*see* **Fig. 13.2A**). A target site for a restriction enzyme, such as *Eco*RI, can be included in the target site as a positive control (*see* **Note 3**).

1. Set up PCR reaction: 5 μL of 10× Stratagene Pfu buffer, 1 μL of 10 mM dNTPs, 1 μL of 10 μM primer pLF, 1 μL of 10 μM primer pER, 1 μL of 10 ng/μL template DNA, 1 μL of 2.5 U/μL Pfu DNA polymerase, and water to 50 μL.
2. Run PCR using the following cycling conditions: initial denaturation at 95°C for 2 min; 30 cycles at 95°C for 30 s, 56°C for 30 s, 72°C for 60 s; final elongation at 72°C for 7 min.
3. In the meantime, pour a 1.2% TAE-buffered agarose gel, supplemented with ethidium bromide (100 ng/mL final concentration).

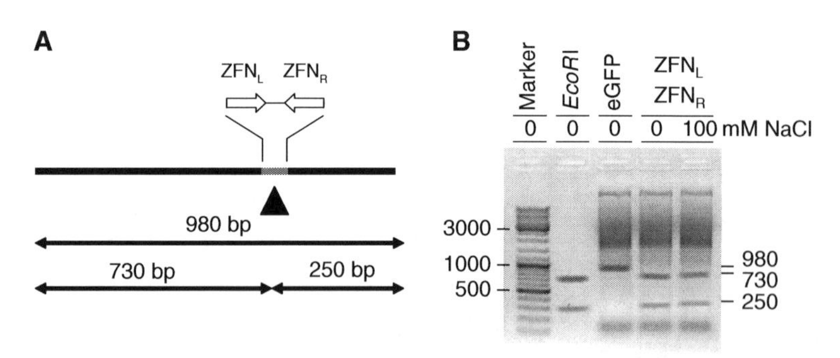

Fig. 13.2. In vitro cleavage assay. (**A**) Schematic. A linear DNA substrate containing the two ZFN target half-sites, ZFN$_L$ and ZFN$_R$, is cleaved into two differently sized products at the site indicated by the *triangle*. The *grey section* indicates the ZFN target site containing an EcoRI site in the spacer. (**B**) Analysis of a cleavage reaction. The linear DNA substrate was incubated with the in vitro-translated ZFNs and the extent of cleavage analyzed by agarose gel electrophoresis. "*Eco*RI" and "eGFP" indicate incubation of the DNA substrate with *EcoRI* (positive control) or in vitro-translated eGFP (negative control). The amount of NaCl in the reaction is indicated on *top* (*see* **Note 4**) and size markers (in base pairs) are shown on both sites of the gel (*see* **Note 5**).

4. Mix the PCR reaction with 10 μL of 6× DNA loading dye and load 20 μL of sample per slot (three slots per sample).
5. Apply 90 V for 30 min to resolve the DNA.
6. After electrophoresis is complete isolate the PCR product and extract the DNA using a gel extraction kit. Elute the DNA in 30 μL of elution buffer (supplied with the kit).
7. Determine the DNA concentration by spectrophotometry and adjust to 200 ng/μL.

3.4. In Vitro Cleavage Assay

As outlined above, the in vitro cleavage assay is performed with unlabelled TNT lysates containing the ZFNs. There is no need to purify the expressed ZFNs.

1. In case the TNT reticulocyte lysates were stored at –20°C, thaw them by hand warming and immediately place on ice.
2. Set up cleavage reaction in a 1.5 mL microcentrifuge tube as follows: 1 μL of 10× NEB4 buffer (New England Biolabs), 1 μL of 1 mg/mL BSA (New England Biolabs), 3 μL of 200 ng/μL DNA target fragment, 1 μL of TNT lysate containing ZFN_L, 1 μL of TNT lysate containing ZFN_R, 1 μL of 1 M NaCl (*see* **Note 4**), and water to 10 μL.
3. Incubate the reaction at 37°C for 90 min.
4. In the meantime pour a 1.2% TAE-buffered agarose gel, supplemented with ethidium bromide (100 ng/mL final concentration).
5. Add 1 μL of 6× DNA loading dye to the cleavage reaction (Step 3) and load the whole reaction on the gel.
6. Apply 90 V for 30 min to separate the DNA fragments.
7. Assess the cleavage reaction by taking a picture of the gel using your gel documentation system (*see* **Fig. 13.2B** and **Note 5**).

4. Notes

1. The apparent molecular weights of ZFN_L and ZFN_R are dissimilar due to variations in the sequence of the *Fok*I domains. The swapping of charged residues in the *Fok*I nuclease domain prevents the formation of ZFN homodimers and was shown to reduce ZFN-associated toxicity (20).
2. The expressed ZFNs can be stored at –20°C for several weeks. However, in order to ensure high ZFN activity, the authors recommend using the protein lysates immediately after in vitro expression in the cleavage reaction.

3. In order to have a positive control for the cleavage reaction, we usually place an *EcoR*I site in the 6-bp spacer separating the two target half-sites (*see* **Fig. 13.2a**). A typical target site thus contains the sequence. 5′-xXXXXXXXXXgaattcNNNNNNNNNn, where X_9 stands for the complementary sequence of the ZFN_L target half-site and N_9 for the ZFN_R target half-site (*see* **Fig. 13.1A**). Because some ZFs tend to reach over to bind an additional nucleotide at the 3′-end of their respective target triplets (15, 24, 25), it is advisable to include the actual 10-bp target half-sites, as indicated by "xX_9" and "N_9n," into the target DNA fragment.
4. The salt concentrations in the reaction can be varied by adding more or less NaCl to the reaction (*see* **Fig. 13.2B**). A high salt concentration, like the one used in this protocol (20 mM Tris–acetate, pH 7.9, 10 mM magnesium acetate, 50 mM potassium acetate, 100 mM NaCl, 1 mM dithiothreitol), is more restrictive and mimics the cellular environment better.
5. In general, agarose gel electrophoresis of an in vitro cleavage assay reveals additional bands (*see* **Fig. 13.2B**). The signals around 3 kb and 100 bp may represent RNA contained in the TNT lysate.

Acknowledgments

The authors thank Anton McCaffrey and Thomas J. Cradick for sharing their know-how on ZFN in vitro cleavage reactions and Tatjana Cornu for carefully reading this manuscript. This chapter is based on work supported by grant LSHB-CT2006-037783 ZNIP of the European Commission's 6th Framework Programme.

References

1. Carroll, D. (2008) Progress and prospects: zinc-finger nucleases as gene therapy agents. *Gene Ther.* **15**, 1463–1468.
2. Cathomen, T. and Joung, J.K. (2008) Zinc-finger nucleases: the next generation emerges. *Mol Ther.* **16**, 1200–1207.
3. Vasquez, K.M., Marburger, K., Intody, Z., and Wilson, J.H. (2001) Manipulating the mammalian genome by homologous recombination. *Proc Natl Acad Sci USA.* **98**, 8403–8410.
4. Maeder, M.L., Thibodeau-Beganny, S., Osiak, A., Wright, D.A., Anthony, R.M., Eichtinger, M., Jiang, T., Foley, J.E., Winfrey, R.J., Townsend, J.A., Unger-Wallace, E., Sander, J.D., Muller-Lerch, F., Fu, F., Pearlberg, J., Gobel, C., Dassie, J.P., Pruett-Miller, S.M., Porteus, M.H., Sgroi, D.C., Iafrate, A.J., Dobbs, D., McCray, P.B., Jr., Cathomen, T., Voytas, D.F., and Joung, J.K. (2008) Rapid "open-source" engineering of customized zinc-finger nucleases for highly

efficient gene modification. *Mol Cell.* **31**, 294–301.
5. Lombardo, A., Genovese, P., Beausejour, C.M., Colleoni, S., Lee, Y.L., Kim, K.A., Ando, D., Urnov, F.D., Galli, C., Gregory, P.D., Holmes, M.C., and Naldini, L. (2007) Gene editing in human stem cells using zinc finger nucleases and integrase-defective lentiviral vector delivery. *Nat Biotechnol.* **25**, 1298–1306.
6. Perez, E.E., Wang, J., Miller, J.C., Jouvenot, Y., Kim, K.A., Liu, O., Wang, N., Lee, G., Bartsevich, V.V., Lee, Y.L., Guschin, D.Y., Rupniewski, I., Waite, A.J., Carpenito, C., Carroll, R.G., Orange, J.S., Urnov, F.D., Rebar, E.J., Ando, D., Gregory, P.D., Riley, J.L., Holmes, M.C., and June, C.H. (2008) Establishment of HIV-1 resistance in CD4+ T cells by genome editing using zinc-finger nucleases. *Nat Biotechnol.* **26**, 808–816.
7. Doyon, Y., McCammon, J.M., Miller, J.C., Faraji, F., Ngo, C., Katibah, G.E., Amora, R., Hocking, T.D., Zhang, L., Rebar, E.J., Gregory, P.D., Urnov, F.D., and Amacher, S.L. (2008) Heritable targeted gene disruption in zebrafish using designed zinc-finger nucleases. *Nat Biotechnol.* **26**, 702–708.
8. Foley, J.E., Yeh, J.R., Maeder, M.L., Reyon, D., Sander, J.D., Peterson, R.T., and Joung, J.K. (2009) Rapid mutation of endogenous zebrafish genes using zinc finger nucleases made by Oligomerized Pool ENgineering (OPEN). *PLoS One.* **4**, e4348.
9. Meng, X., Noyes, M.B., Zhu, L.J., Lawson, N.D., and Wolfe, S.A. (2008) Targeted gene inactivation in zebrafish using engineered zinc-finger nucleases. *Nat Biotechnol.* **26**, 695–701.
10. Morton, J., Davis, M.W., Jorgensen, E.M., and Carroll, D. (2006) Induction and repair of zinc-finger nuclease-targeted double-strand breaks in Caenorhabditis elegans somatic cells. *Proc Natl Acad Sci USA.* **103**, 16370–16375.
11. Bibikova, M., Beumer, K., Trautman, J.K., and Carroll, D. (2003) Enhancing gene targeting with designed zinc finger nucleases. *Science.* **300**, 764.
12. Kim, Y.G., Cha, J., and Chandrasegaran, S. (1996) Hybrid restriction enzymes: zinc finger fusions to Fok I cleavage domain. *Proc Natl Acad Sci USA.* **93**, 1156–1160.
13. Elrod-Erickson, M., Rould, M.A., Nekludova, L., and Pabo, C.O. (1996) Zif268 protein-DNA complex refined at 1.6 A: a model system for understanding zinc finger-DNA interactions. *Structure.* **4**, 1171–1180.
14. Pavletich, N.P. and Pabo, C.O. (1991) Zinc finger-DNA recognition: crystal structure of a Zif268-DNA complex at 2.1 A. *Science.* **252**, 809–817.
15. Dreier, B., Segal, D.J., and Barbas, C.F., 3rd (2000) Insights into the molecular recognition of the 5′-GNN-3′ family of DNA sequences by zinc finger domains. *J Mol Biol.* **303**, 489–502.
16. Beerli, R.R. and Barbas, C.F., 3rd (2002) Engineering polydactyl zinc-finger transcription factors. *Nat Biotechnol.* **20**, 135–141.
17. Bibikova, M., Carroll, D., Segal, D.J., Trautman, J.K., Smith, J., Kim, Y.G., and Chandrasegaran, S. (2001) Stimulation of homologous recombination through targeted cleavage by chimeric nucleases. *Mol Cell Biol.* **21**, 289–297.
18. Smith, J., Bibikova, M., Whitby, F.G., Reddy, A.R., Chandrasegaran, S., and Carroll, D. (2000) Requirements for double-strand cleavage by chimeric restriction enzymes with zinc finger DNA-recognition domains. *Nucleic Acids Res.* **28**, 3361–3369.
19. Miller, J.C., Holmes, M.C., Wang, J., Guschin, D.Y., Lee, Y.L., Rupniewski, I., Beausejour, C.M., Waite, A.J., Wang, N.S., Kim, K.A., Gregory, P.D., Pabo, C.O., and Rebar, E.J. (2007) An improved zinc-finger nuclease architecture for highly specific genome editing. *Nat Biotechnol.* **25**, 778–785.
20. Szczepek, M., Brondani, V., Buchel, J., Serrano, L., Segal, D.J., and Cathomen, T. (2007) Structure-based redesign of the dimerization interface reduces the toxicity of zinc-finger nucleases. *Nat Biotechnol.* **25**, 786–793.
21. Alwin, S., Gere, M.B., Guhl, E., Effertz, K., Barbas, C.F., 3rd, Segal, D.J., Weitzman, M.D., and Cathomen, T. (2005) Custom zinc-finger nucleases for use in human cells. *Mol Ther.* **12**, 610–617.
22. Cornu, T.I., Thibodeau-Beganny, S., Guhl, E., Alwin, S., Eichtinger, M., Joung, J.K., and Cathomen, T. (2008) DNA-binding specificity is a major determinant of the activity and toxicity of zinc-finger nucleases. *Mol Ther.* **16**, 352–358.
23. Handel, E.M., Alwin, S., and Cathomen, T. (2009) Expanding or restricting the target site repertoire of zinc-finger nucleases: the inter-domain linker as a major determinant of target site selectivity. *Mol Ther.* **17**, 104–111.
24. Isalan, M., Choo, Y., and Klug, A. (1997) Synergy between adjacent zinc fingers in sequence-specific DNA recognition. *Proc Natl Acad Sci USA.* **94**, 5617–5621.
25. Mandell, J.G. and Barbas, C.F., 3rd (2006) Zinc finger tools: custom DNA-binding domains for transcription factors and nucleases. *Nucleic Acids Res.* **34**, W516–W523.

Chapter 14

Quantification of Zinc Finger Nuclease-Associated Toxicity

Tatjana I. Cornu and Toni Cathomen

Abstract

The recent development of artificial zinc finger nucleases (ZFNs) for targeted genome editing has opened a broad range of possibilities in biotechnology and gene therapy. The ZFN technology allows a researcher to deliberately choose a target site in a complex genome and create appropriate nucleases to insert a DNA double-strand break (DSB) at that site. Gene editing frequencies of up to 50% in non-selected human cells attest to the power of this technology. Potential side effects of applying ZFNs include toxicity due to cleavage at off-target sites. This can be brought about by insufficient specificity of DNA binding, hence allowing ZFN activity at similar target sequences within the genome, or by activation of the ZFN nuclease domains before the nuclease is properly bound to the DNA. Here, we describe two different methods to quantify ZFN-associated toxicity: the genotoxicity assay is based on quantification of DSB repair foci induced by ZFNs whereas the cytotoxicity is based on assessing cell survival after application of ZFNs.

Key words: Zinc finger nucleases, toxicity, genotoxicity, γ-H2AX, cytotoxicity, cell survival assay.

1. Introduction

Zinc finger nucleases (ZFNs) are artificial enzymes designed to insert a site-specific cut in a complex genome with high specificity (1, 2). The engineered DNA-binding domains consist of three or four small zinc finger (ZF) motifs, which confer specificity of DNA binding. The structure of the transcription factor Zif268, which contains three ZF motifs, revealed that the protein–DNA interaction is brought about by three to four residues located at the amino-terminal end of the zinc finger motif that interacts with three to four bases in the major groove of the DNA double helix (3, 4). The second element in a ZFN is the non-specific cleavage domain of the *Fok*I restriction endonuclease, which is fused via

J.P. Mackay, D.J. Segal (eds.), *Engineered Zinc Finger Proteins*, Methods in Molecular Biology 649,
DOI 10.1007/978-1-60761-753-2_14, © Springer Science+Business Media, LLC 2010

a short linker to the C-terminus of the ZF domain. Because the *Fok*I cleavage domain has to dimerize to become active, two ZFN subunits are typically designed to recognize the target sequence in a tail-to-tail conformation (5). The C-terminal *Fok*I domains (6) are positioned in close proximity to each other on the same face of the DNA helix by separating the two ZFN target half-sites by spacers of 5, 6, 7, or 16 bp (5, 7, 8). Although the efficiency of ZFN cleavage at the target site can be as high as 50% (9, 10), the frequency of off-target cleavage has not been determined yet for any engineered ZFN. Off-target activity can induce high toxicity in human cells (11–16) and can be due to different reasons: (i) Low DNA-binding specificity: the protein does not discriminate between the genuine target sequence and the similar sequences in the genome. (ii) Cleavage at homodimeric target sites or target half-sites: the *Fok*I dimerization domain does not discriminate between a homodimeric and a heterodimeric partner. Moreover, at high ZFN concentrations the formation of an active dimer might be possible in solution with only one subunit properly bound to DNA. (iii) Ambiguity of the linker domain: ZFN subunits cleave at similar sequences that are separated by non-canonical spacers. Different approaches have been employed to overcome those limitations: (i) A novel zinc finger engineering platform called Oligomerized Pool ENgineering (OPEN) has been shown to allow for the generation of highly specific DNA-binding domains that work well in the context of ZFNs (10, 17). (ii) Rational protein modeling combined with computational energy calculations have been used to create variant *Fok*I domains that favor the formation of heterodimers (12, 14). (iii) Linker variants have been identified that permit the formation of active dimers only on defined spacer lengths (7, 8).

Simple assays to assess and quantify ZFN activity and ZFN-associated toxicity have been crucial to advance the ZFN technology to its current state. Here, we describe in detail two simple transfection-based assays to quantify ZFN-associated toxicity. The cytotoxicity assay can be used to determine how ectopic ZFN expression affects cell survival (8, 10, 11, 15). To this end, the fate of transfected HEK293T cells, which over-express a ZFN of interest, are followed over time. To mark transfected cells, an expression plasmid for DsRed-Express (REx), a red fluorescent marker protein, is co-transfected. The cell survival frequency is determined by simply calculating the fraction of REx-positive cells at day 5 as compared to 30 h after transfection (*see* **Fig. 14.1a**).

The genotoxicity assay is based on the formation of nuclear repair foci induced by DNA double-strand breaks (DSBs). One of the earliest responses after the detection of a DSB by cellular DNA repair factors is the phosphorylation of histone H2AX (γ-H2AX) and the formation of repair foci at the site of the break (18). Consequently, ZFN-associated genotoxicity can be

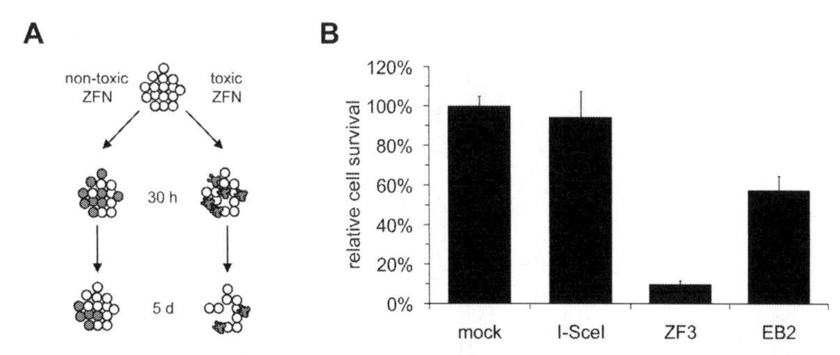

Fig. 14.1. Cytotoxicity assay. (**A**) Cartoon of cytotoxicity assay. HEK 293T cells are co-transfected with expression plasmids encoding a ZFN and the DsRed-Express (REx) fluorescent marker protein (*gray cells*). The "cell survival rate" is calculated as the ratio between REx-positive cells at day 5 as compared to REx-positive cells 30 h after transfection. (**B**) Typical result. The graph shows the cell survival rate after over-expression of a toxic (ZF3-N; (13)) and a non-toxic (EB2-N; (11)) ZFN. "Mock" and "I-SceI" indicate transfection with an empty vector or an I-*Sce*I expression plasmid (12), which is considered to be non-toxic under the conditions used.

determined by quantitatively assessing the level of γ-H2AX in response to over-expression of the nucleases in HT-1080 cells by flow cytometry (11, 12). Alternatively, the number of repair foci per nucleus can be determined by counting the foci after visualization by immunofluorescence, as described in (14).

2. Materials

2.1. Cytotoxicity Assay

1. Plasmids: Expression plasmids for ZFN, as described in (11), pDsRed-Express-N1 was originally generated by Clontech. The plasmids and the corresponding maps can be obtained upon request.
2. HEK 293T cells. These cells are a highly transfectable derivative of HEK 293 cells in which the temperature-sensitive gene for SV40 large T antigen was inserted (ATCC). The cells are maintained in Dulbecco's Modified Eagle Medium (DMEM with high glucose, Gibco/Invitrogen) supplemented with 10% fetal bovine serum (FBS, Gibco/Invitrogen) and penicillin/streptomycin (100× solution, Gibco/Invitrogen).
3. Trypsin solution (0.05% trypsin–EDTA, Gibco/Invitrogen).
4. CO_2 incubator with 5% CO_2 atmosphere, such as Heraeus Hera Cell 240.
5. 2× BES-buffered saline (BBS) for transfection: dissolve 4.28 g of BES, 6.4 g of NaCl, and 0.108 g of Na_2HPO_4 (or 0.161 g of $Na_2HPO_4{\cdot}7H_2O$) in 360 mL of double distilled

water (ddH_2O). Adjust the pH to 6.96 with HCl at room temperature (RT) before adjusting the volume to 400 mL with ddH_2O. Passage solution through a 0.2-μm filter and prepare 5-mL aliquots, which can be stored at –20°C.

6. 2.5 M $CaCl_2$: dissolve 13.5 g of $CaCl_2 \cdot 6H_2O$ (or 7.35 g of $CaCl_2 \cdot 2H_2O$) in 20 mL of ddH_2O. Filter sterilize (0.2 μm) and prepare 1 mL aliquots, which can be stored at –20°C.
7. FACS Buffer: PBS (phosphate-buffered saline) supplemented with 20% FBS.
8. Flow cytometry tubes, e.g., from BD Bioscience.
9. Flow cytometer, such as FacsCalibur (BD Bioscience), and software for analysis, such as CellQuest (BD Bioscience).

2.2. Genotoxicity Assay

1. Plasmids: Expression plasmids for ZFN and EGFP-tubulin, as described in (11). The plasmids and the corresponding maps can be obtained upon request.
2. HT-1080 cells. These cells show minimal background staining for γ-H2AX. The cells are maintained in Dulbecco's Modified Eagle Medium (DMEM with high glucose, Gibco/Invitrogen), supplemented with 10% fetal bovine serum (FBS, Gibco/Invitrogen), and penicillin/streptomycin (100× solution, Gibco/Invitrogen).
3. Lipofectamine 2000 (Invitrogen).
4. Opti-MEM (Gibco/Invitrogen).
5. Ice-cold 100% EtOH.
6. Blocking buffer: PBS with 3% bovine serum albumin.
7. Antibodies: Mouse monoclonal anti-phospho-Histone H2A.X (Ser139), clone JBW301 (Upstate Cell Signaling Solutions), diluted 1:1,000 in blocking buffer. Secondary antibody: Goat Anti-Mouse SFX Kit, conjugated with Alexa Fluor® 594 (Molecular Probes), diluted 1:200 in blocking buffer.
8. Platform shaker.
9. Tabletop centrifuge.

3. Methods

3.1. Cell Survival Assay

1. Seed 2×10^5 HEK293T in 1 mL of culture medium into each well of a 12-well plate and incubate cells for 24 h. Set up each sample at least in duplicate.
2. Replace culture medium carefully 1 h prior to transfection and prepare transfection mix in 1.5-ml microfuge tube: 2 μg of DNA (400 ng of ZFN expression plasmid, 10 ng of

pDsRed-Express-N1, and pUC118 (Stratagene) to 2 μg) with 45 μL of water, 2.5 μL of 2 M $CaCl_2$, and 25 μL of 2× BBS buffer. Mix solution carefully, incubate for 15 min at RT, and add dropwise to the cells. Place cells back into CO_2 incubator for another 24 h.

3. The next day carefully aspirate the supernatant and replace it with fresh cell culture medium.
4. 30 h after transfection, carefully aspirate the medium, wash the cells once with PBS, add 100 μL of trypsin, and place the plate back in the incubator at 37°C for a couple of minutes until the cells detach. Add 200 μL of FACS buffer, resuspend cells by pipetting up and down, transfer 270 μL of the cell solution to a flow cytometry tube, and keep on ice until analysis. Add 1 mL of medium to the remaining cells and place plate in CO_2 incubator.
5. Set up your flow cytometer (*see* **Note 1**).
6. Count 50,000 cells (of Step 4) and evaluate the number of REx-positive cells.
7. Five days after transfection harvest the remaining cells and perform flow cytometry as described above.
8. The relative survival rate is calculated by dividing the number of REx-positive cells after 5 days by the number of positive cells at 30 h post-transfection. These numbers can be normalized to a control experiment involving a non-functional nuclease or I-*Sce*I, as described in (12) (*see* **Fig. 14.1B**).
9. Repeat this assay at least three times in order to generate statistically relevant data.

3.2. Genotoxicity Assay

1. Seed 2 × 10^5 HT-1080 cells in 1 mL of culture medium into each well of a 12-well plate and incubate the cells for 24 h. Cells should reach 90% confluence the next day. Set up each sample at least in duplicate.
2. Replace the culture medium carefully 1 h prior to transfection. Prepare in 1.5-mL microfuge tubes (1): mix 1.6 μg of DNA (400 ng of ZFN expression plasmid, 400 ng of pEGFP-tubulin (*see* **Note 2**), and pUC118 to 1.6 μg) with 100 μL of Opti-MEM (2). Dilute 4 μL of Lipofectamine 2000 in 96 μL of Opti-MEM, mix, and incubate at room temperature for 5 min. Combine carefully the DNA solution with the Lipofectamine 2000 solution, incubate at room temperature for 20 min, and add it dropwise to the cells. Place the cells back into the CO_2 incubator for another 24 h.
3. The next day aspirate the supernatant carefully and replace it with fresh cell culture medium.

4. Harvest the cells 30 h after transfection: wash cells once with PBS, add 100 μL of trypsin, place the plate in the incubator for a couple of minutes until cells detach, resuspend the cells in 1 mL of FACS buffer, and transfer the cell suspension to a 15-ml Falcon tube.
5. Centrifuge for 5 min at 500×*g*, then aspirate the supernatant and resuspend the cells thoroughly in 1 mL of PBS before adding 3 mL of 100% ice-cold EtOH. Vortex immediately!
6. Leave the cell suspension at 4°C for at least 1 h, then transfer to –20°C (at this point it is possible to store the samples for several weeks).
7. For staining, centrifuge the cells for 5 min at 500×*g*, aspirate the supernatant using a yellow tip (*see* **Note 3**), and resuspend the cells in 1 mL of PBS.
8. Centrifuge the cells for 5 min at 500×*g*, aspirate the supernatant using a yellow tip, and resuspend the cells in 300 μL of blocking buffer (*see* **Note 4**).
9. Incubate cells on a rocking shaker for at least 2 h at RT or overnight at 4°C.
10. Centrifuge cells for 5 min at 500×*g* and aspirate the blocking solution using a yellow tip.
11. Add 200 μL of the primary antibody (mouse anti-γ-H2AX) solution and incubate for 2 h at 4°C on a rocking platform shaker.
12. Centrifuge the cells for 5 min at 500×*g*, aspirate the antibody solution, and wash the cells once with 1 mL of blocking buffer.
13. Centrifuge the cells for 5 min at 500×*g* and aspirate the supernatant carefully using a yellow tip.
14. Add 200 μL of the secondary antibody solution and resuspend the cells carefully. Protect the cells from light exposure in the following steps!
15. Incubate the cells for 1 h at 4°C in the dark on rocking platform shaker.
16. Centrifuge the cells for 5 min at 500×*g*, aspirate the antibody solution using a yellow tip, then resuspend the cells in 1 mL of PBS.
17. Centrifuge the cells for 5 min at 500×*g*, aspirate the supernatant using a yellow tip, then resuspend the cell pellet in 250 μL of PBS.
18. Transfer the cell suspension to a new FACS tube and keep the tubes on ice in the dark until analysis.

19. Set up your flow cytometer (*see* **Note 1**).
20. Count 50,000 cells and evaluate the number of EGFP-positive (FL1 channel) and γ-H2AX-positive cells (FL2 channel).
21. To evaluate genotoxicity, the γ-H2AX signal is assessed within the EGFP-positive cell population (*see* **Fig. 14.2a**). The γ-H2AX levels can be presented in either one of two ways: Determine the mean fluorescence intensity (MFI) using CellQuest software and show in combination with

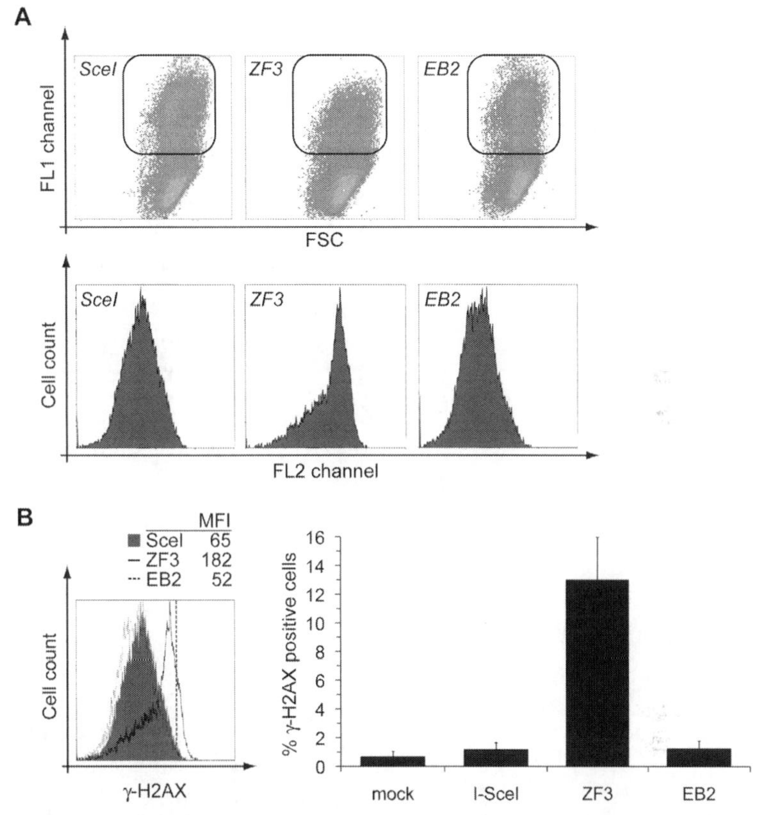

Fig. 14.2. Genotoxicity assay. (**A**) Flow cytometric assessment of cellular γ-H2AX levels. HT-1080 cells were transfected with expression plasmids encoding a toxic (ZF3-N; (13)) or a non-toxic (EB2-N; (11)) ZFN along with an EGFP-tagged tubulin. After 30 h, cells were fixed and permeabilized with EtOH and subsequently stained with a γ-H2AX specific antibody. Transfected (EGFP-positive) cells were identified by plotting FSC vs. FL1 (density plots on *top*) and gated cells were then assessed for γ-H2AX levels by plotting FL2 vs. cell count (histogram plots on the *bottom*). (**B**) Typical result. Cellular γ-H2AX levels can be shown as mean fluorescence intensity (MFI) in combination with an overlayed histogram (*left panel*). Alternatively, a gate can be drawn (*dotted line*; *left panel*) to encompass about 1% of cells transfected with an expression plasmid for a control nuclease, like I-*SceI* (12), which is considered to be non-toxic under the conditions used. The results are then presented as percentage of γ-H2AX-positive cells, which exceed this threshold (*right panel*).

an overlayed histogram (*see* **Fig. 14.2B**, left panel). Alternatively, set a gate that encompasses about 1% of cells transfected with a control plasmid (*see* **Fig. 14.2B**, left panel) and display the intracellular γ-H2AX levels for all samples (*see* **Fig. 14.2B**, right panel).

22. Repeat this assay three times in order to generate statistically relevant data.

4. Notes

1. For a FacsCalibur (BD Bioscience), we typically use the following settings for Voltage and AmpGain: FSC (E-1/4.40), SSC (270/6.23), FL1 (500/1.00), and FL2 (445/1.00). Set compensation between the green and the red channels as "FL1 – 1.2% FL2" and "FL2 – 34.5% FL1" (FSC: forward scatter, SSC: side scatter, FL1: green channel, FL2: red channel). Settings can slightly change, depending on your flow cytometer, the age of the laser, and the cell batch.
2. It is important to use EGFP-tubulin instead of EGFP because the cytoplasmic EGFP will be washed out of the fixed cells once they are permeabilized.
3. Avoid contact of fixed cells with the yellow tip, since they will stick to it. If not possible, coat the tips in FACS solution before use.
4. Keep cells on ice whenever possible.

Acknowledgments

We thank Cem Şöllü for careful reading of the chapter. This chapter is based on work supported by grant CA 311/2 of the Research Priority Programme 1230 (SPP 1230) of the German Research Foundation (DFG).

References

1. Segal, D.J., Cathomen, T., and Joung, J.K. (2008) Zinc-finger nucleases: the next generation emerges. *Mol Ther.* **16**, 1200–1207.
2. Carroll, D. (2008) Progress and prospects: zinc-finger nucleases as gene therapy agents. *Gene Ther.* **15**, 1463–1468.
3. Elrod-Erickson, M., Rould, M.A., Nekludova, L., and Pabo, C.O. (1996) Zif268 protein-DNA complex refined at 1.6 A: a model system for understanding zinc finger-DNA interactions. *Structure.* **4**, 1171–1180.
4. Pavletich, N.P. and Pabo, C.O. (1991) Zinc finger-DNA recognition: crystal structure of a Zif268-DNA complex at 2.1 A. *Science.* **252**, 809–817.
5. Smith, J., Bibikova, M., Whitby, F.G., Reddy, A.R., Chandrasegaran, S., and Carroll, D. (2000) Requirements for double-strand

cleavage by chimeric restriction enzymes with zinc finger DNA-recognition domains. *Nucleic Acids Res.* **28**, 3361–3369.

6. Bitinaite, J., Wah, D.A., Aggarwal, A.K., and Schildkraut, I. (1998) FokI dimerization is required for DNA cleavage. *Proc Natl Acad Sci USA.* **95**, 10570–10575.
7. Bibikova, M., Carroll, D., Segal, D.J., Trautman, J.K., Smith, J., Kim, Y.G., and Chandrasegaran, S. (2001) Stimulation of homologous recombination through targeted cleavage by chimeric nucleases. *Mol Cell Biol.* **21**, 289–297.
8. Handel, E.M., Alwin, S., and Cathomen, T. (2009) Expanding or restricting the target site repertoire of zinc-finger nucleases: the inter-domain linker as a major determinant of target site selectivity. *Mol Ther.* **17**, 104–111.
9. Lombardo, A., Genovese, P., Beausejour, C.M., Colleoni, S., Lee, Y.L., Kim, K.A., Ando, D., Urnov, F.D., Galli, C., Gregory, P.D., Holmes, M.C., and Naldini, L. (2007) Gene editing in human stem cells using zinc finger nucleases and integrase-defective lentiviral vector delivery. *Nat Biotechnol.* **25**, 1298–1306.
10. Maeder, M.L., Thibodeau-Beganny, S., Osiak, A., Wright, D.A., Anthony, R.M., Eichtinger, M., Jiang, T., Foley, J.E., Winfrey, R.J., Townsend, J.A., Unger-Wallace, E., Sander, J.D., Muller-Lerch, F., Fu, F., Pearlberg, J., Gobel, C., Dassie, J.P., Pruett-Miller, S.M., Porteus, M.H., Sgroi, D.C., Iafrate, A.J., Dobbs, D., McCray, P.B., Jr., Cathomen, T., Voytas, D.F., and Joung, J.K. (2008) Rapid "open-source" engineering of customized zinc-finger nucleases for highly efficient gene modification. *Mol Cell.* **31**, 294–301.
11. Cornu, T.I., Thibodeau-Beganny, S., Guhl, E., Alwin, S., Eichtinger, M., Joung, J.K., and Cathomen, T. (2008) DNA-binding specificity is a major determinant of the activity and toxicity of zinc-finger nucleases. *Mol Ther.* **16**, 352–358.
12. Szczepek, M., Brondani, V., Buchel, J., Serrano, L., Segal, D.J., and Cathomen, T. (2007) Structure-based redesign of the dimerization interface reduces the toxicity of zinc-finger nucleases. *Nat Biotechnol.* **25**, 786–793.
13. Alwin, S., Gere, M.B., Guhl, E., Effertz, K., Barbas, C.F., 3rd, Segal, D.J., Weitzman, M.D., and Cathomen, T. (2005) Custom zinc-finger nucleases for use in human cells. *Mol Ther.* **12**, 610–617.
14. Miller, J.C., Holmes, M.C., Wang, J., Guschin, D.Y., Lee, Y.L., Rupniewski, I., Beausejour, C.M., Waite, A.J., Wang, N.S., Kim, K.A., Gregory, P.D., Pabo, C.O., and Rebar, E.J. (2007) An improved zinc-finger nuclease architecture for highly specific genome editing. *Nat Biotechnol.* **25**, 778–785.
15. Pruett-Miller, S.M., Connelly, J.P., Maeder, M.L., Joung, J.K., and Porteus, M.H. (2008) Comparison of zinc finger nucleases for use in gene targeting in mammalian cells. *Mol Ther.* **16**, 707–717.
16. Porteus, M.H. and Baltimore, D. (2003) Chimeric nucleases stimulate gene targeting in human cells. *Science.* **300**, 763.
17. Foley, J.E., Yeh, J.R., Maeder, M.L., Reyon, D., Sander, J.D., Peterson, R.T., and Joung, J.K. (2009) Rapid mutation of endogenous zebrafish genes using zinc finger nucleases made by Oligomerized Pool ENgineering (OPEN). *PLoS ONE.* **4**, e4348.
18. Rogakou, E.P., Boon, C., Redon, C., and Bonner, W.M. (1999) Megabase chromatin domains involved in DNA double-strand breaks in vivo. *J Cell Biol.* **146**, 905–916.

Chapter 15

A Rapid and General Assay for Monitoring Endogenous Gene Modification

Dmitry Y. Guschin, Adam J. Waite, George E. Katibah, Jeffrey C. Miller, Michael C. Holmes, and Edward J. Rebar

Abstract

The development of zinc finger nucleases for targeted gene modification can benefit from rapid functional assays that directly quantify activity at the endogenous target. Here we describe a simple procedure for quantifying mutations that result from DNA double-strand break repair via non-homologous end joining. The assay is based on the ability of the Surveyor nuclease to selectively cleave distorted duplex DNA formed via cross-annealing of mutated and wild-type sequence.

Key words: Zinc finger nuclease (ZFN), designed zinc finger proteins, non-homologous end joining (NHEJ), Surveyor nuclease, Cel 1, genome modification.

1. Introduction

Designed zinc finger nucleases (ZFNs) promise to broadly enable genome engineering of higher eukaryotes. Recent studies have demonstrated the ability of these proteins to catalyze efficient modification of endogenous genes (up to 50% (1)) in diverse species and cell types (1–13). Designed ZFNs may be developed and assayed using a variety of methods (for example *see* (4, 7, 14–17)), but to be useful for genome engineering they must efficiently cleave the target locus in the desired cells. Since context-specific factors such as transcriptional status and chromatin structure can limit the ability of reporters to predict endogenous function, an assay that monitors activity directly at

J.P. Mackay, D.J. Segal (eds.), *Engineered Zinc Finger Proteins*, Methods in Molecular Biology 649,
DOI 10.1007/978-1-60761-753-2_15, © Springer Science+Business Media, LLC 2010

the endogenous target provides the most reliable means for identifying and comparing ZFN candidates.

Genome engineering strategies generally take advantage of one of the two major DNA repair pathways – homology-directed repair (HDR) or non-homologous end joining (NHEJ) – to modify target sequences (18, 19). The choice of strategy constrains options for quantifying modification. HDR-based approaches, for example, copy sequence from a "donor" DNA into the genome and, therefore, may be designed to introduce features that facilitate quantitation such as a new restriction site or size-resolvable insert (5–7). However, screening for HDR-mediated modifications can be cumbersome given the need to construct and validate a new donor with homology arms specific for each ZFN target site. Moreover, assays that PCR amplify the targeted locus must be carefully designed to avoid artifacts caused by the presence of the (typically) large amount of residual donor DNA.

Donor-free strategies for genome engineering utilize the other major DNA repair pathway, NHEJ, to bring about targeted mutation. These approaches yield a heterogeneous mix of minor insertions and deletions that are consequently more difficult to monitor than the discrete modifications introduced via HDR (for example *see* (20)). Such events may be quantified by sequencing the targeted locus, although cost and throughput considerations can render this approach impractical for quantifying low modification levels and/or a large number of samples. Alternatively, one may monitor disruption of an overlapping restriction site (4, 8, 13). This strategy relies on the ability of the minor mutations introduced by NHEJ to confer resistance to digestion by a restriction enzyme whose target overlaps the site of the ZFN-induced break. Although rapid and simple, this approach lacks generality, as it requires fortuitous positioning of a restriction site at the intended target.

In this chapter, we describe an alternative approach for monitoring the heterogeneous mix of mutations introduced via NHEJ. Our assay uses the Surveyor nuclease (Transgenomic) to selectively digest mismatched duplex, which allows one to detect and estimate the mutated fraction of an amplicon pool after cross-annealing with wild-type sequences (21). **Figure 15.1** provides an overview of the assay procedure in the context of a typical gene modification study. After cellular expression of ZFNs (*see* **Notes 1** and **2**) and subsequent mutagenesis, the targeted locus is PCR amplified (**Section 3.1**). The amplicons are then reannealed, which converts any mutations into mismatched duplexes, and digested with the Surveyor nuclease, which selectively cleaves distorted duplex DNA (**Section 3.2**) (22). Finally, digestion products are resolved via PAGE and quantified using a gel imaging station. Data analysis is performed on the digital gel image

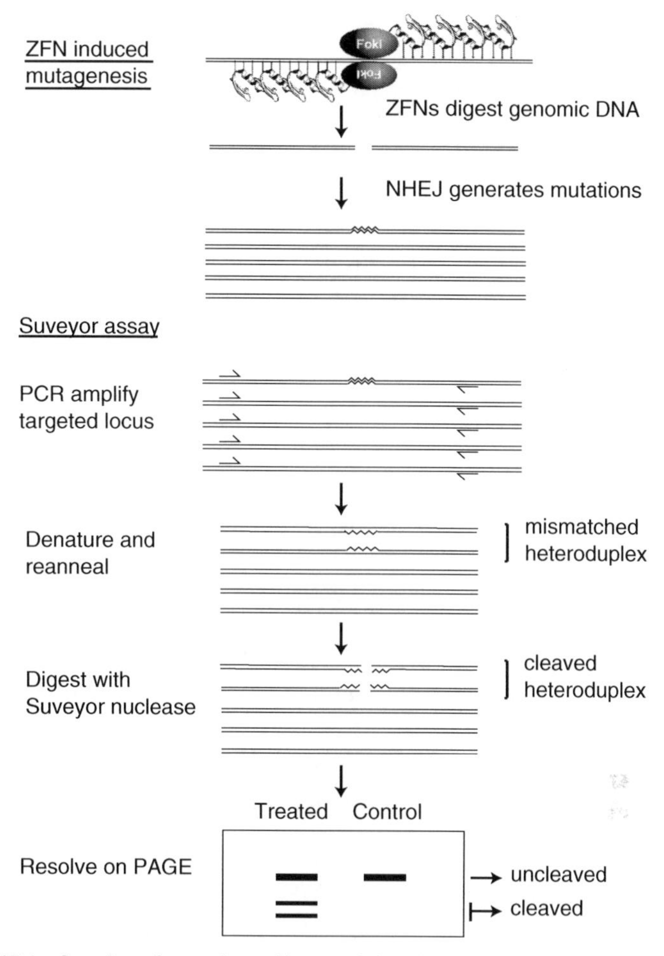

Fig. 15.1. Overview of procedures. *Top panel*: A typical gene modification experiment initiates with ZFN expression in an appropriate cell type. ZFNs cleave their endogenous target, and subsequent repair yields minor mutations in a subset of chromosomes (in this case 20%). A ZFN dimer is pictured schematically at the top of the panel bound to genomic DNA (*straight double lines*) and the resultant cleavage event is depicted by a gap. NHEJ-induced mutations in the genomic DNA are denoted by *jagged double lines*. *Bottom panel*: The Surveyor assay initiates with cell lysis and PCR amplification of the targeted locus. Cross-annealing of mutated and wild-type sequence then converts mutations into mismatched duplexes. The reannealed amplicon is digested with the Surveyor nuclease, which selectively cleaves distorted duplex DNA (22). Digestion products are then resolved via PAGE. *Horizontal half arrows* indicate PCR primers; *straight double lines* and *jagged lines* denote, respectively, wild-type and mutated amplicon sequences. The *rectangle* at bottom schematically depicts a polyacrylamide gel with banding patterns yielded by ZFN-treated and control cells.

(**Section 3.3**). The assay is sensitive down to ~1% gene modification and yields values that correlate with mutation rates obtained from direct cloning and sequencing of the target locus. The assay is rapid, inexpensive, and as convenient as a restriction digest.

2. Materials

2.1. Preparation and PCR Amplification of Genomic DNA

1. QuickExtract DNA extraction solution (Epicentre Biotechnologies, QE09050).
2. AccuPrime taq DNA polymerase (Invitrogen, 12346-086).
3. 96-well PCR reaction plates.
4. PCR thermocycler.
5. PCR primers (for design considerations, *see* **Note 3**).
6. 1% precast agarose gel (Biorad, 161-3063).

2.2. Surveyor Digest and Polyacrylamide Gel Electrophoresis (PAGE)

1. Surveyor nuclease S (Transgenomic, 706025).
2. 10% acrylamide Criterion TBE gel (26-well, BioRad, 345-0053).
3. 10 mg/mL ethidium bromide solution (Sigma-Aldrich, E1510).
4. UV imaging station (e.g., Fluorochem SP, Alpha Innotech).
5. Gel analysis software (e.g., ImageQuant 5.1, GE Healthcare).
6. 5× gel loading buffer: 15% Ficoll-400, 0.05% Orange G (Sigma-Aldrich, 46327 and O3756).
7. Multi flex round gel tips (Sorenson, 13790).
8. 5–50 μL eight-channel pipettor (ThermoFisher, 4610120).
9. 1× TBE gel running buffer: prepare by diluting 10× TBE stock (Fisher BioReagents, BP13334)

3. Methods

3.1. Preparation and PCR Amplification of Genomic DNA

1. Pellet 10^6 ZFN-treated cells at 270×*g* for 5 min.
2. Gently remove the supernatant.
3. Add 100 μL of QuickExtract solution and pipette vigorously (*see* **Note 4**).
4. Transfer the reaction to PCR-compatible tubes and finish the genomic DNA extraction by incubating at 68°C for 15 min, followed by 95°C for 8 min. Then hold at 4°C.
5. Set up a 50 μL PCR reaction using the AccuPrime Taq DNA Polymerase High Fidelity kit: 5 μL of 10× Accuprime buffer II, 2 μL of template DNA from step 4, 2.5 μL of each primer at 10 μM, 0.2 μL of AccuPrime Taq DNA Polymerase High Fidelity, and sterile water to 50 μL (*see* **Note 5**).

6. Run the PCR as follows: 95°C for 5 min; 35× (95°C for 30 s, 60°C for 30 s, 68°C for 40 s); 68°C for 2 min; hold at 4°C.
7. Verify the amplification by running 5 μL of PCR reaction on an agarose gel.

3.2. Surveyor Digest and Gel Electrophoresis

1. Mix 6 μL of 1× Accuprime buffer II with 3 μL of PCR product and incubate as follows: 95°C for 5 min, 95–85°C at –2°C/s, 85–25°C at –0.1°C/s; hold at 4°C. This step melts and randomly reanneals the amplicons, which converts any mutations into mismatched duplex DNA.
2. Store the annealed samples on ice. Add 1 μL of Surveyor nuclease S per sample.
3. Incubate at 42°C for 20 min in a PCR machine.
4. Add to each reaction 3 μL of 5× gel loading buffer.
5. Load the samples on a 10% acrylamide BioRad Criterion TBE gel.
6. Run the gel at 100 V until the Orange G dye front reaches bottom.
7. Stain the gel with 5 μL of ethidium bromide solution in 100 mL of 1× TBE gel running buffer for 10 min, then wash the gel quickly three times with 100 mL of deionized water (<5 s/wash).
8. Capture the gel image with a UV imaging station (for an example gel image, *see* **Fig. 15.2**).

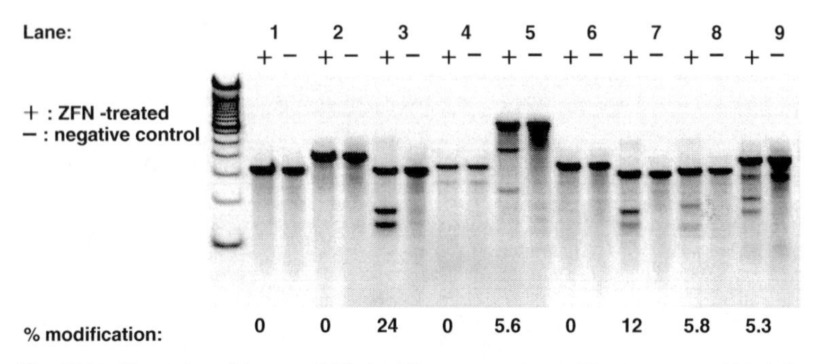

Fig. 15.2. Example gel image yielded by these procedures. The image provides information for estimating mutation rates at nine endogenous loci ("1" to "9"). For each locus, a "+" lane (ZFN-treated cells) and a "–" lane (negative control cells) are supplied. ZFN-dependent bands in lanes 3+, 5+, 7+, 8+, and 9+ yield gene modification estimates of 24, 5.6, 12, 5.8, and 5.3%.

3.3. Data Analysis

1. In order to estimate gene modification levels, we (i) save image files in 16 bit tiff format, (ii) open these files using ImageQuant 5.1, and (iii) generate traces for all lanes

containing visually obvious ZFN-specific bands using the "line" tool widened to encompass the entire lane. This procedure reveals ZFN-driven gene modification as a pair of peaks migrating faster than the parent amplicon and having sizes consistent with cleavage at the ZFN heterodimer target. If the negative control lane contains significant cleavage then this lane is also traced for comparison (*see* **Note 6**). **Figure 15.3A** shows an overlay of lane traces for lanes "8+" (ZFN-treated) and "8–" (negative control) from the gel in **Fig. 15.2**. The ZFN lane trace reveals two additional peaks while the negative control lane lacks these features.

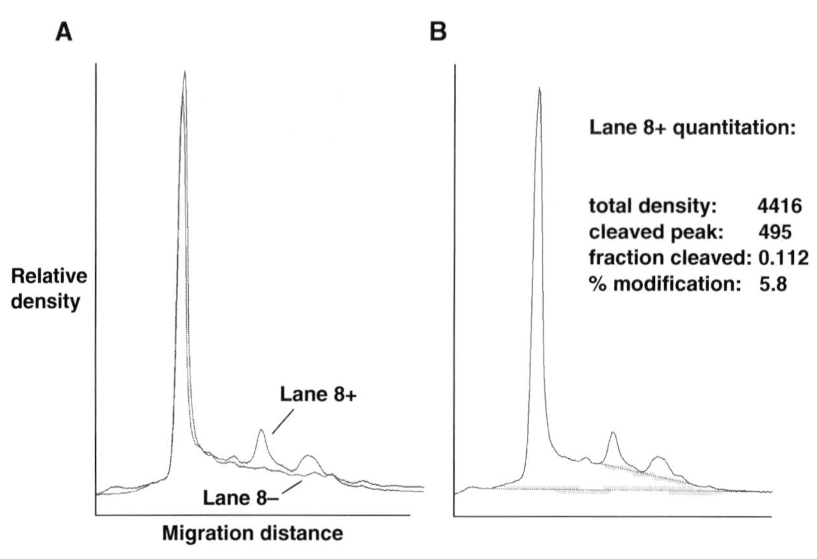

Fig. 15.3. (**A**) Example traces from lane "8+" (ZFN treated) and lane "8–" (negative control) from the gel in **Fig. 15.2**. (**B**) Density profile of lane "8+" with peaks and quantitation. "Total density" and "cleaved peak" refer to density above the lower and upper baselines, respectively.

2. Next, each lane trace is analyzed to determine the fraction of the parent band that has been cleaved at the location of the presumed ZFN-induced modification. In choosing a strategy for lane analysis, an important factor that we consider is the degree of nonspecific cleavage evident in the trace, which can vary depending on the ratio of enzyme to substrate in the digest. If nonspecific cleavage is low, separate baselines are drawn for each parent and cleavage product peak. Each peak is then integrated, and the gene modification level is estimated using the equation

$$\%\text{ gene modification} = 100 \times (1 - (1 - \text{fraction cleaved})^{1/2})$$

where the fraction cleaved is the sum of the cleavage product peaks divided by the sum of the cleavage product and parent peaks. This procedure was used in our prior studies (1, 21).

Often the lane trace will reveal a relatively high level of nonspecific cleavage (up to ~50% of the parent band), which is evident as a large righthand "tail" of the major peak. **Figure 15.3a** provides an example of this behavior. In such cases, the cleavage product peaks ride on top of the nonspecific cleavage tail from the parent amplicon. To quantify traces such as these, we use an analysis strategy that assumes that only the portion of a peak rising above the tail is derived from site-specifically cleaved (and therefore ZFN-modified) parent. Thus lane quantification proceeds as follows: (i) First, a baseline is drawn at background, as judged by the flat, flanking segments of the trace. This is the lower baseline in **Fig. 15.3B**. Density above the baseline is then integrated. The goal of this step is to quantify the total lane density. (ii) Next, baselines for the cleavage product peaks are determined by extrapolating the slope of the tail (upper baseline in **Fig. 15.3B**). Density above this baseline is then integrated and summed. (iii) The fraction cleaved is then calculated by dividing the density from (ii) by the density from (i). (iv) Finally, the apparent gene modification percentage of the original cell pool is calculated using the formula

$$\% \text{ gene modification} = 100 \times (1 - (1 - \text{fraction cleaved})^{1/2})$$

For additional details, *see* **Notes** 7 and **8.**

4. Notes

1. A variety of vectors and delivery methods may be used to obtain sufficient ZFN expression for endogenous target cleavage (1, 3–11, 13, 23). We typically use plasmid vectors derived from pVAX1 (Invitrogen) (5, 6) delivered via nucleofection (Amaxa). We have also achieved robust modification via adenoviral (1), mRNA (3), and integration-deficient lentiviral (8) delivery.
2. We use the K562 cell line for routine screening of ZFNs for human genome modification, as the line is highly transfectable, durable, and easy to maintain in culture.
3. Primers may be designed using a variety of publicly available programs such as Primer3 (24). Input sequences are repeat masked to avoid primer annealing to repeat elements. Settings are chosen to yield primers with melting temperatures of 60°C. Primers are positioned to yield amplicon lengths of between 300 and 400 bp, since digestion products from shorter amplicons can be too faint to accurately quantify, and longer amplicons are prone to excessive nonspecific

digestion by the Surveyor enzyme. ZFN targets are also skewed away from the amplicon center so that a cleavage event by the Surveyor enzyme will yield two distinct product bands during PAGE while maintaining lengths of >100 bp.

4. Although the protocols described here utilize tissue culture samples processed using QuickExtract solution (Epicentre Biotechnologies) as the source of genomic DNA, we note that the Surveyor assay can be used to quantify mutation levels in DNA from any source. For example, in other studies we have successfully used genomic DNA from transgenic fish (3) and isolated using alternative purification kits (21) as template for the assay.
5. We have observed that the lysate can inhibit PCR, so we typically use no more than 2 μL of lysate per 50 μL of reaction.
6. We occasionally observe cleavage products that are not ZFN dependent (*see* **Fig. 15.2** lane "9–"), which we attribute to the presence of polymorphisms or intrinsically distorted sequence. In many such cases, one may still estimate mutation rates if the ZFN-dependent bands migrate away from the ZFN-independent band. PCR primers may also be redesigned to reposition the amplicon to exclude the Surveyor-susceptible sequence.
7. Accurate quantitation rests on four assumptions: (i) complete duplex melting, (ii) random strand reassortment during the reannealing step, (iii) a mutation spectrum of sufficient diversity to prevent significant annealing of identically mutated DNA strands, and (iv) complete cleavage of all mismatches by the Surveyor enzyme. To the extent that conditions deviate from (i) to (iv), the assay will tend to underestimate mutation levels. In practice, we have found that the methods described in this chapter typically agree with mutation levels gauged by sequencing, with underestimates by up to a factor of 2 for a minority of ZFNs.
8. In some instances, a ZFN-treated lane will contain visibly obvious undigested heteroduplex running above the parent amplicon (e.g., the faint band above the parent band in **Fig. 15.2** lanes 3+, 7+, and 8+). If present, this band is quantified separately and the fraction cleaved is calculated as in step 2 of **Section 3.3** using the formula

$$\frac{\text{fraction cleaved} = \text{density from (ii)} + \text{heteroduplex density}}{\text{density from (i)} + \text{heteroduplex density}}$$

Such samples may also be re-assayed using a higher level of Surveyor enzyme or lower level of input DNA in order to achieve complete cleavage.

Acknowledgments

We thank Elo Leung, Xiangdong Meng, Sarah Hinkley, and Lei Zhang for help with the design and assembly of ZFNs; Jianbin Wang and Geoff Friedman for transfections; and Philip Gregory, Susan Abrahamson, and Lei Zhang for helpful comments on the manuscript.

References

1. Perez, E.E., Wang, J., Miller, J.C., et al. (2008) Establishment of HIV-1 resistance in CD4+ T cells by genome editing using zinc-finger nucleases. *Nat Biotechnol.* **26**, 808–816.
2. Cai, C.Q., Doyon, Y., Ainley, W.M., et al. (2009) Targeted transgene integration in plant cells using designed zinc finger nucleases. *Plant Mol Biol.* **69**, 699–709.
3. Doyon, Y., McCammon, J.M., Miller, J.C., et al. (2008) Heritable targeted gene disruption in zebrafish using designed zinc-finger nucleases. *Nat Biotechnol.* **26**, 702–708.
4. Meng, X., Noyes, M.B., Zhu, L.J., Lawson, N.D., and Wolfe, S.A. (2008) Targeted gene inactivation in zebrafish using engineered zinc-finger nucleases. *Nat Biotechnol.* **26**, 695–701.
5. Moehle, E.A., Rock, J.M., Lee, Y.L., et al. (2007) Targeted gene addition into a specified location in the human genome using designed zinc finger nucleases. *Proc Natl Acad Sci USA.* **104**, 3055–3060.
6. Urnov, F.D., Miller, J.C., Lee, Y.L., et al. (2005) Highly efficient endogenous human gene correction using designed zinc-finger nucleases. *Nature.* **435**, 646–651.
7. Maeder, M.L., Thibodeau-Beganny, S., Osiak, A., et al. (2008) Rapid "open-source" engineering of customized zinc-finger nucleases for highly efficient gene modification. *Mol Cell.* **31**, 294–301.
8. Lombardo, A., Genovese, P., Beausejour, C.M., et al. (2007) Gene editing in human stem cells using zinc finger nucleases and integrase-defective lentiviral vector delivery. *Nat Biotechnol.* **25**, 1298–1306.
9. Bibikova, M., Golic, M., Golic, K.G., and Carroll, D. (2002) Targeted chromosomal cleavage and mutagenesis in Drosophila using zinc-finger nucleases. *Genetics.* **161**, 1169–1175.
10. Bibikova, M., Beumer, K., Trautman, J.K., and Carroll, D. (2003) Enhancing gene targeting with designed zinc finger nucleases. *Science.* **300**, 764.
11. Beumer, K., Bhattacharyya, G., Bibikova, M., Trautman, J.K., and Carroll, D. (2006) Efficient gene targeting in Drosophila with zinc-finger nucleases. *Genetics.* **172**, 2391–2403.
12. Wright, D.A., Townsend, J.A., Winfrey, R.J., Jr., et al. (2005) High-frequency homologous recombination in plants mediated by zinc-finger nucleases. *Plant J.* **44**, 693–705.
13. Morton, J., Davis, M.W., Jorgensen, E.M., and Carroll, D. (2006) Induction and repair of zinc-finger nuclease-targeted double-strand breaks in *Caenorhabditis elegans* somatic cells. *Proc Natl Acad Sci USA.* **103**, 16370–16375.
14. Carroll, D., Morton, J.J., Beumer, K.J., and Segal, D.J. (2006) Design, construction and in vitro testing of zinc finger nucleases. *Nat Protoc.* **1**, 1329–1341.
15. Mandell, J.G. and Barbas, C.F., 3rd. (2006) Zinc Finger Tools: custom DNA-binding domains for transcription factors and nucleases. *Nucleic Acids Res.* **34**, W516–W523.
16. Sander, J.D., Zaback, P., Joung, J.K., Voytas, D.F., and Dobbs, D. (2007) Zinc Finger Targeter (ZiFiT): an engineered zinc finger/target site design tool. *Nucleic Acids Res.* **35**, W599–W605.
17. Porteus, M.H. and Baltimore, D. (2003) Chimeric nucleases stimulate gene targeting in human cells. *Science.* **300**, 763.
18. Jasin, M. (1996) Genetic manipulation of genomes with rare-cutting endonucleases. *Trends Genet.* **12**, 224–228.
19. Valerie, K. and Povirk, L.F. (2003) Regulation and mechanisms of mammalian double-strand break repair. *Oncogene.* **22**, 5792–5812.
20. Perez, E., Jouvenot, Y., Miller, J.C., et al (2006) Towards gene knock out therapy for AIDS/HIV: targeted disruption of CCR5 using engineered zinc finger protein nucleases (ZFNs). American Society of Gene Therapy, Baltimore.

21. Miller, J.C., Holmes, M.C., Wang, J., et al. (2007) An improved zinc-finger nuclease architecture for highly specific genome editing. *Nat Biotechnol.* **25**, 778–785.
22. Qiu, P., Shandilya, H., D'Alessio, J.M., O'Connor, K., Durocher, J., and Gerard, G.F. (2004) Mutation detection using Surveyor nuclease. *Biotechniques.* **36**, 702–707.
23. Santiago, Y., Chan, E., Liu, P.Q., et al. (2008) Targeted gene knockout in mammalian cells by using engineered zinc-finger nucleases. *Proc Natl Acad Sci USA.* **105**, 5809–5814.
24. Rozen, S. and Skaletsky, H. (2000) Primer3 on the WWW for general users and for biologist programmers. *Methods Mol Biol.* **132**, 365–386.

Chapter 16

Engineered Zinc Finger Proteins for Manipulation of the Human Mitochondrial Genome

Michal Minczuk

Abstract

Currently there are no effective methods to manipulate or modify particular sequences in mammalian mitochondrial DNA (mtDNA) within cells. The availability of such methods would be of great value for basic mitochondrial research and would aid in development of therapies for diseases linked with mutations in mtDNA. Engineered zinc finger proteins have been used as powerful tools for intervening in nuclear gene expression and modifying nuclear DNA in a sequence-specific manner. Here, basic methods that are helpful in adapting the engineered zinc finger technology for targeting the DNA in mitochondria are presented with the main emphasis on mitochondrial import of zinc finger proteins.

Key words: Mitochondrial import, nuclear export signal, mitochondrial diseases, gene therapy, engineered zinc fingers, zinc finger nuclease.

1. Introduction

Mammalian mitochondria have their own small (16.6 kb) circular genome that encodes essential subunits of the oxidative phosphorylation machinery. Correct expression of mitochondrial DNA (mtDNA) is crucial for mitochondrial function, and therefore mtDNA mutations can have severe consequences for the cell by disrupting oxidative phosphorylation.

Currently, more than 200 point mutations and large-scale rearrangements in human mtDNA are known that are associated with a wide spectrum of clinical manifestations (1, 2). The inability to genetically transform mammalian mitochondria limits gene therapeutic approaches to mtDNA diseases and is a

J.P. Mackay, D.J. Segal (eds.), *Engineered Zinc Finger Proteins*, Methods in Molecular Biology 649,
DOI 10.1007/978-1-60761-753-2_16,

major impediment to basic research on mitochondrial biogenesis. Therefore there is a pressing need to alleviate this obstacle to progress in human mitochondrial biology.

Because delivering wild-type copies of DNA into mitochondria in a heritable manner has not yet been achieved, alternative approaches for the treatment of mtDNA-related diseases are being sought. One of these takes advantage of the fact that each cell harbors hundreds of mtDNA molecules, so that mutant mtDNAs generally co-exist with wild-type mtDNA – a phenomenon known as heteroplasmy. In heteroplasmic cells the phenotype of a pathogenic mtDNA mutation is determined by the ratio of mutant and wild-type genomes, and pathology is only observed when the proportion of mutated mtDNA exceeds a threshold. As cells have a tendency to maintain a fixed number of mtDNA molecules (3), one treatment option is the selective elimination of mutated mtDNA, thus allowing the cell to repopulate cells with wild-type mtDNA molecules and alleviate the defective mitochondrial function that underlies mtDNA diseases (4).

Efficient, rapid, and selective elimination of mtDNA has been achieved by targeting restriction endonucleases (RE) to mitochondria in cultured cells (5–7) and in vivo in mouse models harboring non-pathogenic variants of mtDNA (8, 9). However, there is no appropriate RE for the vast majority of pathological point mutations, which limits the usefulness of this approach. Furthermore, mtDNA deletions, where a large section of the mitochondrial genome is lost, are also a common cause of sporadic mtDNA disease. In these cases the same mtDNA deletion is present in all cells of an affected tissue and many of these deletions are flanked by short direct repeats (10). Mitochondrial DNA deletions are invariably heteroplasmic and tend to require a lower threshold (~60%) than point mutations for a biochemical defect to manifest (10). Importantly, targeting of mtDNA deletions with REs is not possible since the deletion site retains one of the direct repeats and hence is indistinguishable from wild-type mtDNA.

One way to circumvent this problem is to use sequence-specific nucleases that can be designed to cleave any target sequence in the mitochondrial genome. Zinc finger technology allows the engineering of zinc finger proteins (ZFPs) that can bind any predetermined DNA sequence (11). Fusing zinc fingers to a nuclease domain creates a zinc finger nuclease (ZFN) that can cleave DNA adjacent to the specific ZFP binding site, thus providing virtually universal sequence specificity (12). Furthermore, ZFPs can be fused with various other effector domains (e.g., methyltransferases, integrases) in order to target their DNA-modifying activities to specific DNA sequences (11).

We have shown previously that zinc finger technology could be used to target and alter human mtDNA in a sequence-specific manner (13). We developed an efficient method to deliver engineered zinc fingers to mitochondria and demonstrated the selective binding of mitochondria-specific zinc fingers to mtDNA in human cells (13). We have also shown that ZFNs can be efficiently transported into mitochondria. Furthermore, a novel variant of the ZFN that carries two cleavage domains linked in the same protein (single-chain ZNF, *see* **Fig. 16.1B**), expressed in heteroplasmic cells, selectively degraded mutant mtDNA harboring the m.8993T>G mutation responsible for two mitochondrial diseases – maternally inherited Leigh's syndrome and neurogenic muscle weakness, ataxia, and retinitis pigmentosa. Thus we provided proof-of-principle that ZFN technology can be used to selectively deplete mutant mtDNA while sparing the

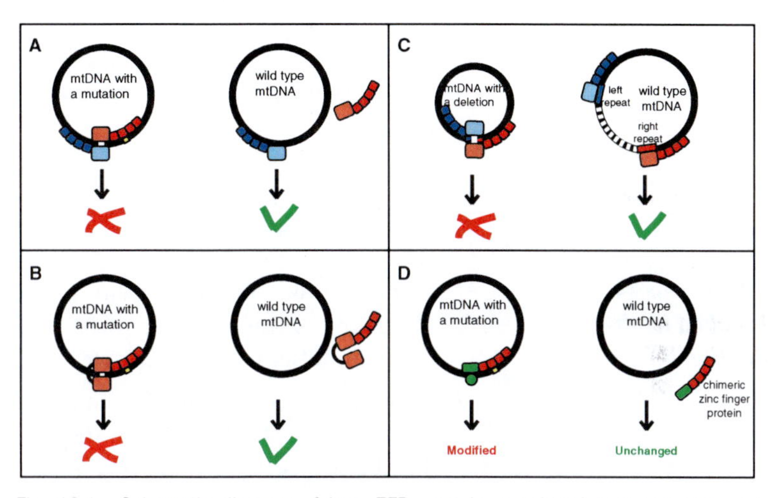

Fig. 16.1. Schematic diagram of how ZFPs can be employed to target and modify the mitochondrial genome with the aim of eliminating mutated mtDNA molecules. A pathogenic mitochondrial point mutation is indicated in yellow. *Dashed line* denotes a region of a pathogenic large-scale deletion (**A**) A ZFN heterodimer is bound to the mutated mtDNA target (*left*). Each of the monomeric ZFNs consists of the *Fok*I nuclease (*large square*) domain linked to a ZFP. One of the ZFNs (*red*) is designed to bind to the mutated mtDNA site (*yellow*), whereas its partner ZFN binds a native sequence on the opposite DNA strand (*blue*). The dimerization of the *Fok*I domains results in a DNA cleavage and the elimination of mutant mtDNA. In the case of wild-type mtDNA (*right*), the mutation-specific ZFN does not bind the target, precluding formation of a heterodimer and DNA cleavage. (**B**) A variant ZFN consisting of two *Fok*I nuclease domains tethered together by a long protein linker and fused to a ZFP (single-chain ZFN) is presented. The ZFP is designed to bind exclusively to the mutated mtDNA site; therefore, only mtDNA molecules harboring the mutation are cleaved (*left*) while the wild-type copies are spared (*right*). (**C**) In order to target a large-scale mitochondrial deletion, a pair of ZFNs is designed to bind to mtDNA sequences on the either side of the deletion junction; binding is accompanied by cleavage (*left*). Wild-type mtDNA will be spared as the binding sites for the ZFN monomers are several kilobases apart, thereby preventing dimerization of the *Fok*I nuclease domains and DNA cleavage (*right*). (**D**) A chimeric ZFP is designed to bind and modify mutated mtDNA molecules (*left*), while the wild-type mtDNA is unchanged (*right*). The modification introduced by the ZFP might, for example, affect the transcription or the replication of mutated mtDNA and hence eliminate the mutant mtDNA from a population of heteroplasmic molecules.

wild type (14). **Figure 16.1** shows further examples of how zinc finger proteins can be applied to eliminate mutated mitochondrial genomes form a heteroplasmic population.

In this chapter the basic techniques for assessing the feasibility of using ZFP fusions for modification of mtDNA are described. These include (**Section 3.1**) guidelines for how to design a DNA construct that allows for efficient import of a zinc finger protein into mitochondria (**Section 3.2**), a method for delivery of plasmid DNA that codes for mitochondrially targeted ZFPs to cells in culture, and (**Section 3.3**) methods for routine assessment of mitochondrial import of zinc finger proteins by immunofluorescence and (**Section 3.4**) cell fractionation. Other methods for testing mitochondrially targeted ZFPs such as EMSA assays, in vitro cleavage assays of zinc finger nucleases, assessment of cytotoxicity of ZFPs are not described here as these topics are covered in other chapters of this book (e.g., *see* **Chapters 6, 13,** and **14**).

2. Materials

2.1. Sequences Required for ZFP Constructs

1. F1β subunit of the human mitochondrial ATP synthase (Protein sequence Genbank Acc. No. NP_001677, nucleotide sequence Genbank Acc. No. NM_001686.3).
2. Protein 2 of murine minute virus (Protein sequence Genbank Acc. No. YP_656489.1, nucleotide sequence Genbank Acc. No. U59501.1).
3. Standard mammalian expression vectors, i.e., pcDNA3.1, pTRACER/CMV/BGH (Invitrogen).

2.2. DNA Preparation and Cell Transfection

1. QiaFilter Plasmid Midi Kit (Qiagen, cat. no. 12243).
2. Human osteosarcoma cell line 143B (ATCC no. CRL-8303).
3. High glucose (4.5 g/L) Dulbecco's Modified Eagle's Medium (DMEM) medium containing pyruvate and L-glutamine, supplemented with 10% fetal calf serum (FCS), 50 μg/mL uridine.
4. 1× PBS, without $CaCl_2$ and $MgCl_2$.
5. 1× trypsin solution in HBSS medium.
6. Coated tissue culture dishes 100 × 20 (58 cm^2) and 140 × 20 (145 cm^2).
7. Round 22 mm cover slips (BDH, cat. no. 406/0187/33).
8. Nucleofector II electroporation system (Lonza, cat. no. AAD-1001).
9. Cell Line Nucleofector Kit V (Lonza, cat. no. VCA-1003).

2.3. Immunofluorescence Experiments

1. MitoTracker Red CMXRos (Invitrogen, M-7512) resuspended in DMSO to 1 μM.
2. Extra fine tips tweezers (Ideal-tek, type 5).
2. 6-well tissue culture plates.
3. 4% (v/v) formaldehyde in PBS (stored at 4°C for up to 1 year).
4. 1% (v/v) Triton X-100 in PBS (stored at RT for up to 1 year).
5. 10% (v/v) FCS in PBS prepared the same day.
6. Primary antibody: anti-HA (clone 3F10, Roche, cat. no. 11867423001).
7. Secondary antibody: anti-rat FITC (Abcam, cat. no. ab6840-1).
8. Vectashield mounting medium (Vector Laboratories, cat. no. H-1000).
9. 1 mm thick microscope slides.
10. Clear nail varnish.

2.4. Isolation of Mitochondria and Proteinase K Treatment

1. Centrifuge for Eppendorf tubes with regulated speed and temperature (e.g., Eppendorf 5417R).
2. 1× WB: 1 mM Tris–HCl, pH 7.0, 130 mM NaCl, 5 mM KCl, and 7.5 mM $MgCl_2$ (*see* **Note 6**).
3. 10× IB: 400 mM Tris–HCl, pH 7.6, 250 mM NaCl, 50 mM $MgCl_2$. Dilutions of 1× IB and 0.1× IB in water are also required (*see* **Note 6**).
4. 1 mL Luer syringe with a rubber-tipped plunger (e.g., B.D. Plastipak, Becton Dickinson, cat. no. 300013).
5. 10 mg/mL proteinase K in water. Store at −20°C. A working solution is prepared the same day of use at 40 μg/mL, kept on ice.
6. 20% (v/v) Triton X-100 in water.
7. Standard equipment and reagents for SDS-PAGE: XCell SureLock® Mini-Cell (Invitrogen, cat. no. EI0002); NuPAGE® Novex 4–12% Bis-Tris gel (Invitrogen, cat. no. NP0301); 20× MOPS buffer system (Invitrogen, cat. no. NP0001); and 2× SDS-PAGE loading buffer: 0.5 M Tris–HCl, pH 6.8, 4.4% (w/v) SDS, 20% (v/v) glycerol, 2% (v/v) 2-mercaptoethanol, and 0.4 mg/mL bromophenol blue.
8. Standard equipment and reagents for Western blotting: Trans-blot SD semi-dry transfer cell (Bio-Rad, cat. no. 170-3848); Protran BA 85 membrane with pore size

0.45 μm (Whatman, cat. no. 10 401 196); semi-dry transfer buffer: 25 mM Tris base (do not adjust pH), 192 mM glycine, 20% methanol; 5% non-fat milk in PBS for membrane blocking and incubation with antibodies; 0.05% (v/v) Tween 20 in PBS; and ECL Plus Western Blotting Detection system (GE Healthcare, cat. no. RPN2132).

10. Primary antibodies:
    (a) anti-HA (clone 3F10, Roche, cat. no. 11867423001).
    (b) anti-TFAM (Abcam, cat. no. ab47517).
    (c) anti-Tom22 mAb (Abcam, cat. no. ab10436).
    (d) anti-GAPDH mAb (Abcam, cat. no. ab8245).
11. Secondary antibodies:
    (a) anti-Rabbit IgG HRP (Promega, cat. no. W401B).
    (b) anti-mouse IgG HRP (Promega, W402B).
    (c) anti-Rat IgG HRP (Santa Cruz Biotechnology, cat. no. sc-2065).

3. Methods

3.1. DNA Constructs Encoding Zinc Finger Proteins

Zinc fingers are predominantly DNA-binding motifs adapted to operate in the nucleus. It has been reported that in many zinc finger proteins the nuclear localization signal (NLS) overlaps with specific DNA-binding residues within the zinc finger motif (15, 16). In order to use engineered ZFPs to manipulate mtDNA, they have to be both effectively targeted to mitochondria and at the same time excluded from the nucleus to avoid their binding to nuclear DNA, which could give rise to toxicity (17). The majority of mitochondrial proteins are encoded in nuclear DNA and many of them are imported from the cytoplasm with the aid of a cleavable N-terminal mitochondrial targeting sequence (MTS). The MTSs vary greatly in length and composition and appear to be individually tailored to different proteins (18).

Fusing an MTS to N-termini can deliver exogenous proteins of various kinds to mitochondria. In our initial attempts to import ZFPs to mitochondria by fusing them with an MTS, we observed that longer arrays of zinc fingers (four fingers and more) are difficult to import into mitochondria. Mitochondrial import was also hampered when a functional domain was fused to ZFP suggesting a possible size exclusion effect (13). As a way to circumvent this problem, in addition to the MTS, we incorporated a nuclear export signal (NES) into a zinc finger fusion protein. The addition of an NES overcame the tendency of zinc fingers to localize in the nucleus, and this method proved effective at directing various ZFPs exclusively to mitochondria (13). Below some guidelines for the design of a DNA construct that allows for efficient import of a ZFP to mitochondria are presented:

1. Among various sequences tested, the MTS from the F1β subunit of the human mitochondrial ATP synthase was found to the most efficient in targeting zinc fingers to mitochondria. We routinely use the first 51 aa of the F1β subunit of the human mitochondrial ATP synthase (*see* **Section 2.1**). This sequence must be located at the N-terminus of a ZFP.
2. The NES that is routinely used in our studies is the one from the nonstructural protein 2 of minute virus of mice: the 18-residue sequence VDEMTKKFGTLTIHD-TEK (residues 79–96). We have found that this NES works efficiently in exporting ZFPs from the nucleus when placed either at the C-terminus or in the middle of a ZFP. However, locating the NES at the N-terminus of a ZFP should be avoided as it might hamper the activity of an MTS.
3. In our experiments ZFPs always contain an epitope tag (or a protein domain) that can be easily detected by specific antibodies. We routinely tag ZFPs with the HA epitope (YPYD-VPDYA). The HA epitope is recommended; we have found that some short tags (e.g., myc), or longer domains (e.g., GFP), which are routinely used to facilitate protein detection in other studies, can hamper the mitochondrial import of ZFPs.
4. In order to express ZFPs in mammalian cells standard expression vectors can be used, for example, pcDNA3.1 and pTRACER/CMV/BGH (Invitrogen).

3.2. Cell Transfection

We use electorporation in order to deliver DNA encoding mitochondrially targeted ZFPs to cells in culture. The electroporation system provided by Lonza (called nucleofection) allows for very high transfection efficiencies and direct entry of DNA to the nucleus. The electroporation can be performed in less than an hour and requires minimal optimization.

1. The plasmid encoding the ZFP is purified using the QiaFilter Midi Kit according to the manufacturer's instructions (*see* **Note 1**)
2. Human osteosarcoma 143B cells are passaged for 24 h before the electroporation; this greatly improves transfection efficiency. In order to obtain 80% confluency the next day about 5×10^6 cells are plated on a 145 cm^2 tissue culture dish.

3. Exponentially growing osteosarcoma 143B cells are electroporated with 1 μg plasmid DNA encoding mitochondrially targeted ZFP. For the 143B cell line, Lonza Nucleofector II programme I-13 is used (*see* **Note 2**). The following cell number and plating methods are used for different applications presented below:
   (a) For immunofluorescence studies 0.5×10^6 cells are electroporated and plated on a 100×20 mm (58 cm^2) tissue culture dish with 10 sterile cover slips placed at the bottom.
   (b) For cell fractionation experiments 2×10^6 cells are electroporated and plated on a 140×20 mm (145 cm^2) tissue culture dish.

3.3. Immunofluorescence Studies of Mitochondrially Expressed ZFPs

In some cases we observed an extensive nuclear localization of mitochondrially targeted ZFPs even though an NES was attached to them. Therefore, it is important to individually test the mitochondrial localization of each ZFP intended to be used to modify mtDNA. It is especially important when a protein contains a functional domain that has never been used in mitochondria before as it may contribute to the undesired nuclear localization of a zinc finger fusion. A quick method of routine assessment of mitochondrial import by immunofluorescent colocalization of a ZFP with mitochondria of fixed cells is provided below.

1. The 143B cells are electroporated as described in **Section 3.2** and grown overnight in the supplemented DMEM medium at 37°C and 5% CO_2 in a humidified incubator.
2. 12–24 h after the electroporation, MitoTracker Red CMXRos is added to a final concentration of 200 nM and the samples are incubated at 37°C for 30 min in a humidified incubator.
3. The cover slips are transferred to a clean 6-well plate (one cover slip to each well) using fine tip tweezers and washed for three times 2 min with 2 mL of PBS.
4. The cells are fixed with 1 mL of 4% formaldehyde in PBS for 15 min at RT, then washed for three times 2 min with 2 mL of PBS.
5. The cells are permeabilized with 1 mL of 1% Triton X-100 in PBS for 5 min and washed again three times 2 min with 2 mL of PBS (*see* **Note 3**).
6. 200 μL of 10% FCS in PBS is carefully layered over a cover slip to form a droplet and incubated for 1 h at RT.
7. The blocking solution is removed and primary rat monoclonal anti-HA antibodies (diluted 1:200 in PBS with 10% FCS) are layered as above and incubated for 1 h. The cover

slips are washed for three times 2 min with PBS with gentle rocking.

8. The cover slips are then incubated with 200 μL of FITC-conjugated anti-rat IgG secondary antibodies (diluted 1:200 in PBS with 10% FCS) for 1 h. After the final incubation the cover slips are washed for three times 5 min with PBS with gentle rocking.
9. A small drop of the mounting medium (*see* **Note 4**) is spotted on a microscope slide and a cover slip is placed on the drop (with the side containing the cells facing down). The cover slip is carefully pressed down; note that it is important to avoid sliding of the cover slip on the microscopic slide. Excess of mounting shield is removed using a medical wipe and the cover slip is immobilized to the slide by applying a nail varnish around the edges.
10. The slide is then analyzed using a confocal microscope. The staining pattern of mitochondria labeled by MitoTracker Red CMXRos and protein labeled with antibodies is compared on digitally overlaid images (*see* **Fig. 16.2 b**).

3.4. Biochemical Assessment of Mitochondrial Import of ZFPs

The immunofluorescence analysis presented above (**Section 3.3**) allows one to detect whether or not a protein colocalizes with mitochondria. The procedure cannot verify, however, if a mitochondrially targeted protein localizes inside the mitochondria. The uptake of proteins with an N-terminal MTS to the mitochondrial matrix through the conventional mitochondrial import pathway is accompanied with the cleavage of the MTS by a mitochondrial processing peptidase (18). This process can be assessed by applying the method presented below that consists of small-scale isolation of mitochondria in conjunction with a treatment of isolated mitochondria with proteinase K in order to degrade the proteins that are located outside the mitochondria. The method of mitochondria isolation has been adopted and optimized for a small-scale from the original method reported by Gaines (19).

1. The 143B cells are transfected as described in **Section 3.2** and grown in the supplemented DMEM medium at 37°C and 5% CO_2 in a humidified incubator (*see* **Note 5**).
2. 48 h post-transfection the cells (3–4 × 10^6) are trypsinized and washed twice in PBS (1/20 of the total cell pellet is saved for further analysis by Western blot – *see* point 11).
3. The pellet is washed twice with 1 mL of 1× WB (*see* **Note 6**). For the last wash the cells are transferred to a 1.5 mL Eppendorf tube.
4. The cell pellet (~50–70 μl) is resuspended in half of the cell pellet volume of 0.1× IB, incubated on ice for 5 min, and homogenized using 15 strokes of a plunger of 1 mL

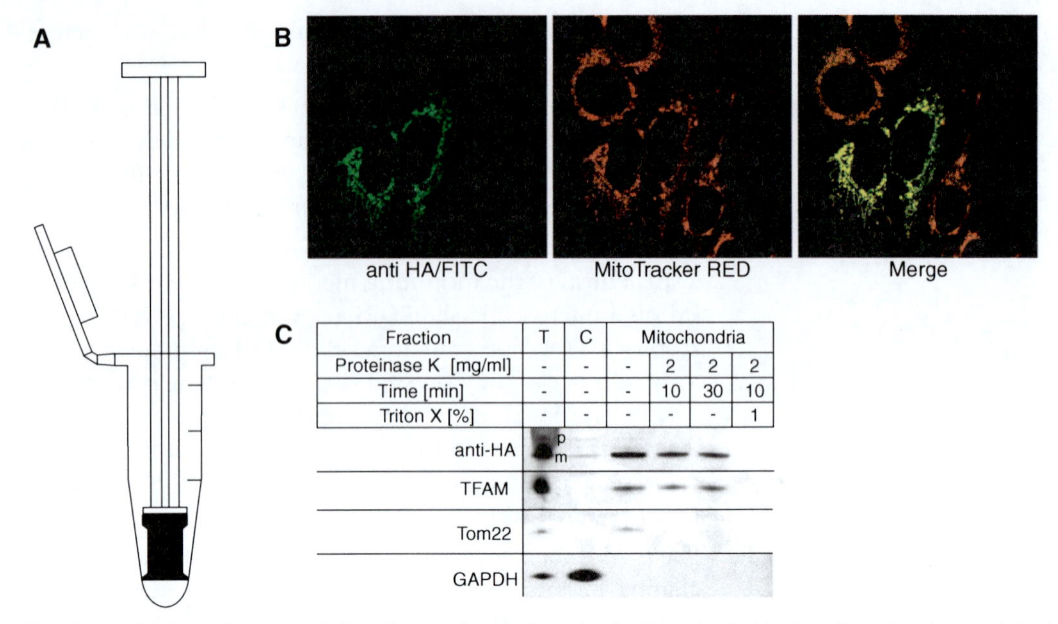

Fig. 16.2. (**A**) Schematic representation of a small-scale 'homogenizer' constructed by inserting of a plunger of 1 mL syringe (with a rubber tip) into a standard 1.5 mL Eppendorf tube. By moving the plunger up and down, suction force is created that efficiently disrupts cells. (**B**) A HA-tagged ZFN was electroporated into 143B cells harboring wild-type mtDNA and its intra-cellular localization was analyzed by immunofluorescence 24 h after transfection. The protein detected with primary antibodies against the HA epitope tag followed by secondary antibodies conjugated to FITC appears in green (*left*). Mitochondria stained with MitoTracker appear in *red* (*middle*). The ZFN exhibits an exclusively mitochondrial staining pattern, as revealed by *yellow* staining on digitally overlaid images (*right*). (**C**) A HA-tagged ZFN was electroporated into 143B cells, the cells were fractionated 48 h post-transfection, and the fractions were analyzed by western blotting using anti-HA mAb. The distribution of the ZFN precursor ('p') and its mature ('m') form in total cell lysate ('T'), cytosolic ('C'), and a mitochondrial fraction treated with proteinase K under various conditions as indicated was compared with the distribution of marker proteins. The precursors of mitochondrial ZFNs found in the mitochondrial fractions were located outside the mitochondria, since they were accessible to protease digestion. In contrast, the mature form of ZFNs was protected and became accessible to proteolysis only after the mitochondria were lysed with Triton X-100. The following endogenous proteins were used as fractionation markers: (a) TFAM: a transcription factor that is localized in the mitochondrial matrix; (b) Tom22: a subunit of the mitochondrial translocase of outer membrane (TOM) complex; and (c) GAPDH: a protein localized in cytoplasm. (Parts of the figure were reproduced from (14) obeying the Oxford University Press copyright policy).

syringe moved up and down in a 1.5 mL Eppendorf tube (*see* **Fig. 16.2a**).

5. The homogenate is immediately mixed with one-ninth of the cell pellet volume of 10× IB and 200 μL of 1× IB is then added to increase the homogenate volume.
6. The cell debris (containing mostly unbroken cells and nuclei) is pelleted by low-speed centrifugations (700×*g* for 3 min at 4°C) and the supernatant is transferred to a new pre-chilled Eppendorf tube. This step is repeated three times.
7. The supernatant is then centrifuged at 18,000×*g* for 2 min. The pellet contains a crude mitochondrial fraction and the

supernatant a cytosolic fraction. The mitochondrial pellet is washed again using 200 μL 1× IB buffer (*see* **Note 7**).

8. The mitochondrial pellet is resuspended in 100 μL of 1× IB (that should result in protein concentration of 1–2 mg/mL) and 9.5 μL aliquots are distributed to six 1.5 mL Eppendorf tubes as presented in **Table 16.1**.

Table 16.1
Treatment of the mitochondrial fraction aliquots (*see* Steps 8–10 of Section 3.4)

Sample no.	1	2	3	4	5	6
Mitochondrial fraction (μL)	9.5	9.5	9.5	9.5	9.5	9.5
Water (μL)	0.5	0.5	–	–	–	–
Proteinase K (μL)	–	–	0.5	0.5	0.5	0.5
Triton-X (μL)	–	–	–	–	–	0.55
Incubation time (min)	0	30	10	20	30	10

9. Water (0.5 μL) is added to the first two tubes and 0.5 μL of proteinase K working solution is added to the remaining four tubes (the final concentration of proteinase K is 2 μg/mL). In addition, 0.55 μL of 20% Triton X-100 is added to one of the tubes already containing proteinase K (*see* **Table 16.1**).
10. The tubes are then incubated at room temperature as presented in **Table 16.1**. At each time point 2× SDS-PAGE loading buffer is added to the reaction and the sample is snap-frozen on dry ice.
11. The protein concentration of the total cell pellet and the cytosolic and mitochondrial fractions is determined using the method of choice (*see* **Note 8**). Then the total, cytosolic and mitochondrial fractions (untreated or treated with proteinase K) are normalized for protein contents and analyzed by Western blot.
12. On the Western blot the difference in size between the mature form, in which the MTS has been cleaved off upon the mitochondrial import, and the precursor ZFP found in the total cell lysate should be visible (*see* **Fig. 16.2C** and **Note 9**).
13. When the isolated mitochondria are incubated with proteinase K the mature form of a ZFP is protected from proteolysis to the same extent as a mitochondrial matrix marker protein (e.g., TFAM). In contrast, the precursor ZFP and proteins associated with the mitochondrial outer

membrane (e.g., Tom22) are degraded by proteinase K present outside the mitochondria. GAPDH can be used as a marker for the cytosolic fraction (*see* **Note 10**).

4. Notes

1. The final plasmid DNA concentration should be equal to or greater than 200 ng/μL as the maximum volume of DNA solution that can be used for the electroporation is 5 μL.
2. The following programmes and Lonza kits were successfully used when other cell lines were transfected with mitochondrially targeted ZFPs: HeLa programme I-13, Cell Line Nucleofector Kit R (Lonza, VCA-1001); Cos-1: programme A-24, Cell Line Nucleofector Kit V; HEK293T: programme A-23, Line Nucleofector Kit V.
3. At this stage fixed and permeabilized cells can be stored at 4°C for up to a month.
4. Mounting medium with DAPI (Vector Laboratories, cat. no. H-1200) can be used in order to visualize nuclei.
5. The fractionation protocol has been optimized for the 143B cells. However, it has been also successfully applied for other mammalian cells lines expressing mitochondrially targeted zinc fingers proteins such as HeLa, Cos-1, and HEK293T.
6. Buffers are pre-chilled to 4°C, contain 2 mM PMSF (Sigma, cat. no. P7626), and 1× proteinase inhibitor cocktail (Roche, EDTA-free, cat. no. 11 873 580 001).
7. For larger scale culture (of more then 5×10^7 cells) further purification of mitochondria can be achieved using a sucrose gradient. The crude mitochondria are resuspended in TES buffer (10 mM Tris–HCl, pH 7.4, 1 mM EDTA, 0.25 M sucrose supplemented with protease inhibitor cocktail (*see* **Note 6**) and 2 mM PMSF), layered on a discontinuous sucrose gradient made by successive layering of 1.5 and 1.0 M sucrose in 10 mM Tris–HCl, pH 7.4, 1 mM EDTA, and centrifuged at 87,390×*g* for 1 h, 4°C in a Beckman Coulter Optima ultracentrifuge in a swing-out rotor MLS50 (Beckman Coulter). The interface fraction between the 1.5 and 1.0 M sucrose layers (mitochondrial fraction) is collected using a 1 mL tip and four volumes of TES buffer is slowly added with gentle vortexing (speed 3 on VORTEX GENIE 2, Scientific Industries). Mitochondria are then collected by centrifugation at 18,000×*g* for

2 min at 4°C and processed as described in **Section 3.4**, point 8.

8. We use a bicinchoninic acid-based colorimetric quantitation method – BCA Protein Assay Kit (Pierce, cat. no. 23227).
9. For some mitochondrially routed ZFPs the precursor protein might not be detected in Western blot. This is probably owing to a very efficient mitochondrial import of these particular constructs. In these cases, a ZFP produced in an in vitro transcription system (e.g., The TNT® T7 Quick Coupled Transcription/Translation System Promega, cat. no. L4610) can be run alongside a mitochondrial fraction in order to compare the masses of the precursor and mature forms.
10. In some of our experiments a fraction of GAPDH has been found to be associated with mitochondria, probably owing to its electrostatic association with the mitochondrial outer membrane as reported previously (20, 21). Mitochondria-bound GAPDH is easily degraded when the mitochondrial fraction is incubated with proteinase K (*see*, for example, Fig. 3 in (13)).

Acknowledgments

This work was supported by the Medical Research Council, UK, and the Federation of European Biochemical Societies Long-Term Fellowship. I would like to thank Michael Murphy for his help with the chapter.

References

1. Taylor, R.W. and Turnbull, D.M. (2005) Mitochondrial DNA mutations in human disease. *Nat Rev Genet.* **6**, 389–402.
2. Schapira, A.H. (2006) Mitochondrial disease. *Lancet.* **368**, 70–82.
3. Tang, Y., Manfredi, G., Hirano, M., and Schon, E.A. (2000) Maintenance of human rearranged mitochondrial DNAs in long-term cultured transmitochondrial cell lines. *Mol Biol Cell.* **11**, 2349–2358.
4. Taylor, R.W., Chinnery, P.F., Turnbull, D.M., and Lightowlers, R.N. (1997) Selective inhibition of mutant human mitochondrial DNA replication in vitro by peptide nucleic acids. *Nat Genet.* **15**, 212–215.
5. Tanaka, M., Borgeld, H.J., Zhang, J., Muramatsu, S., Gong, J.S., Yoneda, M., Maruyama, W., Naoi, M., Ibi, T., Sahashi, K., et al. (2002) Gene therapy for mitochondrial disease by delivering restriction endonuclease SmaI into mitochondria. *J Biomed Sci.* **9**, 534–541.
6. Srivastava, S. and Moraes, C.T. (2001) Manipulating mitochondrial DNA heteroplasmy by a mitochondrially targeted restriction endonuclease. *Hum Mol Genet.* **10**, 3093–3099.
7. Alexeyev, M.F., Venediktova, N., Pastukh, V., Shokolenko, I., Bonilla, G., and Wilson, G.L. (2008) Selective elimination of mutant

mitochondrial genomes as therapeutic strategy for the treatment of NARP and MILS syndromes. *Gene Ther.* **15**, 516–523.

8. Bayona-Bafaluy, M.P., Blits, B., Battersby, B.J., Shoubridge, E.A., and Moraes, C.T. (2005) Rapid directional shift of mitochondrial DNA heteroplasmy in animal tissues by a mitochondrially targeted restriction endonuclease. *Proc Natl Acad Sci USA.* **102**, 14392–14397.
9. Bacman, S.R., Williams, S.L., Hernandez, D., and Moraes, C.T. (2007) Modulating mtDNA heteroplasmy by mitochondria-targeted restriction endonucleases in a 'differential multiple cleavage-site' model. *Gene Ther.* **14**, 1309–1318.
10. Krishnan, K.J., Reeve, A.K., Samuels, D.C., Chinnery, P.F., Blackwood, J.K., Taylor, R.W., Wanrooij, S., Spelbrink, J.N., Lightowlers, R.N., and Turnbull, D.M. (2008) What causes mitochondrial DNA deletions in human cells? *Nat Genet.* **40**, 275–279.
11. Papworth, M., Kolasinska, P., and Minczuk, M. (2006) Designer zinc-finger proteins and their applications. *Gene.* **366**, 27–38.
12. Kim, Y.G., Cha, J., and Chandrasegaran, S. (1996) Hybrid restriction enzymes: zinc finger fusions to Fok I cleavage domain. *Proc Natl Acad Sci USA.* **93**, 1156–1160.
13. Minczuk, M., Papworth, M.A., Kolasinska, P., Murphy, M.P., and Klug, A. (2006) Sequence-specific modification of mitochondrial DNA using a chimeric zinc finger methylase. *Proc Natl Acad Sci USA.* **103**, 19689–19694.
14. Minczuk, M., Papworth, M.A., Miller, J.C., Murphy, M.P., and Klug, A. (2008) Development of a single-chain, quasi-dimeric zinc-finger nuclease for the selective degradation of mutated human mitochondrial DNA. *Nucleic Acids Res.* **36**, 3926–3938.
15. Fernandez-Martinez, J., Brown, C.V., Diez, E., Tilburn, J., Arst, H.N., Jr., Penalva, M.A., and Espeso, E.A. (2003) Overlap of nuclear localisation signal and specific DNA-binding residues within the zinc finger domain of PacC. *J Mol Biol.* **334**, 667–684.
16. Matheny, C., Day, M.L., and Milbrandt, J. (1994) The nuclear localization signal of NGFI-A is located within the zinc finger DNA binding domain. *J Biol Chem.* **269**, 8176–8181.
17. Papworth, M., Moore, M., Isalan, M., Minczuk, M., Choo, Y., and Klug, A. (2003) Inhibition of herpes simplex virus 1 gene expression by designer zinc-finger transcription factors. *Proc Natl Acad Sci USA.* **100**, 1621–1626.
18. Pfanner, N. and Geissler, A. (2001) Versatility of the mitochondrial protein import machinery. *Nat Rev Mol Cell Biol.* **2**, 339–349.
19. Gaines, G.L., 3rd.. (1996) In organello RNA synthesis system from HeLa cells. *Methods Enzymol.* **264**, 43–49.
20. Hartmann, C.M., Gehring, H., and Christen, P. (1993) The mature form of imported mitochondrial proteins undergoes conformational changes upon binding to isolated mitochondria. *Eur J Biochem.* **218**, 905–910.
21. Taylor, S.W., Fahy, E., Zhang, B., Glenn, G.M., Warnock, D.E., Wiley, S., Murphy, A.N., Gaucher, S.P., Capaldi, R.A., Gibson, B.W., et al. (2003) Characterization of the human heart mitochondrial proteome. *Nat Biotechnol.* **21**, 281–286.

Chapter 17

High-Efficiency Gene Targeting in *Drosophila* with Zinc Finger Nucleases

Dana Carroll, Kelly J. Beumer, and Jonathan K. Trautman

Abstract

We describe a method for making targeted double-strand breaks in *Drosophila melanogaster* using zinc finger nucleases (ZFNs). After design and construction of the appropriate coding sequences, synthetic mRNAs for the ZFNs are injected directly into fly embryos. Frequencies of target cleavage and mutagenesis in the range of 1–10% have been achieved at several different loci. A donor DNA carrying desired sequence changes can be incorporated in the injection mix and leads to targeted gene replacement, with particularly good efficiency when the recipient embryos are defective for nonhomologous end joining.

Key words: Gene targeting, Drosophila, DNA repair, zinc finger nucleases, homologous recombination (HR), nonhomologous end joining (NHEJ), embryo injection.

Abbreviations

DSB	Double-strand break
HR	Homologous recombination
NHEJ	Nonhomologous end joining
ZFN	Zinc finger nuclease

1. Introduction

Zinc finger nucleases (ZFNs) have now been in use for quite a number of years as targetable tools for gene manipulation. Many examples are described in this book. The keys to the utility of ZFNs are as follows: (1) DNA recognition by the zinc fingers is very modular, with each finger binding principally to one DNA triplet; (2) because fingers have been identified for many individual triplets and triplet combinations, zinc finger sets can be assembled for many different target sequences; (3) the binding

J.P. Mackay, D.J. Segal (eds.), *Engineered Zinc Finger Proteins*, Methods in Molecular Biology 649,
DOI 10.1007/978-1-60761-753-2_17,

and the cleavage domains are quite separate, so changing binding specificity has little effect on cleavage; (4) cleavage requires dimerization of the cleavage domain, and because the dimer interface is weak, pairs of designed ZFNs are needed to cut a single target; a monomer is not a nuclease; (5) pairs of proteins with three or more fingers each are adequate, in principle, to specify unique sequences in complex genomes.

The first application of ZFNs to a natural genomic target was achieved in the fruit fly, *Drosophila melanogaster* (1, 2). Since that time our lab has generated mutations in five different genes with five different ZFN pairs (3, 4). We have explored experimental parameters that affect the outcomes, including utilization of an introduced donor DNA as a template for repair of the targeted break (2, 3). Most recently we greatly simplified the method for delivery of ZFNs to *Drosophila* (4), taking advantage of procedures for embryo injection that are standard in the research community.

The first step in targeting a new gene is, of course, the choice of that gene. The gene is then searched for plausible ZFN targets. Several approaches are available, and each makes use of existing libraries of zinc fingers. Those fingers can be assembled and tested directly (5, 6), or "improved" through partial randomization and further selection (7, 8). These procedures are described elsewhere in this volume (*see* **Chapters 1, 2** and **3**). It is also now possible to order from Sigma-Aldrich zinc finger sets and ZFN pairs that are made and tested by procedures perfected at Sangamo Biosciences.

All of the ZFNs we have used in *Drosophila* carry three fingers in their binding domains and have been assembled from specific fingers described in the literature for particular DNA triplets, using qualitative assessments of their apparent specificity (5). This approach has proved adequate for *Drosophila* but, for organisms with significantly larger genomes, it may be necessary to achieve increased specificity through increasing the number of zinc fingers and/or applying more stringent selection for specific binding (9, 10).

In this chapter we describe the injection method for ZFN-mediated targeting (4). We assume that the choice of target sequence has been made and that the design, synthesis and possibly testing of the corresponding ZFN coding sequences have already occurred (5). The remaining challenges are effective delivery of the targeting materials to flies and recovery of the desired products. After ZFN cleavage, sequence alterations are produced by inaccurate nonhomologous end joining (NHEJ) or by homologous recombination (HR) with an introduced donor DNA. Both types of products may be desired.

The heat-shock method for ZFN delivery is described in detail in a previous volume of this series, as are more general considerations regarding the use of ZFNs (11). The one addition we would make to the earlier discussion of the heat-shock procedure is that the outcome is affected by the genetic background in which the DSB is induced (12). For example, HR using the donor DNA as a template is reduced in strains deficient in components of that repair pathway, like the Rad51 protein (*spnA* in *Drosophila*). Conversely, the proportion of HR products is increased when NHEJ is impaired, as in mutants lacking DNA ligase IV (*lig4*).

2. Materials

2.1. Fly Stocks

1. For general injections, any wild-type stock will serve as a recipient.
2. To enhance HR, use a stock that carries a null mutation in the *lig4* gene – e.g., $lig4^{169}$ (4, 12, 13). The *lig4* gene lies on the X chromosome. Males carrying this allele and homozygous females are viable and fertile, and stocks seem quite healthy on continuous passage.
3. Other recipients can be used for specific purposes. For example, if the donor DNA has been constructed to carry a visible marker, like the w^+ gene, the recipient must carry the corresponding mutation, w^-.
4. To identify ZFN-induced mutations, candidate flies are crossed to partners that will reveal or facilitate recovery of the new mutations. This is most readily done with stocks carrying deletion mutations (deficiencies in *Drosophila* parlance) that cover the target gene and are carried opposite a marked balancer chromosome. Many such mapped deficiency stocks are available from the Drosophila Stock Center, and additional ones are constructed frequently.

2.2. Molecular Biology

1. Suitable cloning vectors.
   a. We have constructed an entry vector, pENTR-NLS-G-FN, that carries a stuffer fragment that is flanked by *NdeI* and *SpeI* sites and is located between an encoded nuclear localization sequence (NLS) and the FokI cleavage domain (4). Replacement of the stuffer with ZF coding sequences produces a ZFN gene that can be transferred to any desired destination vector (*see* **Note 1**).
   b. A destination vector for in vitro transcription. We use pCS2-DEST, a derivative of pCS2. It carries an SP6 promoter upstream of the insertion site, and an encoded string of A residues (polyA tail) and a *NotI* site downstream.

c. A vector for the donor DNA. We have used pBluescript (Stratagene) and pGEM-T (Promega) with equivalent results. This is only required for gene replacement by HR, not when simple mutagenesis is intended.

2. LR Clonase II Enzyme Mix (Invitrogen).
3. An in vitro transcription kit matched to the destination vector. We typically use AmpliScribe SP6 High Yield Transcription Kit (Epicentre) and the ScriptCap m7G Capping System from the same supplier.
4. E. coli strains
   a. DH5α or other suitable host for general cloning.
   b. DB3.1 (Invitrogen) for destination vectors. It carries the ccdB gene for selection.
6. AquaGenomic DNA extraction reagent (MultiTarget Pharmaceuticals, Salt Lake City, UT; http://www.aquaplasmid.com) for isolating genomic DNA from single flies.
7. Taq DNA polymerase (New England Biolabs) for PCR.
8. Qiagen MinElute columns and Qiagen Plasmid Maxiprep kits (Valencia, CA).
9. Dissecting microscope to observe flies, along with equipment for CO_2 anesthesia (Genesee Scientific, San Diego, CA; http://flystuff.com).
10. Drosophila embryo injection equipment (14).
11. 1× Phosphate Buffered Saline (PBS) (Mediatech).

3. Methods

3.1. DNA and RNA

3.1.1. ZFN RNA

1. Digest both the ZF coding sequences you have produced and pENTR-NLS-G-FN with *Nde*I + *Spe*I (*see* **Note 2**). This assumes you have made the ZF coding sequences by assembly from long oligonucleotides, by mutagenic PCR, or by some other method that yields a double-stranded fragment with an *Nde*I site at the start codon and an *Spe*I site where joining to the cleavage domain will occur. Example sequences are illustrated in a previous publication (11). The following steps apply to both of the two ZFN sequences for a particular target, and should be pursued in parallel.

2. Gel purify the long fragment from both digests. The stuffer in the vector is 633 bp; the vector fragment is 2.9 kb. For three-finger proteins using our standard method and protein framework (5), the insert is 269 bp after digestion.
3. Ligate the ZF coding sequence fragment to the vector, transform into competent DH5α cells, and select on plates containing 50 μg/mL kanamycin.
4. Pick individual colonies and screen for the correct insert by colony PCR. Verify by DNA sequencing.
5. Purify the verified plasmid DNA using any standard miniprep procedure.
6. Transfer the ZFN insert to the pCS2-DEST vector with a Clonase reaction according to the supplier's instructions.
7. Transform the mixture into DH5α cells and select on plates containing 100 μg/mL ampicillin.
8. Screen individual colonies by PCR and verify by DNA sequencing.
9. Grow up cells with verified plasmid and make highly purified DNA with the Qiagen Maxiprep kit.
10. Cut the DNA with *Not*I, which cleaves beyond the 3′ end of the ZFN gene. Check for completeness of digestion by agarose gel electrophoresis (1% agarose works fine).
11. Extract the cut DNA once with phenol/chloroform/isoamyl alcohol (25:24:1, from US Biochemicals) and once with chloroform/isoamyl alcohol (24:1). Add 0.1 volume of 3 M sodium acetate and 2.5 volumes of 100% ethanol. Collect by centrifugation and wash the pellet with 70% ethanol.
12. Dissolve the DNA pellet in Qiagen elution buffer (or RNase-free water; *see* **Note 3**). Determine the DNA concentration with a spectrophotometer, and adjust to 500 μg/mL.
13. Transcribe the resulting linear DNA with SP6 RNA polymerase at 40°C using the AmpliScribe Kit. Treat with RNase-free DNase (provided with the kit) as recommended, then precipitate the RNA with an equal volume of 5 M ammonium acetate. After centrifugation, rinse the pellet with 1 mL of 70% ethanol, dry the pellet, and redissolve in 50 μL of RNase-free water.
14. Determine the RNA concentration with a spectrophotometer. Check the integrity of the RNA by running a 1 μL aliquot on an (1%) agarose gel.
15. Treat ~50 μg of RNA with the AmpliCap reagents for 1 h at 37°C.

16. Place 5 aliquots of the capped RNA (about 10–15 μg each) in separate tubes and add an equal volume of 5 M ammonium acetate. Store at –80° until use.
17. When ready to inject, collect the RNA in one aliquot by centrifugation and redissolve in 7 μL of RNase-free water to a final concentration of ~2 μg/μL.

3.1.2. Donor DNA

1. Design of the donor DNA
   a. The donor DNA must have homology to both sides of the ZFN cleavage site. The extent of homology required for efficient homologous recombination has not been carefully tested, but success has been achieved with a donor having 4.2 kb of total homology and as little as 1 kb in one arm (3, 4).
   b. We have made donor DNAs in which the ZFN recognition sites have been deleted to prevent donor cleavage (2, 3), but this may not be necessary (9).
   c. Markers can be included in the donor DNA to facilitate identification of HR products. For example, insertion of an autonomous w^+ gene or GFP gene would allow visual tracking of the donor.
2. The donor DNA can be made conveniently by PCR amplification and cloning of sequences flanking the genomic ZFN target. How you introduce the desired sequence alteration into the donor will depend on the nature of that alteration, and can be done by standard methods of mutagenesis and cloning.
3. Clone the donor DNA by standard procedures into pBluescript or another vector and verify.
4. Purify the verified plasmid DNA using a Qiagen Maxiprep Kit. Determine the DNA concentration with a spectrophotometer.
5. Take a 90-μg aliquot of DNA and extract once with phenol/chloroform/isoamyl alcohol (25:24:1) and once with chloroform/isoamyl alcohol (24:1). Add 0.1 volume of 3 M sodium acetate and 2.5 volumes of RNase-free 100% ethanol. Collect by centrifugation and wash the pellet with 70% ethanol.
6. Dissolve in 11 μL of Qiagen elution buffer. Determine the DNA concentration with a spectrophotometer.

3.2. Injection

3.2.1. Target Mutagenesis by NHEJ

1. Make an injection mix containing 250 μg/mL of each synthetic ZFN mRNA in 0.1 x Phosphate Buffered Saline (PBS).

2. Inject into 200–300 wild-type embryos by standard procedures (*see* **Note 4**).
3. Rear flies to adulthood.

3.2.2. Target Replacement by HR

1. Make an injection mix containing 250 μg/mL of each synthetic ZFN mRNA, plus the intact circular donor DNA at ~500 μg/mL, in 0.1× PBS (*see* **Note 5**).
2. Inject into 200–300 lig4– embryos by standard procedures (*see* **Note 4**).
3. Rear flies to adulthood.

3.3. Screening for Induced Mutants

1. Cross the injected flies to partners that can reveal new germline mutations in the target.
   a. If the mutant phenotype is known and viable, the cross can be made to existing mutants. For example, we reveal new mutations in the ry gene by crossing to a strain carrying a *ry* deletion that covers the ZFN target (3). Mutants are revealed by eye color in this case.
   b. If the mutant phenotype is unknown or lethal, candidate chromosomes can be isolated opposite corresponding balancers. For example, we recovered new mutations in the *coil* gene by crossing to flies carrying a deletion of the corresponding region on one chromosome 2 and a dominant marker (*CyO*) on a chromosome 2 balancer (4).
   c. If the injection mix included a marked donor DNA, a screen can be employed that takes advantage of donor characteristics. For example, if the donor carries a w^+ gene insertion, crossing to a w^- strain will reveal targets carrying this marker in the germline by the presence of red eyes in the offspring.
2. Characterize the candidate mutations at the molecular level.
   a. Choose individual flies from the F1 or F2 generation and prepare total genomic DNA with the AquaGenomic reagent, according to the supplier's instructions. This is typically done in 96-well plates using a multiple homogenizer (Burkard Scientific, UK).
   b. Amplify a genomic region around the ZFN target by PCR, with appropriate primers. Most NHEJ mutations are small insertions and/or deletions, so products in the range of 200–400 bp work well.
   c. If a donor DNA was included in the injection, analyze PCR products for the presence of the donor. The method will depend on the structure of the donor. We have

endowed donors with unique restriction sites, so analysis by gel electrophoresis after digestion is diagnostic for the donor (2, 3) (*see* **Note 6**).

d. Analyze PCR products for the presence of sequence alterations from NHEJ. We have used each of the following three methods successfully.
   i. If the frequency of mutation is high enough, PCR products from single flies can be sequenced directly, particularly if the candidate chromosome has been isolated opposite a deletion of the ZFN target.
   ii. Mix genomic DNA from three flies prior to PCR. After amplification, analyze with Surveyor nuclease (Transgenomic), which specifically cleaves duplexes containing mismatches. If one of the flies carries a new mutation, it is very likely that the others will have wild-type sequences that will generate the mismatch upon reannealing of mixed PCR products.
   iii. Mix genomic DNA from two flies prior to PCR. Subject the PCR products to high-resolution electrophoresis. Mutant products will typically differ in length from the wild-type target.

4. Notes

1. We use the Gateway system (Invitrogen) to clone new ZFNs for several reasons. Cloning new ZF coding sequences into an entry vector that already contains the nuclease domain and the NLS is convenient and reproducible. Once in that vector, this whole construct can be transferred to any destination vector for a variety of purposes. For example, we have destination vectors for expression in Drosophila, as well as the one for in vitro transcription. The transfer reaction uses bacteriophage lambda integrase (called LR Clonase by Invitrogen) to catalyze site-specific recombination between the entry insert and the destination vector. This reaction is both efficient and specific, and it is less subject to artifact than more traditional restriction enzyme-based cloning. (*See* www.invitrogen.com for a more extensive description of the Gateway system.)
2. For the injection method, it is essential to include a nuclear localization sequence (NLS) in the ZFN coding sequence. The earlier heat-shock procedure did not require it.
3. When working with RNA it is very important to avoid contamination with reagents that contain any source

of ribonucleases. Fresh, unopened reagents (e.g., 100% ethanol) are usually clean enough, and many reagents supplied with commercial kits (e.g., in vitro transcription reagents) are often specified as RNase-free.

4. The yield of fertile adult flies after ZFN mRNA injection is often lower than is typical for DNA injections. Therefore, it may be necessary to inject 300 or more embryos to recover a sufficient number of adults for effective screening.
5. A circular donor DNA is much more efficient than a linear one (4). This may be due to degradation of linear molecules in the embryonic syncytial cytoplasm, or to concatenation by end joining activities. Linear donor DNA was more effective in the heat-shock procedure (2), and it is certainly possible that a useful method for linearization after injection could be devised.
6. The ratio between HR and NHEJ products depends both on the repair capabilities of the recipient flies and the target locus. Using wild-type embryos, we have seen HR products represent 5–15% of recovered mutants at the *ry* locus, but essentially 0% at *y*. In *lig4* mutant embryos, HR products were nearly 100% of new mutants at *ry* and 70% at *y* (4).
7. As we noted earlier (11), off-target cleavage by the ZFNs can be a problem. We have found that some of our designed ZFNs are toxic to flies when expressed too enthusiastically (1), and this is typically a property of a single protein, not exacerbated by its partner. In one case we demonstrated directly that lethality was due to excessive cleavage (3). To ameliorate the toxicity of single ZFNs, two groups have redesigned the dimer interface to prevent homodimerization, while still allowing heterodimerization with the designed partner (15, 16). We recommend starting with unmodified cleavage domains, as targeting efficiency is negatively affected by the modifications in some cases. If one of a pair of ZFNs proves toxic, dimer interface modifications can be incorporated.

Acknowledgments

We are grateful to members of our lab who have contributed to development of these procedures over several years. Our work has been supported by the US National Institutes of Health through Research Grant R01 GM078571 to D.C. and through core facilities funded in part by the University of Utah Cancer Center Support Grant.

References

1. Bibikova, M., Golic, M., Golic, K.G., and Carroll, D. (2002) Targeted chromosomal cleavage and mutagenesis in *Drosophila* using zinc-finger nucleases. *Genetics.* **161**, 1169–1175.
2. Bibikova, M., Beumer, K., Trautman, J.K., and Carroll, D. (2003) Enhancing gene targeting with designed zinc finger nucleases. *Science.* **300**, 764.
3. Beumer, K., Bhattacharyya, G., Bibikova, M., Trautman, J.K., and Carroll, D. (2006) Efficient gene targeting in Drosophila with zinc finger nucleases. *Genetics.* **172**, 2391–2403.
4. Beumer, K.J., Trautman, J.K., Bozas, A., et al. (2008) Efficient gene targeting in Drosophila by direct embryo injection with zinc-finger nucleases. *Proc Natl Acad Sci USA.* **105**, 19821–19826.
5. Carroll, D., Morton, J.J., Beumer, K.J., and Segal, D.J. (2006) Design, construction and in vitro testing of zinc finger nucleases. *Nat Protoc.* **1**, 1329–1341.
6. Wright, D.A., Thibodeau-Beganny, S., Sander, J.D., et al. (2006) Standardized reagents and protocols for engineering zinc finger nucleases by modular assembly. *Nat Protoc.* **1**, 1637–1652.
7. Maeder, M.L., Thibodeau-Beganny, S., Osiak, A., et al. (2008) Rapid "Open-Source" engineering of customized zinc-finger nucleases for highly efficient gene modification. *Mol Cell.* **31**, 294–301.
8. Meng, X., Noyes, M.B., Zhu, L.J., Lawson, N.D., and Wolfe, S.A. (2008) Targeted gene inactivation in zebrafish using engineered zinc-finger nucleases. *NaT Biotechnol.* **26**, 695–701.
9. Urnov, F.D., Miller, J.C., Lee, Y.-L., et al. (2005) Highly efficient endogenous gene correction using designed zinc-finger nucleases. *Nature.* **435**, 646–651.
10. Doyon, Y., MaCammon, J.M., Miller, J.C., et al. (2008) Heritable targeted gene disruption in zebrafish using designed zinc-finger nucleases. *NaT Biotechnol.* **26**, 702–708.
11. Carroll, D., Beumer, K.J., Morton, J.J., Bozas, A., and Trautman, J.K. (2008) Gene targeting in *Drosophila* and *Caenorhabditis elegans* with zinc-finger nucleases *in* "Chromosomal Mutagenesis". In: (Davis, G., and Kayser, K.J., Eds.), Methods in molecular biology. Vol. 435, pp. 63–77. Humana Press, Totowa.
12. Bozas, A., Beumer, K.J., Trautman, J.K. and Carroll, D. (2009) Genetic analysis of zinc-finger nuclease-induced gene targeting in Drosophila. *Genetics* **182**, 641–651.
13. McVey, M., Radut, D., and Sekelsky, J.J. (2004) End-joining repair of double-strand breaks in *Drosophila melanogaster* is largely DNA ligase IV independent. *Genetics.* **168**, 2067–2076.
14. Spradling, A.C. (1986) P element-mediated transformation. In Drosophila: A Practical Approach (Roberts, D.B., Ed.), pp. 175–197. IRL Press, New York.
15. Miller, J.C., Holmes, M.C., Wang, J., et al. (2007) An improved zinc-finger nuclease architecture for highly specific genome cleavage. *Nat Biotechnol.* **25**, 778–785.
16. Szczepek, M., Brondani, V., Buchel, J., Serrano, L., Segal, D.J., and Cathomen, T. (2007) Structure-based redesign of the dimerization interface reduces the toxicity of zinc-finger nucleases. *Nat Biotechnol.* **25**, 786–793.

Chapter 18

Using Zinc Finger Nucleases for Efficient and Heritable Gene Disruption in Zebrafish

Jasmine M. McCammon and Sharon L. Amacher

Abstract

While the experimental tools developed for zebrafish have continued to advance the organism as a laboratory model, techniques for reverse genetics remain somewhat limited in scope. Zinc finger nucleases (ZFNs), chimeric fusions between DNA-binding zinc finger proteins and the non-specific cleavage domain of the FokI endonuclease, hold great promise for targeted mutagenesis in zebrafish, as demonstrated by two recent publications (Doyon et al., 2008, *Nat Biotechnol.* 26, 702–708; Meng et al., 2008, *Nat Biotechnol.* 26, 695–701). Because ZFNs can be designed to recognize a unique sequence in the genome, they can specifically bind and cleave a target locus, creating a double-strand break (DSB) that is repaired by one of two major DNA repair pathways. Repair by one of these pathways, non-homologous end joining, is often mutagenic, allowing one to screen for induced mutations in the target locus. By injecting into zebrafish embryos RNA encoding ZFNs that target three different loci, two groups have shown that ZFNs work efficiently to induce somatic and germline mutations (reviewed in (3)). We review here protocols for injection of ZFN-encoding mRNA into zebrafish embryos, screening of injected fish for induced mutations, and subsequent recovery of the induced mutations.

Key words: Zinc finger nuclease, zebrafish, reverse genetics, targeted mutagenesis, *golden*, *no tail*, non-homologous end joining.

1. Introduction

The zebrafish has proven to be a valuable model organism; however, technology for reverse genetics in the zebrafish remains somewhat limited. Although it is possible to screen a mutagenized zebrafish stock for lesions within a target locus (4, 5), zebrafish researchers have been unable to specifically target a locus of interest for directed modification. The use of ZFNs for targeted

J.P. Mackay, D.J. Segal (eds.), *Engineered Zinc Finger Proteins*, Methods in Molecular Biology 649,
DOI 10.1007/978-1-60761-753-2_18,

gene modification in other model systems such as *Drosophila* and *Caenorhabditis elegans* (6, 7) has demonstrated the potential for application of ZFN technology to other models such as zebrafish.

ZFNs are a chimeric fusion between zinc finger motifs and the non-specific cleavage domain of FokI endonuclease (8). Each zinc finger motif basically recognizes a 3-bp sequence; therefore, motifs can be pieced together in a modular fashion to create a zinc finger protein (ZFP) that binds a specific and unique genome sequence (9). Because the FokI cleavage domain requires dimerization for catalytic activity, pairs of ZFNs must be designed to bind adjacent sequences with a 5- to 6-bp spacer domain (10). Once bound, the ZFNs cleave the intervening sequence, creating a double-strand break (DSB) that is repaired via one of two major pathways. These are homology-directed repair, which uses a homologous template such as a sister chromatid or an exogenously supplied donor to direct repair, and non-homologous end joining (NHEJ), an error-prone mechanism that introduces small insertions and/or deletions when ends are improperly ligated together without use of homology (11). In this way, ZFNs can be considered mutagenic.

The basic experimental overview for a ZFN-mediated gene disruption project in zebrafish is outlined in **Fig. 18.1**. Clearly, ZFN design is a critical component for any application of ZFN technology. Since we obtained ZFNs for our study through collaboration with Sangamo Biosciences, we do not cover design methodology, but instead refer the reader to the ZFN design chapters in this volume (*see* **Chapters 1, 2, 3, 4** and **11**), as well as several recent articles that detail the design approaches taken by academic laboratories (12, 13) and to website tools and protocols developed by Wolfe and Lawson for zebrafish applications (14). In addition, we briefly describe a yeast-based assay that efficiently pre-selects ZFNs that recognize and cleave the designated sequence in vivo (1); this assay may be generally useful in the field, especially when combined with modular design approaches. Our chapter concentrates on how to use ZFNs for highly efficient mutagenesis in zebrafish (steps 3–5 in **Fig. 18.1**).

To show that ZFNs can induce targeted mutations in zebrafish, we tested their ability to disrupt two loci, *golden* (*gol*) and *no tail* (*ntl*) (1). These are two well-characterized genes with classic mutant phenotypes: *gol* mutants lack pigment 2 days post-fertilization (15, 16), while *ntl* mutants lack posterior mesoderm and notochord (17, 18). Because of the inherent problems in having a nuclease with mutagenic potential lurking in the genome, we wanted ZFN exposure to be transient, so we transcribed RNA-encoding *gol*- and *ntl*-targeting ZFNs and injected it directly into one-cell zebrafish embryos. To immediately assess whether the ZFNs were functional, we developed a transient assay in which ZFN RNA was injected into embryos that are heterozygous for

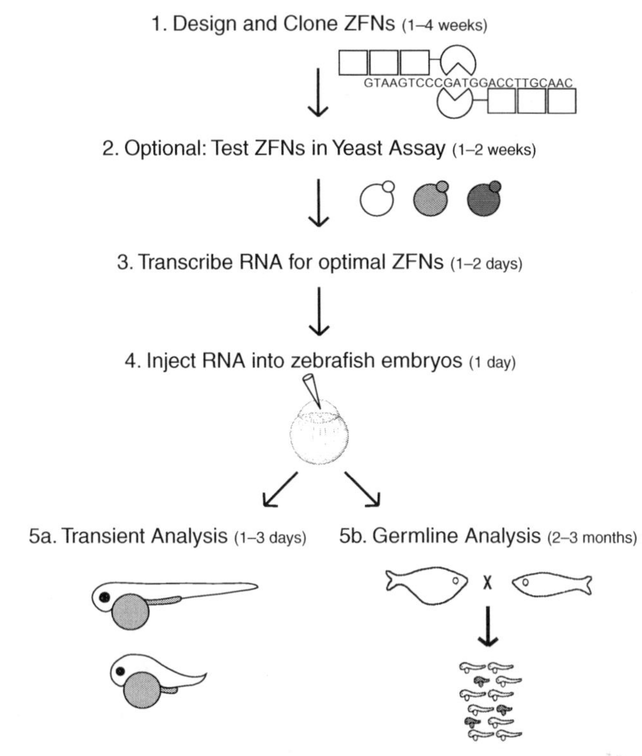

Fig. 18.1. Experimental overview. *Step 1:* Design and clone ZFNs. Considerations include number of zinc fingers (3 or 4, thereby creating a 9- or 12-bp recognition site for each ZFN) and FokI variant (wild-type or high-fidelity versions, *see* **Note 2**). *Step 2:* Test ZFNs in a yeast assay. In this cleavage-based assay, described in further detail in **Fig. 18.5**, one can identify ZFNs that efficiently bind and cleave their targets in a chromatin-based setting. In addition, one can also monitor whether particular ZFN pairs affect yeast survival, which correlates with toxicity when injected in zebrafish embryos. *Step 3:* Transcribe RNA for selected ZFNs. *Step 4:* Inject RNA into zebrafish embryos. *Steps 5a and 5b:* These can be done simultaneously. In the transient assay, it may be useful to use embryos that are heterozygous for the target locus, if an allele is available. Screening for mosaic phenotype in transient analysis takes 1 day for *ntl*-targeting ZFN RNA-injected embryos and 2 days for *gol*-targeting ZFN RNA-injected embryos. The Cel-1 assay (*see* step 6 of **Section 3.4** and **Note 10**) takes 1–2 days. In situ hybridization takes 2–3 days. Germline analysis takes longer because the injected embryos have to be raised to sexual maturity (2–3 months). The subsequent complementation crosses (3 days), genotyping (1–2 days), and sequencing of alleles (4 days) take less time once embryos can be obtained from the mature founder fish.

a previously characterized mutant allele of the test loci; however, our results indicate that the mutagenic efficiency is sufficiently high that one does not need an existing mutant allele to successfully isolate ZFN-induced alleles. We also injected ZFN-encoding RNAs into wild-type embryos and raised them to sexual maturity to evaluate germline mutagenesis frequency. Using a combination of morphological analysis, molecular genotyping, and genetics, we determined that ZFNs efficiently induce somatic and germline mutations (1). We review in detail our methods for the published study, pointing out where one might take protocol deviations depending upon the ZFNs utilized and the nature of the target site.

2. Materials

2.1. RNA Synthesis (step 3 in Fig. 18.1)

1. ELIMINase (Decon Labs).
2. Sharp precision barrier tips (Denville).
3. EagI restriction enzyme (NEB), store at –20°C.
4. mMessage mMachine kit (Ambion), store at –20°C.
5. Nuclease-free water, e.g., UltraPure distilled water, DNase, RNase free (Invitrogen).
6. Phenol:chloroform:isoamyl alcohol (25:24:1) (Fisher), store at 4°C.
7. Chloroform.
8. 3 M sodium acetate.
9. Nuclease-free ethanol, store at –20°C.
10. RNasin (Roche), store at –20°C.

2.2. Zebrafish Husbandry

1. Strains: AB wild type, *golden*^{b1} homozygous adults, *no tail*^{b195} heterozygous adults.
2. Mating tanks.
3. Optional: Hank's premix: 0.137 M NaCl, 5.4 mM KCl, 0.25 mM Na_2HPO_4, 0.44 mM KH_2PO_4, 1.3 mM $CaCl_2$, 1.0 mM $CaCl_2$, 1.0 mM $MgSO_4$. Store premix at 4°C. On the day of the in vitro fertilization experiment, make 10 mL of Hank's solution by mixing 9.9 mL of Hank's premix with 0.1 mL of freshly made 0.42 M $NaHCO_3$ (0.35 g $NaHCO_3$ diluted to 10 mL with RO water).
4. Optional: Blunt tipped tweezers (e.g., Millipore MF filter forceps, XX6200006).
5. Optional: Sponges/packing foam with 2–3 cm slits cut in them.
6. Optional: Tricaine stock: 0.4 g 3-amino benzoic acid ethyl ester (Sigma), 0.8 g Na_2HPO_4 (anhydrous) in 100 mL RO water. Add 4.2 mL of stock to 100 mL fish water to anesthetize adult fish. Store stock at 4°C.
7. Optional: 35 mm × 10 mm culture dishes (Falcon 353003).
8. Optional: Metal weighing spatulas with rounded ends (e.g., Fisherbrand #21-401-5).
9. Optional: Kimwipes.

10. Optional: Capillary micropipets, 25 and 50 μL (Drummond Scientific, 5-000-2050).
11. Optional: Paper towels.
12. Optional: Drummond short length microcaps micropipets (20 μL; No.: 1-000-0200).

2.3. Embryo Microinjection (step 4 in Fig. 18.1)

1. Capillaries for injection needles, borosilicate with filament. O.D. 1.2 mm, I.D. 0.94 mm. 10-cm length (Sutter Instruments).
2. Sutter Instrument Co. Flaming/Brown Micropipette Puller (Model P-87) or comparable puller.
3. Phenol red, 1% stock solution.
4. Olympus SZ-60 dissecting microscope with transillumination base or comparable microscope.
5. Pressure injector (Applied Scientific Instrumentation, #MMPI-3), including back pressure unit and micropipette holder, or comparable injection apparatus.
6. An apparatus to hold embryos during injection. We used plexiglass molds, 3.5 cm × 12 cm, with V-shaped grooves cut every 1 cm. The grooves were 1 mm wide and 0.5 mm deep. However, there are many other alternatives (19).

2.4. Phenotype Characterization: Transient Assay (step 5a in Fig. 18.1)

1. Tricaine: For embryos add 15 drops of stock (*see* step 6 of **Section 2.2**) to a 35 mm × 10 mm culture dish of fish water. Final concentration is approximately 0.004%.
2. Pokers: 2-pound test fishing line or eyelash hairs glued into small glass capillary tubes. Poker stems can be enfolded in laboratory tape to facilitate gripping.
3. Camera/microscope setup: Zeiss Axioplan 2 with Axiocam digital camera or comparable microscope and camera.
4. Depression slides ("Hanging drop" slides, Fisher Scientific #12-560A).
5. Methylcellulose, 3% solution.
6. Paraformaldehyde fixative: 4% solution in 1× PBS, pH~7.4. Store at 4°C for up to 1 week.
7. PBST: 1× PBS plus 0.1% Tween-20.
8. Forceps: Dumont #5 tweezers, rustless.
9. Proteinase K (Roche): 10 mg/mL stock in sterile H_2O.
10. Hybridization buffer: 25 mL formamide (pure grade, Fisher), 12.5 mL 20× SSC, 50 μL 50 mg/mL heparin, 500 μL 50 mg/mL yeast tRNA, 250 μL 20% Tween-20, 460 μL 1 M citric acid, sterile water to 50 mL. Store at –20°C. Keeps for 6 months.

11. Blocking solution: 2.5 g BSA powder (Sigma, store powder at 4°C) in 50 mL PBST. Store at 4°C for up to 3 weeks.
12. Anti-digoxigenin antibody (Roche), store at 4°C.
13. Coloration buffer: 100 mM Tris–HCl, 50 mM $MgCl_2$, 100 mM NaCl, 0.1% Tween-20.
14. NBT and BCIP (Roche), store at –20°C.

2.5. Germline Mutagenesis Analysis

The materials used in **Section 3.5** overlap with those in **Sections 2.2** and **2.6** (step 5b in **Fig. 18.1**).

2.6. Genotyping

1. Thermopol buffer (New England Biolabs): 20 mM Tris–HCl, 10 mM $(NH_4)_2SO_4$, 10 mM KCl, 2 mM $MgSO_4$, 0.1% Triton X-100. Store at –20°C.
2. Primers (Bioneer): For *gol*: 5′-ATCTGATATGGCCATGTCCAACATCG-3′ and 5′-GGAACAATCCCATACGCTCCTGCAG-3′. For *ntl*: 5′-ACGAATGTTTCCCGTGCTCAGAGCC-3′ and 5′-GCTGAAAGATACGGGTGCTTTCATCCAGTGCG-3′.
3. TOPO TA cloning kit (Invitrogen).
4. Competent cells (e.g., TOP10, Invitrogen).
5. LB media: Add 25 g LB Miller Broth to 1 L water and autoclave for 20 min.
6. Ampicillin plates: Add 25 g LB Miller broth and 15 g agar to 1 L water, then autoclave for 20 min. Once cool to touch, add ampicillin (final concentration: 50 μg/mL) to liquid and pour into Petri dishes. Store poured plates upside-down at 4°C.
7. Mini-prep kit (Qiagen).
8. BsrDI (NEB), store at –20°C, or appropriate enzyme for the designated target locus.
9. Agarose gel electrophoresis materials.

3. Methods

3.1. RNA Synthesis

1. Select the zinc finger sequence (*see* **Note 1**) and FokI nuclease variant (*see* **Note 2**) to be used in the study and subclone ZFNs into appropriate plasmids to use for RNA synthesis (*see* **Note 3**) (step 3 in **Fig. 18.1**).
2. Prepare an RNase-free area and RNase-free micropipettors by wiping the area and pipettors down with ELIMINase.

3. Make the template for the mMessage mMachine transcription reaction by linearizing ~8–12 μg of plasmid DNA encoding ZFNs with the appropriate enzyme. In our case, we used 2 μL EagI, a restriction enzyme with a unique cut site after the BGH polyA signal in the pVAX vector, in 50 μL reaction for 4 h at 37°C.
4. Stop the reaction by purifying with phenol:chloroform extraction (*see* **Note 4**). Add 50 μL of nuclease-free water to digest and then add 100 μL of phenol:chloroform (1:1). Mix by vortexing and centrifuge at top speed for 4 min at room temperature. Extract the top aqueous layer to a clean microcentrifuge tube. Add 1 volume of chloroform to the extracted layer, mix, and centrifuge as before. Extract the top aqueous layer to a clean tube.
5. Perform an ethanol precipitation to concentrate the DNA template. Add 1/10th volume of 3 M sodium acetate and 2 volumes of ice-cold 100% ethanol, then mix, and freeze at –20°C for at least 20 min (can go overnight). Centrifuge at top speed for 10 min at room temperature, noting where the pellet will form, aspirate the liquid, add 100 μL of cold 70% ethanol, and centrifuge for 1 min to make the pellet stick. Carefully aspirate 70% ethanol to avoid disturbing or moving pellet. Evaporate the remaining ethanol by leaving the tube caps open on the bench top for ~5 min. Resuspend the pellet in 6 μL of nuclease-free water.
6. Set up a transcription reaction. Thaw the 10× reaction buffer and 2× NTP/CAP, vortex, and spin down. Keep the 2× NTP/CAP on ice but leave the 10× reaction buffer at room temperature. In a clean microcentrifuge tube at room temperature, add 10 μL of 2× NTP/CAP, 2 μL of 10× reaction buffer, 5.5 μL of linearized DNA template (recommended ~1 μg), 0.5 μL of RNase inhibitor, and 2 μL of T7 (or other appropriate polymerase for sense transcription) enzyme mix. Incubate for 3 h at 37°C. Add 1 μL of DNase to the transcription reaction to destroy the template and incubate at 37°C for 15 min.
7. Stop the reaction by precipitating with LiCl solution included with the kit (*see* **Note 5**). Mix and store at –20°C for at least 20 min or overnight. Centrifuge at top speed for 15 min at 4°C, noting where the pellet of RNA will form. Aspirate the liquid, being careful not to disturb the pellet. Add 100 μL of ice-cold 70% ethanol and centrifuge for 5 min. Carefully remove the ethanol and allow the pellet to air dry for 3–4 min. Resuspend the RNA in 15 μL of nuclease-free water and store at –20°C.

8. Determine the RNA concentration with a 1:100 dilution (1 μL RNA and 99 μL water) by measuring the absorbance with a UV spectrophotometer. Multiply the 260 nm absorbance reading by 4 to determine the concentration in μg/μL.

3.2. Zebrafish Husbandry

1. Maintain adult strains at 28.5°C on a 14/10 h light/dark cycle.
2. Obtain embryos from natural spawning. The afternoon before the desired embryo collection day, separate male and female fish in holding tanks without food. Shortly after the lights turn on the next morning, combine three females and three males of the desired genotype in mating tanks with mesh bottoms to prevent the adults from consuming the embryos once they are made. Alternatively, in vitro fertilization can be used to generate embryos (*see* **Note 6**). If there is a concern that fish strains may carry single nucleotide polymorphisms (SNPs) in the loci of interest, thereby affecting subsequent analyses (i.e., step 6 of **Section 3.4**), then embryos can be generated from genotyped individuals.

3.3. Embryo Microinjection

1. Prepare injection needles using a micropipette puller and borosilicate capillaries with filaments (*see* **Note 7**). The parameters for our micropipette puller to generate injection needles are as follows: heat 573, pull 150, velocity 100, time 50, pressure 110 (step 4 in **Fig. 18.1**).
2. Prepare the injection solution at 5-fold serial dilutions. Dilute the RNA with RNAse-free water and 0.1% phenol red as an injection tracer. Typically the range for the *golden* and *no tail* experiments was 80.4–1 ng/nL. For experiments with other ZFNs, the effective range may depend upon the number of zinc fingers in the ZFN and the FokI endonuclease variant used. Load 1 μL of injection mix into needles.
3. Inject 3–5 nL of the mRNA solution into 1–2 cell embryos (heterozygous embryos for transient assays and wild-type embryos for transient or germline assays) into a cell or the yolk (*see* **Note 8**).
4. Remove infertile embryos during the blastula stages (3–4 h post-fertilization, hpf).

3.4. Phenotype Characterization: Transient Assay

To analyze *gol*-targeting ZFN-injected embryos (step 5a in **Fig. 18.1**):

1. At 2 days post-fertilization (dpf), immobilize *golden*^{b1} heterozygous embryos injected with *gol*-targeting ZFNs in 0.004% tricaine. Carefully manipulate individual embryos using pokers and evaluate each eye for the appearance of

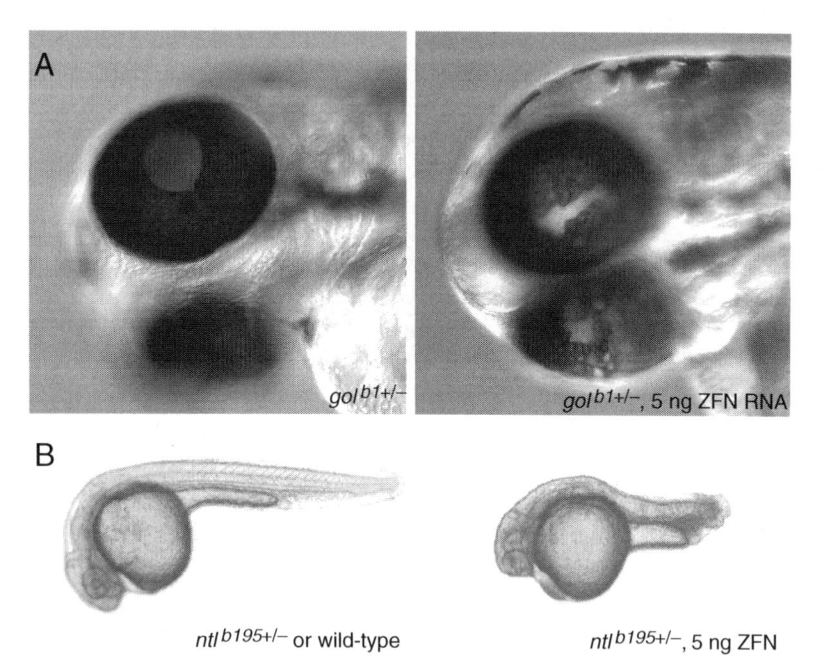

Fig. 18.2. Mosaic analysis of ZFN-induced mutations at the *gol* and *ntl* loci. (**A**) A wild-type eye (*left panel*) and an eye of a gol^{b1} heterozygote injected with 5 ng of *gol*-targeting ZFN RNA (*right panel*) at 2 dpf. Note the unpigmented patches of cells in both eyes of the injected embryo while the wild-type eye is uniformly darkly pigmented. (**B**) A wild-type embryo (*left panel*) and a ntl^{b195} heterozygote injected with 5 ng of *ntl*-targeting ZFN RNA (*right panel*). The injected embryo exhibits similar defects as a *ntl* null mutant: reduced or lacking notochord, misshapen somites, and reduced tail mesoderm.

patches of unpigmented cells, using a standard dissecting microscope with transillumination (*see* **Fig. 18.2a**).

2. Mount immobilized embryos on a drop of 3% methylcellulose in a depression slide to photograph.

To analyze *ntl*-targeting ZFN-injected embryos:

3. At 1 dpf, using a standard dissecting microscope with transillumination, evaluate *no* $tail^{b195}$ heterozygous embryos injected with *ntl*-targeting ZFNs for *ntl*-like appearance: reduced or missing posterior mesoderm, misshapen somites, and a lack of notochord cells (*see* **Fig. 18.2B**).
4. Immobilize the embryos in 0.004% tricaine and photograph as described above (*see* step 2 of **Section 3.4**).
5. To characterize subtler notochord defects, fix 18–22 hpf ZFN-injected ntl^{b195} heterozygous embryos in 4% paraformaldehyde overnight at 4°C in microcentrifuge tubes (20 embryos/tube) for in situ hybridization. There are several published protocols for in situ hybridization; our protocol is essentially the same as described by Thisse and Thisse for the zebrafish expression screen project (detailed protocol available online at http://zfin.org/

ZFIN/Methods/ThisseProtocol.html), thus we only summarize the protocol here. Rinse the embryos in PBST and dechorionate with forceps. Wash 5 × 5 min with PBST. Permeabilize the embryos by incubating in 10 μg/μL proteinase K in PBST for 2–8 min, depending upon the stage. Refix the embryos in 4% PFA for 20 min. Wash 5 × 5 min with PBST. Incubate the embryos in hybridization buffer at 65°C for 1–4 h. Add 20–300 μL of 1:200 ntl riboprobe (*see* **Note 9**) in hybridization buffer at 65°C overnight. After incubation, perform several washes at 65°C: 5 min with 66% hybridization buffer/33% 2× SSC buffer, 5 min with 33% hybridization buffer/66% 2× SSC, 5 min with 2× SSC, 5 min with 1× SSC, 20 min with 0.2× SSC, 2 × 20 min with 0.1× SSC. Then, perform several room temperature washes: 5 min with 66% 0.1× SSC/33% PBST, 5 min with 33% 0.1× SSC/66% PBST, 5 min with PBST. Incubate the embryos for 1 h in blocking solution. Incubate for 2 h at RT or 4°C overnight in 1:5,000 anti-digoxigenin antibody in blocking solution. Perform 5 × 15 min washes in PBST, followed by 3 × 5 min washes in coloration buffer. Prepare the coloration solution: 4.5 μL NBT and 3.5 μL BCIP per 1 mL of coloration buffer. Add 500 μL of coloration solution per tube and store the tubes protected from the light while the color reaction proceeds. Periodically check the staining intensity in the notochord. For detection of *ntl* transcripts, color development is usually complete after 1–2 h, but for different riboprobes, time will vary depending upon the specific transcript being detected and riboprobe quality. Stop the reaction by washing twice with water and store the embryos in PBST at 4°C.

6. To evaluate ZFN mutagenesis at the molecular level, one can assay for heterozygosity at the test locus by using a mismatch-sensitive endonuclease, Cel-1 (*see* **Note 10**). Readers are referred to the chapter by Sangamo Biosciences (*see* **Chapter 15**).

3.5. Germline Mutagenesis Analysis

1. Inject wild-type embryos with RNA encoding *ntl*-targeting ZFNs and raise healthy larvae that develop swim bladders (scored using a dissecting microscope at 5 dpf) to adulthood (step 5b in **Fig. 18.1**).
2. When injected potential founders are 2- to 3-months old, perform complementation crosses with ntl^{b195} heterozygous fish by natural crosses or in vitro fertilization (*see* **Note 6**).
3. Score the resulting progeny for a *ntl* mutant phenotype (*see* **Fig. 18.3**). To estimate the fraction of mutant gametes in the germline, double the percentage of mutant embryos observed in the complementation cross, since only half the

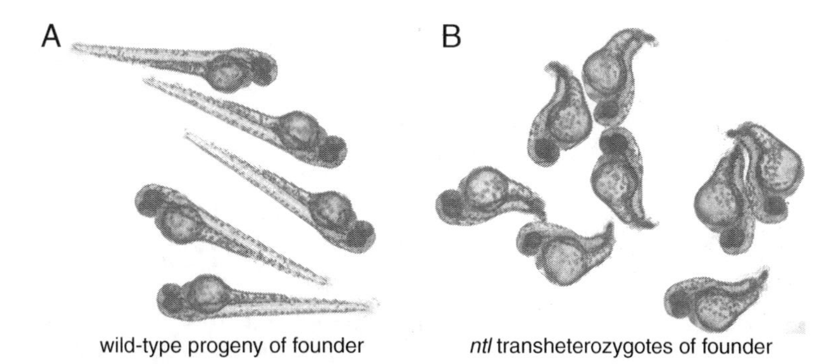

Fig. 18.3. Germline analysis of ZFN-induced mutations at the *ntl* locus. Wild-type (**A**) and *ntl* (**B**) progeny resulting from a complementation cross between a *ntl*^{b195} heterozygous fish and a *ntl*-targeting ZFN-injected founder fish.

ZFN-injected embryos inherited a *ntl*^{b195}allele. Genotyping can be used to confirm the actual percentage of ZFN-induced mutant chromatids among progeny of the complementation cross (*see* **Section 3.6**).

3.6. Genotyping

BsrDI (or equivalent enzyme) digest

1. To confirm that the phenotypes observed are a result of ZFN-induced mutagenic repair by NHEJ at the desired site, one can assay for insertions and/or deletions at the cut site by restriction digestion. First, isolate genomic DNA from *ntl*-like embryos. Place individual embryos in 50 μL of 1× thermopol buffer. Heat the samples to 98°C for 10 min, then cool to 55°C. Add 5 μL of 10 mg/mL proteinase K and incubate at 55°C for 1 h, flicking the tubes after 30 min to help break up embryos. Heat to 98°C for 10 min to inactivate proteinase K. Centrifuge the samples for 3 min at 18,407 RCF to pellet undigested tissue. Remove the supernatant containing the genomic DNA to clean tubes.
2. PCR amplify the 226-bp region for *ntl* using the primers listed in step 2 of **Section 2.6**. Use 1 μL of genomic DNA in a 10-μL PCR reaction (PCR: 94°C for 2 min; 94°C for 30 s, 55°C for 30 s, 72°C for 30 s, for 30 cycles; 72°C for 2 min, 4°C hold). Due to a 1.5-kb insertion in the *ntl*^{b195} allele, the *ntl*^{b195} chromatids are not amplified using this protocol, allowing one to assay only the chromatid inherited from the ZFN mRNA-injected parent.
3. Digest the PCR product with BsrDI in a 20-μL reaction at 65°C for 1 h (*see* **Note 11**).
4. Run the digest on a 2% agarose TBE gel. Expected band sizes: 226 bp for the mutant allele, 176 bp + 50 bp for the wild-type allele (*see* **Note 12, Fig. 18.4**).

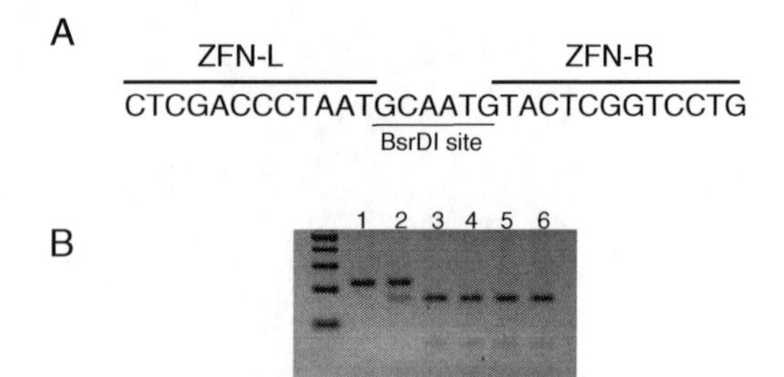

Fig. 18.4. Genotyping of ZFN-induced mutations at the *ntl* locus by screening for loss of sensitivity to BsrDI. (**A**) The sequence of the ZFN target site of the *ntl* locus. Note that a recognition site for the restriction enzyme BsrDI is found in the spacer domain between the left and right ZFN-binding sites (ZFN-L and ZFN-R, respectively). The spacer region is where the ZFN pair will induce a DSB and where mutagenic repair by NHEJ will likely confer loss of restriction enzyme recognition site. (**B**) Gel analysis of BsrDI-digested PCR products of the region surrounding the *ntl* ZFN-binding sites. Template DNA used for the PCR was genomic DNA isolated from progeny of complementation crosses between ZFN-injected founders and a ntl^{b195} heterozygote. Due to a 1.5-kb insertion in the ntl^{b195} allele, ntl^{b195} chromatids are not amplified using this protocol. A product obtained from a genotypically wild-type embryo will be completely sensitive to BsrDI digestion and result in 176- and 50-bp products (*lanes* 3–6). A product from a genotypically homozygous mutant embryo should be resistant to BsrDI cleavage resulting in a 226-bp product (*lane* 1). Product from a heterozygous embryo should have all three band sizes (*lane* 2).

Sequencing

5. To subclone the PCR product into the TOPO TA cloning pCRII vector, combine 4 μL of the undigested PCR product with 1 μL of salt solution and 1 μL of vector for 10 min at room temperature.
6. To transform bacteria, add 1 μL of the ligation reaction to 100 μL of competent cells and incubate the cells on ice for 20 min before heating at 42°C for 45 s. Recover the cells on ice for 2 min, add 900 μL of LB medium, and shake the tube at 37°C for 45 min.
7. Plate 50 μL of cells on selective media plates and grow overnight at 37°C.
8. Select multiple colonies for sequencing with a T7 sequencing primer. If mutations are ZFN-induced, there should be short insertions, deletions, or a combination thereof at the site of ZFN binding.

4. Notes

1. There are a variety of approaches used to generate a zinc finger protein (ZFP) that can recognize a unique stretch of base pairs (usually 9–12) in the genome. Some

design approaches use bacterial hybrid selection strategies, in which the "best" ZFP is built via selection in bacteria to find a combination of single ZFPs that best recognize the target sequence (2, 20, 21). This strategy has been used successfully to build ZFNs that efficiently recognize and cleave their intended target sequences in the zebrafish genome (2). The ZFNs used in our study (1) were assembled by Sangamo Biosciences using their proprietary platform (now available commercially through Sigma-Aldrich). One potential drawback of modular design approaches is that successful binding of the ZFP to the target in vitro does not necessarily mean that the ZFP (or the ZFN constructed from it) will recognize the target in vivo. In our published study, we describe a yeast-based assay that accurately predicts which ZFN pair combinations will efficiently cleave the intended target sequence in the context of zebrafish chromatin (1, *see* **Fig. 18.5** for a summary). The yeast strain, YSC1048-645440 yeast parental strain – BY4741, can be purchased from Open Biosciences, and yeast plasmids and protocols are available from Sangamo Biosciences after receipt of a simple MTA.

2. Several FokI nuclease variants are available (22, 23). In most cases, we used an obligate heterodimer high-fidelity version of the FokI cleavage domain (23). This version has point mutations at the interaction interface to ensure that only ZFN pairs binding as heterodimers can create an active version of the FokI cleavage domain, not homodimers binding at off-target sites. When we used the wild-type FokI in zebrafish, we found that while these ZFNs induced more mutations at the target site, they were also more toxic. We observed more embryos exhibiting developmental defects after injection with ZFNs containing wild-type FokI than with the same amount of ZFNs containing the high-fidelity FokI variant.

3. The expression vector used can vary depending upon the ease of subcloning as well as personal preference. We used pVAX for most of our experiments; however, a pCS2+-based vector was used in the study by Meng et al. (2). In most cases, pairs of ZFNs were placed in the same plasmid, spaced by a viral 2A peptide ribosomal stutter sequence, which should give equal amounts of both transcripts (24). However, we also made singly transcribed RNA from plasmids containing only one ZFN-coding sequence, quantified the amount of RNA using a UV spectrophotometer, and co-injected equal amounts of RNA for paired ZFNs. We did not observe a significant difference in either case and mostly used 2A fusion plasmids for convenience.

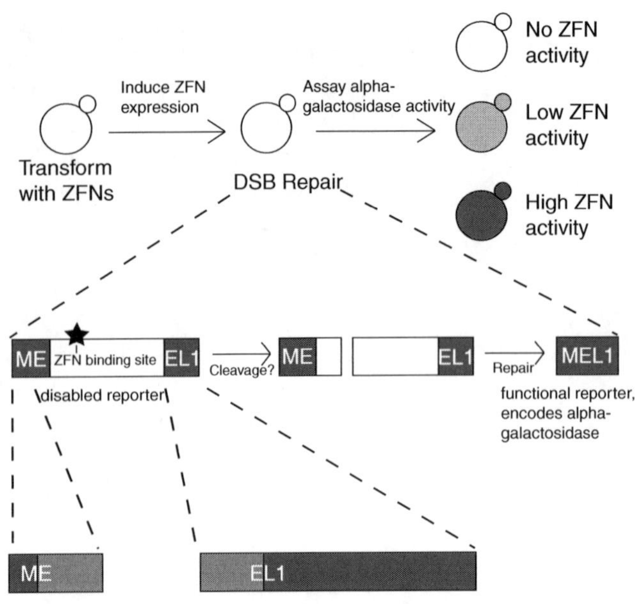

Fig. 18.5. Yeast-based assay to identify ZFNs that are active in vivo. Because successful binding in vitro does not necessarily identify ZFNs that will efficiently cleave DNA in vivo, investigators at Sangamo Biosciences developed a budding yeast-based assay to more accurately identify ZFNs that work well in the context of chromatin (1). The assay exploits the observation that a DSB between two short direct repeats in yeast cells will trigger resection of intervening heterologous sequence and will restore chromosome integrity through single-strand annealing of the direct repeats. Constructs containing a secreted form of α-galactosidase (encoded by the yeast *MEL1* gene) were disabled by insertion of the ZFN target sequence flanked by direct repeats of *MEL1* and transformed into the yeast genome. In the magnified view of the *MEL1* reporter, the disabled *MEL1* arms (5′-749 bp, 3′-1,810 bp) are shown in two-tone gray, with light gray indicating the position of the direct repeats (450 bp). The yeast cells were then transformed with plasmids carrying ZFNs under the control of an inducible promoter. Functional restoration of the "broken" MEL1 open reading frame (and production of α-galactosidase) after induction of ZFN expression indicates that the ZFN pair recognizes and cleaves the intended target sequence. A quantitative measure of how efficiently they do so is determined by assaying α-galactoside activity colorimetrically. The ZFNs that were most active in the yeast assay while not significantly affecting yeast cell survival were prioritized for analysis in zebrafish embryos.

4. Phenol:chloroform extraction is essential. We do not recommend alternatives, such as heat inactivation of the digest reaction, as this method leads to decreased transcription efficiency in our hands. Note that phenol is corrosive. Use appropriate protection.
5. We sometimes phenol:chloroform extract RNA after LiCl precipitation, which tends to decrease RNA yield. We did not note a significant increase in toxicity when injecting embryos without this additional phenol:chloroform extraction step.
6. The in vitro fertilization method is described in detail in *The Zebrafish Book* (25) and available online (http://

zfin.org/zf_info/zfbook/chapt2/2.8.html), so we only briefly summarize here. The reagents and materials for the procedure are listed as "optional" in **Section 2.2**. To procure sperm, anesthetize males in 0.004% tricane. Once gills stop rapidly moving, remove a single male with a spoon, being careful to scoop from head to tail to avoid damaging gills, rinse the fish quickly in aquarium water, and place the male belly up in the slit of a damp sponge. Dry off the genital region by blotting with a Kimwipe (exposure to water will activate gametes). Under a dissecting microscope, hold a capillary tube against genital pore while gently stroking sides of male with blunt ended tweezers. Transfer collected sperm from one to several males to ice-cold Hank's solution until the solution is cloudy. To procure eggs, anesthetize females in 0.004% tricane. Remove and rinse as described earlier, but transfer to a 1-cm thick pile of paper towels. Using the spoon, gently roll the female across the paper towels to dry her body, then transfer her to a 35 mm × 10 mm Petri dish, laying her on her side. With damp fingertips, brace her back with the index finger of one hand while gently pressing her abdomen with the index finger of the other hand to express the eggs. If the female has eggs they will come out easily. Good eggs will be yellow in color and come out compactly, while bad eggs are more milky white and watery. Immediately add 10–20 μL of sperm stored in Hank's solution and mix gently. Activate fertilization by adding ~0.5 mL of fish water. After about 1 min, fill dish with fish water.

7. Needles were not specially treated to remove RNase; however, care was taken to wear gloves when pulling needles and when loading RNA into needles.

8. To obtain *ntl*-like embryos in transient assays, we injected heterozygous *ntl* embryos with RNA-encoding ZFNs containing the high-fidelity version of FokI. However, we could also obtain *ntl*-like embryos by injecting wild-type embryos if we used RNA-encoding ZFNs containing the wild-type version of FokI (*see* **Note 2**). Many NHEJ mutagenic repair sequences were recovered from these *ntl*-like embryos (1). However, biallelic mutations were also likely induced with the obligate heterodimer variant of FokI, as wild-type embryos injected with *ntl* ZFNs carrying this variant occasionally grew up as adult fish with tail truncations (1).

9. Riboprobe synthesis is described online (http://zfin.org/ZFIN/Methods/ThisseProtocol.html) and briefly summarized here. Linearize the template with an appropriate restriction enzyme and purify with phenol:chloroform extraction. To make the antisense probe, combine in a

20 μL reaction: 1 μg of linearized template, 4 μL of transcription buffer, 2 μL of NTP-DIG-RNA labeling mix, 1 μL of RNase inhibitor, and 1 μL of an appropriate RNA polymerase (i.e., T7, T3, or SP6). Incubate for 2 h at 37°C. Digest the DNA template with DNase for 15 min at 37°C. Stop the reaction by adding 1 μL of 0.5 M EDTA. Precipitate the RNA with 2.5 μL of 4 M LiCl and 75 μL of cold 100% ethanol. After cold incubation, centrifuge at 4°C, remove the supernatant, centrifuge again with 70% ethanol, remove the supernatant, and resuspend the pellet in 100 μL of sterile water. Store at –20°C or –70°C.

10. Note that in order for an assay with a mismatch-sensitive endonuclease to work, one will need non-polymorphic strains, at least in a 200- to 300-bp region containing the ZFN-binding site. Lack of existing polymorphism ensures that a PCR product from the region will be cleaved only when heterozygosity has been induced by ZFN-directed cleavage and NHEJ-mediated repair (instead of cleavage due to preexisting SNPs). Thus, we recommend that the region surrounding the target be screened for segregating SNPs before ZFNs are injected and progeny raised. Mismatch-sensitive endonuclease analysis is particularly useful to detect mutagenic events in the case where the ZFN target site does not contain a restriction enzyme site (*see* **Note 11**).
11. When selecting ZFNs-binding sites, it is helpful to choose one where the 5- to 6-bp spacer region is a recognition site for a restriction enzyme; loss of this site is a likely indicator of mutagenic NHEJ repair. One can then identify mutagenic events at the molecular level without needing a preexisting mutant allele over which to genetically screen for loss of complementation.
12. Because of the high percentage of germline mutations, one does not need a preexisting mutant allele to identify germline mutations when targeting a novel locus with ZFNs. In addition, with careful planning of ZFN-binding sites around a restriction enzyme recognition site, one can also genotype for germline mutations.

Note added in proof

After this manuscript was reviewed, another group published the successful application of ZFN-mediated mutagenesis in zebrafish (26).

References

1. Doyon, Y., McCammon, J.M., Miller, J.C., Faraji, F., Ngo, C., Katibah, G.E., et al. (2008) Heritable targeted gene disruption in zebrafish using designed zinc-finger nucleases. *Nat Biotechnol.* **26**, 702–708.
2. Meng, X., Noyes, M.B., Zhu, L.J., Lawson, N.D., and Wolfe, S.A. (2008) Targeted gene inactivation in zebrafish using engineered zinc-finger nucleases. *Nat Biotechnol.* **26**, 695–701.
3. Amacher, S.L. (2008) Emerging gene knockout technology in zebrafish: zinc-finger nucleases. *Brief Funct Genomic Proteomic.* **7**, 460–464.
4. Stemple, D.L. (2004) TILLING – a high-throughput harvest for functional genomics. *Nat Rev Genet.* **5**, 145–150.
5. Moens, C.B., Donn, T.M., Wolf-Saxon, E.R., and Ma, T.P. (2008) Reverse genetics in zebrafish by TILLING. *Brief Funct Genomic Proteomic.* **7**, 454–459.
6. Bibikova, M., Golic, M., Golic, K.G., and Carroll, D. (2002) Targeted chromosomal cleavage and mutagenesis in *Drosophila* using zinc-finger nucleases. *Genetics.* **161**, 1169–1175.
7. Morton, J., Davis, M.W., Jorgensen, E.M., and Carroll, D. (2006) Induction and repair of zinc-finger nuclease-targeted double-strand breaks in *Caenorhabditis elegans* somatic cells. *Proc Natl Acad Sci USA.* **103**, 16370–16375.
8. Kim, Y.G., Cha, J., and Chandrasegaran, S. (1996) Hybrid restriction enzymes: zinc finger fusions to Fok I cleavage domain. *Proc Natl Acad Sci USA.* **93**, 1156–1160.
9. Pabo, C.O., Peisach, E., and Grant, R.A. (2001) Design and selection of novel Cys2His2 zinc finger proteins. *Annu Rev Biochem.* **70**, 313–340.
10. Smith, J., Berg, J., and Chandrasegaran, S. (1999) A detailed study of the substrate specificity of a chimeric restriction enzyme. *Nucleic Acids Res.* **27**, 674–681.
11. Valerie, K. and Povirk, L.F. (2003) Regulation and mechanisms of mammalian double-strand break repair. *Oncogene.* **22**, 5792–5812.
12. Maeder, M.L., Thibodeau-Beganny, S., Osiak, A., Wright, D.A., Anthony, R.M., Eichtinger, M., et al. (2008) Rapid "open-source" engineering of customized zinc-finger nucleases for highly efficient gene modification. *Mol Cell.* **31**, 294–301.
13. Ramirez, C.L., Foley, J.E., Wright, D.A., Müller-Lerch, F., Rahman, S.H., Cornu, T.I., et al. (2008) Unexpected failure rates for modular assembly of engineered zinc fingers. *Nat Methods.* **5**, 374–375.
14. For resources and reagents related to ZFN design and application in zebrafish developed by the Wolfe and Lawson labs (*2*), we refer readers to their respective lab websites (http://labs.umassmed.edu/WolfeLab and http://lawsonlab.umassmed.edu). The available resources and tools include the plasmids and bacterial selection strain for bacterial 1-hybrid selections to make ZFNs, as well as an online tool to help identify good ZFN target sites (http://pgfe.umassmed.edu/ZFPsearch.html).
15. Streisinger, G., Coale, F., Taggart, C., Walker, C., and Grunwald, D.J. (1989) Clonal origins of cells in the pigmented retina of the zebrafish eye. *Dev Biol.* **131**, 60–69.
16. Lamason, R.L., Mohideen, M.A., Mest, J.R., Wong, A.C., Norton, H.L., Aros, M.C., et al. (2005) SLC24A5, a putative cation exchanger, affects pigmentation in zebrafish and humans. *Science.* **310**, 1782–1786.
17. Halpern, M.E., Ho, R.K., Walker, C., and Kimmel, C.B. (1993) Induction of muscle pioneers and floor plate is distinguished by the zebrafish *no tail* mutation. *Cell.* **75**, 99–111.
18. Schulte-Merker, S., van Eeden, F.J.M., Halpern, M.E., Kimmel, C.B., and Nüsslein-Volhard, C. (1994) *no tail (ntl)* is the zebrafish homologue of the mouse T (Brachyury) gene. *Development.* **120**, 1009–1015.
19. Agarose or Agar Troughs: http://zfin.org/zf_info/zfbook/chapt5/5.1.html Other alternatives: http://www.springerprotocols.com/Full/doi/10.1385/1-59259-678-9:125?encCode=QkVDOjUyMTo5LTg3Ni05NTI5NS0x&tokenString=165AlvaizdUK49gbSjNE4w==
20. Durai, S., Bosley, A., Abulencia, A.B., Chandrasegaran, S., and Ostermeier, M. (2006) A bacterial one-hybrid selection system for interrogating zinc finger-DNA interactions. *Comb Chem High Throughput Screen.* **9**, 301–311.
21. Thibodeau-Beganny, S. and Joung, J.K. (2007) Engineering Cys2His2 zinc finger domains using a bacterial cell-based two-hybrid selection. *Method Mol Biol.* **408,** 317–334.
22. Szczepek, M., Brondani, V., Büchel, J., Serrano, L., Segal, D.J., and Cathomen, T. (2007) Structure-based redesign of the dimerization interface reduces the toxicity

of zinc-finger nucleases. *Nat Biotechnol.* **25**, 786–793.
23. Miller, J.C., Holmes, M.C., Wang, J., Guschin, D.Y., Lee, Y.L., Rupniewski, I., et al. (2007) An improved zinc-finger nuclease architecture for highly specific genome editing. *Nat Biotechnol.* **25**, 778–785.
24. Provost, E., Rhee, J., and Leach, S.D. (2007) Viral 2A peptides allow expression of multiple proteins from a single ORF in transgenic zebrafish embryos. *Genesis.* **45**, 625–629.
25. Westerfield, M. (2000) The zebrafish book. A guide for the laboratory use of zebrafish (Danio rerio). 4th ed. University of Oregon Press, Eugene.
26. Foley, J.E., Yeh, J.R., Maeder, M.L., Reyon, D., Sander, J.D., Peterson, R.T., and Joung, J.K. (2009) Rapid mutation of endogenous zebrafish genes using zinc finger nucleases made by Oligomerized pool ENgineering (OPEN). *PLoS One 4*, e4348.

Chapter 19

A Transient Assay for Monitoring Zinc Finger Nuclease Activity at Endogenous Plant Gene Targets

Justin P. Hoshaw, Erica Unger-Wallace, Feng Zhang, and Daniel F. Voytas

Abstract

Advances in plant biology have been frustrated by the lack of an efficient means to create targeted mutations. Zinc finger nucleases (ZFNs) hold much promise for overcoming this limitation: they can be used to generate targeted gene knockouts through imprecise repair of broken chromosomes by non-homologous end joining (NHEJ), or they can stimulate the introduction of specific DNA sequence changes through homologous recombination. Critical to the function of ZFNs is their ability to access and cleave chromosomal target sites. Numerous factors may obscure cleavage, including packaging of DNA into chromatin, DNA methylation, or the presence of other proteins at the target site. Here we describe a transient assay that rapidly assesses ZFN function at chromosomal targets in plant cells. The assay monitors the ability of a ZFN to introduce mutations by imprecise repair through NHEJ, resulting in the loss of a restriction endonuclease recognition sequence. The requirement for the restriction endonuclease recognition sequence coincident with the ZFN spacer region has thus far not been a limiting factor in identifying ZFN target sites in genes of interest suitable for this assay.

Key words: Plant transformation, *Arabidopsis*, mutagenesis, non-homologous end joining.

1. Introduction

As in many organisms, the use of ZFNs in plants to create targeted DNA modifications is often a laborious and time-consuming process. Because of this, surrogate assay systems are widely used to first assess ZFN activity and to predict the likelihood that the desired genome modifications can be achieved. Such assays include bacterial two-hybrid assays that monitor target site recog-

J.P. Mackay, D.J. Segal (eds.), *Engineered Zinc Finger Proteins*, Methods in Molecular Biology 649, DOI 10.1007/978-1-60761-753-2_19, © Springer Science+Business Media, LLC 2010

nition by zinc finger arrays in *Escherichia coli* (1) or assays that measure ZFN function either in vitro (2) or in yeast (3, 4). A drawback to all of these surrogate assay systems is that they do not evaluate ZFN function at the target locus in the organism of choice. DNA-bound proteins, DNA methylation, or chromosome packaging could obscure a given ZFNs ability to find, recognize, and/or cleave its intended target. Here we describe an assay that surveys imprecise repair of ZFN-induced chromosome breaks to determine whether a specific ZFN can recognize and cleave its chromosomal target site. Mutations introduced by imprecise repair at the cleavage site by NHEJ can result in the loss of the restriction endonuclease recognition sequence located between the binding sites for the two zinc finger arrays (*see* **Fig. 19.1a**). Loss of the restriction site is measured by PCR amplifying the target site and then characterizing restriction endonuclease-insensitive amplicons by DNA sequencing for mutations at the cleavage site. This assay was first described for use in *Arabidopsis* plants (5), and we have adapted the approach to allow rapid assessment of ZFN function in *Arabidopsis* protoplasts following transient ZFN expression. It should be easy to modify this assay for use in other organisms.

Potential ZFN target sites are analyzed for the presence of a restriction endonuclease recognition sequence within the spacer region. We have found that the majority of ZFN target sites have an appropriate 4 bp restriction endonuclease recognition sequence within the spacer region. Because we find the assay so useful, we increasingly select ZFN target sites that contain suitable restriction sequences as part of the ZFN design process. The construction of ZFNs has been described elsewhere (5–5 and *see* **Chapters 1** and **14**). Once the zinc finger arrays are assembled, plasmids optimized for plant expression of both left and right ZFNs are constructed and introduced into *Arabidopsis* protoplasts using polyethylene glycol (PEG)-mediated transformation (6). Although untested in our lab, other DNA delivery methods developed for transient plant cell expression may also be used. Forty-eight hours after transformation, DNA is prepared and digested with the restriction endonuclease that recognizes the ZFN spacer sequence. This enriches for the population of DNA fragments that have sustained a mutation at the cut site. A fragment encompassing the ZFN target site is then PCR amplified using flanking primers. The amplicon is treated with the restriction enzyme and uncleaved PCR products are cloned and sequenced. Although some undigested DNA makes it through this step, most of the cloned PCR products have short insertions or deletions at the cut site that indicate activity of the ZFN.

This protocol is divided into six different methods. The goal of the first three methods is to obtain high-quality *Arabidopsis* protoplasts that are suitable for transformation. This involves

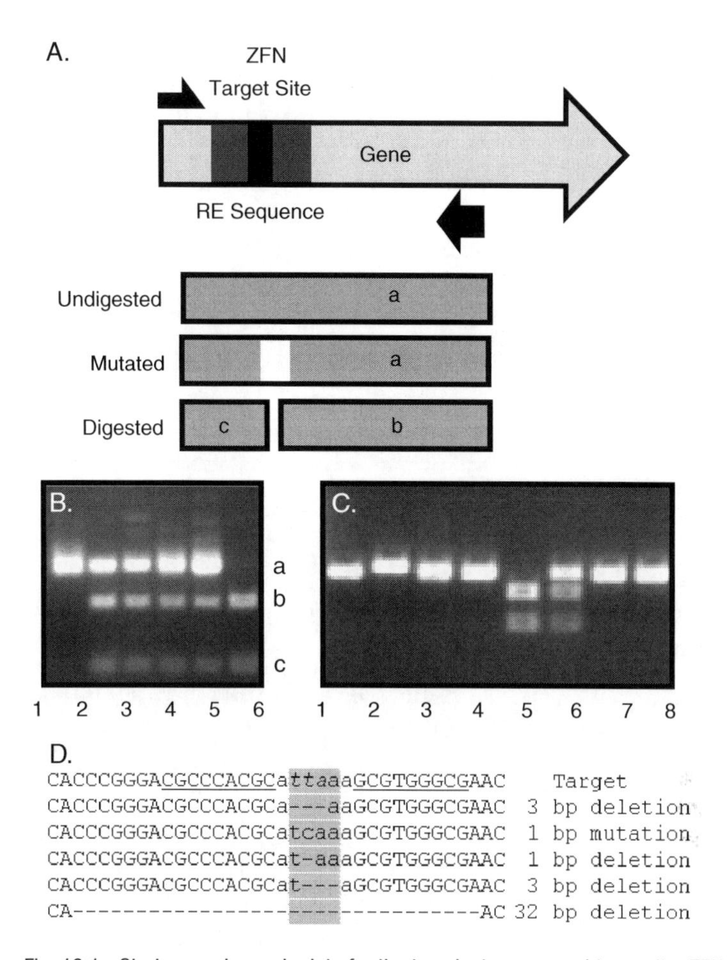

Fig. 19.1. Strategy and sample data for the transient assay used to monitor ZFN function in plant cells. (**A**) The large *arrow* depicts a typical target gene. The ZFN target site is shaded in *gray* and the restriction endonuclease cleavage sequence between the left and right ZFN recognition sites is shown in *black*. Flanking the target site are two half *arrows* that denote PCR primers 1 and 2. The PCR product resulting from amplification with the primers is digested with the restriction endonuclease, and three possible outcomes are shown: PCR fragments without mutations are digested by the restriction enzyme generating fragments *b* and *c*; PCR fragments with ZFN-induced mutations and those that are not digested are indistinguishable (fragment *a*). The two classes of fragments comprising fragment *a* can be distinguished by cloning individual PCR products and digesting with the restriction endonuclease (*see* panel **c**). (**B**) Sample data for the agarose gel described in step 5 of **Section 3.6**. After PCR amplifying the restriction endonuclease-digested genomic DNA, the PCR products are then digested with the same restriction endonuclease. Lane 1, undigested PCR product control; lanes 2–4, digested PCR product (note the presence of all three fragments illustrated in panel **a**); lane 5, a fully digested control that lacks mutations in the ZFN target site. (**C**) Individual clones obtained by cloning fragment *a* in panel **b** are digested with the restriction endonuclease and run on an agarose gel. Lanes 1–4 and 7–8 identify clones with mutations at the cut site. The slight variations in size are due to the fact that some samples have deletions or insertions due to imprecise repair of the ZFN cut site. Lanes 5–6 are wild-type clones derived from undigested genomic DNA; lane 6 is a partial digest. (**D**) Sample DNA sequences illustrating ZFN-induced mutagenesis. The various mutants are aligned with the unmodified target sequence that has the left and right ZFN sites *underlined*. The ZFN spacer sequence has an *Mse*I recognition sequence in *italics*.

(**Section 3.1**) growing seedlings under aseptic conditions and then (**Section 3.2**) harvesting and digesting the tissue with enzymes to release plant cells that lack cell walls (i.e., protoplasts). The protoplasts are then purified by centrifugation to remove cellular debris (**Section 3.3**). **Section 3.4** involves polyethylene glycol-mediated transformation of the protoplasts to introduce the ZFN-encoding plasmids. Finally, (**Section 3.5**) DNA is isolated from the treated protoplasts and then (**Section 3.6**) assayed for mutations introduced by non-homologous end joining at the ZFN cleavage site.

2. Materials

2.1. Seed Sterilization and Plating

1. 50% bleach: mix equal volumes of ddH_2O and bleach (household chlorine bleach) and store at room temperature.
2. 0.1% agarose: prepare 0.1% agarose in 500 mL of ddH_2O, autoclave, and store at room temperature.
3. Half-strength Murashige and Skoog (MS) plates: Dissolve half a package of MS salts (Caisson Laboratories) and 10 g of sucrose in 800 mL of ddH_2O, and adjust the volume to 1 L. Add 7.5 g of agarose, autoclave, pour into large Petri dishes with beveled ridges (150 mm diameter × 15 mm height, Fisher), and store at 4°C.
4. Custom plate rack for growing seedlings: MS plates are positioned vertically while seeds are germinating. This allows the roots to grow across the agar surface. A plate rack for this purpose can be created by taking a medium to large piece of Styrofoam and cutting out rows one-half inch deep and sufficiently wide to snuggly hold the MS plates.
5. Fluorescent lights: shop lights can be used that are positioned 18–24 in. above a bench or shelf. Standard fluorescent bulbs are suitable for germination and plant growth.

2.2. Digestion of Plant Tissue to Release Protoplasts

1. Dissecting forceps, 7 in. (Fisher).
2. Metal scalpel, 3.8 cm (Fisher).
3. Nuncon cell culture plates: 90 mm diameter × 20 mm height (Fisher).
4. K3 medium major salts: 250 mM KNO_3, 10 mM $(NH4)_2SO_4$, 10 mM $MgSO_4{\cdot}7H_2O$, 12 mM KH_2PO_4, 30 mM NH_4NO_3. Prepare by dissolving 25 g of KNO_3, 1.34 g of $(NH4)_2SO_4$, 2.5 g of $MgSO_4{\cdot}7H_2O$, 2 g of KH_2PO_4, and 2.5 g of NH_4NO_3 in 800 mL of ddH_2O. Adjust the volume to 1 L, autoclave, and store at 4°C.

5. 625 mM $CaCl_2 \cdot 2H_2O$ solution: dissolve 92.3 g of $CaCl_2 \cdot 2H_2O$ in 800 mL of ddH_2O, adjust the volume to 1 L, autoclave, and store at 4°C.
6. 10 mM Fe EDTA solution: prepare by dissolving 3.75 g of Na_2EDTA completely in 800 mL of ddH_2O. Then add 2.75 g of $FeSO_4 \cdot 7 \cdot H_2O$, adjust to pH 3.0, adjust the volume to 1 L, autoclave, and store at 4°C.
7. B5 vitamins: 0.8 mM nicotinic acid, 3 mM thiamine HCl, 0.5 mM pyridoxine HCl, 55 mM myo-inositol. Prepare by dissolving 0.1 g of nicotinic acid, 1.0 g of thiamine HCl, 0.1 g of pyridoxine HCl, and 10 g of myo-inositol in 800 mL of ddH_2O. Adjust the volume to 1 L and store at 4°C.
8. MS minors: 10 mM H_3BO_3, 10 mM $MnCl_2 \cdot 4H_2O$, 32 mM $ZnSO_4 \cdot 7H_2O$, 0.5 mM KI, 0.1 mM $Na_2MoO_4 \cdot 2H_2O$, 0.01 mM $CuSO_4 \cdot 5H_2O$, 0.01 mM $CoCl_2 \cdot 6H_2O$. Prepare by dissolving 0.62 g of H_3BO_3, 1.98 g of $MnCl_2 \cdot 4H_2O$, 0.92 g of $ZnSO_4 \cdot 7H_2O$, 0.083 g of KI, 0.025 g of $Na_2MoO_4 \cdot 2H_2O$, 0.0025 g of $CuSO_4 \cdot 5H_2O$, and 0.0022 g of $CoCl_2 \cdot 6H_2O$ in 800 mL of ddH_2O. Adjust the volume to 1 L, autoclave, and store at 4°C.
9. Naphthelene acetic acid (NAA) solution: dissolve 0.1 g of NAA in dilute KOH and adjust the volume to 100 mL with ddH_2O. Filter sterilize and store at 4°C for up to 6 months.
10. Benzylamino purine (BAP) solution: dissolve 0.1 g of BAP in dilute HCl and adjust the volume to 100 mL with ddH_2O. Filter sterilize and store at 4°C for up to 6 months.
11. K3/S1 medium with enzymes: combine 100 mL of K3 major salts, 10 mL of $CaCl \cdot 2H_20$ solution, 10 mL of Fe EDTA solution, 10 mL of B5 vitamins, 1 mL of MS minors, 136.8 g of sucrose, 3 mL of NAA solution, 1 mL of BAP solution, 2.5 g of cellulase R10 (Karlan) (*see* **Note 1**), 0.05 g of macerozyme R10 (Karlan), and ddH_2O to 800 mL. Once sucrose is dissolved, adjust the volume to 1 L, adjust to pH 5.7, filter sterilize, and aliquot 40 mL into sterile 50 mL disposable centrifuge tubes. Store at –20°C.
12. Air incubator set at 30°C.

2.3. Protoplast Collection

1. Pyrex funnels with 60° angle bowls, 65 mm top diameter, 150 mm stems, and 206 mm overall length (Fisher).
2. 1 and 10 mL graduated pipettes.
3. Rubber tubing: 7 cm in length with an inside diameter of 4.8 mm.
4. VWR Cheesecloth Wipers (VWR).

5. Scissors.
6. Aluminum foil.
7. Kimax brand Babcock bottles for milk testing: 8% bottle with "Sealed 3" marking (Fisher).
8. Kimwipes (Kimtech Science).
9. 70% ethanol.
10. Medium-sized, cast-iron, tripod-base support stand (Fisher).
11. Two-prong extension clamp for tripod-base support stand (Fisher).
12. Flexaframe castaloy hook connector for joining two-prong extension clamp to tripod-base support stand (Fisher).
13. K3/S1 medium: combine 100 mL K3 major salts, 10 mL $CaCl{\cdot}2H_20$ solution, 10 mL Fe EDTA solution, 10 mL B5 vitamins, 1 mL MS minors, 136.8 g sucrose, 3 mL NAA solution, 1 mL BAP solution, and ddH_2O to 800 mL. Once sucrose is dissolved, adjust volume to 1 L, adjust to pH 5.7, filter sterilize, and store at 4°C.

2.4. Protoplast Transformation

1. MMg medium: 0.4 M mannitol, 15 mM $MgCl_2$, 4 mM MES. Dissolve 10.92 g of mannitol, 0.46 g of $MgCl_2$, and 0.15 g of MES in 150 mL of ddH_2O. Adjust to pH 5.7, filter sterilize, and store at 4°C.
2. Plasmid DNA encoding the ZFN to be tested. Each sample should have approximately 10 μL at a concentration of 2 μg/μL for the left and right array.
3. Plasmid DNA encoding a green fluorescent protein (GFP). Use approximately 10 μL at a concentration of 2 μg/μL.
4. 0.8 M mannitol: dissolve 29.14 g of mannitol in 200 mL of ddH_2O. Store at room temperature.
5. 1 M $CaCl_2$: dissolve 22.2 g of $CaCl_2$ in 200 mL of ddH_2O. Store at room temperature.
6. 40% polyethylene glycol (PEG) solution: dissolve 4 g of PEG 4,000 (Fluka, #81240), 3 mL of ddH_2O, 2.5 mL of 0.8 M mannitol, and 1 mL of 1 M $CaCl_2$, filter sterilize, and store at 4°C.
7. W5 solution: 154 mM NaCl, 125 mM $CaCl_2$, 5 mM KCl, 2 mM MES. Dissolve 4.5 g of NaCl, 9.2 g of $CaCl_2$, 0.19 g of KCl, 0.2 g of MES in 500 mL of ddH_2O. Adjust to pH 5.7, filter sterilize, and store at 4°C.
8. K3/S1 medium (as in step 13 of **Section 2.3**).
9. Nuncon cell culture plates: 90 mm diameter × 20 mm height (Fisher).

2.5. Protoplast DNA Isolation

1. 2× CTAB: 27.44 mM hexadecyl trimethyl ammonium bromide (CTAB), 1.4 M NaCl, 20 mM EDTA, 100 mM Tris–HCl, pH 8.0, and 1.4 M NaCl. Dissolve 1 g of CTAB (EM Science) in 5.0 mL of 1 M Tris–HCl, pH 8.0 (Invitrogen), 2 mL of 0.5 M EDTA, and 14 mL of 5 M NaCl, adjust volume to 50 mL with ddH_2O, and store at room temperature.
2. Chloroform (caution: hazardous).
3. Isopropanol (caution: hazardous).
4. 70% ethanol.
5. TE buffer: 10 mM Tris–HCl, pH 7.5, 1 mM EDTA.

2.6. DNA Digestion

1. Specified restriction enzyme that has a target site located in the spacer region of the ZFN target site.
2. Corresponding buffer for the restriction enzyme.
3. Primers 1 and 2 to amplify and sequence the ZFN target site.
4. HotStarTaq Plus Master Mix Kit (Qiagen).
5. Razor blades.
6. MinElute PCR Purification Kit (Qiagen).
7. 2% agarose gel: prepare 2% agarose solution in 500 mL of ddH_2O. Microwave until completely melted, mixing the solution every minute while heating. Add 10 μL of 10 mg/mL ethidium bromide solution, mix gently, and pour the gel.
8. Qiaex II Agarose Gel Extraction Kit (Qiagen).
9. TOPO TA Cloning Kit: pCR 2.1-TOPO Vector (Invitrogen).
10. One Shot TOP10 Electrocomp Cells (Invitrogen). Store at –80°C.
11. 5-bromo-4-chloro-3-indolyl-beta-D-galactopyranoside (Xgal): dissolve 0.02 g of Xgal in 1 mL of dimethyl sulfoxide (DMSO), filter sterilize, and put into a sterile Eppendorf tube covered with aluminum foil. Store at –20°C.
12. Kanamycin: dissolve 1 g of kanamycin in 20 mL of ddH_2O. Filter sterilize, aliquot out into Eppendorf tubes, and store at –20°C.
13. Kanamycin plates: dissolve 10 g of tryptone, 5 g of yeast extract, 5 g of NaCl, and 15 g bacteriological agar in 1 L of ddH_2O. Autoclave the medium and when cool add 1 mL of kanamycin and 1 mL of Xgal.
14. Selective LB medium: dissolve 10 g of tryptone, 5 g of yeast extract, and 5 g of NaCl in 1 L of ddH_2O. Autoclave the medium and when cool add 1 mL of kanamycin.

15. 2 mL 96-well culture plates (Costar).
16. AirPore tape sheets (Qiagen).

3. Methods

3.1. Seed Sterilization and Plating

1. Weigh out two aliquots of 0.05 g of *Arabidopsis thaliana* seeds (approximately 4,000) on weighing paper (*see* **Note 2**). Carefully transfer each seed aliquot to an Eppendorf tube. Each tube has sufficient seeds for two MS plates (*see* **Note 3**).
2. Add 1 mL of 50% bleach solution to each tube and make sure the tube is tightly sealed.
3. Vortex the tubes for 5 min at a medium speed to make sure that the seeds are constantly suspended in solution. All subsequent steps should be performed in a tissue culture hood to ensure that the sterile seeds do not become contaminated.
4. Centrifuge the tubes in a microcentrifuge for 10 s.
5. Remove the bleach solution with a micropipettor and a sterile pipette tip.
6. Wash the seeds with 1 mL of sterile ddH_2O, vortex briefly, centrifuge, and remove the liquid. Wash the seeds two additional times. Remove the water using fresh, sterile pipette tips each time.
7. Add 1.2 mL of 0.1% agarose to the Eppendorf tubes. Resuspend the seeds evenly in the agarose by pipetting up and down several times.
8. Plate the seeds on large MS plates using a 1,000 μL micropipettor. Disperse the seeds in long narrow rows across the plate with about half an inch between each row. Plating the seeds in rows makes it easier to remove the seedlings in small clumps rather than one matte of intertwined seedlings.
9. Seal the plates with Parafilm and place them vertically in the custom plate racks (*see* **Fig. 19.2**). Position the racks under fluorescent lights at room temperature. The top of the plates should be 6–12 in. from the lights. Tilting the plates vertically prevents the roots from growing into the medium and makes the seedlings easier to harvest. Seedlings should be ready to harvest in 4–8 days (*see* **Note 4**).

3.2. Digestion of Plant Tissue to Release Protoplasts

1. Sterilize the forceps and scalpel using either an autoclave or a glass bead sterilizer.
2. Thaw two 40-mL aliquots of K3/S1 medium with enzymes at room temperature. When completely thawed, transfer 20 mL into each of four cell culture plates.

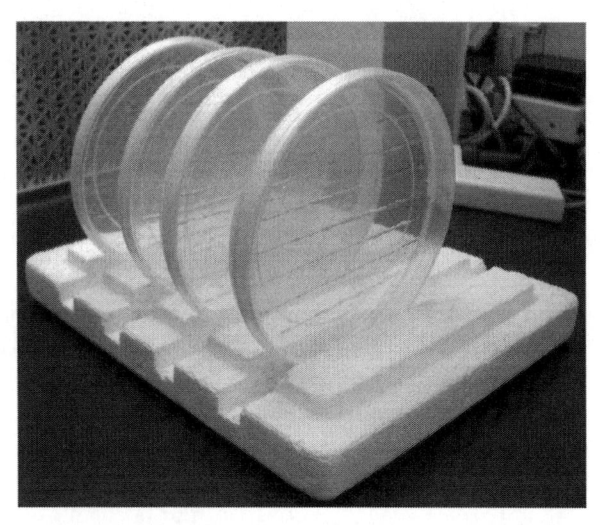

Fig. 19.2. The custom plate rack is shown holding plates with MS medium and newly planted *Arabidopsis* seeds.

3. Collect the seedlings from one of the four MS plates using the sterilized forceps. Take care that no agarose adheres to the roots of the seedlings. Transfer the seedlings onto the sterile side of a cell culture plate lid. While holding a group of seedlings with the forceps, use the scalpel to chop them into 0.5 cm pieces.
4. Place the cut plant tissue into one of the culture plates with K3/S1 medium with enzymes. Plant tissue can be added until a majority of the surface has been occupied. Make sure that the plant tissue is evenly spread out and there are no clumps.
5. Repeat steps 3 and 4 with the other three MS plates with seedlings.
6. Wrap the plates with Parafilm and incubate at 30°C for 16 h.

3.3. Protoplast Collection

1. Remove the cotton filter from a 1 mL glass pipette. Carefully break off the narrow tip of the pipette. Insert about 1.5 cm of one end of the pipette into the 7 cm rubber tubing. Insert the end of the funnel into the other end of the rubber tubing. Cut a 20 cm × 20 cm sheet of cheesecloth that is two layers thick. Fold the cheesecloth twice to make a smaller square. Place the cheesecloth into the funnel and open it so that the cheesecloth fits inside the funnel. Cut off any excess material. Dampen the cheesecloth with water. Wrap the apparatus with aluminum foil using one sheet for the funnel and another for the pipette. Autoclave the funnel apparatus to sterilize.

2. Place a small, approximately 5 cm × 5 cm, square of foil over the top of each of three Babcock bottles and fold down around the neck of each bottle. Autoclave to sterilize.
3. Wipe down the tissue culture hood with Kimwipes and 70% ethanol to sterilize. Remove the Parafilm from the plates with the digesting plant material. Gently agitate the plates for about 1 min to dislodge as many protoplasts into the medium as possible.
4. Bring the autoclaved funnel and bottles into the hood. Remove the foil from the base of the funnel and attach the Babcock funnel to the stand using the clamp. Now remove the rest of the foil. Take off the foil from one Babcock bottle, but keep the foil in a sterile area. Slide the funnel's pipette tip into the Babcock bottle until the tip is about 3 cm from the bottom.
5. Using a 10-mL pipette, add 5 mL of K3/S1 medium to the funnel to moisten the cheesecloth. Use the same pipette to collect the protoplasts. It may be necessary to pipette up and down to release the cells and break up clumps of tissue. Tilt the plate so that the cells are on one side and wash the surface of the plate with liquid to free any cells stuck to the bottom of the plate. After pipetting, release the cells into the center of the funnel. Avoid filling the funnel too full, because the cheesecloth can wick media up and over the sides of the funnel and contaminate the sample.
6. With a new 10-mL pipette, add 5 mL of additional K3/S1 medium to wash any remaining cells out of the cheesecloth. Remove the Babcock bottle from the bottom of the funnel. Add additional K3/S1 medium until you reach the top mark on the Babcock bottle. This can be done using the same pipette and releasing the medium slowly into the Babcock bottle while tilted. Replace the foil on the Babcock bottle opening.
7. Centrifuge the Babcock bottles in a table-top centrifuge for 10 min at 100×*g*. After centrifugation, the divisions (markings) on the Babcock bottle can be used to estimate the yield of protoplasts. One Babcock division approximates 250,000 protoplasts and is sufficient for one transformation experiment.
8. Wash the protoplasts to remove undigested plant tissue. Add 10 mL of K3/S1 medium to a new bottle (*see* **Note 5**). Using a 1-mL pipette remove the cells from the top of the centrifuged Babcock bottle and put into the new Babcock bottle. Again add K3/S1 medium to the top mark on the bottle.

9. Centrifuge for 10 min at 100×*g* in the table-top centrifuge. A different Babcock bottle filled with water and foil wrapped around the neck can serve as a counterbalance for this step.
10. Repeat steps 8 and 9 two additional times.

3.4. Protoplast Transformation

1. In the tissue culture hood, add 100 μL of MMg medium for each Babcock division to the corner of a sterile tissue culture plate that is slightly tilted. MMg medium helps to stabilize the protoplasts that were weakened during purification.
2. Remove the protoplasts from the top of the Babcock bottle and add them to the MMg medium. For each division required, remove 100 μL of cells and liquid from the top of the Babcock bottle. Mix gently by slowly pipetting up and down using a micropipettor with a 1-mL sterile tip.
3. Add 200 μL of the protoplast solution to individual Eppendorf tubes, making sure to keep the protoplasts evenly suspended during the transfer process.
4. Add 60 μg of DNA per tube (*see* **Note 6**). Mix by inverting several times and wait 2 min.
5. Add 100 μL of PEG solution to each sample. Mix every 10 min for 30 min by inverting the tube several times.
6. Add 800 μL of W5 solution and mix by inverting. Wait for 5 min.
7. Centrifuge for 3 min at 150×*g* in a microcentrifuge.
8. Remove the supernatant. Add 1 mL of W5 solution to resuspend the protoplasts. Wait 5 min. Centrifuge as in step 7 and repeat the wash one more time.
9. Add 1 mL of K3/S1 medium to the washed protoplasts. Transfer the solution onto the Nuncon cell culture plates and add 4 mL of K3/S1 medium (*see* **Note 7**).
10. Parafilm the plates and place at room temperature for 2 days.
11. The health of the cells can be monitored using a microscope. Healthy cells are full and round with visible chloroplasts (*see* **Note 8**).

3.5. Protoplast DNA Isolation

1. Collect approximately 5 mL of protoplasts in three Eppendorf tubes using a pipette. Remove the supernatant after centrifuging at 5,500×*g* in a microcentrifuge for 15 min.
2. Add 750 μL of 2× CTAB to one of the microcentrifuge tubes and resuspend the cells by pipetting up and down. Use this solution to resuspend the cells in the other

tubes. All of the cells from one sample should now be in one tube.

3. Incubate at 65°C for 30 min and mix every 10 min.
4. Add 750 μL of chloroform. Mix by inverting the tube for 5 min.
5. Centrifuge at 9,500×*g* in a microcentrifuge for 10 min.
6. Transfer the top layer to a new tube. Add 700 μL of isopropanol and mix by inverting the tube several times.
7. Centrifuge again.
8. Wash the pellet twice with 1 mL of 70% ethanol, centrifuge after each wash.
9. Air dry the DNA pellet for 7–10 min.
10. Dissolve the DNA in 30 μL of TE buffer. DNA can be kept at 4°C for short-term use or –20°C for longer periods.

3.6. DNA Digestion

1. Digest the genomic DNA (*see* **Note 9**). Use 10 μL of DNA, 5 μL of the appropriate restriction enzyme buffer, 4 μL of restriction enzyme, and 31 μL of ddH_2O for a total of 50 μL. Let the digestion go overnight at the temperature specific for the enzyme being used. Performing this digestion before the PCR step will remove wild-type DNA targets and enrich for targets with mutations.
2. PCR the digested DNA (*see* **Note 10**). **Figure 19.1A** provides a visual for how to set up the placement of the primers in order to obtain an easily distinguishable banding pattern. Use 4 μL of DNA, 50 μL of HotStarTaq Plus Master Mix, 2 μL of primer 1, 2 μL of primer 2, and 42 μL of ddH_2O. Heat the lid of the PCR machine to 105°C. Add reactions, hot start for 15 min at 95°C, and then carry out 40 cycles at 94°C for 30 s, 55°C for 30 s, 72°C for 1 min, and an extension at 72°C for 10 min.
3. Purify the PCR product using a MinElute PCR Purification Kit.
4. Digest the purified product again using 10 μL of DNA, 2.5 μL of buffer, 1 μL of enzyme, and 11.5 μL of ddH_2O for 3 h at the preferred temperature for the restriction enzyme. This step further removes targets with wild-type sequences.
5. Following digestion the DNA is separated by electrophoresis in a 2% agarose gel (*see* **Fig. 19.1B**). The larger, restriction enzyme-insensitive fragments are excised and purified from the gel using a Qiaex II Agarose Gel Extraction Kit.
6. This population of fragments is cloned using a TOPO TA Cloning Kit. Use 4 μL of PCR product, 1 μL of salt solution, and 1 μL of pCR 2.1-TOPO vector.

7. Transform 2 μL of the ligation product into one vial of One Shot TOP10 Electrocomp Cells. Incubate on ice for 30 min before heat shocking the cells for 30 s at 42°C in a water bath. Add 250 μL of SOC medium (supplied with the cells) to the cells, shake for 1 h at 37°C, and spread 100 μL of the transformation onto a kanamycin plate. Incubate the plate overnight at 37°C.
8. Pick white colonies from the plate with a 200-μL pipette tip and put into a 2-mL 96-well culture plate containing 1 mL of selective LB medium in each well. Seal the plate using an AirPore tape sheet and incubate overnight in a 37°C shaker at 250 rpm.
9. PCR amplify the cloned fragments directly, using 1 μL of overnight culture, 0.5 μL of primer 1, 0.5 μL of primer 2, 12.5 μL of Master Mix, and 10.5 μL of ddH_2O. Use the same PCR conditions as in step 2, but with 35 cycles.
10. Verify the presence of restriction enzyme-insensitive fragments by digesting an aliquot of the PCR reaction, followed by electrophoresis on a 2% agarose gel (*see* **Fig. 19.1C**). Digestion reactions should include 5 μL of PCR product, 1.5 μL of buffer, 0.5 μL of enzyme, and 8 μL of ddH_2O. Digest the samples for 3 h at the appropriate temperature for the enzyme.
11. For those clones containing restriction enzyme-insensitive inserts, the DNA is submitted for sequence analysis. Some clones may produce partially digested products (e.g., **Fig. 19.1C**, lane 6) and such clones should not be sent for sequence analysis.
12. Sequences are aligned to the reference target DNA sequence using a DNA analysis software program (e.g., Vector NTI or Sequencher). Mutations at the ZFN cleavage site should be apparent (*see* **Fig. 19.1D**).

4. Notes

1. Cellulase R10 should be used. Cellulose RS is too strong and will inhibit the healthy growth of protoplasts.
2. If you do not wear gloves for this step it reduces the amount of static electricity and the seeds are easier to handle.
3. One Eppendorf tube of sterilized seeds will yield sufficient plant material for approximately five transformation

experiments after 5 days of growth. If the seeds grow for 8 days, the plate(s) will yield approximately 10 transformation experiments.

4. If the MS plates with seeds become contaminated with fungi or bacteria they will need to be discarded. To overcome contamination, the seeds can be left in the bleach solution for a longer amount of time. Also, adding 0.05% Plant Preservative Mixture (Plant Cell Technology, Inc.) to the half-strength MS solid medium can reduce contamination. If the problem persists, it may be worth using a different batch of seeds. Seeds collected under high humidity often have high levels of microbial contaminants.
5. If you tilt the bottle to about a 50° angle, you can easily use a 10-mL pipette to fill the bottle. Go slowly and almost touch the tip of the pipette to the inside rim of the bottle.
6. Remember to include a positive and negative control for your experiment. A plasmid encoding GFP works well as a positive control and can be mixed with the ZFN DNA sample. About a third of the cells should contain all three plasmids, and transformation can be monitored by observing GFP fluorescence. GFP may be visible after as little as 6 h. Do not introduce any DNA into the negative control.
7. If you find that the cells do not look very healthy (e.g., many small or shriveled cells) you can try adding extra protoplasts to the transformed cells after plating in order to give the cells a greater density. Healthy growth is often density dependent.
8. If after 1 day a majority of the cells appear to be dying, you can isolate the DNA at that time. Depending on the toxicity of the ZFN, the transformed protoplasts can survive for up to a week. In our experience, 48 h seems to be the optimal time for DNA isolation, as it provides sufficient time to allow the ZFN to cause double-strand breaks and the host to repair the break.
9. The restriction enzyme site should be in the middle of the ZFN site in order to obtain the most mutants. Deletions of up to 50 bp have been recorded, but smaller mutations of about 4 bp are much more common. The use of an efficient enzyme can greatly aid in the isolation of mutants, because it will more robustly cleave the unmodified DNA targets.
10. Position the PCR primers so that the digested amplification products give you distinguishable bands on the gel. A difference of at least 50 bp is necessary to detect a size difference by standard agarose gel electrophoresis.

References

1. Hurt, J.A., Thibodeau, S.A., Hirsh, A.S., Pabo, C.O., and Joung, J.K. (2003) Highly specific zinc finger proteins obtained by directed domain shuffling and cell-based selection. *Proc Natl Acad Sci USA.* **100**, 12271–12276.
2. Carroll, D., Morton, J.J., Beumer, K.J., and Segal, D.J. (2006) Design, construction and in vitro testing of zinc finger nucleases. *Nat Protoc.* **1**, 1329–1341.
3. Doyon, Y., McCammon, J.M., Miller, J.C., Faraji, F., Ngo, C., Katibah, G.E., Amora, R., Hocking, T.D., Zhang, L., Rebar, E.J., et al. (2008) Heritable targeted gene disruption in zebrafish using designed zinc-finger nucleases. *Nat Biotechnol.* **26**, 702–708.
4. Townsend, J., Wright, D.A., Winfrey, R.J., Fu, F., Maeder, M., Joung, J.K., and Voytas, D.F. (2009) High-frequency modification of plant genes using engineered zinc finger nucleases. *Nature.* **495**, 442–445.
5. Lloyd, A., Plaisier, C.L., Carroll, D., and Drews, G.N. (2005) Targeted mutagenesis using zinc-finger nucleases in *Arabidopsis. Proc Natl Acad Sci USA.* **102**, 2232–2237.
6. Yoo, S.D., Cho, Y.H., and Sheen, J. (2007) Arabidopsis mesophyll protoplasts: a versatile cell system for transient gene expression analysis. *Nat Protoc.* **2**, 1565–1572.

Chapter 20

Validation and Expression of Zinc Finger Nucleases in Plant Cells

Andriy Tovkach, Vardit Zeevi, and Tzvi Tzfira

Abstract

Zinc finger nucleases (ZFNs) can be designed to target virtually any long stretch of DNA sequence. Their expression in living cells has been shown to lead to gene targeting via homologous recombination, site-specific mutagenesis, and targeted DNA integration in various species. A variety of assays have been developed to test ZFN activity both in vitro and in vivo, and an assortment of vectors have been constructed to facilitate the analysis and expression of ZFNs in mammalian, and specifically human cells, as well as in other model organisms. Here we describe a set of protocols and vectors that were specifically designed to analyze ZFN activity in plant cells. Our assays provide the user with versatile tools and simple protocols for *in-planta* analysis of ZFN activity on transiently delivered and stably integrated mutated plant reporter (GUS)-encoding genes. Specifically designed for maximum compatibility with a generalized plant expression system, our vector system also allows easy assembly of ZFN plant transformation vectors for gene-targeting experiments in plants.

Key words: Gene targeting, binary vectors, plant transformation, multigene vectors.

1. Introduction

Genome editing via gene targeting (GT) is one of the most sought after technologies for basic research and biotechnology in plants. GT in plant cells (and in many other species) is limited by the domination of the cell's non-homologous end joining (NHEJ) DNA repair mechanisms over homologous recombination (HR) (1–3). Thus, foreign DNA molecules, delivered via *Agrobacterium*-mediated genetic transformation, microbombardment, or PEG-mediated transformation of plasmid DNA into plant cells, often integrate at random locations across the genome

J.P. Mackay, D.J. Segal (eds.), *Engineered Zinc Finger Proteins*, Methods in Molecular Biology 649,
DOI 10.1007/978-1-60761-753-2_20,

and not at specific (i.e., targeted) locations. Different experimental methods have been designed to select for rare GT events or to redirect foreign DNA molecules to specific genomic locations. Thus, for example, strong positive and negative selection schemes have been used to achieve reproducible GT in rice (4, 5), and the expression of yeast RAD54 in *Arabidopsis* plants has been shown to lead to increased HR rates in this model species (6).

Induction of site-specific genomic double-strand breaks (DSBs) has also been shown to enhance HR at specific genomic locations (7), and various reports have demonstrated that expression of restriction enzymes and other endonucleases can lead to increased HR-mediated GT in model plants (8–10). Expression of restriction enzymes in plant cells can also lead to site-specific mutagenesis and the incorporation of foreign DNA at specific genomic locations (e.g., 11–13). Nevertheless, the very limited number of sequence-specific rare-cutting restriction enzymes greatly limits their potential use for the targeting of native genomic sequences in living cells. Zinc finger nucleases (ZFNs) acting as synthetic restriction enzymes may provide a valuable alternative to natural and engineered rare-cutting restriction enzymes, since they can be specifically designed to cleave virtually any long stretch of double-stranded DNA sequence (for recent reviews *see* 14–16). ZFNs have been successfully used for the production of genomic DSBs in various species, including plants, leading to a variety of outcomes (e.g., enhanced HR, site-specific mutagenesis, and targeted gene addition, 17–25).

Realizing the power of ZFNs as a novel emerging technology, a collection of procedures, web tools, plasmids, and assays has been developed to facilitate the design and assembly of novel ZFNs and the analysis and validation of ZFN activity both in vitro and in vivo. Thus, for example, various tools and vectors for the design and assembly of novel ZFNs are available from Carlos Barbas's laboratory (www.zincfingertools.org) and the Zinc Finger (ZF) Consortium (www.zincfingers.org). More recently, a unique strategy for constructing multifinger arrays (designated Oligomerized Pool ENgineering or OPEN) has been described as a rapid and public tool for the assembly of novel ZFNs (26). Other resources include, for example, a bacterial cell-based reporter assay to analyze the DNA-binding activities of zinc finger domains (27, and *see* **Chapter 2**), a cell culture-based assay useful for validation of ZFN activity in mammalian cells (28, and *see* **Chapters 1, 13,** and **14**), and a digestion assay for in vitro analysis of ZFN digestion activity (29, and *see* **Chapter 15**).

Here we describe a collection of vectors and assays that are useful for the validation and expression of ZFNs in plants. **Section 3.1** describes the construction of a reporter-based vector for validating ZFN activity in plant cells. We used the GUS reporter system, which is commonly used for visualization of gene expression in plant cells. The *gusA* gene encodes β-glucuronidase (GUS) which cleaves the X-Gluc (5-bromo-4-chloro-3-indolyl-beta-D-glucuronic acid) colorless substrate to colorless glucuronic acid and a blue precipitate of chloro-bromoindigo. Other reporter genes (e.g., GFP) can also be used, but the advantage of GUS lies in the simplicity of the staining and detection procedures and its applicability to many plant species. **Section 3.2** describes the assembly of an *Agrobacterium* core binary vector. This vector is to be used in the *in-planta* validation assays (i.e., *Agrobacterium*'s transferred DNA (T-DNA) repair assay and transgene repair assay, described in **Sections 3.2** and **3.3**), and will also serve as part of the whole-plant DNA repair assay (described in **Section 3.4**). Finally, in **Section 3.5**, we describe the assembly of multi-ZFN plant transformation vectors that can facilitate the transfer and use of ZFNs for GT experiments in plant species. The simplicity of our assays and the compatibility of our ZFN validation and expression vectors with a versatile and extensive family of plant transformation vectors (30–32) allow tailored cloning of the ZFN validation and transformation vectors to a specific need.

2. Materials

2.1. Equipment and Consumables

1. Environmentally controlled shaker (37°C) for culturing *Escherichia coli*.
2. Environmentally controlled incubator (37°C) for growing *E. coli*.
3. Environmentally controlled shaker (28°C) for culturing *Agrobacterium*.
4. Environmentally controlled incubator (28°C) for growing *Agrobacterium*.
5. Spectrophotometer for measuring optical density of bacterial cultures.
6. Small-volume (1.5–2 mL) refrigerated centrifuge.
7. Disposable tall 90-mm plant culture dishes.
8. Disposable 50-mL sterile plastic tubes.
9. Plasmid DNA extraction kit (e.g., cat. no. K3030, Bioneer).
10. PCR fragment purification kit (e.g., Illustra GFX PCR DNA and Gel Band Purification Kit, cat. no. 28-9034-70, GE Healthcare).
11. Thin-walled PCR tubes.
12. 10-mL single-use disposable syringes.

13. $1^1/_2$ gauge single-use sterile needles.
14. Razor blades.
15. Sterile 10 × 10 cm filter papers.
16. Standard equipment and reagents for agarose gel electrophoresis.
17. Power supply.
18. Heated water bath (42°C).
19. No. 11 scalpel blades and scalpel handles for plant tissue culture.
20. Polymerase chain reaction (PCR) thermocycler.
21. Dry/wet heat block.

2.2. Media, Antibiotics, and Chemicals

1. 1,000× stock solutions of antibiotics: 50 mg/mL kanamycin, 100 mg/mL ampicillin, 50 mg/mL spectinomycin, 50 mg/mL streptomycin, and 1,000 mg/mL carbenicillin in double-distilled water (ddH_2O). Filter sterilize and maintain for exclusive use for plant tissue culture, aliquot, and store at –20°C up to 30 days.
2. 1 M acetosyringone (3,5-dimethoxy-4-hydroxyacetophenone, prod. no. D134406, Sigma-Aldrich): dissolve 196.2 mg of acetosyringone in 1 mL of dimethyl sulfoxide (DMSO), aliquot, and store at –20°C up to 30 days.
3. Luria broth (LB) liquid medium: 5 g yeast extract, 10 g tryptone, and 10 g NaCl in 1 L of ddH_2O. Autoclave and store at room temperature.
4. LB solid medium: same as LB liquid medium, with the addition of 15 g/L agar before autoclaving. Store at room temperature.
5. Murashige and Skoog (MS) basal salts with vitamins and glycine (cat. no. MSP0109, Caisson Laboratories).
6. Plant cell culture agar (Sigma-Aldrich).
7. 2-(*N*-morpholine)-ethane sulfonic acid (MES), low moisture (Sigma-Aldrich).
8. 1 mg/mL naphthalene acetic acid (NAA, PhytoTechnology Laboratories).
9. 1 mg/mL benzylaminopurine (BAP, PhytoTechnology Laboratories).
10. X-Gluc (Gold Biotechnology).
11. Silwet L-77 (Lehle Seeds).
12. Agrobacterium Induction Medium: 10.5 g/L K_2HPO_4, 4.5 g/L KH_2PO_4, 1 g/L $(NH_4)_2SO_4$, 0.5 g/L Na-citrate,

1 g/L glucose, 1 g/L fructose, 4 g/L glycerol, 1 mM $MgSO_4$, 10 mM MES; adjust pH to 5.7 with KOH. Autoclave and store at room temperature (*see* **Note 1**).

13. Agrobacterium Infiltration Medium: 10 mM $MgSO_4$, 10 mM MES; adjust pH to 5.7 with KOH. Autoclave and store at room temperature (*see* **Note 1**).
14. Tobacco Callus Induction Medium: full-strength Gamborg's B5 medium (PhytoTechnology Laboratories), 30 g/L sucrose, 0.2 mg/L BAP, 2.0 mg/L NAA; adjust pH to 5.7 with KOH. Supplement the medium with 10 g/L plant cell culture agar. Autoclave and store at room temperature.
15. Tobacco Callus Selection Medium: supplement the tobacco callus induction medium with 300 mg/L carbenicillin and the appropriate selection antibiotics (e.g., 25 mg/L kanamycin for nptII-based selection).
16. Tobacco Callus Growth Medium: full-strength MS medium (PhytoTechnology Laboratories), 0.5 g/L MES, 30 g/L sucrose, 8 g/L plant cell culture agar, adjust pH to 5.7 with KOH. Autoclave and store at room temperature.
17. Triton X-100.
18. X-Gluc Staining Buffer: 0.2 M sodium phosphate buffer, pH 7.5, 10 mM ethylenediaminetetraacetic acid (EDTA), 0.1% (v/v) Triton X-100 in ddH_2O. Filter sterilize and store at –20°C.
19. X-Gluc Staining Solution: dissolve 25 mg of X-Gluc in 1 mL of dimethylformamide (DMF) and add to 50 mL of freshly thawed X-Gluc Staining Buffer.
20. 2.5 mM (each) dNTP mix for PCR.
21. Molecular biology-grade agarose.
22. TaKaRa EX Taq DNApolymerase at a concentration of 5 U/μL and EX Taq 10× reaction buffer (Pan Vera Corp.).
23. Cloned Pfu Taq DNA polymerase at a concentration of 2.5 U/μL and cloned Pfu Taq DNA polymerase 10× reaction buffer (Stratagene).
24. T4-DNA ligase at a concentration of 3 U/μL (Promega).
25. rAPid alkaline phosphatase (Roche).
26. Restriction endonucleases *Spe*I, *Xho*I, *Nco*I, *Age*I, *Not*I, and *Bam*HI (New England Biolabs).
27. Rare-cutting restriction endonucleases PI-*Psp*I, I-*Sce*I, *Asc*I, and I-*Ceu*I (New England Biolabs).

2.3. Plasmids, Bacterial Strains, and Primers

1. *Nicotiana tabacum* plants – grown in the greenhouse or growth chamber.
2. Sterile tissue culture-grown *N. tabacum* plants (*see* **Note 1**).
3. Arabidopsis thaliana plants and seeds.
4. Agrobacterium tumefaciens strain EHA105.
5. pSAT4.35S.NLS-FokI, a high-copy cloning plasmid for assembly and expression of ZFNs (*see* **Note 2**). All plasmids described in this chapter are available from the author for non-commercial use upon request.
6. pRCS2, binary plant transformation vector.
7. pRCS2.ocs.nptII, binary plant transformation vector with the kanamycin resistance gene.
8. pSAT6-MCS, pSAT6A-MCS, pSAT5-MCS, pSAT4-MCS, and pSAT6-rbcP-MCS-rbcT satellite plant expression vectors.
9. Chemically competent cells of *E. coli* strain DH5α.
10. pRTL2-GUS, GUS-coding sequence DNA template (33).
11. A pair of primers for the construction of a mutated GUS reporter gene. Primers should be designed to carry a TGA (stop) codon within the 6-bp spacer of the ZFN target site (TS). The primers' general structure should be GUS amplification forward primer (generic GUS-ZFN-TS-F-*Kpn*I, which is comprised of a ZFN-TS and a *Kpn*I recognition site for cloning) 5′-GGGGTACCATGNNNNNNNNNTCCTGANNNNNNNNNTTACGTCCTGTAGAAACCCC, and GUS amplification reverse primer (GUS-R-*Bam*HI, which carries a *Bam*HI recognition site for cloning) 5′-CGGGGTACCATGTTACGTCCTGTAGAAACCCC.
12. Plant-inducible promoter amplification primers (e.g., hsp-F 5′-TAGTTAACCGGTAAGCTTGCTGCAGCTTTG and hsp-R 5′-TAACATCCATGGTGTCGTTGCTTTTCGGGAG, which were designed to be used for amplification of the hsp18.2 heat-shock-inducible promoter from Arabidopsis and contains *Age*I and *Nco*I recognition sites for cloning).
13. pSAT-F sequencing primer (5′-TCTTCGCTATTACGCCAGC). To be used for sequencing of newly assembled pSAT vectors.
14. pSAT-R sequencing primer (5′-ACGACAGGTTTCCCGACTGG). To be used for sequencing of newly assembled pSAT vectors.

15. 35S-F sequencing primer (5′-CGACAGTGGTCCCA AAGATGG). To be used for sequencing of newly assembled ZFNs in pSAT4.35S.NLS-*Fok*I.

3. Methods

Rules for selecting target sites and methods for constructing novel ZFNs can be found in several excellent papers (28, 34, 35). In addition, data sets describing the DNA-binding domains for various triplets (34, and *see* **Chapter 1**), publicly available web-based programs (www.zincfingers.org and www.zincfinger-tools.org), and even collections of zinc finger libraries can also be instrumental during the construction of novel ZFNs. The construction of ZFNs with novel specificities is beyond the scope of the current chapter and users can choose from a wide variety of vectors useful for the assembly and in vitro analysis of their enzymes (e.g. 16, 26–28, 35, 36, and *see* **Chapters 1–4**). We provide the user with the option of using our pSAT4.35S.NLS-*Fok*I zinc finger cloning vector, which was specifically designed to allow for cloning compatibility and flexibility with binary plant transformation vectors, as illustrated in **Fig. 20.1**. Further details on the use of pSAT4.35S.NLS-*Fok*I and a general strategy for the construction of zinc finger proteins (ZFPs) from overlapping oligonucleotides can be found in **Chapter 12** (37) of this issue and in references (38, 39). Briefly, newly assembled DNA-binding domains, which are comprised of several fingers (i.e., ZFPs), can be fused as *Xho*I–*Spe*I fragments between a nuclear localization signal and *Fok*I domain in pSAT4.35S.NLS-*Fok*I (*see* **Fig. 20.1**). We used this strategy for the assembly of ZFN3, a ZFN enzyme capable of binding the 9-bp-long sequence 5′-GTTGGTGCT, creating pSAT4.35S. ZFN3. Alternatively, ZFNs that are assembled in other systems can be transferred as complete coding sequences into the *Nco*I and *Bam*HI sites of pSAT4.35S.NLS-*Fok*I. We used this strategy to transfer the QQR ZFN-coding sequence from pHS::QQR-QEQ/2300 (25), yielding pSAT4.35S.QQR. It should be noted that the pSAT4.35S.NLS-*Fok*I plasmid is also compatible with the pET28.XH ZFP expression vector (39), allowing for single-step transfer of ZFN-coding sequences into *E. coli* expression vectors, which may be useful for in vitro analysis of ZFN activity (39). Here we provide a set of procedures for the construction of plant transformation vectors that are useful for the validation and expression of ZFNs in plant cells, using ZFN3 and QQR as model enzymes.

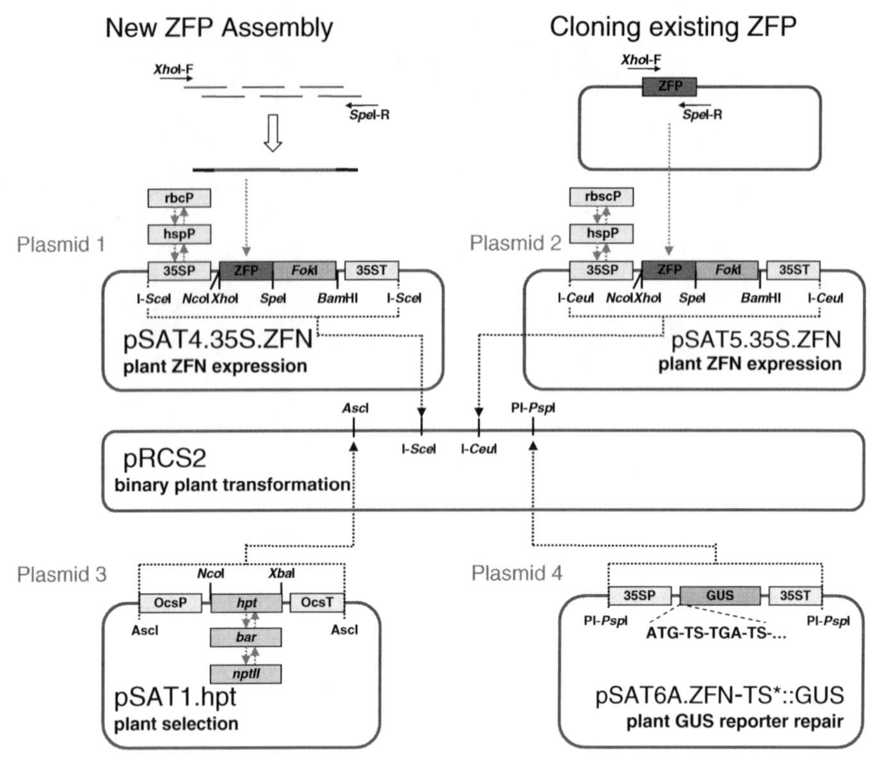

Fig. 20.1. Structure and key features of the ZFN validation and expression vector system. Our vector system is based on the assembly of ZFN expression cassettes (plasmids 1 and 2) and/or plant selection expression cassette (plasmid 3) and/or GUS reporter cassette (plasmid 4) onto the plant binary vector pRCS2. Plant ZFN expression vectors serve as acceptor vectors for zinc finger proteins (ZFPs) assembled from overlapping primers (illustrated on plasmid 1) or by PCR amplification of existing ZFPs (illustrated on plasmid 2). The regulatory elements on plant ZFN expression vectors can be replaced by virtually any other regulatory sequence [illustrated here by the exchange of 35S promoter (35SP) with those of rbcP or hspP], allowing for either controlled or constitutive expression of ZFNs in target cells. Various plant selection cassettes (e.g., *hpt*, *npt*II, and *bar*) can be used for the production of transgenic tissues and/or whole plants (illustrated by plasmid 3). Using different combinations of ZFN, GUS reporter, and selection marker expression cassettes, it is possible to assemble the various plant transformation vectors for the validation and expression of ZFN in plant cells, as described in the text.

Our three validation assays are based on activation of a mutated GUS reporter gene, engineered to carry a TGA (stop) codon within the 6-bp spacer of the ZFN target site (*see* **Fig. 20.2**), constructed on a reporter repair plasmid (*see* plasmid 4, **Fig. 20.1**). The T-DNA repair assay calls for the construction of a single binary plasmid composed of a constitutive ZFN expression cassette and the reporter repair plasmid (i.e., mount plasmids 1 and 4 on the single pRCS-based binary vector, **Fig. 20.1**). The transgene repair assay requires the assembly of two binary vectors: one carrying a plant selection marker and the reporter repair plasmid (i.e., mounting plasmids 3 and 4 on the single pRCS-based binary vector, **Fig. 20.1**) and the other carrying only the constitutive ZFN expression cassette (i.e., cloning plasmid 1 on a separate pRCS-based binary vector, **Fig. 20.1**),

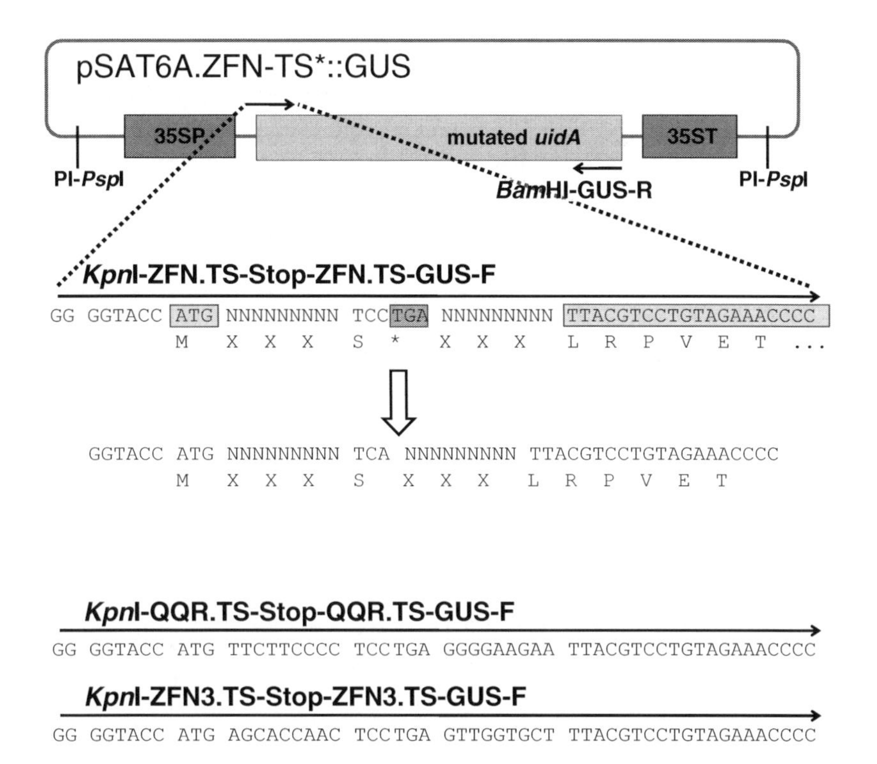

Fig. 20.2. Key features of the GUS reporter vector. The plasmid carries a plant expression cassette engineered for constitutive expression of a mutated *uidA* (GUS) gene. A stop (TGA) codon was engineered within the 6-bp spacer of the ZFN target site, leading to premature termination of *uidA* translation in plant cells. Digestion and successive dis-repair of the broken ZFN recognition site by the plant's NHEJ machinery may lead to deletion and/or alteration of the stop codon and to functional GUS expression. The generic forward primer (*Kpn*I-ZFN.TS-Stop-ZFN.TS-GUS-F) used for amplification and cloning of a mutated GUS-coding sequence is presented. Also presented are the *Kpn*I-QQR.TS-Stop-QQR.TS-GUS-F and *Kpn*I-ZFN3.TS-Stop-ZFN3.TS-GUS-F primers used for the construction of QQR-TS*::GUS- and ZFN3-TS*::GUS-expressing reporter vectors, respectively. 35SP and 35ST, cauliflower mosaic virus 35S constitutive promoter and terminator, respectively; TS*, a stop codon-containing ZFN target site.

Finally, the whole-plant repair assay calls for replacing the constitutive promoter on the ZFN expression cassette with an inducible one (e.g., the heat-shock-inducible promoter, hspP, **Fig. 20.1**) and mounting it together with a plant selection marker and the reporter repair plasmid on a single binary plasmid (i.e., mounting plasmids 1, 3, and 4 on a single pRCS binary vector, **Fig. 20.1**). Each validation tool is then used to test ZFN activity in plant tissues or whole transgenic plants, as we describe below using ZFN3 and QQR as model enzymes. In addition, our vector system facilitates the replacement of promoters on ZFN expression cassettes and the construction of dual ZFN monomer expression cassette transformation vectors (i.e., by mounting ZFN expression cassettes from modified plasmids 1, 2 and a selection marker from plasmid 3 onto a single binary vector, **Fig. 20.1**) which may be useful for GT experiments in plant species, as we describe below.

3.1. Construction of a GUS Reporter Repair Plasmid

Our validation assays consist of a mutated GUS reporter gene, which is activated upon expression of ZFNs in living plant cells. A key component of our validation assay is a plant expression cassette which drives the expression of the mutated GUS reporter gene under the control of the constitutive cauliflower mosaic virus 35S promoter (35SP) and terminator (35ST) (*see* plasmid 4, **Fig. 20.1**). The mutated GUS-coding sequence can be amplified using a set of primers, in which the forward primer has been designed to carry a stop codon within the ZFN recognition site (*see*, for example, primers *Kpn*I-QQR.TS-Stop-QQR.TS-GUS-F and *Kpn*I-ZFN3.TS-Stop-ZFN3.TS-GUS-F, **Fig. 20.2**). Digestion of the mutated GUS reporter gene and its reconstruction by the plant's NHEJ machinery can be measured by the activation of GUS expression. Here we describe the assembly of pSAT5.ZFN-TS*::GUS, a key component of our validation assays.

1. Design a GUS amplification forward primer (i.e., GUS-ZFN-TS-F-*Kpn*I). The primer should carry a stop codon within the 6-bp spacer of a ZFN recognition target site as illustrated in **Fig. 20.2**.
2. Dilute the GUS-ZFN-TS-F-*Kpn*I and GUS-R-*Bam*HI primers to 20 μM.
3. Prepare a PCR mixture with a total volume of 50 μL containing 2 μL of GUS-ZFN-TS-F-*Kpn*I, 2 μL of GUS-R-*Bam*HI, 4 μL of dNTPs (2.5 mM each), 2 U of *Pfu* Taq DNA polymerase, 5 μL of *Pfu* Taq 10× reaction buffer, and 5 ng of pRTL2-GUS DNA template.
4. Perform PCR with the following program: denaturation at 95°C for 5 min followed by 35 cycles of 45 s at 95°C, 30 s at 55°C, and 2 min at 72°C.
5. Run the PCR product on a 1.0% agarose gel.
6. Excise the ca 1.7-bp mutated GUS-coding sequence band from the gel and purify it using GFX PCR purification kit. Elute the sample in 50 μL of ddH_2O.
7. Digest the mutated GUS-coding sequence with *Kpn*I and *Bam*HI and repurify the digested fragment using GFX PCR purification kit.
8. Digest the pSAT6A-MCS with *Kpn*I and *Bam*HI, separate the linear vector backbone in a 1% agarose gel, and purify it using GFX PCR purification kit.
9. Prepare a ligation cocktail with a total volume of 15 μL containing approximately 50 ng of linearized pSAT6A-MCS target vector, 150 ng mutated GUS-encoding PCR fragment, 0.5 μL T4-DNA ligase (concentrated to 3 U/μL), and 1.5 μL of 10× ligation buffer.

10. Perform the ligation for 2 h at room temperature and use the entire ligation reaction for transformation into chemically competent *E. coli* strain DH5α cells.
11. Plate the bacterial culture on LB solid plates supplemented with 100 μg/mL ampicillin.
12. Pick selected colonies, extract DNA, and identify recombinant colonies by digestion with *Kpn*I and *Bam*HI. This will produce the generic pSAT6A.ZFN-TS*::GUS plasmid.
13. Verify the accuracy of the ZFN target site by DNA sequencing of recombinant plasmids, using the 35S-F sequencing primer. We used this procedure for the construction of pSAT6A.ZFN3-TS*::GUS and pSAT6A.QQR-TS*::GUS using the GUS-ZFN3-TS-F-*Kpn*I and GUS-QQR-TS-F-*Kpn*I primers, respectively (*see* **Fig. 20.2** and **Note 3**).

3.2. Construction of an Agrobacterium T-DNA Repair Vector

We recommend using *Agrobacterium*-mediated genetic transformation to deliver the mutated GUS and ZFN expression cassettes. Our pSAT-based transformation system allows mounting several components onto the T-DNA region of a single *Agrobacterium*'s binary vector (*see* **Fig. 20.1**). Here we describe the construction of our basic binary vector which will be used to deliver components of our validation assay as single T-DNA molecules.

1. Digest the pRCS2 binary vector with PI-*Psp*I at 65°C, separate the linearized vector in a 0.7% agarose gel, and purify it using GFX PCR purification kit (*see* **Note 4**).
2. Treat the linearized vector with rAPid alkaline phosphatase for 7–10 min at 37°C and deactivate the enzyme by heat treatment for 3 min at 75°C.
3. Digest the pSAT6A.ZFN-TS*::GUS (e.g., pSAT6A.QQR-TS*::GUS or pSAT6A.ZFN3-TS*::GUS) with PI-*Psp*I at 65°C, separate the mutated GUS expression cassette from the vector backbone in 0.7% agarose gel, and purify it using GFX PCR purification kit (*see* **Note 4**).
4. Prepare a ligation cocktail with a total volume of 15 μL containing approximately 50 ng of linearized pRCS2, 150 ng of mutated GUS expression cassette, 0.5 μL T4-DNA ligase (3 U/μL) and 1.5 μL 10× ligation buffer.
5. Perform the ligation overnight at 4°C, let sit at room temperature for an additional hour, and use the entire ligation reaction for transformation into chemically competent *E. coli* cells (*see* **Note 5**).

6. Plate the bacteria on LB solid medium supplemented with 50 μg/mL spectinomycin and 50 μg/mL streptomycin (*see* **Note 6**).
7. Pick selected colonies, extract DNA, and identify recombinant colonies by digestion with PI-*Psp*I. This will produce the generic vector pRCS2.[ZFN-TS*::GUS]. We used this procedure for the construction of pRCS2.[QQR-TS*::GUS].
8. Digest the generic pRCS2.[ZFN-TS*::GUS] binary vector with I-*Sce*I, separate the linearized vector in 0.7% agarose gel, and purify it using GFX PCR purification kit (*see* **Note 4**).
9. Treat the linearized vector with rAPid alkaline phosphatase for 7–10 min at 37°C and deactivate the enzyme by heat treatment for 3 min at 75°C.
10. Digest the generic vector pSAT4.35S.ZFN with I-*Sce*I, separate the mutated GUS expression cassette from the vector backbone in 0.7% agarose gel, and purify it using GFX PCR purification kit (*see* **Note 4**).
11. Prepare a ligation cocktail with a total volume of 15 μL containing approximately 50 ng of linearized generic pRCS2.[ZFN-TS*::GUS] target vector, 150 ng of ZFN expression cassette (from generic pSAT4.35S.ZFN), 0.5 μL T4-DNA ligase (3 U/μL), and 1.5 μL of 10× ligation buffer.
12. Perform the ligation overnight at 4°C, let sit at room temperature for an additional hour, and use the entire ligation reaction for transformation into chemically competent *E. coli* cells (*see* **Note 5**).
13. Plate the bacteria on LB solid medium supplemented with 50 μg/mL spectinomycin and 50 μg/mL streptomycin (*see* **Note 6**).
14. Pick selected colonies, extract DNA, and identify recombinant colonies by digestion with I-*Sce*I. This will produce the generic T-DNA repair vector pRCS2.[ZFN][ZFN-TS*::GUS]. We used this procedure for the construction of pRCS2.[QQR][QQR-TS*::GUS] by cloning the QQR constitutive expression cassette from pSAT4.35SP.QQR into pRCS2.[QQR-TS*::GUS].

3.3. In-Planta T-DNA Repair Assay

This assay allows the rapid and easy validation of ZFN activity in living plant cells. The assay relies on the transient transfer of a T-DNA molecule from *Agrobacterium* into plant cells but does not necessarily require the integration of that molecule into the plant genome. Thus, repair of the mutated GUS gene can be assayed within several days of infection of the target plant tissue. Here we describe the procedure for T-DNA repair in tobacco leaf disks. The procedure can be adapted to other plant species for which a transient *Agrobacterium* transformation procedure has been established.

1. Transform the generic T-DNA repair vector p*RCS*2.[ZFN][ZFN-TS*::GUS] and control vector pRCS2.[ZFN-TS*::GUS] into chemically competent EHA105 *Agrobacterium* cells (*see* **Note 7**).
2. Plate and maintain the cells on solid LB, supplemented with 200 μg/mL spectinomycin and 200 μg/mL streptomycin (*see* **Note 6**).
3. Grow *Agrobacterium* cells overnight at 28°C in 10-mL liquid LB medium supplemented with 200 μg/mL spectinomycin and 200 μg/mL streptomycin (*see* **Note 6**).
4. Collect the bacterial cells by centrifugation for 10 min at 3,000×*g* and resuspend in 4 mL of *Agrobacterium* Induction Medium supplemented with 100 μM acetosyringone, 200 μg/mL spectinomycin, and 300 μg/mL streptomycin. Measure OD_{600} of resulting bacterial suspension and dilute it further with supplemented Induction Medium in order to obtain 4 mL of culture with an OD_{600} of approximately 0.3.
5. Allow the *Agrobacterium* culture to grow at 28°C to an OD_{600} of 0.6.
6. Collect the bacterial cells by centrifugation for 10 min at 3,000×*g* and resuspend in 4 mL *Agrobacterium* Infiltration Medium supplemented with 200 μM acetosyringone.
7. Pick healthy young tobacco plants and infiltrate the *Agrobacterium* cells into the abaxial sides of the leaves using a 10-mL syringe (*see* **Note 8**).
8. Mark the infected regions on the leaves and let the plants recover in the growth room for 48–56 h.
9. Detach the infected leaves from the plants, slice the infected areas with a razor blade, and stain with X-Gluc solution for 24 h at 37°C. ZFN activity can be monitored by the appearance of the blue precipitate chloro-bromoindigo. We transformed pRCS2.[QQR][QQR-TS*::GUS] and pRCS2.[QQR-TS*::GUS] into *Agrobacterium* strain EHA105 and used them for detection of QQR activity in tobacco plants, as shown in **Fig. 20.3**.

3.4. Transgene Repair Assay

The transgene repair assay provides another level of confidence for the expression and function of ZFNs in living plant cells. In this assay, ZFN activity is measured by the reconstruction of a mutated GUS reporter gene following its integration into the

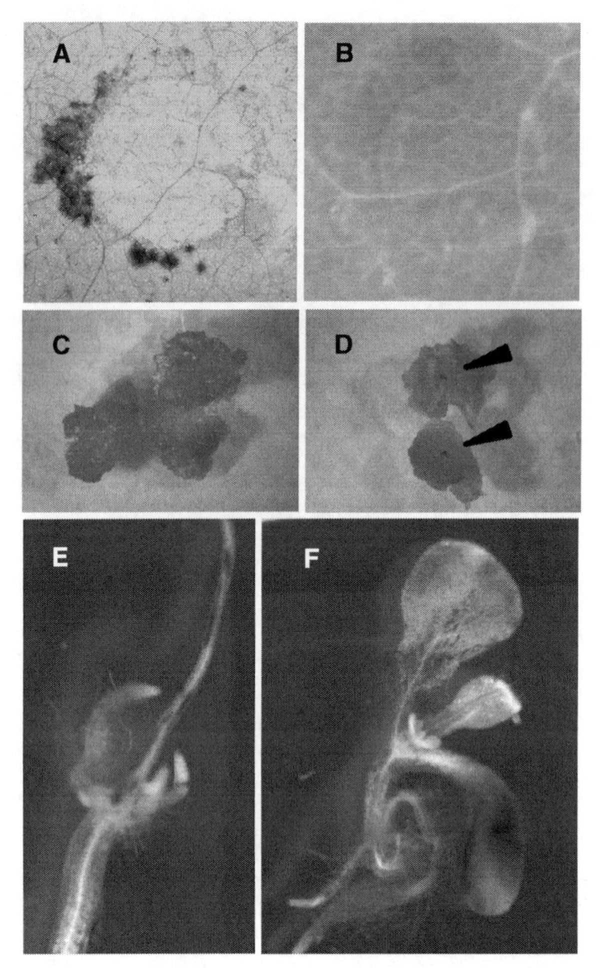

Fig. 20.3. Validation of ZFN activity in plant cells (**a** and **b**). T-DNA repair assay of QQR. Tobacco leaves were infected with QQR-TS*::GUS transformation vector with (**A**) or without (**B**) the QQR-expressing cassette on the T-DNA molecule. The reconstruction of GUS expression indicates the successful expression and digestion activity of QQR in plant cells in the presence of QQR expression only (**c** and **d**). Transgene repair assay of ZFN3. Activation of the ZFN3-TS*::GUS-expressing cassette in transgenic tobacco calli (**C**) by inoculation with a ZFN3-expressing binary vector (**D**). While no GUS expression could be observed in the transgenic callus which does not express ZFN3 (**C**), reconstruction of the GUS gene could be observed following delivery of ZFN3 into the callus cells (*arrowheads*, **d**). (**e** and **f**) Whole-plant DNA repair assay of QQR. Heat-shock-induced activation of QQR activity in various organs of transgenic *Arabidopsis* seedlings engineered to carry the constitutive QQR-TS*::GUS-expressing cassette and the heat-shock-inducible QQR expression cassette. Expression of reconstructed GUS gene can be observed in root tissues (**E**) and new true leaves (**F**) of germinating transgenic *Arabidopsis* seeds. Images reproduced with permission from (39).

plant genome. Here we describe the steps for implementing the assay in transgenic tobacco calli. The procedure can be adapted to other plant species for which a procedure for stable *Agrobacterium* transformation and regeneration of transgenic tissue has been established.

1. Use the above-described cloning methods for the construction of a stable plant transformation vector for the transgene repair assay by mounting the GUS-reporter

repair cassette (using PI-*Psp*I) together with a plant selection expression cassette (using *AscI*) *onto* a single binary vector. This will produce the generic transgene repair vector pRCS2.[KAN][ZFN-TS*::GUS] (*see* **Note 9**).

2. Use the above-described cloning methods for the construction of the ZFN-expressing binary vector by mounting the ZFN expression cassette (using I-*Sce*I) onto a pRCS2 vector. This will produce the generic ZFN expression vector pRCS2.[ZFN].
3. Transform the generic transgene repair vector pRCS2.[KAN][ZFN-TS*::GUS] into chemically competent EHA105 *Agrobacterium* cells (*see* **Note 7**).
4. Plate and maintain the cells on solid LB medium, supplemented with 200 μg/mL spectinomycin and 200 μg/mL streptomycin (*see* **Note 6**).
5. Grow and induce the *Agrobacterium* cells as described in Steps 2–6 of **Section 3.3**.
6. Cut sterile tobacco leaves into 1.5-cm^2 pieces and suspend them in the *Agrobacterium* cell suspension for 10 min. Using a sterile scalpel blade, gently wound the leaf segments a few times while in suspension.
7. Blot the leaf segments dry on sterile filter paper and transfer to plates with Callus Induction Medium.
8. Incubate the leaf segments in the dark for 2–3 days at 28°C.
9. Transfer the leaf segments to a 50-mL plastic screw-cap sterile tube, add 30 mL sterile ddH_2O supplemented with 300 μg/mL carbenicillin, and shake gently to wash off the bacteria.
10. Discard the ddH_2O and repeat the washing procedure twice more.
11. Blot the leaf disks dry on sterile filter paper and transfer to Callus Induction Medium supplemented with 300 mg/L carbenicillin and 25 mg/L kanamycin.
12. Subculture the leaf segments every 2–3 weeks and monitor for regeneration of resistant calli, typically 3–4 weeks after infection.
13. Distribute the regenerated calli evenly on a fresh plate of callus induction medium supplemented with 300 mg/L carbenicillin and 25 mg/L kanamycin and grow until they reach the desired size (typically approximately 5–8 mm in diameter).
14. Transform the generic ZFN-expressing binary vector pRCS2.[ZFN] into chemically competent EHA105 *Agrobacterium* cells, grow and maintain the cells in the

presence of 200 μg/mL spectinomycin and 200 μg/mL streptomycin as described above.

15. Grow and induce the *Agrobacterium* cells as described in Steps 3–6 of **Section 3.3**.
16. Slice calli clusters into halves, stab them once or twice with a sterile needle, and inoculate them with *Agrobacterium* cell suspension.
17. Transfer the callus clusters to Callus Growth Medium without antibiotics and culture for 2–3 days at 28°C.
18. Stain the clusters with X-Gluc Staining Solution for 24 h at 37°C. ZFN activity can be monitored by the appearance of the blue precipitate chloro-bromoindigo. We used the pRCS2.oc*s.nptII* vector (31), which carries the kanamycin resistance gene under the control of the octopine synthase promoter and the pSAT6A.ZFN3-TS*::GUS vector for the assembly of pRCS2.[KAN][ZFN3-TS*::GUS]. We also used pRCS2 and pSAT4.35S.ZFN3 for the construction of pRCS2.[ZFN3]. We used these vectors to validate ZFN3 activity in transgenic calli, as shown in **Fig. 20.3**.

3.5. Whole-Plant DNA Repair Assay

ZFN activity may vary in different plant tissues and organs. Transgenic plants in which the expression of ZFNs is induced by inducible promoters can be used to validate and study ZFN activity in various organs and at different developmental stages of the plant. Here we describe the construction and use of a whole-plant DNA repair assay in *Arabidopsis* using a heat-shock-inducible promoter. The procedure can be adapted to other plant species for which a procedure for stable *Agrobacterium* transformation and regeneration of whole transgenic plants has been established.

1. Select an inducible, tissue-specific, or developmental promoter suitable for driving the expression of the ZFN being tested in transgenic plants. We used the *Arabidopsis* heat-shock-inducible promoter hsp18.2 (40) to drive the expression of QQR in transgenic *Arabidopsis* plants following a short heat-shock induction step.
2. Design a pair of primers for amplification of the selected promoter, including *Age*I and *Nco*I sites. In our example, we amplified the hsp promoter from pHS::QQR-QEQ/2300 (25) using *hsp*-F (forward) and *hsp*-R (reverse) primers.
3. Dilute the *hsp*-F and *hsp*-R primers to 20 μM.
4. Prepare a PCR mixture with a total volume of 50 μL containing 2 μL of *hsp*-F, 2 μL of *hsp*-R, 4 μL of dNTPs (2.5 mM each), 2 U of *Pfu* Taq DNA polymerase, 5 μL of *Pfu* Taq 10× reaction buffer, and 10 ng of template DNA (in our example, pHS::QQR-QEQ/2300).

5. Perform PCR with the following program: denaturation at 95°C for 5 min followed by 35 cycles of 45 s at 95°C, 30 s at 55°C, and 30 s at 72°C (*see* **Note 10**).
6. Run the PCR product on a 1.0% agarose gel and purify it using a GFX PCR purification kit.
7. Digest the pSAT4.35S.ZFN (in our example, pSAT4.35S.QQR) with *Age*I and *Nco*I, separate the plasmid from the ~890-bp tandem 35S promoter in a 1.0% agarose gel, and purify it using GFX PCR purification kit.
8. Prepare a ligation between the pSAT4.35S.ZFN-digested plasmid and the amplified promoter. Perform the ligation, transformation into *E. coli* cells, selection, and verification of recombinant colonies as described above.
9. Verify the accuracy of the promoter sequence by DNA sequencing of recombinant plasmids using the pSAT-F sequencing primer. We used this procedure for the construction of pSAT4.hspP.QQR in which the expression of QQR ZFN is driven by the *Arabidopsis* heat-shock-inducible promoter.
10. Use the above-described cloning methods to add the induced ZFN expression cassette (*using I-Sce*I) to the stable plant transformation vector. We cloned the heat-shock induced QQR from pSAT4.hspP.QQR into pRCS2.[KAN][QQR-TS*::GUS] and produced pRCS2.[KAN][hsp.QQR][QQR-TS*::GUS].
11. Transform the whole-plant repair vector (e.g., pRCS2.[KAN][hsp.QQR][QQR-TS*::GUS]) into chemically competent EHA105 *Agrobacterium* cells, grow and maintain the cells in the presence of 200 μg/mL spectinomycin and 200 μg/mL streptomycin as described above.
12. Use the *Agrobacterium* cells for transformation of *Arabidopsis* plants (as described in Ref. 41) and select several transgenic lines.
13. Transgenic plants and tissues can be induced at various developmental stages and stained with X-Gluc Staining Solution for 24 h at 37°C to monitor ZFN activity. We typically induce our plants for 90–150 min at 42°C and let them recover for an additional 24–72 h prior to staining, as shown in **Fig. 20.3**.

3.6. Construction of Multi-ZFN Transformation Vectors

Co-transformation of several plasmids is not an efficient method for delivering multiple genes into plant cells. Thus, the coordinated delivery of several ZFN monomers, together with a selectable reporter marker, can pose a technical challenge for plant species. Our pSAT vector construction system was designed to facilitate the assembly of several expression cassettes onto a single binary plant transformation plasmid (*see* **Fig. 20.1**), and here we describe the construction of multi-ZFN plant transformation vectors.

1. Select the promoter(s) suitable for driving the expression of your target pair of ZFNs in transgenic plants. For the purpose of overexpression, we recommend combining the tandem 35S promoter (35SP) and the rubisco small subunit promoter (rbcP) and their corresponding terminators (35ST and rbcT, respectively) on a single transformation vector. The use of other constitutive (e.g. 31), or inducible promoters, should also be considered. In our example, we used 35SP and rbcP to mount a pair of ZFN monomers on a single binary vector for the purpose of constructing a dual-gene plant transformation cassette for targeting experiments in plants. Common cloning strategies can be deployed for the construction of modified pSAT4.35S.NLS-*Fok*I vectors.
2. Transfer the rbcP as an *Age*I–*Nco*I fragment from pSAT6.rbcP.MCS.rbcT into the same sites of pSAT4. 35S.NLS-*Fok*I. This will replace the tandem 35SP with rbcP (*see* **Note 11**).
3. Next, transfer the rbcT as a *Bam*HI–*Not*I fragment from pSAT6.rbcP.MCS.rbcT into the same sites of the resultant plasmid. This will replace the 35ST with rbcT and produce pSAT6.rbcP.NLS-*Fok*I (*see* **Note 12**).
4. Transfer the entire rbcP.NLS-*Fok*I expression cassette as an *Age*I–*Not*I fragment from pSAT6.rbcP.NLS-*Fok*I into pSAT5-MCS to produce pSAT5.rbcP.NLS-*Fok*I.
5. Construct pairs of ZFN-expressing plasmids based on pSAT4.35SP.NLS-*Fok*I and pSAT5.rbcP.NLS-*Fok*I.
6. Use the above-described cloning methods for the construction of a stable plant transformation vector that is useful for GT experiments, by mounting the ZFN expression cassettes (using successive cloning by I-*Sce*I and I-*Ceu*I) together with a plant selection expression cassette (using *Asc*I) onto a single binary vector (*see* **Note 9**).

4. Notes

1. The *Agrobacterium* induction and infiltration media and the *Arabidopsis* infection method are based on the protocol by Lee and Yang (42). Other *Agrobacterium* strains, tobacco plants, and *Arabidopsis* ecotypes can also be used. EHA105 and the plant material described in this chapter are available for non-commercial use upon request.

2. For further details on the pSAT- and pRCS-based vector systems, (*see* Refs. 30, 31, 43).
3. The GUS reporter repair plasmids can also be used as substrates in various in vitro ZFN digestion analyses (e.g. 39, 44).
4. Short incubation times (i.e., less than 30 min) are advisable in order to minimize the possibility of non-specific digestion and degradation of the vector. This may lead to partial digestion, but the linear and digested form of pRCS2 is distinguishable from the undigested form on a 0.7% agarose gel.
5. Use ultra-competent *E. coli* cells to ensure high transformation efficiency and recovery of recombinant colonies.
6. It is important to select and maintain pRCS2-based vectors on both antibiotics due to the relatively high resistance background in many *E. coli* and other bacterial species.
7. For step-by-step preparation of *Agrobacterium*-competent cells (*see* Ref. 45). Other *Agrobacterium* strains (e.g., LBA4404, GV3101, and GV2260) can also be used.
8. Do not apply high pressure as it may damage the infected tissue. Alternatively, *Arabidopsis* leaves can be used. For a detailed protocol on transient gene expression via agro infiltration (*see* Ref. 42).
9. You can use previously constructed pRCS vectors which already carry selectable marker expression in their T-DNA region (e.g., pRCS2.ocs.*nptII*, pRCS2.ocs.*hpt*, or pRCS2.ocs.*bar*, which carry the genes encoding kanamycin, hygromycin, and basta resistance, respectively (31)).
10. Adjust the extension time at 55°C to allow amplification of the selected promoter.
11. Use PCR to amplify and clone the selected promoter. Use the pSAT-F sequencing primer to verify the accuracy of the construct.
12. Use PCR to amplify and clone the selected promoter. Use the pSAT-R sequencing primer to verify the accuracy of the construct.

Acknowledgments

We thank Dr. G.N. Drews for the gift of pHS::QQR-QEQ/2300. In our lab, our work is supported by grants from the Biotechnology Research and Development Corporation (BRDC), The Consortium for Plant Biotechnology Research (CPBR), the Binational Agricultural Research and Development Fund (BARD), Dow Agrocience, and University of Michigan startup funds.

References

1. Britt, A.B. and May, G.D. (2003) Re-engineering plant gene targeting. *Trends Plant Sci.* **8**, 90–95.
2. Siebert, R. and Puchta, H. (2002) Efficient repair of genomic double-strand breaks by homologous recombination between directly repeated sequences in the plant genome. *Plant Cell.* **14**, 1121–1131.
3. Tzfira, T., Li, J., Lacroix, B., and Citovsky, V. (2004) *Agrobacterium* T-DNA integration: molecules and models. *Trends Genet.* **20**, 375–383.
4. Terada, R., Urawa, H., Inagaki, Y., Tsugane, K., and Iida, S. (2002) Efficient gene targeting by homologous recombination in rice. *Nat Biotechnol.* **20**, 1030–1034.
5. Terada, R., Johzuka-Hisatomi, Y., Saitoh, M., Asao, H., and Iida, S. (2007) Gene targeting by homologous recombination as a biotechnological tool for rice functional genomics. *Plant Physiol.* **144**, 846–856.
6. Shaked, H., Melamed-Bessudo, C., and Levy, A.A. (2005) High frequency gene targeting in *Arabidopsis* plants expressing the yeast RAD54 gene. *Proc Natl Acad Sci USA.* **102**, 12265–12269.
7. Puchta, H. (2002) Gene replacement by homologous recombination in plants. *Plant Mol Biol.* **48**, 173–182.
8. Puchta, H., Dujon, B., and Hohn, B. (1993) Homologous recombination in plant cells is enhanced by in vivo induction of double strand breaks into DNA by a site-specific endonuclease. *Nucleic Acids Res.* **21**, 5034–5040.
9. Puchta, H., Dujon, B., and Hohn, B. (1996) Two different but related mechanisms are used in plants for the repair of genomic double-strand breaks by homologous recombination. *Proc Natl Acad Sci USA.* **93**, 5055–5060.
10. Chiurazzi, M., Ray, A., Viret, J.F., Perera, R., Wang, X.H., Lloyd, A.M., and Signer, E.R. (1996) Enhancement of somatic intrachromosomal homologous recombination in Arabidopsis by the HO endonuclease. *Plant Cell.* **8**, 2057–2066.
11. Chilton, M.-D.M. and Que, Q. (2003) Targeted integration of T-DNA into the tobacco genome at double-strand breaks: new insights on the mechanism of T-DNA integration. *Plant Physiol.* **133**, 956–965.
12. Tzfira, T., Frankmen, L., Vaidya, M., and Citovsky, V. (2003) Site-specific integration of *Agrobacterium tumefaciens* T-DNA *via* double-stranded intermediates. *Plant Physiol.* **133**, 1011–1023.
13. Salomon, S. and Puchta, H. (1998) Capture of genomic and T-DNA sequences during double-strand break repair in somatic plant cells. *EMBO J.* **17**, 6086–6095.
14. Porteus, M.H. and Carroll, D. (2005) Gene targeting using zinc finger nucleases. *Nat Biotechnol.* **23**, 967–973.
15. Durai, S., Mani, M., Kandavelou, K., Wu, J., Porteus, M.H., and Chandrasegaran, S. (2005) Zinc finger nucleases: custom-designed molecular scissors for genome engineering of plant and mammalian cells. *Nucleic Acids Res.* **33**, 5978–5990.
16. Carroll, D., Morton, J.J., Beumer, K.J., and Segal, D.J. (2006) Design, construction and in vitro testing of zinc finger nucleases. *Nat Protoc.* **1**, 1329–1341.
17. Bibikova, M., Carroll, D., Segal, D.J., Trautman, J.K., Smith, J., Kim, Y.G., and Chandrasegaran, S. (2001) Stimulation of homologous recombination through targeted cleavage by chimeric nucleases. *Mol Cell Biol.* **21**, 289–297.
18. Urnov, F.D., Miller, J.C., Lee, Y.L., Beausejour, C.M., Rock, J.M., Augustus, S., Jamieson, A.C., Porteus, M.H., Gregory, P.D., and Holmes, M.C. (2005) Highly efficient endogenous human gene correction using designed zinc-finger nucleases. *Nature.* **435**, 646–651.
19. Porteus, M.H. and Baltimore, D. (2003) Chimeric nucleases stimulate gene targeting in human cells. *Science.* **300**, 763.
20. Beumer, K., Bhattacharyya, G., Bibikova, M., Trautman, J.K., and Carroll, D. (2006) Efficient gene targeting in *Drosophila* with zinc-finger nucleases. *Genetics.* **172**, 2391–2403.
21. Moehle, E.A., Rock, J.M., Lee, Y.L., Jouvenot, Y., Dekelver, R.C., Gregory, P.D., Urnov, F.D., and Holmes, M.C. (2007) Targeted gene addition into a specified location in the human genome using designed zinc finger nucleases. *Proc Natl Acad Sci USA.* **104**, 3055–3060.
22. Santiago, Y., Chan, E., Liu, P.Q., Orlando, S., Zhang, L., Urnov, F.D., Holmes, M.C.,

Guschin, D., Waite, A., Miller, J.C., Rebar, E.J., Gregory, P.D., Klug, A., and Collingwood, T.N. (2008) Targeted gene knockout in mammalian cells by using engineered zinc-finger nucleases. *Proc Natl Acad Sci USA.* **105**, 5809–5814.

23. Wright, D.A., Townsend, J.A., Winfrey, R.J., Jr., Irwin, P.A., Rajagopal, J., Lonosky, P.M., Hall, B.D., Jondle, M.D., and Voytas, D.F. (2005) High-frequency homologous recombination in plants mediated by zinc-finger nucleases. *Plant J.* **44**, 693–705.
24. Perez, E.E., Wang, J., Miller, J.C., Jouvenot, Y., Kim, K.A., Liu, O., Wang, N., Lee, G., Bartsevich, V.V., Lee, Y.L., Guschin, D.Y., Rupniewski, I., Waite, A.J., Carpenito, C., Carroll, R.G., Orange, J.S., Urnov, F.D., Rebar, E.J., Ando, D., Gregory, P.D., Riley, J.L., Holmes, M.C., and June, C.H. (2008) Establishment of HIV-1 resistance in CD4+ T cells by genome editing using zinc-finger nucleases. *Nat Biotechnol.* **26**, 808–816.
25. Lloyd, A., Plaisier, C.L., Carroll, D., and Drews, G.N. (2005) Targeted mutagenesis using zinc-finger nucleases in *Arabidopsis*. *Proc Natl Acad Sci USA.* **102**, 2232–2237.
26. Maeder, M.L., Thibodeau-Beganny, S., Osiak, A., Wright, D.A., Anthony, R.M., Eichtinger, M., Jiang, T., Foley, J.E., Winfrey, R.J., Townsend, J.A., Unger-Wallace, E., Sander, J.D., Muller-Lerch, F., Fu, F., Pearlberg, J., Gobel, C., Dassie, J.P., Pruett-Miller, S.M., Porteus, M.H., Sgroi, D.C., Iafrate, A.J., Dobbs, D., McCray, P.B., Jr., Cathomen, T., Voytas, D.F., and Joung, J.K. (2008) Rapid "open-source" engineering of customized zinc-finger nucleases for highly efficient gene modification. *Mol Cell.* **31**, 294–301.
27. Wright, D.A., Thibodeau-Beganny, S., Sander, J.D., Winfrey, R.J., Hirsh, A.S., Eichtinger, M., Fu, F., Porteus, M.H., Dobbs, D., Voytas, D.F., and Joung, J.K. (2006) Standardized reagents and protocols for engineering zinc finger nucleases by modular assembly. *Nat Protoc.* **1**, 1637–1652.
28. Porteus, M. (2008) Design and testing of zinc finger nucleases for use in mammalian cells. *Methods Mol Biol.* **435**, 47–61.
29. Mani, M., Kandavelou, K., Dy, F.J., Durai, S., and Chandrasegaran, S. (2005) Design, engineering, and characterization of zinc finger nucleases. *Biochem Biophys Res Commun.* **335**, 447–457.
30. Tzfira, T., Tian, G.-W., Lacroix, B., Vyas, S., Li, J., Leitner-Dagan, Y., Krichevsky, A., Taylor, T., Vainstein, A., and Citovsky, V. (2005) pSAT vectors: a modular series of plasmids for autofluorescent protein tagging and expression of multiple genes in plants. *Plant Mol Biol.* **57**, 503–516.
31. Chung, S.M., Frankman, E.L., and Tzfira, T. (2005) A versatile vector system for multiple gene expression in plants. *Trends Plant Sci.* **10**, 357–361.
32. Dafny-Yelin, M. and Tzfira, T. (2007) Delivery of multiple transgenes to plant cells. *Plant Physiol.* **145**, 1118–1128.
33. Carrington, J.C., Freed, D.D., and Leinicke, A.J. (1991) Bipartite signal sequence mediates nuclear translocation of the plant potyviral NIa protein. *Plant Cell.* **3**, 953–962.
34. Mandell, J.G. and Barbas, C.F., 3rd (2006) Zinc Finger Tools: custom DNA-binding domains for transcription factors and nucleases. *Nucleic Acids Res.* **34**, W516–W523.
35. Sander, J.D., Zaback, P., Joung, J.K., Voytas, D.F., and Dobbs, D. (2007) Zinc Finger Targeter (ZiFiT): an engineered zinc finger/target site design tool. *Nucleic Acids Res.* **35**, W599–W605.
36. Liu, Q., Xia, Z., Zhong, X., and Case, C.C. (2002) Validated zinc finger protein designs for all 16 GNN DNA triplet targets. *J Biol Chem.* **277**, 3850–3856.
37. Zeevi, V., Tovkach, A., and Tzfira, T. (2010) Artificial zinc finger nucleases for DNA cloning. *Methods Mol Biol.* (in press).
38. Zeevi, V., Tovkach, A., and Tzfira, T. (2008) Increasing cloning possibilities using artificial zinc finger nucleases. *Proc Natl Acad Sci USA.* **105**, 12785–12790.
39. Tovkach, A., Zeevi, V., and Tzfira, T. (2009) A toolbox and procedural notes for characterizing novel zinc finger nucleases for genome editing in plant cells. *Plant J.* 57, 747–757.
40. Takahashi, T. and Komeda, Y. (1989) Characterization of two genes encoding small heat-shock proteins in *Arabidopsis thaliana*. *Mol Gen Genet.* **219**, 365–372.
41. Zhang, X., Henriques, R., Lin, S.S., Niu, Q.W., and Chua, N.H. (2006) *Agrobacterium*-mediated transformation of *Arabidopsis thaliana* using the floral dip method. *Nat Protoc.* **1**, 641–646.
42. Lee, M.W. and Yang, Y. (2006) Transient expression assay by agroinfiltration of leaves. *Methods Mol Biol.* **323**, 225–229.
43. Goderis, I.J., De Bolle, M.F., Francois, I.E., Wouters, P.F., Broekaert, W.F., and Cammue, B.P. (2002) A set of modular plant transformation vectors allowing flexible insertion of up to six expression units. *Plant Mol Biol.* **50**, 17–27.
44. Cathomen, T., Segal, D.J., Brondani, V., and Muller-Lerch, F. (2008) Generation and

functional analysis of zinc finger nucleases. *Methods Mol Biol.* **434**, 277–290.
45. Tzfira, T., Jensen, C.S., Wangxia, W., Zuker, A., Altman, A., and Vainstein, A. (1997) Transgenic *Populus*: a step-by-step protocol for its *Agrobacterium*-mediated transformation. *Plant Mol Biol Rep.* **15**, 219–235.

Chapter 21

Non-*Fok*I-Based Zinc Finger Nucleases

Miki Imanishi, Shigeru Negi, and Yukio Sugiura

Abstract

The design of functional proteins is one of the most challenging areas of protein research. We have constructed zinc finger peptides with metal-dependent hydrolytic abilities by mutating the zinc ligands in classical zinc fingers, without the need to add a *Fok*I or other DNA cleavage domain. The designed peptides acquired DNA cleavage ability successfully, retaining the proper zinc finger folding and DNA targeting ability. We have also succeeded in site-specific DNA cleavage in the presence of cerium ions by introducing a lanthanide ion-binding loop as a linker of zinc finger motifs.

Key words: Sp1, catalytic zinc fingers, hydrolysis, DNA cleavage, cerium-binding peptide.

1. Introduction

Zinc can play two roles in metalloproteins, a structural or catalytic role (1). In Cys2His2-type zinc finger motifs, zinc is a structural metal ion that is necessary for formation of the zinc finger structure required for specific DNA recognition (2, 3). On the other hand, a catalytic zinc ion directly interacts with substrate molecules. One of the significant differences between structural and catalytic zinc ions is the zinc coordination geometry. A catalytic zinc site contains a vacant ligand site(s) and the zinc-bound water that fills the fourth coordination site is a critical component for catalytic activity (4, 5). Our previous research suggested that the abilities to bind zinc and fold into a compact structure are retained even in mutant peptides in which a zinc-coordinating residue is substituted and also that the unsaturated zinc sites seem to be occupied by water molecules (6). Based on

J.P. Mackay, D.J. Segal (eds.), *Engineered Zinc Finger Proteins*, Methods in Molecular Biology 649,
DOI 10.1007/978-1-60761-753-2_21, © Springer Science+Business Media, LLC 2010

such information, catalytically active zinc finger mutant peptides were created (7–9). Among them, the hydrolysis ability of the His4-type zinc finger peptide was the highest.

In this chapter, the preparation and cleavage assay of catalytic zinc fingers are described. First, as a representative of catalytic zinc finger peptides, we describe how to prepare and assay the His4-type zinc finger peptide based on the second finger of Sp1 (**Sections 3.1** and **3.2**). Then, we describe zinc finger mutant proteins consisting of three tandemly connected catalytic zinc finger motifs (**Sections 3.3**, **3.4** and **3.5**). Last, we describe artificial endonucleases created using a cerium-binding loop as a linker between zinc finger domains (10) (**Sections 3.6** and **3.7**).

2. Materials

2.1. Preparation of Zinc Finger Peptides with Catalytic Zinc Sites (His4-Type Zinc Finger Peptide) by Peptide Synthesis

1. Peptide synthesis reagents, to be stored at −20°C: *N*-fluorenylmethoxycarbonyl (Fmoc)-protected amino acids (Peptide Institute, Inc.); TGS-RAM resin (Shimadzu); *N*-hydroxybenzotriazole (HOBt); 1-[bis(dimethylamino) methylene]-1*H*-benzotriazolium (HBTU); trifluoroacetic acid (TFA).
2. Peptide synthesis reagents, to be stored at room temperature: *N,N*-diisopropylethylamine (DIEA); piperidine (PIP); methanol; diethyl ether; *N,N*-dimethylformamide (DMF) (peptide synthesis grade). To keep DMF anhydrous, add activated molecular sieves 4A.
3. Cleavage Buffer: 85% (v/v) TFA, 5% (v/v) thioanisole, 5% (v/v) ethanedithiol, 1.5% (v/v) triisopropylsilane, in water.
4. Cosmosil 5C_{18}-AR-300 (10 mm × 250 mm, Nacalai Tesque) or Cosmosil 5C_{18}-ARII (10 mm × 250 mm, Nacalai Tesque) column.
5. Elution Buffer for HPLC: H_2O/TFA (100:0.1) and acetonitrile/TFA (100:0.1). Filter and store at room temperature.

2.2. Hydrolysis of 4-Nitrophenyl Acetate (NA) and Phosphodiester (Bis(4-Nitrophenyl) Phosphoric acid, BNP) by Catalytic Zinc Finger Peptides

1. Reaction Buffer: 20 mM HEPES (pH 8.0), 100 mM NaCl, 3.5% acetonitrile.
2. $ZnCl_2$ (99.999% purity).
3. NA stock solution: 4-Nitrophenyl acetate (NA) in acetonitrile (50 mM).
4. BNP stock solution: Bis(4-nitrophenyl)phosphoric acid (BNP) in acetonitrile (50 mM).

2.3. Construction of Plasmid DNAs Encoding Zinc Finger Proteins

1. Template Plasmid: pEV-Sp1 (530–623) (11).
2. QuickChange Site-Directed Mutagenesis Kit (Stratagene). *E. coli* strain XL1-Blue super competent cells are included and should be stored at −80°C (*see* **Note 1**).
3. Complementary Oligonucleotides for QuickChange: PAGE purification grade (*see* **Note 2**). Following the instructions of the QuickChange Site-Directed Mutagenesis Kit, put the mutation to be introduced in the middle of DNA fragment. Dissolve to a concentration of 100 μM in 5 mM Tris–HCl, pH 8.0. Store at −20°C.
4. 0.1 g/mL ampicillin in Milli-Q purified water. Store at −20°C.
5. LB medium: 1% polypepton (w/v), 0.5% yeast extract (w/v), 1% NaCl (w/v). Adjust pH to 7.0 with 2 M NaOH. Autoclave and store at room temperature.
6. 1.5% LB–agar plates containing 0.1 mg/mL ampicillin: Add 1.5 g of agar (powder) to 100 mL of LB medium and autoclave. When the temperature of the autoclaved LB–agar solution cools to about 50°C, add 100 μL of 0.1 g/mL ampicillin and mix well. Pour it into dishes. Store at 4°C.
7. Plasmid extraction kit.

2.4. Preparation of Zinc Finger Mutant Proteins

1. *E. coli* strain BL21(DE3) pLysS competent cells. Divide into 10 μL aliquots and store at −80°C.
2. LB medium, 800 mL, autoclaved.
3. Antibiotics stocks: 0.1 g/mL ampicillin in Milli-Q purified water; 0.1 g/mL carbenicillin in Milli-Q purified water; 0.1 g/mL chloramphenicol in EtOH. Store at −20°C.
4. 1 M isopropylthiogalactoside (IPTG) in water. Store at −20°C.
5. 100 mM $ZnCl_2$ in water (add HCl if necessary). Autoclave before storage at room temperature.
6. Sonicator for cell disruption.
7. 0.45-μm HT Tuffryn membrane filter.
8. Cation exchange column for low-pressure liquid chromatography: High S column (Bio-Rad).
9. 10× phosphate-buffered saline (PBS) stock: 1.37 M NaCl, 27 mM KCl, 100 mM Na_2HPO_4, 18 mM KH_2PO_4 (pH 7.6). Autoclave before storage at room temperature.
10. 1 M DTT in water. Store at −20°C.
11. Buffers for cation exchange chromatography: Buffer A: 1 mM DTT in 1× PBS. Buffer B: 2 M NaCl, 1 mM DTT in 1× PBS. Filter each buffer through 0.2 μm nylon

membrane filters (Nalgene), degas using an aspirator, and store at 4°C.

12. Reagents and materials for SDS-PAGE analysis.
13. Ultrafilter devices (e.g., Amicon Ultra 10,000 MWCO, Millipore).
14. Ultrafiltration Buffer: 20 mM Hepes buffer, pH 7.5.

2.5. Cleavage of Plasmid DNA by Zinc Finger Mutant Proteins

1. pUC19 (Takara Bio Inc.).
2. Oligonucleotides: Upper strand: 5′-GATCTGGGGCGGGGCCT-3′, Lower strand: 5′-AATTAGGCCCCGCCCCA-3′ in the case of the GC-box DNA (underlined) as a target. Dissolve in 5 mM Tris–HCl, pH 8.0, to be 100 μM.
3. Enzymes for DNA subcloning (store at −20°C): *Eco*RI; *Bam*HI; calf intestinal phosphatase (CIP); T4 polynucleotide kinase (PNK); 100 mM ATP in 1 mM Tris–HCl, pH 7.5; Takara ligation kit Ver 2 (Takara Bio Inc.); DNA purification kit (e.g., Promega); 2× PCR Master Mix (Promega).
4. 1× Tris–Borate–EDTA (TBE) buffer: Mix 109 g Tris base, 55 g boric acid, and 9.8 g Na_2ETDA. Bring the volume to 10 L by adding dH_2O.
5. 10 mg/mL ethidium bromide solution.
6. 1% agarose gel: Add agarose in TBE buffer to be 1% (w/v). Heat with microwave until agarose completely melts and the solution becomes transparent by mixing gently several times. Cool at room temperature to ~50°C. Pour the molten gel in a tray and put a comb.

2.6. Preparation of Artificial Zinc Finger-Type Nuclease with Cerium-Binding Sequence

1. Oligonucleotides (**Table 21.1**) (PAGE purification grade), T7 promoter primer and T7 terminator primer dissolved in 5 mM Tris–HCl, pH 8.0, to be 100 μM. Store at −20°C.
2. Template Plasmids: pEV-Sp1 (566–623) (11) and pEV-GLI (99–160) (12).
3. Enzymes for DNA subcloning (store at −20°C): *Nde*I; *Eco*RI; CIP.
4. Cation exchange column for chromatography: High S (Bio-Rad), Mono S or RESOURCE S (GE Healthcare).
5. Buffer for cation exchange: same as described in Step 11 of **Section 2.4**.
6. Gel filtration column (e.g., Superdex75, GE Healthcare).
7. TN Buffer: 10 mM Tris–HCl, pH 8.0, 50 mM NaCl, 1 mM DTT.

Table 21.1
Sequences of oligonucleotides used for preparation of artificial zinc finger-type nucleases with cerium-binding sequence and for DNA cleavage reactions

Primer name	Oligonucleotide sequences
P1-rev	5′-ATGTAGCCGTCGCCGTCCTTGTCTTCGCCGGTGTGGGTCTTGATATGTTT-3′
P1G-rev	5′-ATGTAGCCGTCGCCGTCCTTGTCGCCTTCGCCGGTGTGGGTCTTGATATGTTT-3′
P1-fwd	5′-GGACGGCGACGGCTACATCAGCGCAGCAGAACCATATATGTGTGAGCAC-3′
P1G-fwd	5′-GGACGGCGACGGCTACATCAGCGCAGCAGAAGGCCCATATATGTGTGAGCAC-3′
GCGLI upper	5′-TATCGAATTAGGCCCCGCCCCATCTTGGGTGGTCAGATCCACTA-3′
GCGLI lower	5′-TAGTGGATCTGACCACCCAAGATGGGGCGGGGCCTAATTCGATA-3′

2.7. Detection of Sequence-Specific DNA Cleavage

1. Oligonucleotides (**Table 21.1**) (PAGE purification grade) dissolved in 5 mM Tris–HCl, pH 8.0, to be 100 μM. Store at −20°C.
2. 1× TBE buffer (*see* Step 4 of **Section 2.5**).
3. T4 polynucleotide kinase (PNK).
4. 10 μCi/μL [^{32}P]-γATP. *Caution*: this is radioactive. Refer to the guidelines at each research facility.
5. MicroSpin G-25 Columns (GE Healthcare).
6. Reagents for phenol/chloroform protein extraction and ethanol DNA precipitation. *Caution*: phenol is corrosive. Use appropriate protection.
7. Cleavage Buffer: 10 mM Tris–HCl, pH 8.0, 50 mM NaCl, 1 mM dithiothreitol, 20 μg/μL calf thymus DNA, and the ^{32}P-end-labeled substrate DNA fragment (~8 nM, 40 kcpm).
8. 500 μM $(NH_4)_2[Ce(NO_3)_6]$ (e.g., Sigma-Aldrich).
9. Reagents and materials for analysis on a 15% polyacrylamide/8 M urea sequencing gel using TBE buffer.
10. 1× Sample Buffer: 50 mM NaOH, 8 M urea, 0.05% bromophenol blue, 0.05% xylene cyanol in 1× TBE buffer.

3. Methods

Synthesized zinc finger peptides can be stored as lyophilized powder at −80°C and partly dissolved with buffer just before experiments. On the other hand, *E. coli*-expressed zinc finger mutant proteins with altered zinc ligands and zinc finger-type nucleases with cerium-binding peptides are not as stable. It is important to promptly purify and use fresh peptides.

3.1. Preparation of Zinc Finger Peptides with Catalytic Zinc Sites (His4-Type Zinc Finger Peptide) by Peptide Synthesis

Herein a Shimadzu's peptide synthesizer PSSM8 is used for preparing the Sp1 finger2 His4 peptide. According to the Shimadzu's program for peptide synthesis, Steps 3–5 of **Section 3.1** are run automatically to synthesize the intended peptide.

1. Resin preparation: Weigh 50 mg TGS-RAM resin (0.21 mmol/g) into a reaction vessel of the automatic synthesizer. Add 1.5 mL DMF to the dried resin, stir gently for 1 min, wait for 30 min, and then remove the solvent via vacuum filtration.
2. Reagent preparation: Prepare Reagent 1 (HBTU and HOBt), Reagent 2 (DIEA), piperidine (PIP), and all Fmoc amino acids according to the PSSM8 program recipe before starting the synthesis. Connect all regents to the PSSM8 synthesizer equipment.
3. Removal of Fmoc group on the resin: Wash the resin with DMF for 3 min. Add DMF with 30% PIP to the resin. Bubble the resulting mixture for 4 min. Repeat this step once again. Then wash the resin with DMF for 1 min five times.
4. Condensation reaction: Bubble 420 μL of DMF containing 105 μmol (44.68 mg) of Fmoc-Glu(OtBu)-OH, 105 μmol of HBTU, 105 μmol of HOBt, and 210 μmol of DIEA for 3 min, then add the mixture to the resin, and bubble for 30 min. After this step, remove the solvent.
5. Cycling: Repeat the above Steps 3–4 of **Section 3.1** (deprotection/coupling cycle) to obtain a resin having a bound, protected peptide. In each successive cycle, substitute the Fmoc-Glu(OtBu)-OH in Step 4 of **Section 3.1** with 105 μmol of the Fmoc amino acid in the following order; Fmoc-Gly-OH, Fmoc-Thr(tBu)-OH, Fmoc-His(Trt)-OH, Fmoc-Thr(tBu)-OH, Fmoc-Arg(Pbf), Fmoc-Lys (Boc)-OH, Fmoc-His(Trt)-OH, Fmoc-Arg(Pbf), Fmoc-Gln(Trt)-OH, Fmoc-Leu-OH, Fmoc-Glu(OtBu)-OH, Fmoc-Asp(OtBu)-OH, Fmoc-Ser(tBu)-OH, Fmoc-Arg (Pbf), Fmoc-Thr(tBu)-OH, Fmoc-Phe-OH, Fmoc-Arg(Pbf), Fmoc-Lys(Boc)-OH, Fmoc-Gly-OH, Fmoc-His(Trt)-OH, Fmoc-Tyr(tBu)-OH, Fmoc-Ser(tBu)-OH, Fmoc-Trp-OH, Fmoc-Thr(tBu)-OH, Fmoc-His(Trt)-OH, Fmoc-Met-OH, Fmoc-Phe-OH, Fmoc-Pro-OH, and Fmoc-Arg(Pbf).
6. Washing: Wash the obtained resin with methanol and diethyl ether twice. Dry under reduced pressure for 1 h.

7. Final cleavage of side chain-protecting group and from the resin: Add 2 mL of Cleavage Buffer to the resin and keep it at room temperature for 6 h, resulting in cleavage of the peptide from the resin. Stir the mixture gently two or three times during the cleavage step. Remove the resin from mixture by filtration. Add about 10 mL of diethyl ether to the filtrate. Collect the precipitate and dry in vacuo to obtain the crude product of the peptide.
8. Purification of crude peptide: Purify the crude product by HPLC using a reversed-phase column (Cosmosil $5C_{18}$-AR-300 or Cosmosil $5C_{18}$-ARII). Elute the purified product according to a linear concentration gradient method with H_2O/TFA and acetonitrile/TFA, detecting absorbance at 230 nm. Collect the main peaks and measure their molecular weight by MALDI-TOF mass spectrometry to determine the fractions containing the correct peptide. Collect all corresponding fractions and lyophilize them to obtain the peptide.

3.2. Hydrolysis of 4-Nitrophenyl Acetate (NA) and Phosphodiester (Bis(4-Nitrophenyl) Phosphoric Acid, BNP) by Catalytic Zinc Finger Peptides

1. Dissolve the zinc finger peptide in Reaction Buffer and determine the concentration by measurement of absorbance at 280 nm (ca. 100 μM).
2. Add 1.2 molar equivalents of $ZnCl_2$ to the zinc finger peptide solution and incubate for 1 h at 25°C.
3. Add NA (**Fig. 21.1**) or BNP (**Fig. 21.2**) solution to the solution of the Zn(II)–peptide complex (final concentration; [NA] or [BNP] = ca. 1 mM).

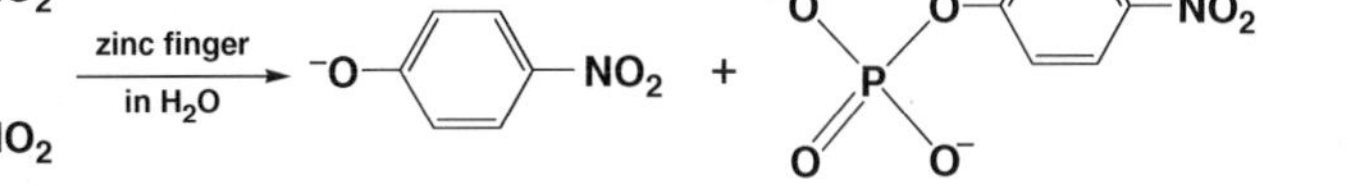

$$O_2N-C_6H_4-OAc \xrightarrow[\text{in } H_2O]{\text{zinc finger}} O_2N-C_6H_4-O^- + CH_3CO_2^-$$

Fig. 21.1. Hydrolytic reaction of NA by zinc finger nuclease.

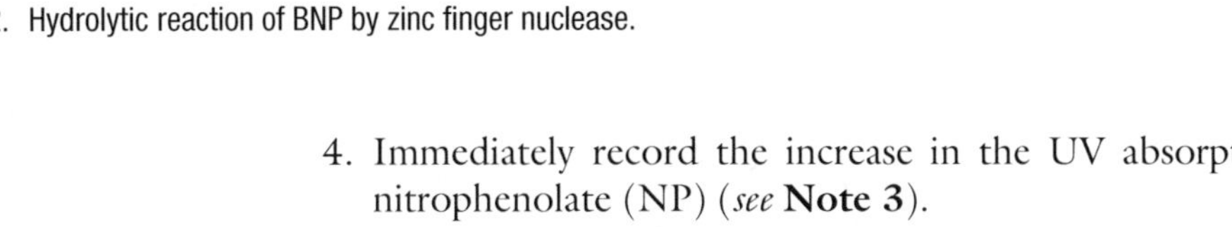

Fig. 21.2. Hydrolytic reaction of BNP by zinc finger nuclease.

4. Immediately record the increase in the UV absorption of nitrophenolate (NP) (*see* **Note 3**).
5. Calculate the rate constant in accordance with following description. In this experimental condition, the initial rate

linearly increases with the concentration of NP. The initial rate (v_{in}) is expressed by equation [**1**]:

$$v_{in}=k_{obs}[NP] \quad [1]$$

where k_{obs} s^{-1} is the pseudo-first-order rate constant. Even in the absence of the catalyst, hydrolysis is observed, indicating that OH^- and solvent can promote hydrolysis. The value k_{obs} includes all the catalytic species such as the zinc complex and bases. Therefore, k_{obs} can be expressed using equation [2]:

$$k_{obs}=k[cat]+k_{OH}[OH^-]+k_0 \quad [2]$$

where k $M^{-1}s^{-1}$ and k_{OH} $M^{-1}s^{-1}$ are the second-order rate constants of the catalyst and OH^-; respectively, and k_0 s^{-1} is the first-order rate constant of solvolysis.

3.3. Construction of Plasmid DNAs Encoding Zinc Finger Mutant Proteins Including Catalytic Zinc Fingers

1. Amplify Template Plasmid DNA with the Complementary Oligonucleotides using the QuickChange Site-Directed Mutagenesis Kit.
2. Add 1 μL of kit-supplied DpnI to the reaction mix and incubate for 1 h at 37°C.
3. Transform 1 μL of the mix into kit-supplied XL-1 Blue super-competent cells. Plate the cells on 1.5% LB–agar plates containing 0.1 mg/mL ampicillin.
4. Pick a colony into 3 mL LB medium containing 0.1 mg/mL ampicillin. Shake at 37°C, 200 rpm, overnight (about 10–16 h).
5. Purify plasmid DNA from the *E. coli* overnight culture using a commercial grade plasmid extraction kit.
6. Cut out the zinc finger-coding region containing the introduced mutations and reinsert the fragment into the original plasmid (*see* **Note 4**)
7. Confirm the DNA sequence of the inserted zinc finger-coding region.

3.4. Preparation of Zinc Finger Mutant Proteins

1. Transform 1 ng of plasmid encoding each zinc finger mutant into *E. coli* strain BL21(DE3) pLysS competent cells and spread on 1.5% LB–agar plates containing antibiotics (0.1 mg/mL ampicillin for pEV-Sp1). Incubate the plates at 37°C overnight.
2. Pick a colony into 3 mL of LB medium containing 0.1 mg/mL ampicillin. Shake at 37°C, 200 rpm, overnight (about 10–16 h) (*see* **Note 5**).
3. Transfer the overnight culture to 800 mL of LB medium containing 0.05 mg/mL carbenicillin. Shake at 37°C, 200 rpm until A_{600} becomes ~0.6.

4. Cool the culture, then add 0.1 mM IPTG and 0.1 mM $ZnCl_2$. Shake at 20°C, 100 rpm, overnight (about 12–18 h).
5. Harvest the cells by centrifuging at 2,600×*g* for 10 min.
6. Resuspend cells with 25 mL of 1× PBS and transfer to a small centrifuge tube. Centrifuge at 2,600×*g* for 10 min.
7. Resuspend cells with 25 mL of 1× PBS. Sonicate for five rounds of 2 min each (40% duty cycle) (*see* **Note 6**).
8. Centrifuge at 4°C, 35,000×*g* for 30 min.
9. Filter the supernatant (soluble fraction) using a 0.45-μm HT Tuffryn membrane.
10. Purify the peptide using a low-pressure cation exchange FPLC chromatography. The peptide is eluted at about 30% Buffer B using a salt gradient from 0 to 100% Buffer B (0–2 M NaCl) over the equivalent of 20 column volumes. Fractions of 5 mL are collected.
11. Check 10 μL of each fraction by SDS-PAGE.
12. Concentrate and exchange the buffer by ultrafiltration in Ultrafiltration Buffer.

3.5. Cleavage of the Plasmid DNA by Zinc Finger Mutant Proteins

1. Construction of a plasmid with the target site for zinc finger mutant proteins:
   a. Phosphorylation and annealing of oligonucleotides coding the target sequence: Mix 1 μL of each 100 μM oligonucleotide, 10 μL 10× T4 PNK Buffer, 1 μL of 100 mM ATP, 1 μL of T4 PNK, and 86 μL of Milli-Q-purified water. Incubate at 37°C for 1 h, followed by incubation at 95°C for 10 min and gradually cooling down to room temperature over 1 h.
   b. Preparation of vector DNA for ligation with the above DNA fragment: Mix 3 μg of pUC19 DNA, 10× *Eco*RI buffer, 0.5 μL of *Bam*HI, 0.5 μL of *Eco*RI, and Milli-Q-purified water to be 100 μL. Incubate at 37°C for 1 h. Add 1 μL of CIP to the reaction mixture and incubate at 37°C for 1 h. Then purify the treated vector DNA using a DNA purification kit with 60 μL of the kit-supplied elution buffer (5 mM Tris–HCl, pH 8.0).
   c. Mix well 1 μL of 20-fold diluted annealed DNA fragment and 2 μL of purified vector DNA, and add 3 μL of Takara ligation kit Mighty Mix (or equivalent). Incubate at 25°C for 5 min. Transform 1 μL of the reaction mixture into 10 μL of *E. coli* DH5α competent cells. Spread onto 1.5% LB–agar plates containing 0.1 mg/mL ampicillin.
   d. Check insertion by colony PCR using 2× PCR Master Mix and appropriate PCR primers (*see* **Note 7**).

2. Cleavage reaction of zinc finger peptides with plasmid DNA.
   a. Mix zinc finger mutant proteins ($\sim 10^{-7}$–10^{-6} M) and plasmid DNA (25 nM) in 5 mM HEPES, pH 7.5, 50 mM NaCl. Incubate at 37°C for 48 h.
   b. Run samples on a 1% agarose gel at 100 V.
   c. Stain with ethidium bromide solution (0.5 μg/mL) and visualize with UV light.

3.6. Preparation of Artificial Zinc Finger-Type Nucleases with Cerium-Binding Sequence, Sp1(P1)GLI and Sp1(P1G)GLI

P1 (DKDGDGYISAAE) is an analogue of a 12-residue calcium-binding loop, to which a lanthanide ion strongly and specifically binds (13). Sp1(P1)GLI and Sp1(P1G)GLI (10) consist of Sp1(zf23) and GLI(zf45) connected by P1 and Gly-P1-Gly peptides, respectively.

1. Construction of plasmid DNAs encoding Sp1(P1)GLI or Sp1(P1G)GLI (*see* **Note 8**):
   a. Amplify Sp1(zf23)- and GLI(zf45)-coding fragments by standard PCR using the following primers and templates:
      i. Sp1(zf23) fragment: T7 promoter and P1 reverse primer, pEV-Sp1 (566–623) (11) as a template.
      ii. GLI(zf45)fragment: P1 forward and T7 terminator primer, pEV-GLI (99–160) (12) as a template.
   b. Amplify the fragment encoding Sp1(P1)GLI using T7 promoter and terminator primers and amplified Sp1(zf23) and GLI(zf45) fragments as templates.
   c. Digest the resultant fragments with *Nde*I and *Eco*RI and insert into pEV3b. The sequence of pEV-Sp1(P1)GLI should be confirmed by DNA sequencing.
   d. pEV-Sp1(P1G)GLI can be created in the same way.
2. Expression and purification of the peptide: Following Steps 1–12 in **Section 3.4**, dilute the fractions containing the zinc finger peptide 5-fold with Buffer A. Further purify the peptide using FPLC and a Mono S or RESOURCE S column. The peptide is eluted using a salt gradient from 5 to 100% Buffer B (0.1–2 M NaCl) over the equivalent of 10 column volumes. Collect 0.25 mL fractions. Finally purify the peptide by gel filtration using a Superdex75 column with TN Buffer.

3.7. Detection of Sequence-Specific DNA Cleavage

1. DNA labeling with [^{32}P]-γATP. *Caution*: this and all subsequent steps contain materials that are radioactive. Mix water, 1 pmol of a PAGE-purified oligonucleotide, 5 μL of 10× PNK buffer, 1 μL of PNK, and 1 μL of [^{32}P]-γATP to a final volume of 50 μL. Incubate at 37°C for 1 h. Remove the unreacted [^{32}P]-γATP using a MicroSpin G-25 column. Add 1 pmol of an unlabeled complementary

oligonucleotide. Heat at 95°C for 10 min and cool to room temperature for about 2 h to complete annealing. Purify the annealed DNA fragment using phenol/chloroform extraction and ethanol precipitation. Dissolve the end-labeled DNA in water to be 10 kcpm/μL. Store at −20°C.

2. Cleavage reaction: Incubate various concentrations of peptides (2.5–10 μM) in Cleavage Buffer containing the ^{32}P-end-labeled substrate DNA fragment and $(NH_4)_2[Ce(NO_3)_6]$ (final concentrations of 5–20 μM) at 37°C for 2 h (*see* **Note 9**). Stop the reaction by adding an equivolume of phenol/chloroform. After protein extraction and ethanol precipitation, dissolve the cleaved products in 4 μL of 1× Sample Buffer (*see* **Note 10**)
3. Prepare a Maxam–Gilbert sequencing reaction of the labeled DNA fragments as a sequence ladder sample using standard protocols.
4. Heat the samples at 95°C for 2 min.
5. After a pre-run of 15% polyacrylamide/8 M urea sequencing gel in 1× TBE buffer, flush the well, and set a comb. Apply the samples and run the gel at 3,000 V until the bromophenol blue and xylene cyanol separate well (*see* **Note 11**).
6. Visualize the radioactive image using X-ray film or a phosphorimager.

4. Notes

1. As suggested in the instruction of the Kit, kit-included cells with high competency are recommended rather than custom-made cells.
2. The grade of oligonucleotides is not important. However, desalting-grade oligonucleotides sometimes lack several bases and may result in additional time and money to get mutant plasmids with correct sequences.
3. The molar extinction coefficients at 400 nm for NP are 1,700 $M^{-1}cm^{-1}$ (pH 6.0), 4,200 (pH 6.5), 8,700 (pH 7.0), 12,800 (pH 7.5), 16,200 (pH 8.0), 17,300 (pH 8.5), 17,800 (pH 9.0), and 18,200 (pH 9.5).
4. This step eliminates incorporation of incorrect nucleotides into the vector sequences and you can check DNA sequences only for the zinc finger-coding region in the next step.

5. You can skip this step and pick a colony directly to 800 mL of LB medium, though it takes more than 10 h to reach A_{600}=0.6. Using an overnight culture, it takes about 3 h to reach A_{600}=0.6.
6. The time depends on the power of the sonicator. We use a transparent container and continue sonicating until the cells are no longer opaque. Be careful not to increase the sample temperature by continuous sonication.
7. Sometimes two kinds of plasmids, with and without insertion, exist in the same colony on the plate. So, it is best to use a pair of primers that should anneal to close to each ends of the inserted region and to compare the length of amplified DNA fragments between with and without the insertion. Or, you can use one primer that is complementary to the insert and one that is complementary to the vector, so that you don't amplify linear DNA insert that is just lying around on the plate. In the latter case, sequencing should be confirmed to avoid contamination of the original plasmid without insertion.
8. The P1G peptide contains Gly residues at the both ends of P1 sequence in order to examine the effects of flexibility of the joints of the zinc finger domain and the cerium-binding peptide. The sequence-specific DNA cleavage ability of Sp1(P1G)GLI is higher than that of Sp1(P1)GLI.
9. When the incubation time is 24 h, non-specific cleavage bands can be detected not only at the spacer region between Sp1 and GLI target sites but also within the binding sites for Sp1 and GLI. The Ce(IV)-bound P1 peptide itself does not cleave DNA in a sequence-specific manner even after 24-h incubation.
10. Measure the counts per minute of 1 μL for each sample and transfer to new tubes a volume that will maintain the same counts per minute between all samples.
11. In a 15% denaturing sequencing gel, the mobilities of bromophenol blue and xylene cyanol are close to those of 10- and 40-mer oligonucleotides, respectively.

References

1. Auld, D.S. (2001) Zinc coordination sphere in biochemical zinc sites. *Biometals.* **14**, 271–313.
2. Klug, A. and Rhodes, D. (1987) 'Zinc fingers': a novel protein motif for nucleic acid recognition. *Trends Biochem Sci.* **12**, 464–469.
3. Klug, A. (1999) Zinc finger peptides for the regulation of gene expression. *J Mol Biol.* **293**, 215–218.
4. Vallee, B.L. and Auld, D.S. (1992) Active zinc binding sites of zinc metalloenzymes. *Matrix Suppl.* **1**, 5–19.
5. Vallee, B.L. and Auld, D.S. (1992) Functional zinc-binding motifs in enzymes and DNA-binding proteins. *Faraday Discuss Chem Soc.* **93**, 47–65.
6. Nomura, A. and Sugiura, Y. (2002) Contribution of individual zinc ligands to metal binding and peptide folding of

zinc finger peptides. *Inorg Chem.* **41**, 3693–3698.
7. Nomura, A. and Sugiura, Y. (2004) Hydrolytic reaction by zinc finger mutant peptides: successful redesign of structural zinc sites into catalytic zinc sites. *Inorg Chem.* **43**, 1708–1713.
8. Nomura, A. and Sugiura, Y. (2004) Sequence-selective and hydrolytic cleavage of DNA by zinc finger mutants. *J Am Chem Soc.* **126**, 15374–15375.
9. Negi, S., Itazu, M., Imanishi, M., Nomura, A., and Sugiura, Y. (2004) Creation and characteristics of unnatural CysHis(3)-type zinc finger protein. *Biochem Biophys Res Commn* **325**, 421–425.
10. Nakatsukasa, T., Shiraishi, Y., Negi, S., Imanishi, M., Futaki, S., and Sugiura, Y. (2005) Site-specific DNA cleavage by artificial zinc finger-type nuclease with cerium-binding peptide. *Biochem Biophys Res Commn* **330**, 247–252.
11. Yokono, M., Saegusa, N., Matsushita, K., and Sugiura, Y. (1998) Unique DNA binding mode of the N-terminal zinc finger of transcription factor Sp1. *Biochemistry.* **37**, 6824–6832.
12. Shiraishi, Y., Imanishi, M., and Sugiura, Y. (2004) Exchange of histidine spacing between Sp1 and GLI zinc fingers: distinct effect of histidine spacing-linker region on DNA binding. *Biochemistry.* **43**, 6352–6359.
13. Wojcik, J., Goral, K., Pawlowski, A., and Bierzynski, A. (1997) Isolated calcium-binding loops of EF-hand proteins can dimerize to form a native-like structure. *Biochemistry.* **28**, 680–687.

Section IV

Beyond Transcription and Cleavage

Chapter 22

Designing and Testing Chimeric Zinc Finger Transposases

Matthew H. Wilson and Alfred L. George

Abstract

Transposons have been effectively utilized as non-viral gene delivery systems that are capable of promoting stable transgene expression in mammalian and human cells. Two specific transposon systems, *Sleeping Beauty (SB)* and *piggyBac (PB)*, have been successfully modified by the addition of a zinc finger (ZF) DNA-binding domain as a strategy for directing transposon integration near ZF target sites in a host cell genome. Site-directed integration could improve transposon-mediated gene transfer by limiting positional effects on transgene integration, limiting genotoxicity, and improving the overall safety in gene therapy applications. In this chapter, we describe methodology for creating and characterizing functional chimeric ZF transposases and experimental approaches for testing their transpositional activity in human cells.

Key words: Transposon, transposase, zinc finger, site-directed integration, human cells, *sleeping beauty*, *piggyBac*.

1. Introduction

Transposon systems have been used for pre-clinical gene therapy approaches in vivo and ex vivo (1–9). Transposon systems contain two components: a transposase enzyme and a transposon segment of DNA. Genomic integration occurs by a "cut and paste" mechanism whereby transposon DNA is excised, usually from plasmid DNA for gene transfer applications, and then integrated into a target genomic sequence in the host cell. In their native state, the site selection for integration is uncontrolled, and this can lead to potential safety issues in the context of gene therapy (10, 11). Zinc finger (ZF) DNA-binding domains can enable protein binding to specific genomic target sequences. The fusion of a site-specific ZF DNA-binding domain to a transposase

J.P. Mackay, D.J. Segal (eds.), *Engineered Zinc Finger Proteins*, Methods in Molecular Biology 649,
DOI 10.1007/978-1-60761-753-2_22, © Springer Science+Business Media, LLC 2010

enzyme is one proposed strategy to enable site-directed integration of transposons near a zinc finger target site within the genome (10, 12–15).

The addition of a protein domain to a native transposase has the potential to impair transposase function (12, 13, 15). For example, *SB* transposase may exhibit decreased or completely abolished enzymatic activity when modified (10, 12, 13, 15). Re-engineering *piggyBac* transposase by fusing proteins to either the amino- or the carboxyl-terminus has been more successful in producing functional enzyme (10, 14, 16). Testing two major activities (transposon excision and genomic integration) of re-engineered transposases is necessary prior to investigating their utility in gene transfer experiments. Confirming protein expression of chimeric transposase enzymes by Western blot analysis is also important to properly interpret the results of functional assays.

Recombinant transposases should be tested for the ability to excise transposon sequences from a plasmid followed by repairing (rejoining) the cleaved donor DNA molecule ends. This activity can be assessed by detecting successful removal of transposon DNA from a transposon containing plasmid when transiently expressed in mammalian cells (10, 12, 17). This type of assay relies upon isolation of plasmid DNA from transfected cells and PCR amplification of a region spanning the transposon. Absence of the transposon inferred by a shorter amplicon than observed with intact plasmid demonstrates excisional activity. Further characterization of transposase activity is accomplished by demonstrating successful transposon integration into the host cell genome. A simple approach can be used to quantify the number of resistant colonies formed after transfer of a transgene conferring antibiotic resistance into cultured cells (1, 10, 12).

After fusion of a protein domain such as a ZF to a transposase, one has to confirm that the chimeric enzyme retains function. Hence, transposon excision analysis is used to measure release of a transgene from the transposon by transposase activity, whereas transposition assays are used to quantify and characterize the success of genomic integration of the transposon. This chapter will summarize the basic methods for creating and characterizing a functional chimeric ZF-*SB* transposase and outline the protocols for testing its activity in human cells.

2. Materials

2.1. Generation of Chimeric ZF Transposases

1. Oligonucleotide primer sequences for recombinant polymerase chain reaction (PCR) fusion of a ZF DNA-binding domain to the amino-terminus of a hyperactive *SB* transposase *(SB12)* (*see* **Table 22.1**).

Table 22.1.

Addition of Sp1

Sp1 forward	GATGCGTATGCCTACGGCCGACCGGT ATGTACCCGTACGATGTTCCAGATT ACGCTAAAGACAGTGAAGGAAGGG GCTC
Sp1 reverse	CTTGGCTGATTTCTTTTGATTTTCCTC CGGTACCACTGTCCAGGGGCAAAGT
SB forward	ACTTTGCCCCTGGACAGTGGTACCGGA GGAAAATCAAAAGAAATCAGCCAAG
SB reverse	CAGTGGCTTCTTCCTTGCTGAG
Addition of ZNF202	
ZNF202 forward	GATGCGTATGCCTACGGCCGACCGG TATGTACCCGTACGATGTTCCAGAT TACGCTACTCCATCAGTGGAGAAA CCCT
ZNF202 reverse	CTTGGCTGATTTCTTTTGATTTTCCCA TACTAGTTCCAGAACCACCAGAGCC TCCGGAACCTCCAGAGCCGGAGGT CTTTTCTGAGTGGGT
SB forward	ACCCACTCAGAAAAGACCTCCGGCTCT GGAGGTTCCGGAGGCTCTGGTGGT TCTGGAACTAGTATGGGAAAATCAA AAGAAATCAGCCAAG
SB reverse	CAGTGGCTTCTTCCTTGCTGAG
Linker 1-top	GGGCTCTGGAGGTTCCGGAGGCTCTG GTGGTTCTGGAACTAGTGGTAC
Linker 1-bottom	CACTAGTTCCAGAACCACCAGAGCCTC CGGAACCTCCAGAGCCCGTAC

Primer sequences are presented 5′–3′, left to right, adapted from (12).

2. pCMV-SB12 transposase expression plasmid (*see* **Fig. 22.1**). Templates for all constructs or plasmids can be obtained from the authors.
3. High-fidelity thermostable DNA polymerase (e.g., Expand Taq, Roche Applied Science).
4. Standard reagents for PCR and cloning; calf alkaline phosphatase and restriction enzymes (*Kpn*I, *Bst*ZI, *Bsr*GI).

2.2. Western Blot of Recombinant Chimeric ZF Transposases

1. Human embryonic kidney (HEK293) cells (American Type Culture Collection).
2. HEK293 Medium: Minimum essential medium with 2 mM L-glutamine, 10% fetal bovine serum, 50 U/mL penicillin, and 50 μg/mL streptomycin, sterile filtered.

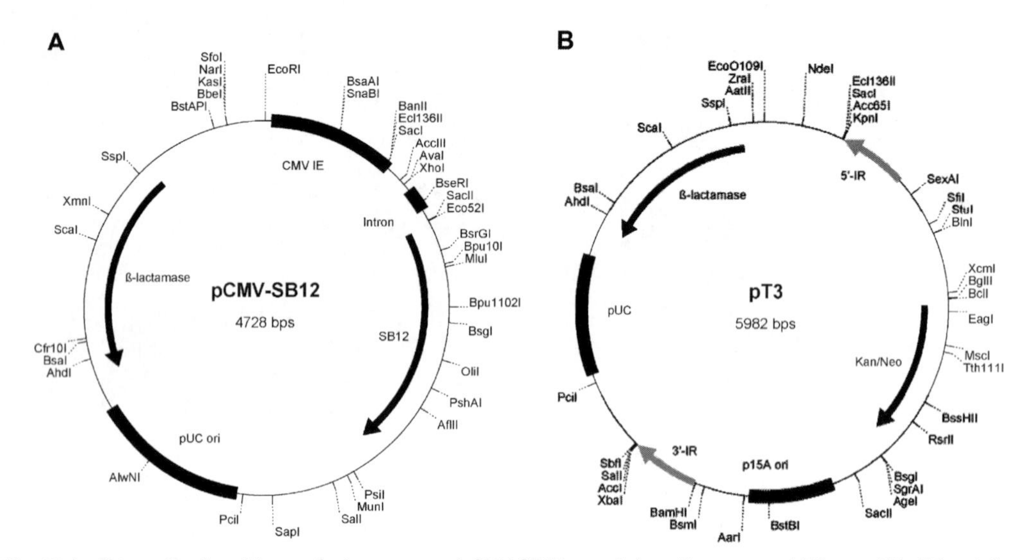

Fig. 22.1. Schematic plasmid maps for transposase (pCMV-SB12, panel **a**) and transposon (pT3, panel **b**). Abbreviations: CMV IE, immediate early promoter for cytomegalovirus; SB12, hyperactive *SB* transposase; pUC, origin of replication; β-lactamase, ampicillin resistance gene; 5′-IR, 5′ inverted repeat element; 3′-IR, 3′ inverted repeat element; Kan/Neo, kanamycin/neomycin resistance gene; p15A, origin of replication.

3. FuGene-6 transfection reagent (Roche Applied Science).
4. RIPA Lysis Buffer: 50 mM Tris–HCl, pH 7.5, 150 mM NaCl, 1% IPEGAL, 0.5% deoxycholate, 0.1% SDS, 1× protease inhibitor cocktail (Complete Mini, Roche Applied Science).
5. 10% SDS-PAGE gels, nitrocellulose membranes, and related equipment.
6. Rat polyclonal hemagglutinin epitope (HA) antibody (Amersham).
7. Secondary antibody against rat IgG conjugated with horseradish peroxidase (Amersham).
8. Tris-buffered saline, pH 7.6 (Invitrogen), with 0.1% Tween-20 (Sigma) (TBS-T) with 5% (w/v) non-fat dry milk (blocking agent).
9. Chemiluminescence detection system (e.g., Thermo Fisher Scientific).

2.3. Transposon Excision Assay

1. HEK293 cells transfected with transposase and transposon plasmids.
2. ZF-SB12 transposase expression plasmid (created in **Section 3.1**) and pT3 transposon plasmid (*see* **Fig. 22.1**).
3. Miniprep spin column kit (e.g., QIAprep, Qiagen).

Table 22.2.

Excision assay primers

Round 1 forward	CAGCTTGTCTGTAAGCGGATG
Round 1 reverse	TACCGCCTTTGAGTGAGCTGATAC
Round 2 forward	TACCGCCTTTGAGTGAGCTGATAC
Round 2 reverse	GTCAGTGAGCGAGGAAGCGGAAGAG
AmpR forward	ATCAGTGAGGCACCTATCTCAG
AmpR reverse	TTCCGTGTCGCCCTTATTCCC

Primer sequences are presented 5′–3′, left to right, adapted from (12).

4. Nested PCR primers (*see* **Table 22.2**).
5. PCR reagents: Taq polymerase and PCR grade nucleotide mix (10 mM of each nucleotide) from Roche Applied Science. 10× NH_4 PCR buffer with 3 mM $MgCl_2$ (final concentration) from Bioline USA.

2.4. Transposition Assay

1. Trypsin (0.25% w/v) in 1 mM EDTA solution (Invitrogen).
2. G418: final concentration of 800 μg/mL in HEK293 Medium (Invitrogen).
3. Formaldehyde (10% v/v) in phosphate-buffered saline (PBS).
4. Methylene blue (0.5% w/v) in PBS, filtered.

3. Methods

3.1. Generation of Chimeric ZF Transposases

1. Use recombinant "double-overlap" PCR to create chimeric transposases having the DNA-binding domain of Sp1 (amino acids 534–623), ZFs 3–8 of the human ZF 202 (ZNF202), or ZF(s) of interest fused to the N-terminus of the *SB12* transposase (primer sequences provided in **Table 22.1** and (12)) (*see* **Notes 1–3**). Use PCR to generate a DNA-binding domain with an overlapping sequence corresponding to the N-terminus of the *SB12* transposase. Use PCR to generate a fragment containing the N-terminus of *SB12* with an overlapping sequence to enable fusion to the DNA-binding domain. Gel-purify the initial PCR products. Next, perform a second round of PCR with the purified DNA-binding domain and *SB12* fragments using the most external primers to generate a fused product for cloning into

the expression vector after restriction digestion with *Bst*ZI and *Bsr*GI.

2. When constructing N-terminal chimeras, include a SV40 T-antigen nuclear localization signal (MPKKKRKV) followed in-frame by an epitope tag (if desired), the corresponding ZF DNA-binding domain, and then the transposase cDNA.
3. Digest the recombinant PCR products and subclone them into pCMV-*SB12*. Digest with *Kpn*I, dephosphorylate the cut ends using calf intestine alkaline phosphatase, and ligate a double-stranded linker encoding 19 amino acid residues (linker 1: GTGSGGSGGSGGSGTSGTG, **Table 22.1**) (*see* **Note 3**).
4. Verify all DNA constructs by DNA sequencing to exclude polymerase errors and to demonstrate an intact open reading frame.

3.2. Western Blot of Recombinant Chimeric ZF Transposases

1. Culture HEK293 cells in HEK293 Medium at 37°C and 5% CO_2. Transfect HEK293 cells at 30–50% confluence in a 60-mm dish with FuGENE-6 and 3 μg of transposase plasmid.
2. Three days after transfection, lyse the cells in RIPA Lysis Buffer containing a 1× protease inhibitor cocktail.
3. Determine the protein concentrations (e.g., by Bradford assay). Electrophorese equivalent amounts of protein on a 10% SDS-PAGE gel.
4. Transfer gel to a nitrocellulose membrane and probe in TBS-T with blocking agent with a polyclonal anti-HA antibody (1:1,000 dilution) followed by incubation with a secondary antibody conjugated to horseradish peroxidase (1:5,000 dilution). A more detailed description of the Western blot procedure is offered in other chapters of this book (*see* **Chapters 6**, **7**, and **11**).
5. Detect epitope-tagged proteins by enhanced chemiluminescence detection. An example of expressed chimeric transposase enzymes is given in **Fig. 22.2** (*see* **Note 4**).

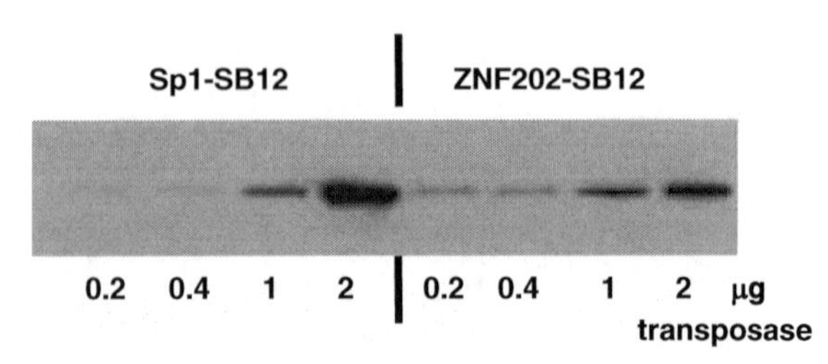

Fig. 22.2. Western blot analysis of chimeric Sp1-SB12 and ZNF202-SB12 transposases expressed in HEK293 cells. The bottom numbers represent the amount of transposase plasmid DNA used in transfections. (Reproduced from (12) with permission.)

3.3. Transposon Excision Assay

1. Transiently transfect HEK293 cells in a 60-mm dish with FuGENE-6 and plasmids containing the transposase (200 ng) and the transposon DNA segment (2 μg).
2. Three days after transfection, isolate plasmid DNA from cells using a QIAprep spin column kit according to the manufacturer's protocol in a total volume of 30 μL.
3. Using 5 μL of isolated plasmid DNA as a template, perform nested PCR using the primer sequences in **Table 22.2** to amplify a 585-bp product from the transposon plasmids that have undergone excision and repair as opposed to the intact ~3.5 kb transposon segment as described in (12) (*see* **Fig. 22.2a** and **Note 5**). The recovery of plasmid DNA from mammalian cells is very low; therefore nested (i.e., two rounds) PCR is necessary for further amplification of products in order to detect excision. Add 5 μL of lysate to the PCR mixture (1× NH_4 PCR buffer, 3 mM $MgCl_2$, 0.2 mM dNTP, 10 pmol forward and reverse primer each, and 0.5 μL Taq polymerase). The PCR conditions are as follows: 94°C for 5 min, 60 cycles of "94°C for 30 s, 64°C for 30 s, 72°C for 20 s," followed by 72°C for 10 min.
4. Dilute the PCR product from the first round 1:50 with sterile water and use 5 μL for the second round of PCR with nested primers and 1.5 mM $MgCl_2$; the composition of the PCR mixture is otherwise the same. Perform nested PCR at 94°C for 5 min, 35 cycles of "94°C for 30 s, 64°C for 30 s, 72°C for 5 s," followed by 72°C for 10 min.
5. Use isolated plasmid DNA for PCR amplification of the ampicillin resistance gene (**Table 22.2**) within the plasmid backbone to normalize for the amount of template in each reaction.
6. Visualize the PCR products using 1% agarose gel electrophoresis. An example of an excision assay is illustrated in **Fig. 22.3**.

3.4. Transposition Assay

1. Transfect HEK293 cells in a 60-mm dish with both transposase and transposon (which contains a neomycin resistance gene) plasmids using FuGene-6.
2. Two days after transfection, split cells to varying densities (1:15 and 1:150 dilutions) and place in medium containing 800 μg/mL G418 for selection.
3. After 2 weeks of selection, fix colonies of cells using 10% formaldehyde in PBS, then stain with methylene blue, and count. The number of colonies visualized is a proxy for transposition activity. More colonies equal higher transposase activity. Counting the number of colonies provides

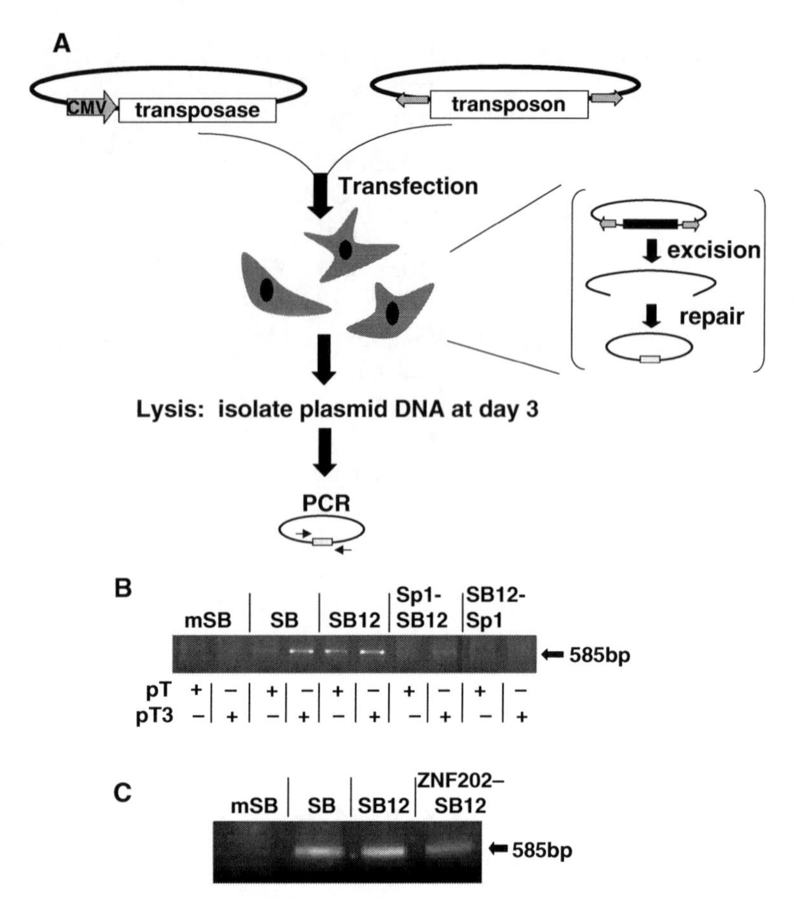

Fig. 22.3. Excision activity of chimeric *SB12* transposases. (**A**) Schematic of an excision assay. Cells are transfected with plasmids encoding transposase and transposon and subsequently lysed at day 3 post-transfection after the cell host machinery has repaired the site of transposon excision from plasmid DNA. Recovered plasmid is then used for a PCR reaction to amplify over the excised area in the transposon plasmid to monitor for excision. (**B**) Representative excision assay for transposase chimeras comprising Sp1 fused to *SB12*. *mSB*, catalytically inactive transposase; *SB12*, hyperactive transposase variant; *Sp1*, Sp1 DNA-binding domain either on N-terminus (Sp1-SB12) or on C-terminus (SB12-Sp1), *pT*, native transposon; *pT3*, hyperactive transposon. (**C**) Representative excision assay of ZNF202-SB12 chimera performed in the presence of *pT3* transposon. (Reproduced from (12) with permission.)

an estimate of the number of cells that underwent a transposition event (i.e., integrated the transposon and permitted selection). Examples of fixed and stained plates used in a transposition assay are shown in **Fig. 22.4** (*see* **Notes 6** and 7). At least three separate experiments should be performed to do a side-by-side comparison of native (i.e., unaltered) transposase to the designed chimeric ZF transposase to ensure that ZF fusion has not altered transposase activity. The colony count assay does not discriminate site-directed integration from random integration events (*see* **Note 8**).

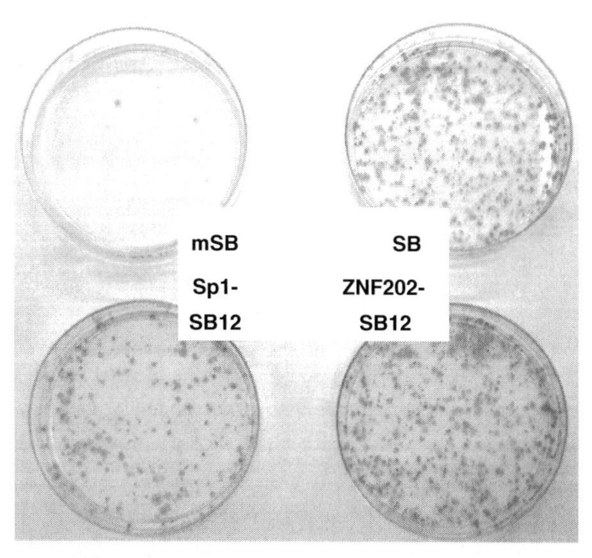

Fig. 22.4. Representative transposition assay comparing catalytically inactive *SB* (*mSB*) and native *SB* to N-terminal ZF fusion chimeras Sp1-SB12 and ZNF202-SB12. The number of colonies (i.e., colony count) equates with transposase activity. More active enzymes more efficiently integrate the neomycin resistance gene into the cellular genome and therefore permit more cells to grow in the selection antibiotic G418.

4. Notes

1. Fusion of a ZF DNA-binding domain to a transposase may impair transposase activity. In previous studies, addition of a ZF element to native *SB* has resulted in no detectable activity (12, 13, 15). In order to retain activity with *SB*, it would appear that hyperactive transposases (e.g., *SB12*) and transposons are necessary (12, 13, 15). The native *piggyBac* transposase has been successfully modified using a ZF element with detectable activity in mammalian cells (14).
2. Fusion of ZFs to the N-terminus preserves detectable transposase activity with both *SB12* and *piggyBac* (10, 12–15). C-terminal addition to *SB12* abolishes enzyme activity (12, 13, 15). Although a ZF has not been added to the C-terminus of *piggyBac*, one report has shown preserved *piggyBac* activity with C-terminal addition of another protein domain (16).
3. An optimal amino acid linker sequence separating the ZF and transposase reading frames may need to be determined. Various linkers have been successfully used in creating chimeric *SB12* and *piggyBac* transposases (10, 12–16). The linker needs to be flexible and of a length to ensure activity of both the ZF and the transposase domains.

4. The addition of a ZF element to a transposase can alter the expression level of the transposase enzyme (15). Codon optimization can be used to increase the expression level of a transposase enzyme if necessary.
5. Depending on the transposon system and specific plasmids used, the primers used for the excision assay may vary. Optimization of PCR conditions is necessary when monitoring for transposon excision in cultured cells.
6. The *SB* transposon system exhibits a phenomenon termed overproduction inhibition, in which increased transposase expression levels can inhibit transposition (5, 10, 12, 18). The addition of a ZF element to the *SB12* transposase may alter this phenomenon (12). When using *SB* as a gene transfer system, it is advisable to do a titration of transposase and transposon plasmids to determine which ratio gives optimal gene transfer. The *piggyBac* transposon system has conflicting reports regarding overproduction inhibition and may or may not exhibit this phenomenon in different cell types under varying conditions (10, 14, 16, 19).
7. Transposition assays require dilution of the transfected cells to such an extent that plating in G418 will permit visualization of colony formation for quantification. Various dilutions are therefore required to ensure that colony formation can be observed. Colony formation will depend on the inherent activity of the transposon system evaluated.
8. Independent assays (as described elsewhere in this book) should be used to verify the DNA-binding activity of the ZF elements used in creating functional transposase chimeras (*see* **Chapters 3**, **6**, **7**, **11**, and **27**).

Acknowledgments

M.H.W. would like to acknowledge support from a Career Development Award from the Department of Veterans Affairs.

References

1. Ivics, Z., Hackett, P.B., Plasterk, R.H., and Izsvak, Z. (1997) Molecular reconstruction of Sleeping beauty, a Tc1-like transposon from fish, and its transposition in human cells. *Cell.* **91**, 501–510.
2. Aronovich, E.L., Bell, J.B., Belur, L.R., Fransden, J.L., Gunther, R., Koniar, B., Erickson, D.C.C., Mcivor, R.S., Hackett, P.B., and Whitley, C.B. (2004) Long-term expression of Sleeping Beauty transposon in the murine model of mucopolysaccharidosis (MPS) type VII. *Mol Ther.* **9**, S331.
3. Belur, L.R., Frandsen, J.L., Dupuy, A.J., Ingbar, D.H., Largaespada, D.A., Hackett, P.B., and Mcivor, R.S. (2003) Gene insertion and long-term expression in lung mediated by

the Sleeping Beauty transposon system. *Mol Ther.* **8**, 501–507.
4. Dupuy, A.J., Clark, K., Carlson, C.M., Fritz, S., Davidson, A.E., Markley, K.M., Finley, K., Fletcher, C.F., Ekker, S.C., Hackett, P.B., Horn, S., and Largaespada, D.A. (2002) Mammalian germ-line transgenesis by transposition. *Proc Nat Acad Sci USA.* **99**, 4495–4499.
5. Geurts, A.M., Yang, Y., Clark, K.J., Liu, G.Y., Cui, Z.B., Dupuy, A.J., Bell, J.B., Largaespada, D.A., and Hackett, P.B. (2003) Gene transfer into genomes of human cells by the sleeping beauty transposon system. *Mol Ther.* **8**, 108–117.
6. Hackett, P.B., Clark, K., Davidson, A., Dupay, A., Largaspaeda, D., and Ekker, S.C. (2000) Insertional mutagenesis and gene tagging with the sleeping beauty transposon system in zebrafish. *Dev Biol.* **222**, 278.
7. Yant, S.R., Meuse, L., Chiu, W., Ivics, Z., Izsvak, Z., and Kay, M.A. (2000) Somatic integration and long-term transgene expression in normal and haemophilic mice using a DNA transposon system. *Nat Genet.* **25**, 35–41.
8. Huang, X., Wilber, A.C., Bao, L., Tuong, D., Tolar, J., Orchard, P.J., Levine, B.L., June, C.H., Mcivor, R.S., Blazar, B.R., and Zhou, X.Z. (2006) Stable gene transfer and expression in human primary T cells by the Sleeping Beauty transposon system. *Blood.* **107**, 483–491.
9. Singh, H., Manuri, P.R., Olivares, S., Dara, N., Dawson, M.J., Huls, H., Hackett, P.B., Kohn, D.B., Shpall, E.J., Champlin, R.E., and Cooper, L.J. (2008) Redirecting specificity of T-cell populations for CD19 using the Sleeping Beauty system. *Cancer Res.* **68**, 2961–2971.
10. Wilson, M.H., Coates, C.J., and George, A.L., Jr. (2007) PiggyBac Transposon-mediated Gene Transfer in Human Cells. *Mol Ther.* **15**, 139–145.
11. Yant, S.R., Wu, X.L., Huang, Y., Garrison, B., Burgess, S.M., and Kay, M.A. (2005) High-resolution genome-wide mapping of transposon integration in mammals. *Mol Cell Biol.* **25**, 2085–2094.
12. Wilson, M.H., Kaminski, J.M., and George, A.L., Jr. (2005) Functional zinc finger/sleeping beauty transposase chimeras exhibit attenuated overproduction inhibition. *FEBS Lett.* **579**, 6205–6209.
13. Ivics, Z., Katzer, A., Stuwe, E.E., Fiedler, D., Knespel, S., and Izsvak, Z. (2007) Targeted Sleeping Beauty transposition in human cells. *Mol Ther.* **15**, 1137–1144.
14. Wu, S.C., Meir, Y.J., Coates, C.J., Handler, A.M., Pelczar, P., Moisyadi, S., and Kaminski, J.M. (2006) piggyBac is a flexible and highly active transposon as compared to sleeping beauty, Tol2, and Mos1 in mammalian cells. *Proc Natl Acad Sci USA.* **103**, 15008–15013.
15. Yant, S.R., Huang, Y., Akache, B., and Kay, M.A. (2007) Site-directed transposon integration in human cells. *Nucleic Acids Res.* **35**, e50.
16. Cadinanos, J. and Bradley, A. (2007) Generation of an inducible and optimized piggyBac transposon system. *Nucleic Acids Res.* **35**, e87.
17. Liu, G.Y., Aronovich, E.L., Cui, Z.B., Whitley, C.B., and Hackett, P.B. (2004) Excision of Sleeping Beauty transposons: parameters and applications to gene therapy. *J Gen Med.* **6**, 574–583.
18. Zayed, H., Izsvak, Z., Walisko, O., and Ivics, Z. (2004) Development of hyperactive Sleeping Beauty transposon vectors by mutational analysis. *Mol Ther.* **9**, 292–304.
19. Wang, W., Lin, C., Lu, D., Ning, Z., Cox, T., Melvin, D., Wang, X., Bradley, A., and Liu, P. (2008) Chromosomal transposition of PiggyBac in mouse embryonic stem cells. *Proc Natl Acad Sci USA.* **105**, 9290–9295.

Chapter 23

Seeing Genetic and Epigenetic Information Without DNA Denaturation Using Sequence-Enabled Reassembly (SEER)

Jason R. Porter, Sarah H. Lockwood, David J. Segal, and Indraneel Ghosh

Abstract

Virtually all methods for reading the sequence of bases in DNA rely on the ability to denature double-stranded DNA into single strands and then use Watson–Crick base-pairing rules to hybridize the strands with high specificity to another DNA primer or probe. However, nature frequently uses an alternative method, reading the sequence information directly from double-stranded DNA using sequence-specific DNA-binding proteins. Here we describe methods for the construction and testing of sequence probes based on engineered zinc finger DNA-binding proteins. Background is reduced using split-reporter molecules, and signal is amplified using enzymatic reporters. The resulting *se*quence-*e*nabled *re*assembly (SEER) probes can be configured to detect DNA sequence (genetic) or DNA methylation (epigenetic) information.

Key words: Double-stranded DNA, diagnostics, DNA sequence, methylation, fluorescent detection, beta-lactamase, luciferase, green florescent protein, engineered zinc finger proteins.

1. Introduction

The direct detection of specific DNA sequences represents a powerful tool for the diagnosis of human disease and the development of new molecular-based therapies. The direct identification of a given DNA sequence of interest can allow for the detection of known pathogenic alterations, including mutations, recombination, DNA sequence repeats, and pathogenically linked DNA chemical modification. Typically, sequence-specific detection requires the use of either labeled ssDNA probes, which require a single-stranded, non-native DNA target, or the use

J.P. Mackay, D.J. Segal (eds.), *Engineered Zinc Finger Proteins*, Methods in Molecular Biology 649,
DOI 10.1007/978-1-60761-753-2_23,

of synthetic small molecule or nucleotide-based reagents where selectivity remains challenging (1–4). Recently, we developed a new method that we call Sequence-Enabled Reassembly (SEER), which offers a selective protein-based alternative to conventional hybridization-based detection approaches (5, 6). Because of their reported modularity and high specificity, zinc finger DNA-binding proteins (ZFPs) are well suited for the development of protein-based sequence-specific DNA detection reagents. With the use of ZFPs, sequence-specific reassembly has been an effective method for the rapid and direct detection of specific dsDNA sequences in addition to the site-specific determination of dsDNA chemical modification (7, 8). SEER has also been used as an effective method to rapidly determine DNA-binding domain specificity as an alternative to EMSAs (9).

SEER combines the use of custom-designed ZFPs with the controllable signal generation of split-protein reassembly to create a "turn-on" sequence-specific DNA biosensor (*see* **Fig. 23.1**). Zinc finger (ZF)-based SEER reagents consist of two different ZFPs, each tethered to inactive fragments of a dissected signal-generating protein or enzyme. Several proteins and enzymes have been utilized as SEER signal-generating domains including green fluorescent protein (GFP), β-lactamase, and firefly luciferase (9). Signal generation and dsDNA detection are achieved by the simultaneous binding of both ZFPs to their binding sites, which serve to bring the two inactive signaling domain fragments into close proximity, facilitating reassembly, and resulting in signal generation only in the presence of a target dsDNA sequence. Through the use of available ZFP design strategies, SEER reagents can be potentially generated with the capability to detect virtually any dsDNA sequence of interest (10–12). Two three-finger ZFPs in SEER would allow for the recognition of 18-bp sites, providing the requisite specificity to target a unique site within the human genome. Alternatively, non-ZFPs, such as the

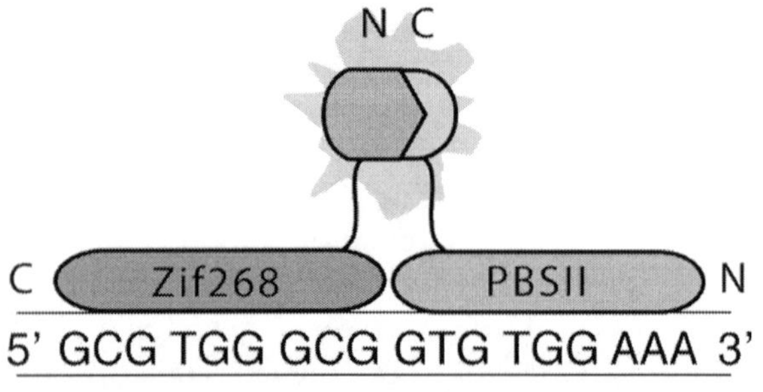

Fig. 23.1. Detection of DNA by Sequence-Enabled Reassembly (SEER). Schematic of SEER. Two zinc finger proteins (Zif268 and PBSII) bind simultaneously, resulting in signaling domain reassembly. Note the often-confusing "antiparallel" orientation of the ZFP with its binding site. When the ZFP-binding site is written 5′ to 3′, the ZFP binds with its C-terminus at the 5′ end and its N-terminus at the 3′ end.

methyl-CpG-binding domain MBD2, can be utilized in conjunction with a ZFP to site specifically determine dsDNA methylation (7–9).

In this chapter we describe the construction of three unique zinc finger-based SEER reagents incorporating the reporter domains GFP (SEER-GFP) or firefly luciferase (SEER-LUC, **Section 3.1**), or β-lactamase (SEER-LAC, **Section 3.2**). Three in vitro methods of analysis are described. Often the SEER fusion proteins are insoluble when expressed in bacteria. **Section 3.3** describes a dialysis method of analysis that starts with insoluble SEER-GFP proteins. **Section 3.4** circumvents the solubility issue by using a cell-free method, which is capable of detecting specific DNA sequences or DNA methylation in less than 2 h starting with in vitro-translated SEER-LUC proteins. **Section 3.5** avoids insolubility by appending maltose-binding protein (MBP) to the SEER fusions, enabling analysis using purified soluble SEER-LAC proteins.

2. Materials

2.1. Generation of Fragmented GFP or Firefly Luciferase–ZFP Fusions

1. The plasmid pETDuet-SEER-GFP: a modified pETDuet-1 prokaryotic expression vector (Novagen) containing two distinct multiple cloning sites (MCSs), each placed under control of an IPTG-inducible T7 promoter for the simultaneous expression of two different protein products. The vector pETDuet-SEER-GFP contains both halves of fragmented GFP (NGFP and CGFP), each appended to a 15-amino acid linker (*see* **Fig. 23.2A**). MCS-1 places the N-terminal hexahistidine tag in frame with an engineered ZFP (ZF1) followed by a 15-amino acid linker (GGGGSGGGGSGGGGT) and the C-terminal fragment of GFP (CGFP, residues 158–238). MCS-2 places the N-terminal fragment of GFP (NGFP, residues 1–157) followed by a 15-amino acid linker (GGSGGGGSGGGGYPG) in frame with an engineered ZFP (ZF2), which is in turn followed by an S-tag affinity purification sequence (SGKETAAAKFERQHMDSSTSSAA). This vector can be obtained from the authors.
2. The plasmid pETDuet-SEER-LUC: a modified pETDuet-1 vector containing both halves of fragmented firefly luciferase (NFluc and CFluc), each appended to a 15-amino acid linker (*see* **Fig. 23.2B**). MCS-1 places the N-terminal hexahistidine tag in frame with an engineered ZFP (ZF1) followed by a 15-amino acid linker (GGGGSGGGGSGGGGT) and the N-terminal fragment of firefly luciferase (NFluc,

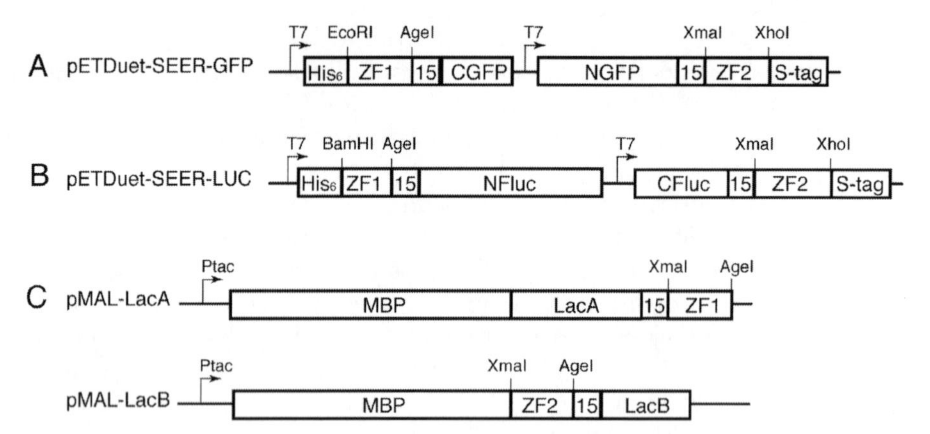

Fig. 23.2. SEER vectors for construction of fragmented GFP, firefly luciferase, and β-lactamase-linked zinc finger fusions. T7: IPTG-inducible T7 promoter; His_6: hexahistidine tag; ZF: engineered ZFP; 15: 15-aa linker; S-tag: S-tag affinity purification sequence; Ptac: IPTG-inducible prokaryotic promoter; MBP: maltose-binding protein. (**A**) Schematic of the dual expression vector pETDuet-SEER-GFP for generating ZF-fragmented GFP fusions. (**B**) Schematic of the dual expression vector pETDuet-SEER-LUC for generating ZF-fragmented luciferase fusions. (**C**) Schematic of the pMAL-LacA and pMAL-LacB expression vectors for generating ZF-fragmented β-lactamase fusions.

residues 2–416). MCS-2 places the C-terminal fragment of firefly luciferase (CFluc, residues 398–550) followed by a 15-amino acid linker (GGSGGGGSGGGGSPG) in frame with an engineered ZFP (ZF2), which in turn is followed by an S-tag affinity purification sequence (SGKETAAAKFER-QHMDSSTSSAA). This vector can be obtained from the authors.

3. ZFPs. Methods enabling the generation of ZFPs capable of targeting a wide range of DNA sequences are described in other chapters in this volume (*see* **Chapters 1**, **2**, **3**, **4**, and **11**). The methods in this chapter assume that appropriate ZFP sequences have been obtained (*see* **Note 1**).

4. ZFP-specific PCR primers:
   a. To place a ZFP N-terminal to CGFP or NFluc, the following forward (MCS-1 Fwd) and reverse (MCS-1 Rev) primer templates can be used:
      i. MCS1-Fwd (CGFP): 5′-gcgatagaattcGnnnnnnnnnnnnnnnnn
      ii. MCS1-Fwd (NFluc): 5′-gcgataggatccGnnnnnnnnnnnnnnnnn
      iii. MCS1-Rev: 5′-gcctaaaccggtnnnnnnnnnnnnnnnnnn

where the underlined sequences are *Eco*RI (MCS-1 Fwd (CGFP)) or *Bam*HI (MCS-1 Fwd (NFluc)) and *Age*I (MCS-1 Rev) restriction endonuclease sites and the variable regions correspond to the 5′ and 3′ coding regions of the ZF motif to be amplified. The capitalized "G" in MCS-1 is necessary for keeping the CGFP/NFluc-zinc finger coding region in frame with the N-terminal hexahistidine purification tag (*see* **Note 2**).

b. To place an engineered ZFP C-terminal to NGFP or CFluc, the following forward (MCS-2 Fwd) and reverse (MCS-2 Rev) primer templates can be used:
   i. MCS2-Fwd: 5′-gcgatacccgggnnnnnnnnnnnnnnnn
   ii. MCS2-Rev: 5′-gcctaactcgagnnnnnnnnnnnnnnnn

   where the underlined sequences are *Xma*I/*Sma*I (MCS-2 Fwd) and *Xho*I (MCS-2 Rev) restriction endonuclease sites and the variable regions correspond to 5′ and 3′ coding regions of the ZFP to be amplified.

5. Sequencing primers: Up1 and Dn1 (5′-atgcgtccggcgtaga-3′ and 5′-gattatgcggccgtgtacaa-3′) and Up2 and Dn2 (5′-ttgtacacggccgcataatc-3′ and 5′-gctagttattgctcagcgg-3′) can be used to sequence MCS-1 and MCS-2, respectively.
6. Standard equipment and reagents for agarose gel electrophoresis.
7. PCR reagents: Taq DNA Polymerase (New England BioLabs), supplied with 10× ThermoPol-Reaction Buffer, 10 mM dNTPs (Stratagene), PCR purification kit (e.g., QIAquick PCR purification kit, Qiagen).
8. Cloning reagents: T4 DNA ligase (400,000 U/mL) supplied with 10× ligase buffer, and restriction endonucleases (New England BioLabs).
9. Competent cells (e.g., *Escherichia coli* XL-1 Blue, Stratagene).

2.2. Generation of Fragmented β-Lactamase ZF Fusions

1. The plasmids pMAL-LacA and pMAL-LacB. These are modified pMAL-c2X prokaryotic expression vectors (New England BioLabs) that place the N-terminal maltose-binding protein (MBP) affinity purification tag in frame with either a LacA–ZFP or a LacB–ZFP fusion (*see* **Fig. 23.2C**). MBP allows for the purification of soluble ZFP-β-lactamase fusions in a single step using amylose resin. pMAL-LacA places the N-terminal MBP tag in frame with the N-terminal fragment of β-lactamase (LacA, residues 26–196), followed by a 15-amino acid linker (LQGGSGGGGSGGGGT), followed by an engineered ZFP. pMAL-LacB places the N-terminal MBP tag in frame with an engineered ZFP, followed by a 15-amino acid linker (GGGGSGGGGSGGGGT), followed by the C-terminal fragment of β-lactamase (LacB, residues 198–290). These vectors can be obtained from the authors.
2. ZFP-specific PCR primers:

a. To place a ZFP C-terminal to LacA or N-terminal to LacB the following forward (Lac Fwd) and reverse (Lac Rev) primers can be used:

i. Lac Fwd: 5′-gagtctcccgggnnnnnnnnnnnnnnnnn-3′

ii. Lac Rev: 5′-gcgtataccggtnnnnnnnnnnnnnnnnn-3′

where the underlined sequences are *Xma*I/*Sma*I (Lac Fwd) and *Age*I (Lac Rev) restriction endonuclease sites and the variable regions correspond to the 5′ and 3′ coding regions of the ZF motif to be amplified.

3. Sequencing primers: MalF (5′- ggtcgtcagactgtcgatgaagcc-3′) and MalR (5′- cgccagggttttcccagtcacgac-3′) can be used to sequence ZF split β-lactamase fusions. MalF anneals within the MBP coding region, just 5′ to the start of the fusion protein. MalR anneals 3′ to the end of the fusion protein.
4. PCR reagents, cloning reagents, and competent cells as in **Section 2.1**.

2.3. In Vitro SEER-GFP Assay (Dialysis Method)

1. Vector pETDuet-SEER-GFP containing appropriate ZFPs (*see* **Section 3.1**).
2. DNA target oligonucleotides: target oligonucleotides containing the ZFP-binding sites can be generated from two complementary ssDNA oligonucleotides that are annealed to create a dsDNA oligonucleotide target. Alternatively, long oligonucleotides containing complementary regions ("hairpin oligos") can be used to produce unimolecular dsDNA targets (*see* **Note 1**).
3. Cells: *E. coli* BL-21 DE3 (Stratagene). These cells provide tight repression of fusion protein expression before IPTG induction, while providing high expression levels after the addition of IPTG.
4. 2×YT medium: Dissolve 31 g of 2×YT medium powder (Sigma) in 1 L of distilled water and autoclave. Dissolve ampicillin in water to 100 mg/mL, sterilize using a 0.2-μm filter, and store at –20°C. The antibiotic is added to 2×YT medium at 100 μg/mL just before use.
5. Isopropyl-beta-D-thiogalactopyranoside (IPTG): Dissolve IPTG in water to 0.5 M, sterilize using a 0.2-μm filter, and store at –20°C. IPTG is added to 2×YT medium to a final concentration of 1 mM in induce protein expression.
6. Zinc Buffer B (ZBB): 10 mM Tris–HCl, pH 7.5, 100 mM NaCl, 1 mM DTT, 100 μM $ZnCl_2$.
7. Empty PD-10 columns (GE Healthcare).
8. Ni-NTA agarose resin (Qiagen).
9. Standard equipment and reagents for SDS-PAGE.

10. 10× *Bam*HI buffer (New England BioLabs).
11. Sonicator (e.g., Branson 450 sonifier).
12. 0.1–0.5 mL 3.5-kDa MWCO dialysis cassettes (e.g., Slide-A-Lyzer, Thermo Scientific).
13. Spectrofluorometer (e.g., Photon Technology International spectrofluorometer).

2.4. In Vitro SEER-LUC Assay (Cell-Free Method)

1. The vector pETDuet-SEER-LUC containing appropriate ZFPs (*see* **Section 3.1**).
2. PCR reagents as described in **Section 2.1**.
3. DNA target oligonucleotides as described in **Section 2.3**.
4. In vitro transcription primers. Primers should be generated that incorporate 5′ T7 promoter and Kozak sequences and a 3′ stem loop into the PCR product used as the template for in vitro transcription (*see* **Note 3**). These primers will be referred to as MCS1-Kozak-Fwd and MCS1-SL-Rev or MCS2-Kozak-Fwd and MCS2-SL-Rev below.
5. RiboMAX Large Scale RNA Production System-T7 (Promega), supplied with T7 polymerase Enzyme Mix, 5× T7 Transcription Buffer, 25 mM rNTPs, and RQO1 RNase-free DNase.
6. RNA purification kit (e.g., Illustra ProbeQuant G-50 Micro Columns, GE Healthcare).
7. Flexi-Rabbit Reticulocyte Lysate kit (Promega) supplied with rabbit reticulocyte lysate, 2.5 M KCl, and three amino acid mixtures: 1 mM minus methionine, 1 mM minus leucine, and 1 mM minus cysteine.
8. RNasin Plus (Promega).
9. $ZnCl_2$ (100 μM) in water.
10. Steady-Glo Luciferase Assay System (Promega).
11. One-mL glass single-tube luminometer vials.
12. Luminometer (e.g., Turner TD-20e Luminometer or a $20/20^n$ Luminometer, Turner Biosystems).

2.5. In Vitro SEER-LAC Assay (Solution Method)

1. The vectors pMAL-LacA and pMAL-LacB containing appropriate ZFPs (*see* **Section 3.2**).
2. PCR reagents as described in **Section 2.1**.
3. DNA target oligonucleotides as described in **Section 2.3**.
4. Cells, LB medium, and IPTG as described in **Section 2.3** (*see* **Note 4**).
5. Zinc Buffer A (ZBA): 10 mM Tris–HCl, pH 7.5, 90 mM KCl, 1 mM $MgCl_2$, 100 μM $ZnCl_2$, 5 mM DTT.

6. Empty Econo-Pac chromatography columns (Bio-Rad).
7. Amylose resin (New England BioLabs).
8. Maltose.
9. Glycerol.
10. Nitrocefin (1 mM), a colorimetric β-lactamase substrate: Dissolve 1 mg of nitrocefin (Calbiochem) in 300 μL dimethyl sulfoxide (DMSO) and vortex. Add 1.7 mL of 100 mM phosphate buffer, pH 7.0, to total volume of 2 mL. Nitrocefin is light sensitive. Store protected from light at –20°C for up to 2 weeks.
11. CCF2-FA (100 μM), a fluorescent β-lactamase substrate: Just before use, dissolve CCF2-FA (Invitrogen) to 100 μM in phosphate-buffered saline, pH 7.4. CCF2-FA is light sensitive.
12. Sonicator (e.g., Branson 450 sonifier).
13. Standard equipment and reagents for SDS-PAGE.
14. 96-well plates, flat bottom, clear (colorimetric), or black (fluorescence).
15. Microplate reader: (e.g., for colorimetric assays, a μQuant spectrophotometer, Bio-Tek; for fluorescent assays, a Spectra Max Gemini fluorometer, Molecular Devices).

3. Methods

3.1. Generation of Fragmented GFP or Firefly Luciferase–ZFP Fusions

1. Set up a PCR reaction using an engineered ZFP as the template for amplification: 200 ng of template, 2.5 μL of 10 mM dNTPs, 10 μL of 10× ThermoPol-Reaction Buffer, 2.5 μL of each 10 pmol/μL primer MCS1-Fwd (CGFP)/MCS1-Rev, MCS1-Fwd (NFluc)/MCS1-Rev or MCS2-Fwd/Rev, 0.25 μL of Taq Polymerase and ddH_2O to a final volume of 100 μL. Run the PCR using the following conditions: 1 cycle at 95°C for 5 min; 30 cycles of 95°C for 30 s, 56°C for 1 min, and 72°C for 1 min; 1 cycle at 72°C for 7 min; final cycle at 4°C until further use. Verify the purity and correct size of the PCR DNA product by agarose gel electrophoresis.
2. Purify the PCR DNA product using PCR purification kit and elute with 50 μL of the kit-supplied elution buffer.
3. Digest the eluted PCR DNA and a corresponding vector with *Eco*RI (pETDuet-SEER-GFP) or *Bam*HI (pETDuet-SEER-LUC) and *Age*I to clone into MCS-1, or *Xma*I/*Sma*I and *Xho*I to clone into MCS-2 (*see* **Note 5**).

4. Ligate the digested PCR DNA insert and pETDuet vector using T4 DNA ligase.
5. Transform the ligation product into XL1-Blue cells.

3.2. Generation of Fragmented β-Lactamase ZF Fusions

1. Set up a PCR reaction using an engineered ZFP as the template for amplification: 200 ng of template, 2.5 μL of 10 mM dNTPs, 10 μL of 10× ThermoPol-Reaction Buffer, 2.5 μL of each primer (Lac Fwd and Lac Rev) at 10 pmol/μL, 0.25 μL of Taq Polymerase, and ddH_2O to a final volume of 100 μL. Run the PCR using the following conditions: 1 cycle at 95°C for 5 min; 30 cycles of 95°C for 30 s, 56°C for 1 min, and 72°C for 1 min; 1 cycle at 72°C for 7 min; final cycle at 4°C until further use. Verify the purity and correct size of the PCR DNA product by agarose gel electrophoresis.
2. Purify the PCR product using a PCR purification kit and elute with 50 μL of the kit-supplied elution buffer.
3. Digest eluted PCR DNA and a corresponding vector (pMAL-LacA or pMAL-LacB) with *Xma*I/*Sma*I and *Age*I, following standard procedures.
4. Ligate the digested PCR DNA insert and pMAL vector using T4 DNA ligase.
5. Transform the ligation product into XL1-Blue cells.

3.3. In Vitro SEER-GFP Assay (Dialysis Method)

1. Transform *E. coli* BL-21 DE3 cells with the pETDuet-SEER-GFP plasmid containing the ZFP–GFP fusions.
2. Inoculate a 50-mL culture of 2×YT medium supplemented with 100 μg/mL ampicillin and incubate overnight at 37°C with shaking.
3. Use the overnight culture to inoculate 1 L of 2×YT medium supplemented with 100 μg/mL ampicillin and 100 μM $ZnCl_2$. The initial OD_{600} should be about 0.05. Incubate at 37 °C with shaking until an OD_{600} of 0.5–0.8 is reached.
4. Induce simultaneous expression of the ZFP–GFP fusions by the addition of 2 mL of 0.5 M IPTG. Continue incubation at 37°C with shaking for an additional 3 h.
5. Pellet the cells in a 50-mL conical tube by centrifugation at 3,000×*g*.
6. Discard the supernatant as waste. If desired, the pelleted cells can be stored at −80°C until needed, then thawed on ice.
7. Resuspend the cells in 40 mL of ZBB. Sonicate on ice. Pellet the inclusion bodies by centrifugation at 18,000×*g* for 40 min at 4°C.

8. Both fusion proteins are typically insoluble and found in inclusion bodies and should be purified under denaturing conditions (*see* **Note 6**). Solubilize the inclusion bodies in 40 mL of ZBB supplemented with 6 M urea followed by incubation on ice for 1 h.
9. Dilute the resulting solution to 4 M urea by the addition of 20 mL of ZBB, followed by centrifugation at 18,000×*g* for 20 min.
10. Add 2 mL of Ni-NTA agarose resin to an empty PD-10 column to produce 1 mL of packed Ni-NTA resin. Equilibrate the resin with 20 mL of ZBB supplemented with 4 M urea.
11. Resuspend the equilibrated Ni-NTA resin from Step 10 in the supernatant from Step 9 and let equilibrate at 4°C for 1 h while rocking.
12. Isolate the Ni-NTA resin by re-applying the resuspended Ni-NTA resin to the PD-10 column.
13. Wash the Ni-NTA resin with 10 mL of ZBB supplemented with 4 M urea.
14. Elute with ZBB supplemented with 4 M urea and imidazole using the following elution scheme: 4 × 0.5 mL volumes each containing 2, 10, 20, and 50 mM imidazole, followed by 10 × 0.5 mL volumes containing 500 mM imidazole. Be sure to maintain a pH of 7.5 at high concentrations of imidazole.
15. Analyze 10 μL of the fractions by SDS-PAGE. NGFP-ZF2 should elute in the 2 mM imidazole fractions whereas CGFP-ZF1 should elute in the 50–500 mM fractions (a mixture of both proteins can typically be found in the 10–20 mM fractions) (*see* **Note 7**).
16. Fractions of CGFP-ZF1 that contain small amounts of NGFP-ZF2 can be further purified by dialyzing into ZBB with 4 M urea and 2 mM imidazole and re-purifying on the same PD-10 column.
17. Elute with the same scheme as in Step 14. NGFP-ZF2 should elute in 2 mM imidazole, whereas CGFP-ZF1 should elute in the previously observed fractions. Protein concentrations can be obtained by UV absorbance at 280 nm (*see* **Note 8**).
18. Double-stranded DNA targets can be generated from bi- or unimolecular ssDNA oligonucleotides in 1 × *Bam*HI buffer as follows: heat to 95°C for 7 min, cool to 56°C at a rate of 1°C/min, equilibrate at 56°C for 5 min, cool to 25°C at a rate of 1°C/min.

19. In a 250-μL reaction volume, combine 5 μM of each ZFP–GFP fusion with 2.5 μM of target oligonucleotide in ZBB supplemented with 4 M urea. It may also be useful to analyze a range of target oligonucleotide concentrations including a no-DNA target control to assess background levels of GFP reassembly.
20. Inject the 250-μL reaction mixture from Step 19 into a 3.5-kDa MWCO dialysis cassette and place in 300 mL of ZBB supplemented with 2 M urea. Dialyze uncovered at 4°C for 4 h.
21. Perform additional dialysis in a stepwise manner in ZBB containing 1 M urea, ZBB containing 0.5 M urea, then twice into ZBB with no urea. Dialyze uncovered at 4°C for 4 h at each step.
22. Remove the reaction mixture from the dialysis cassette and incubate for 7–24 h at 4°C to allow for chromophore maturation.
23. Measure the GFP fluorescence at excitation 468 nm and emission 505 nm in the presence and absence of dsDNA targets to evaluate DNA sequence-dependant GFP reassembly (*see* **Fig. 23.3A** and **Note 9**).

3.4. In Vitro SEER-LUC Assay (Cell-Free Method)

1. Generate PCR products of each MCS of a pETDuet-SEER-LUC containing ZFP–LUC fusions (*see* **Note 10**). Set up the PCR reaction as follows: 200 ng of template, 2.5 μL of 10 mM dNTPs, 10 μL of 10× ThermoPol-Reaction Buffer, 2.5 μL of each 10 pmol/μL primer (Up1 and Dn1 for MCS-1; Up2 and Dn2 for MCS-2), 0.25 μL of Taq Polymerase, and ddH_2O to a final volume of 100 μL. Run the PCR using the following conditions: 1 cycle at 95°C for 5 min; 30 cycles of 95°C for 30 s, 56°C for 1 min, and 72°C for 1 min; 1 cycle at 72°C for 7 min; a final cycle at 4°C until further use. Verify the purity and correct size of PCR DNA product by agarose gel electrophoresis.
2. Purify PCR DNA product using PCR purification kit and elute with 50 μL of the kit-supplied elution buffer.
3. Generate PCR products for in vitro transcription using the following PCR reaction mixture: 200 ng of PCR template generated form Step 1 (i.e., generated from either MCS-1 or MCS-2), 2 μL of 10 mM dNTPs, 10 μL of 10× ThermoPol-Reaction Buffer, 5 μL of each 10 pmol/μL primer (MCS1-Kozak-Fwd and MCS1-SL-Rev, or MCS2-Kozak-Fwd and MCS2-SL-Rev), 1 μL of Taq Polymerase, and ddH_2O to a final volume of 100 μL. Run the PCR using the following

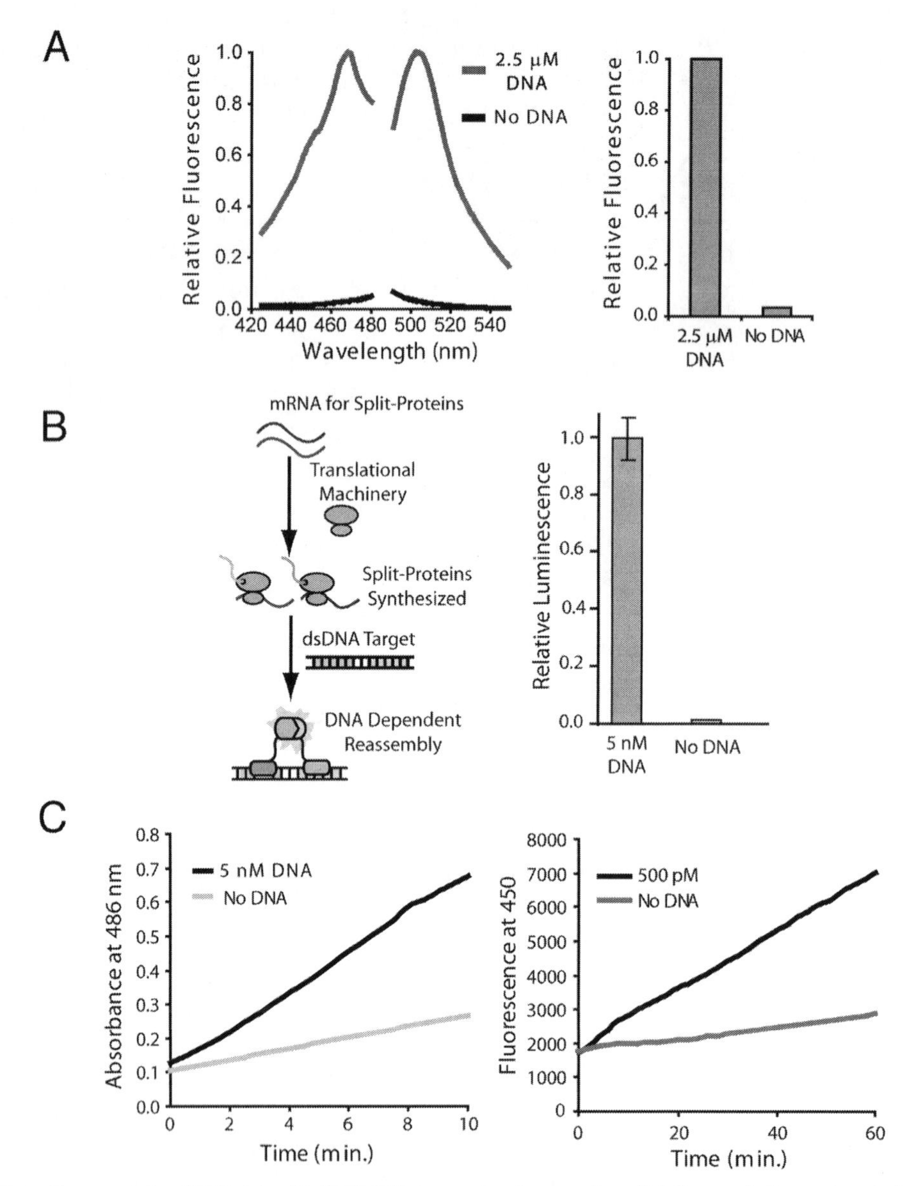

Fig. 23.3. Representative SEER data. (**A**) SEER-GFP assay signal generation. Relative excitation and emission in the presence or absence of target dsDNA (*left*). *Bar graph* showing relative fluorescence emission at 505 nm (*right*). (**B**) Cartoon representation of dsDNA detection using SEER-LUC in a cell-free protein expression assay (*left*). *Bar graph* showing relative luminescence in the presence and absence of target dsDNA (*right*). (**C**) Colorimetric and fluorescence analysis of dsDNA-mediated SEER-LAC in the presence and absence of target dsDNA. Absorbance is taken at 486 nm over a period of 10 min (*left*). When the fluorogenic substrate CCF2-FA is used, β-lactamase reassembly can be observed by monitoring the increase in emission at 450 nm upon excitation at 409 nm in the presence or absence of target dsDNA (*right*).

conditions: 1 cycle at 95°C for 5 min; 30 cycles of 95°C for 30 s, 56°C for 1 min, and 72°C for 3 min; 1 cycle at 72°C for 7 min; a final cycle at 4°C until further use. Verify the purity and correct size of the PCR product by agarose gel electrophoresis.

4. Purify PCR DNA product using PCR purification kit and elute with 50 μL of elution buffer.
5. Generate mRNA by in vitro transcription using the RiboMAX Large Scale RNA Production System-T7 and the following reaction mixture: 10 μL of 5× T7 Transcription Buffer, 15 μL of 25 mM rNTPs, 4 μg of PCR DNA template generated in Step 3, 5 μL of T7 Enzyme Mix, and nuclease-free H_2O to a final volume of 50 μL. Incubate at 37°C for 3 h.
6. Add 4 μL of RQO1 RNase-free DNase and incubate for 15 min at 37°C.
7. Purify the in vitro-transcribed mRNA using an RNA purification kit. Verify the purity and correct size of mRNA product by agarose gel electrophoresis. Determine the mRNA concentration by UV absorbance at 260 nm.
8. Prepare dsDNA targets as described in Step 18 **Section 3.3**.
9. Create a 1 mM amino acid mixture using the components in the Flexi-Rabbit Reticulocyte Lysate kit by combining equal amounts of minus methionine, minus leucine, and minus cysteine (1 mM each) amino acid mixtures.
10. Prepare an in vitro translation reaction as follows: 16.5 μL of Rabbit Reticulocyte Lysate, 0.5 μL of 1 mM amino acid mixture from Step 9, 0.7 μL of 2.5 M KCl, 0.5 μL of RNasin Plus, 1 pmol of mRNA corresponding to each ZFP–LUC fusion, 2.5 μL of 100 μM $ZnCl_2$, 2.5 μL of 50 nM dsDNA targets, and nuclease-free H_2O to a final volume of 25 μL (*see* **Note 11**).
11. Incubate the in vitro translation mixture at 30°C for 90 min.
12. Place 20 μL of translation mixture into a 1 mL glass luminometer vial. Add 80 μL of Steady-Glo Luciferase Assay Reagent, mix gently, and incubate for 1 min at room temperature (*see* **Note 12**).
13. Measure the luminescence in the presence and absence of dsDNA target to evaluate dsDNA target-mediated firefly luciferase reassembly (*see* **Fig. 23.3B** and **Note 13**).

3.5. In Vitro SEER-LAC Assay (Solution Method)

1. Transform *E. coli* BL-21 DE3 cells with plasmids pMAL-LacA or pMAL-LacB containing ZFP–LAC fusions.

2. Inoculate a 5 mL culture in LB supplemented with 100 μg/mL ampicillin and incubate overnight at 37°C with shaking.
3. Use the overnight culture to inoculate 100 mL of LB supplemented with 100 μg/mL ampicillin, and 100 μM $ZnCl_2$. The initial OD_{600} should be about 0.05. Incubate at 37°C with shaking until an OD_{600} of 0.5–0.8 is reached.
4. Induce simultaneous expression of the ZFP–LAC fusions by the addition of 200 μL of 0.5 M IPTG. Continue incubation at 37°C with shaking for an additional 2–5 h.
5. Pellet the cells in a 50-mL conical tube by centrifugation at 3,000×*g*.
6. Discard the supernatant as waste. If desired, the pelleted cells can be stored at -80°C until needed, then thawed on ice.
7. Resuspend the cells in 5 mL of ZBA. Sonicate on ice for 2 min at 50% duty cycle (10 s pulse, 20 s pause).
8. Pellet the debris by centrifugation at 18,000×*g* for 20 min at 4°C.
9. Filter the supernatant using a 0.2 μm syringe filter.
10. Add 3 mL of amylose resin to an empty Econo-Pac column to produce 2 mL of packed amylose resin. Equilibrate the resin with 20 mL of ZBA.
11. Apply the supernatant from Step 9 to the amylose resin. Collect the flow through.
12. Re-apply the flow through to the same Econo-Pac column (*see* **Note 14**).
13. Wash the column with 20 mL of ZBA containing 2 M NaCl (*see* **Note 15**).
14. Wash the column with 20 mL of ZBA twice.
15. Elute with 5 mL of ZBA containing 10 mM maltose. Collect 10 0.5 mL fractions.
16. Analyze 20 μL of each fraction by SDS-PAGE. Protein concentrations can be determined by measuring UV absorbance at 280 nm (*see* **Notes 8** and **16**).
17. Add 100% glycerol to a final concentration of 50% v/v glycerol. Store the ZFP–LAC fusions at –20°C until needed.
18. Prepare dsDNA targets as described in Step 18 of **Section 3.3**.

19. For colorimetric assays: Prepare a 180-μL reaction containing 125 nM of each ZFP–LAC fusion protein and 5 nM dsDNA target in ZBA. Incubate for 30 min at room temperature. After incubation, add 20 μL of 1 mM nitrocefin to achieve a final concentration of 100 μM nitrocefin and a final volume of 200 μL. Measure absorbance at 486 nm every 30 s over a period of 10 min. Signal generation is defined as the rate of nitrocefin hydrolysis between 0.5 and 2 min after nitrocefin addition. Activity is determined by subtracting the background rate of hydrolysis for in the absence of DNA (*see* **Fig. 23.3C left**).
20. For fluorometric assays: Prepare a 90-μL reaction containing 125 nM of each ZFP–LAC fusion protein and 5 nM dsDNA target in ZBA. Incubate for 30 min at room temperature. After incubation add 10 μL of 100 μM CCF2-FA to achieve a final concentration of 10 μM CCF2-FA and a final volume of 100 μL. Following a 30 min incubation in the dark at room temperature, measure the hydrolysis of CCF2-FA in the presence and absence of target dsDNA to evaluate dsDNA target-mediated β-lactamase reassembly. Fluorescence measurements are acquired at an excitation of 409 nm and emission of 450 and 520 nm every 2 min for 20 min. Signal generation is defined as the rate of increase in fluorescence at 450 nm or the rate of decrease in fluorescence at 520 nm. Activity is determined by subtracting the background rate of hydrolysis in the absence of DNA (*see* **Fig. 23.3c right**).

4. Notes

1. When considering DNA targets, one should keep in mind that ZFPs bind C-terminal to N-terminal when the target site is written 5′ to 3′ (*see* **Fig. 23.1A**). The ZFPs and targets should be designed in a manner such that the binding site of ZF2 (Zif268 in **Fig. 23.1A**) is upstream of the binding site for ZF1 (PBSII in **Fig. 23.1A**). This will place the signaling domains at the termini of the two ZFPs facing toward each other upon binding, resulting in a favorable orientation for reassembly. The ZF2- and ZF1-binding sites should be contiguous, with no nucleotides inserted between them.
2. The vector design resulted in a 1-bp out-of-frame shift in MCS1 of pETDuet. This frame shift places the inserted

gene out-of-frame with start ATG and hexahistidine tag. Insertion of the extra "G" in the forward primer between the restriction endonuclease site and the DNA-binding domain being inserted places the zinc finger-GFP/luciferase fusion back in frame with the start ATG and hexahistidine tag.

3. Because the templates for PCR are prokaryotic expression vectors, the PCR primers must include a Kozak sequence in order to introduce a eukaryotic translation initiation site into the resulting in vitro-transcribed mRNA. A T7 promoter sequence must be placed upstream of the Kozak sequence to allow for transcription initiation in the T7 polymerase-based in vitro transcription system. The reverse primer should place a hairpin loop downstream of the PCR product. The introduction of a 3′ stem loop results in greater mRNA stability during in vitro translation (9).
4. In most previous publications, we have used Luria Broth (LB) medium and the antibiotic carbenicillin (a more stable analog of ampicillin, Omega Scientific). The choice of growth medium should not affect the experiment.
5. The N-terminal portion of firefly luciferase contains an internal *Eco*RI site. Take care when designing primers for cloning DNA-binding domains into MCS1 of pETDuet-SEER-LUC. Use of *Bam*HI will provide a unique site for cloning.
6. Typically ZF–GFP fusion proteins are insoluble and are found primarily in inclusion bodies when expressed in *E. coli*. However, SDS-PAGE analysis of both the soluble and the insoluble fractions should be performed to ensure that proper purification procedures are used. In the case when non-ZF DNA-binding domains are used proteins may be expressed in the soluble fraction, in which case purification under native conditions should be performed. Conditions for native hexahistidine tag purification can be found in the Qiagen QIA*expressionist* hexahistidine-tagged protein expression manual.
7. Due to the high number of histidine residues found in the NGFP–ZF2 fusion, this fusion is capable of weak binding to Ni-NTA resin. In the case where non-zinc finger DNA-binding domains are used, purification using an S-tag column may be needed. The hexahistidine-tagged ZF1–CGFP fusion binds tightly to Ni-NTA and will elute with higher concentrations of imidazole.
8. Extinction coefficients can be calculated from the equation: $\varepsilon = (n\text{W}\times 5{,}500) + (n\text{Y}\times 1{,}490) + (n\text{C}\times 125)$, where

n is the number of each residue and the stated values are the Trp, Tyr, and Cys amino acid molar absorptivities at 280 nm. Concentration = A_{280}/ε (for a path length of 1 cm). Alternatively, concentration can be determined by comparison to known mass standards on the SDS-PAGE (e.g., 0.5, 1, 2, and 4 μg of bovine serum albumen).

9. A Photon Technology International (PTI) spectrofluorometer is typically used. Slit widths are typically set to 5 nm for excitation and 10 nm for emission.
10. Purity of in vitro-transcribed mRNA is extremely important if optimal translation and signal generation is expected. Truncated or mRNA products of incorrect size can result in decreased protein expression levels and can result in decrease signal generation due to possible dominant negative effects of partially expressed ZFP–LUC fusions. The use of two PCR steps typically eliminates this concern. If incorrect mRNA products are present, gel extraction and purification can be performed on either or both PCR products to ensure that the proper template is used for in vitro transcription.
11. Initially, concentrations of 1 pmol/reaction of mRNA are used for each ZFP–LUC fusion along with 5 nM dsDNA target. Amounts as low as 0.05 pmol/reaction of each mRNA have been used. Varying mRNA concentrations in either equal or disproportionate amounts can result in increased sensitivity, particularly when extremely low amounts of dsDNA are being tested for detection. The method described calls for the dsDNA target to be present during in vitro translation; alternatively the dsDNA target can be added after translation and allowed to equilibrate at room temperature for 30 min followed by substrate addition.
12. We have found that a ratio of 1:4 translation mixture:substrate works best for Steady-Glo Luciferase Assay Reagent. This ratio can be optimized or reduced depending on the application needed.
13. Luminescence measurements are typically performed using a 3 s delay followed by a 10 s integration time.
14. A higher concentration of ZFP–LAC fusion protein can be obtained by passing the clarified extracts over the amylose resin twice. This step can be eliminated if it is determined to be not required, or if contamination with bacterial proteins seems significant.
15. When expressed as soluble MBP fusion proteins, ZFP purification can be contaminated with non-specifically bound *E. coli* nucleic acid (DNA and/or RNA). An initial

high-salt wash with 2 M NaCl greatly reduces the amount of contaminant nucleic acid found in purified sample.

16. Nucleic acid contamination of purified protein solutions can be determined by comparing the 260 nm:280 nm UV absorbance. A ratio less than 1 is typically seen for solutions containing greater than 95% protein.

References

1. Fechter, E.J., Olenyuk, B., and Dervan, P.B. (2005) Sequence-specific fluorescence detection of DNA by polyamide-thiazole orange conjugates. *J Am Chem Soc.* **127**, 16685–16691.
2. Johnson, M.D., 3rd and Fresco, J.R. (1999) Third-strand in situ hybridization (TISH) to non-denatured metaphase spreads and interphase nuclei. *Chromosoma.* **108**, 181–189.
3. Levsky, J.M. and Singer, R.H. (2003) Fluorescence in situ hybridization: past, present and future. *J Cell Sci.* **116**, 2833–2838.
4. Rucker, V.C., Foister, S., Melander, C., and Dervan, P.B. (2003) Sequence specific fluorescence detection of double strand DNA. *J Am Chem Soc.* **125**, 1195–1202.
5. Ooi, A.T., Stains, C.I., Ghosh, I., and Segal, D.J. (2006) Sequence-enabled reassembly of beta-lactamase (SEER-LAC): A sensitive method for the detection of double-stranded DNA. *Biochemistry.* **45**, 3620–3625.
6. Stains, C.I., Porter, J.R., Ooi, A.T., Segal, D.J., and Ghosh, I. (2005) DNA sequence-enabled reassembly of the green fluorescent protein. *J Am Chem Soc.* **127**, 10782–10783.
7. Porter, J.R., Stains, C.I., Segal, D.J., and Ghosh, I. (2007) Split beta-lactamase sensor for the sequence-specific detection of DNA methylation. *Anal Chem.* **79**, 6702–6708.
8. Stains, C.I., Furman, J.L., Segal, D.J., and Ghosh, I. (2006) Site-specific detection of DNA methylation utilizing mCpG-SEER. *J Am Chem Soc.* **128**, 9761–9765.
9. Porter, J.R., Stains, C.I., Jester, B.W., and Ghosh, I. (2008) A general and rapid cell-free approach for the interrogation of protein-protein, protein-DNA, and protein-RNA interactions and their antagonists utilizing split-protein reporters. *J Am Chem Soc.* **130**, 6488–6497.
10. Dreier, B., Beerli, R.R., Segal, D.J., Flippin, J.D., and Barbas, C.F., 3rd (2001) Development of zinc finger domains for recognition of the 5′-ANN-3′ family of DNA sequences and their use in the construction of artificial transcription factors. *J Biol Chem.* **276**, 29466–29478.
11. Dreier, B., Fuller, R.P., Segal, D.J., Lund, C.V., Blancafort, P., Huber, A., Koksch, B., and Barbas, C.F., 3rd (2005) Development of zinc finger domains for recognition of the 5′-CNN-3′ family DNA sequences and their use in the construction of artificial transcription factors. *J Biol Chem.* **280**, 35588–35597.
12. Segal, D.J., Dreier, B., Beerli, R.R., and Barbas, C.F., 3rd (1999) Toward controlling gene expression at will: selection and design of zinc finger domains recognizing each of the 5′-GNN-3′ DNA target sequences. *Proc Natl Acad Sci USA.* **96**, 2758–2763.

Chapter 24

Zinc Finger-Mediated Live Cell Imaging in *Arabidopsis* Roots

Beatrice I. Lindhout, Tobias Meckel, and Bert J. van der Zaal

Abstract

Following the elucidation of recognition codes, artificial zinc finger (ZF) domains can now be assembled to create custom-made DNA-binding proteins in which the alpha helix of each zinc finger mediates an interaction with 3 or 4 bp of DNA. A module of consecutive zinc finger domains, designated a polydactyl zinc finger (PZF) domain, is thus capable of binding an extended number of base pairs of DNA. Besides the multitude of utilities of PZF domains addressed in other chapters, we have shown that they can also be used for live cell imaging of repetitive DNA sequences in *Arabidopsis*, as well as in mouse cells by generating and expressing PZF:GFP fusion proteins (1). Here we provide a detailed protocol for the construction of such PZF:GFP reporter proteins using our established cloning vehicles, together with a protocol for their expression in plants in order to achieve in vivo labelling of repetitive DNA. Furthermore we provide an accurate quantification method for GFP signals using fluorescent beads (FluoSpheres). Single-molecule precision can be obtained using any confocal setup once the fluorescent beads have been calibrated against purified GFP. The methods can easily be adapted to meet the demands for other situations or for other experimental systems.

Key words: Cys_2His_2 zinc finger domain, live cell imaging, GFP quantification, *Arabidopsis*, CLSM, single-molecule microscopy, FluoSpheres, polydactyl zinc finger.

1. Introduction

The study of the architecture of chromosomes or of more specific chromosomal loci has thus far heavily leaned on fluorescent in situ hybridization techniques. Although the resolution of the images generated by these techniques for even single-copy sequences is often excellent and in principle any DNA probe can be used, this can only be achieved when the material is fixed. Apart from having to work with dead material, ruling out the possibility to

J.P. Mackay, D.J. Segal (eds.), *Engineered Zinc Finger Proteins*, Methods in Molecular Biology 649,
DOI 10.1007/978-1-60761-753-2_24, © Springer Science+Business Media, LLC 2010

study any dynamic behaviour of the objects of interest, fixation will undoubtedly change the native in vivo structure of the chromosomal areas involved. Some way out of these dilemmas has become possible via the expression of fluorescently labelled proteins which naturally bind to the chromatin, sometimes even at a specific location (2, 3). Alternatively, the integration of a construct encoding a fluorescently labelled repressor that binds to a repetitive array of its operator sequence can be a useful strategy (4, 5). Unfortunately, the latter method has very limited practicality since a target DNA sequence first needs to be introduced into the genome, either together with a gene encoding the fluorescently tagged repressor or separately.

Since PZFs can be constructed for a large variety of target sequences and can be fused with, in principle, any other protein domain in a modular fashion, they provide a tool for studying dynamic behaviour of chromosomes. Recently we have demonstrated that it is possible to visualize centromeric repeats in living root cells of the plant *Arabidopsis thaliana* as well as in mouse cells by designing three-fingered PZF domains fused with GFP (PZF:GFP fusions) that are aimed at recognizing a nine-base-pair (bp) motif in the centromere (1). The visualization is mediated by a local accumulation of GFP fluorescence via the binding of the PZF domain to a repetitively organized binding sequence and is therefore mainly dependent on the copy number and the spatial coordination of the binding sites. In addition, parameters such as DNA-binding affinity of the PZF domain and PZF:GFP reporter protein concentration are also relevant. Here we provide a detailed protocol for the construction of typical PZF:GFP reporter proteins, exemplified by the generation and visualization of a construct facilitating the imaging of a 9-bp sequence (5′-GTTGCGGTT-3′) present in *Arabidopsis* centromeres. Other combinations of ZF domains for alternative target sequences can simply be constructed using the same cloning system. For expression of PZF:GFPs in other biological systems, which obviously demand other types of expression signals, the outlined principles are of general use.

We also describe how to quantify the fluorescence intensity of PZF:GFP reporter proteins with a precision that allows us to relate a specific signal to the number of bound PZF:GFP molecules. Provided the signal intensity of a single GFP is known for the microscope setup used to image the domains, the number of PZF:GFP molecules per domain can easily be calculated. The problem, however, is that single GFP molecules are not detectable on commercial confocal systems. To overcome this limitation, we use fluorescent styrol beads as intensity standards on the confocal setup. These fluorescent beads can be calibrated against the intensity of single GFP molecules on a setup providing single-molecule sensitivity. Our approach to combine two setups with different sensitivity allows quantifying signal intensities with near single-molecule precision on setups lacking single-molecule detection capabilities.

Several methods for constructing PZF domains are described in other chapters in this volume (*see* **Chapters 1, 2, 3, 4, 8,** and **11**) or have been published elsewhere. In this chapter, we provide a detailed protocol using the cloning vehicle and wide host range vector that we developed, into which PZF domains can easily be introduced (1, 6). In **Section 3.1**, the PZFs are constructed using the previously described cloning vehicle pSKN-SgrAI, into which each individual ZF-encoding DNA fragment is introduced in the form of two annealed oligomers with an *Sgr*AI-compatible overhang. The procedure involves a repetitive protocol to add new ZF domains to the N-terminus of the growing PZF domain. The order of DNA recognition of PZFs is such that the addition of each ZF domain to the N-terminus will specify another 3 bp of DNA template in the 3′ direction (*see* **Fig. 24.1**). The PZF domain for targeting *Arabidopsis* centromeres is prepared via the sequential cloning of the ZFs that recognize GTT, GCG, and GTT, respectively, and thus targets the sequence 5′-GTTGCGGTT-3′. The cloning vehicle pSKN-SgrAI and destination vector pRF-GFP both contain two incompatible *Sfi*I restriction sites that allow the directional cloning and transfer of any PZF domain constructed in pSKN-SgrAI (*see* **Note 1**). **Section 3.2** describes the transformation of the constructs into

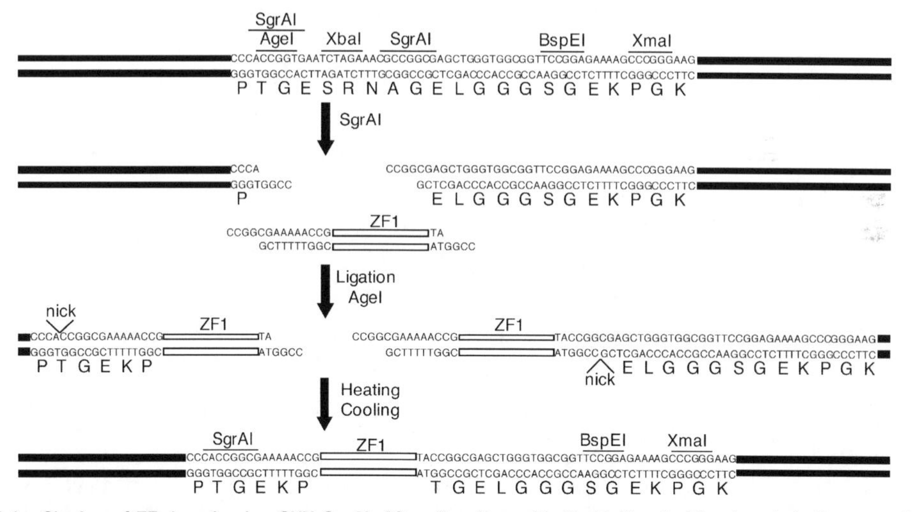

Fig. 24.1. Cloning of ZF domains in pSKN-SgrAI. After digestion with *Sgr*AI, the double-stranded oligo encoding ZF1 (recognizing GTT in the centromere repeat) is ligated to the vector ends. Addition of *Age*I during ligation selects against the antisense orientation. The resulting linear plasmid with a ZF domain attached to each side is incubated at 95°C to remove the non-covalently bound strands, and subsequent annealing via a gradual decrease in temperature circularizes the plasmid which now has one ZF domain enclosed. For the centromere-targeted PZF domain, the procedure starting with *Sgr*AI digestion is repeated twice in order to add ZF2 (GCG) and ZF3 (GTT).

Arabidopsis and the selection of primary transformants. **Section 3.3** describes the procedure for live cell imaging, and **Section 3.4** describes the quantification of bound PZF:GFP proteins using FluoSpheres as a reference.

2. Materials

2.1. Construction of pRF-180:GFP for Live Cell Imaging of Arabidopsis Centromeres

1. Molecular cloning reagents: Competent *Escherichia coli* DH5α cells; antibiotics carbenicillin, kanamycin, and rifampicin.
2. *Agrobacterium tumefaciens* strain Agl1, containing rifampicin and carbenicillin bacterial selection markers.
3. Restriction enzymes: *Sgr*AI, *Sfi*I, *Pvu*II, and *Age*I.
4. 10× STE buffer: 1 M NaCl, 100 mM Tris–HCl, pH 8, 10 mM EDTA.
5. TE buffer: 10 mM Tris–HCl, pH 7.4, 1 mM EDTA.
6. DNA precipitation reagents: 10 M NH_4OAc; 2-propanol; 70% ethanol.
7. Protein extraction reagents: buffer-saturated phenol; chloroform.
8. Vector pSKN-SgrAI (6), containing a carbenicillin-resistance bacterial selection marker.
9. Pairs of complementary oligonucleotides with CCGG overhangs, encoding ZF domains for binding to the triplets GTT and GCG (*see* **Table 24.1**).
10. DNA gel extraction kit that facilitates elution of DNA in small volumes (e.g. Zymoclean, ZYMO Research, Cat D4001).
11. T4 DNA ligase and ligase buffer (New England Biolabs, Cat. M0202).
12. Vector pRF-GFP (1), a wide host range plant expression vector containing a kanamycin-resistance bacterial and plant selection marker.
13. Rapid DNA Dephos & Ligation Kit (Roche, Cat. 04898117001).
14. Universal electroporation cuvettes (2 mm yellow cap) (Eurogentec, Cat. CE-0002-50).
15. Electroporation system (Bio-Rad Gene Pulser or any equivalent type).
16. LB medium (10 g/L Bacto-Tryptone, 5 g/L Bacto-Yeast extract, 10 g/L NaCl).

Table 24.1
Sets of oligomers (FW and RV) encoding ZF domains recognizing DNA triplets GTT and GCG

Triplet	Oligonucleotide sequence from 5′ to 3′
GTT-FW	CCGGCGAGAAACCCTACGCCTGCCCGGAATGCG GGAAGTCGTTTAGCACCTCCGGTAGCCTGGT GAGGCATCAGCGTACGCATA
GTT-RV	CCGGTATGCGTACGCTGATGCCTCACCAGGCTAC CGGAGGTGCTAAACGACTTCCCGCATTCCGG GCAGGCGTAGGGTTTCTCG
GCG-FW	CCGGCGAGAAACCGTATGCGTGTCCCGAATGCG GTAAAAGCTTTAGCCGTAGCGATGACCTTGT GCGTCACCAGAGGACCCATA
GCG-RV	CCGGTATGGGTCCTCTGGTGACGCACAAGGTCA TCGCTACGGCTAAAGCTTTTACCGCATTCG GGACACGCATACGGTTTCTCG

2.2. Plant Growth Conditions and Transformation

1. *Arabidopsis* plants of ecotype *Columbia-0*.
2. For floral dip: 5% sucrose solution in water containing 200 μL/L Silwet L-77 (Lehle Seeds).
3. Seed sterilization reagents: 70% ethanol; 1% sodium hypochlorite solution, usually corresponding to 0.25× of a commercial household bleach solution, depending on the brand.
4. Top agar: 0.1% agarose in water, autoclaved.
5. Primary Transformant Selective Medium: a version of Murashige & Skoog medium with half the concentration of macro elements that can easily be made with a commercially available medium stock (Duchefa, Cat. M0233), using 0.8% Daishin agar (Melford, Cat. D1004) for solidification. After sterilization and just prior to pouring, it must be supplemented with antibiotics Timentin (100 mg/L, to prevent growth of *Agrobacterium*, Duchefa, Cat. T0190), Nystatin (100 mg/L, to prevent growth of filamentous fungi, Duchefa, Cat. N0138), and kanamycin (50 mg/L, for selection of transformed seedlings by the *NPTII* marker gene, Duchefa, Cat. K0126).
6. T2 Selective Medium: prepared similarly as Primary Transformant Selective Medium, but it is supplemented with sucrose (10 g/L, for swifter plant development), and contains only 100 mg/L Timentin and 30 mg/L kanamycin as antibiotics.

2.3. Live Cell Imaging

1. Standard microscopy materials: object glasses, coverslips, forceps.
2. Confocal laser scanning microscope (e.g. Leica TCS SP, Leica) with filters for GFP.

2.4. Quantification of GFP Signals

1. Fluorescent styrol spheres with a diameter of 20 nm (FluoSpheres, F8787, Invitrogen).
2. Recombinant EGFP (rEGFP, BioVision).
3. Spin coater (e.g. model P6700, Specialty Coating Systems).
4. Piranha Solution: 3:1 mixture of concentrated sulfuric acid and 30% hydrogen peroxide solution.
5. Phosphate-buffered saline (PBS), pH 7.4.
6. Polyvinyl alcohol (PVA, Roth Chemicals).
7. Microscope setup capable of imaging the fluorescence signal of single GFP molecules: Inverted microscope (Axiovert 100TV, Zeiss), ×100 objective (NA 1.4, Zeiss), 488-nm laser (ArKr-laser, Spectra Physics), nitrogen-cooled CCD-camera system (Spec-10:400B/LN, Princeton Instruments).
8. Matlab (The MathWorks) for automated quantification of fluorescent signals.

3. Methods

3.1. Construction of ZF Modules for In Vivo Binding to a Repetitive Sequence

1. Anneal the oligomers. Stock concentrations of oligomers are prepared at 100 μM in 0.25× TE. For each complementary set encoding a particular ZF domain, take 10 μL from the stock solution and add 10 μL of 10× STE and water to a volume of 100 μL in a standard 1.5-mL reaction tube. Firmly clamp the tubes containing the mixture, preferably in a rack that can be submerged in a water bath. Heat a water bath to 95°C and incubate the tubes for 2 min. The water is then allowed to cool down gradually. Do not use excessive amounts of hot water (not more than approximately 400 mL), as this unnecessarily lengthens the procedure. Once cooled below 40°C, the tubes can be put on ice or stored at –20°C until use.
2. Digest vector pSKN-SgrAI with *Sgr*AI. Per ligation reaction, 100–200 ng, of digested vector is recommended. Since the *Sgr*AI enzyme has star activity, over-digestion should be avoided. Check a sample for complete digestion on an agarose gel prior to starting ligation reactions.

3. Prepare a ligation reaction in a total volume of 10 μL as follows:
   x μL *Sgr*AI digested vector (100–200 ng)
   1 μL annealed oligomers (encoding the "GTT" ZF in the Step 1 of **Section 3.1** as ZF1)
   1 μL ligation buffer
   0.5 μL *Age*I (to select against antisense orientation)
   1 μL T4 DNA ligase
   x μL water to a final volume of 10 μL

   Ligate for at least 2 h at room temperature. Analyze the ligation reaction on a 0.8% agarose gel. This step is required to remove the large excess of free (unligated) oligomers that would otherwise block the re-annealing reaction (*see* Step 4 of **Section 3.1**). The vector with a ligated double-stranded ZF oligomer at each of its *Sgr*AI ends should be the major band (around 3,000 bp). Cut out and extract the DNA (Zymoclean kit, or other equivalent method) and elute or dissolve it in a maximum volume of 20 μL.
4. Add a maximum of 18 μL of the extracted ligation product to 2 μL 10× STE and Milli-Q purified water to a total reaction volume of 20 μL (i.e. 1× STE final concentration). Consider dividing the mixture into two tubes in order to have one as backup material. Incubate at 95°C for 2 min in a water bath (*see* Step 1 of **Section 3.1**) so that base pair bonds will become unstable, allowing non-covalently bound oligomer sequences to be lost and the subsequent annealing of covalently bound intra-molecular cohesive ends to occur. Allow the mixture to cool gradually for the proper annealing of the ZF-encoding complementary strands. Put on ice (or at –20°C until use). Transform part of the mixture (up to 10 μL) containing circularized plasmid molecules into competent *E. coli* cells following standard procedures.
5. Check the obtained colonies for the presence and proper orientation of ZF1 via restriction digestion with *Pvu*II and *Age*I and subsequent gel analysis. Positive clones will contain a band of 656 bp. Despite the presence of *Age*I during the oligomer ligation reaction, a significant fraction of the inserts will probably still be in the wrong orientation (due perhaps to the fact that nicked sites lacking a 5′ phosphate group are imperfect targets for the enzyme), which leads to the presence of an *Age*I site which will cut within the 656 bp fragment.
6. To add the next ZF domain, repeat Steps 2–5 of **Section 3.1** for two successive rounds using the annealed oligomers encoding the "GCG" ZF (ZF2)

and, subsequently, the "GTT" ZF (ZF3). To check the colonies for the presence of the correct plasmid (Step 5 of **Section 3.1**), consider that the *Pvu*II fragment will increase 84 bp with each finger (i.e. 740 bp for ZF1 and ZF2 together and 824 bp for the final plasmid containing ZF1, ZF2, and ZF3). Sequences verify the final construct containing the three ZF domains (*see* **Notes 2** and **3**).

7. Digest the final pSKN-SgrAI construct containing the three-finger PZF domain with *Sfi*I (note that this enzyme cuts at 55°C). Excise the three-finger PZF fragment from the gel (325 bp) and extract the DNA using Zymoclean.
8. Digest the wide host range plant expression vector pRF-GFP with *Sfi*I (55°C). After digestion, remove the heat stable *Sfi*I by extraction with phenol/chloroform (phenol is corrosive, *see* **Note 4**). Add 0.25× TE buffer to the water phase containing the DNA for a final volume of 100 μL. For differential precipitation of vector DNA, add 45 μL of 10 M NH4OAc and 90 μL of 2-propanol and spin down the DNA by centrifugation at maximum speed for 15 min. Wash the DNA pellet with 70% ethanol, dry the pellet briefly, and dissolve in 0.25× TE at a concentration of about 10–20 ng/μL.
9. Dephosphorylate the plasmid DNA and ligate the extracted three-finger fragment using a Rapid DNA Dephos&Ligation kit. Transform to the obtained plasmid, designated pRF-180:GFP, into *E. coli* and check the colonies via restriction analysis.
10. Introduce pRF-180:GFP into *Agrobacterium tumefaciens* strain Agl1 via electroporation. Briefly, starting with a 100 μL inoculum from a small preculture, *Agrobacterium* cells are grown overnight under vigorous shaking at 29°C in 100 mL LB medium with the appropriate antibiotics – rifampicin (10 μg/L) and carbenicillin (75 μg/L) for strain Agl1 – to an OD600 of about 1.0. After chilling on ice, cells are harvested by centrifugation, washed for three times in 10 mL Milli-Q purified water, one time with 10 mL of 10% glycerol, and finally resuspended in 500–750 μL 10% glycerol. Aliquots of 40 μL are put in 1.5-mL reaction tubes and used directly or after being snap frozen in liquid nitrogen and stored at –80°C. About 10 ng of plasmid DNA (in up to 5 μL water) is mixed with 40 μL cells (on ice) and the mixture is transferred to a pre-chilled electroporation cuvette followed by a single pulse (2.5 kV, 25 μF capacity, 200 Ω for a Bio-Rad Gene Pulser). After 1 h incubation at 29°C in 1 mL LB medium, aliquots are plated on agar medium with rifampicin (10 μg/L), carbenicillin (75 μg/L), and kanamycin (100 μg/L, for

selection of plasmid presence) (*see* **Note 5**). Colonies should be visible after 2–3 days of growth at 29°C.

3.2. Transformation of Arabidopsis and Selection of Primary Transformants

1. *Arabidopsis* seedlings are grown under standard greenhouse conditions (21°C, 16 h day, 8 h night). When the plants start to flower, the primary flower stalk is clipped off to increase the number of flower stalks that are produced per plant. Approximately 1 week later, an *Agrobacterium*-mediated transformation is carried out by means of the floral dip method. Hereto, cells from an overnight 50-mL liquid culture of a kanamycin-resistant *Agrobacterium* strain harbouring a plasmid of interest (Step 10 of **Section 3.1**) are collected by centrifugation and resuspended in water containing 5% sucrose and 200 μL/L Silwet L-77 (*see* **Note 6**) to an OD600 of 0.1. The suspension is put in an open container and the flowering parts of the complete soil-grown *Arabidopsis* plants are briefly submerged into the liquid. Optionally, a control can be included in which *Arabidopsis* plants are transformed with an *Agrobacterium* strain containing the empty pRF-GFP construct, thus lacking a PZF domain. Just after floral dipping, the plants are put in plastics bags which are kept closed for 24 h to maintain high moisture content. Bags are removed afterwards.
2. About 4–5 weeks after the floral dip, dry seeds can be harvested from the plants and transformed seedlings can be selected. Per plant, 20–100 mg of seeds can be expected.
3. All of the following tissue culture steps are conducted in a laminar flow cabinet. To sterilize seeds, up to 500 mg of harvested seeds are transferred to an empty 50-mL disposable conical tube. Wash the seeds briefly (1 min) with 70% ethanol. Replace the ethanol solution with 1% hypochlorite solution and sterilize seeds for 10 min at room temperature. Remove the 1% hypochlorite solution and wash repeatedly with 40 mL sterile water (at least three times). Remove the water and add 10 mL of top agar per 100 mg (= about 5,000) seeds. Mix well to resuspend the seeds evenly.
4. Prepare plates for primary transformant selection using 15-cm Petri dishes containing 60 mL solidified Primary Transformant Selective Medium (*see* Step 5 of **Section 2.2**). Add 10 mL of the seed suspension to each plate, thus using up to five plates per full 50-mL tube. Spread the seeds evenly by swirling the plate. Leave the plates with their lids open in a laminar flow cabinet until the top agar has dried up. Close the plates with only two small straps of tape; do not seal the whole plate. Vernalize by putting the plates at 4°C for 2 days, then transfer plates to a growth chamber. Primary transformants can be distinguished after 7–10 days of

growth under standard tissue culture growth conditions (21°C, 16 h day, 8 h night). Per plate (= per 5,000 seeds), 5–50 primary transformants are routinely obtained.

5. One can use primary transformants for microscopical observations (*see* **Section 3.3**), but for future observations and for establishing seed stocks one needs to grow several primary transformants under greenhouse conditions and let them set seed. It is wise to start-up several lines, since expression levels may vary considerably between independent transformants.
6. Sterilize a small sample of T2 seeds, derived from the primary transformants, in a 1.5-mL Eppendorf reaction tube essentially as described but with up to 1 mL volumes (*see* Step 3 of **Section 3.2**) and sow them on 9-cm Petri dishes with 25 mL solidified T2 Selective Medium (*see* Step 6 of **Section 2.2**). Ideally, seeds are sown on a straight line and the plates are placed vertically in the growth chamber. In this way, the segregation for the resistance marker can be analysed. The roots, where GFP fluorescence is best observed, will grow on top of the medium, facilitating the preparation of samples for microscopy. As a result of using the pRF-GFP plasmid, PZF:GFP expression is driven by a ribosomal protein 5A (*RPS5A*) promoter, which is highly active in meristems. The developmental stage of T2 seedlings at which fluorescence is best observed in primary root tips lies between 5 and 10 days after germination. Depending on one's interest, meristematic cells in aerial parts of the plants are better observed at later stages.

3.3. Live Cell Imaging

1. Within a root tip, a high expression level of the *RPS5A* promoter-driven GFP-tagged ZFs is combined with a near complete lack of interfering auto-fluorescence of the tissue. Therefore, young root tips of T2 transformants are most suitable for live cell imaging. The root tip is placed on a microscope slide in a drop of water and a coverslip is placed on top.
2. Live cell imaging can be conducted with any standard confocal laser scanning microscope with appropriate GFP filters. The centromeric spots will be visible using a ×40 magnification objective (*see* **Fig. 24.2**).

3.4. Quantification Using Fluorescent Beads (FluoSpheres) as Reference

3.4.1. Recording of FluoSphere and rEGFP Calibration Samples

1. Object glasses and coverslips need to be cleaned by submersing them in Piranha solution for 10 min, washing them

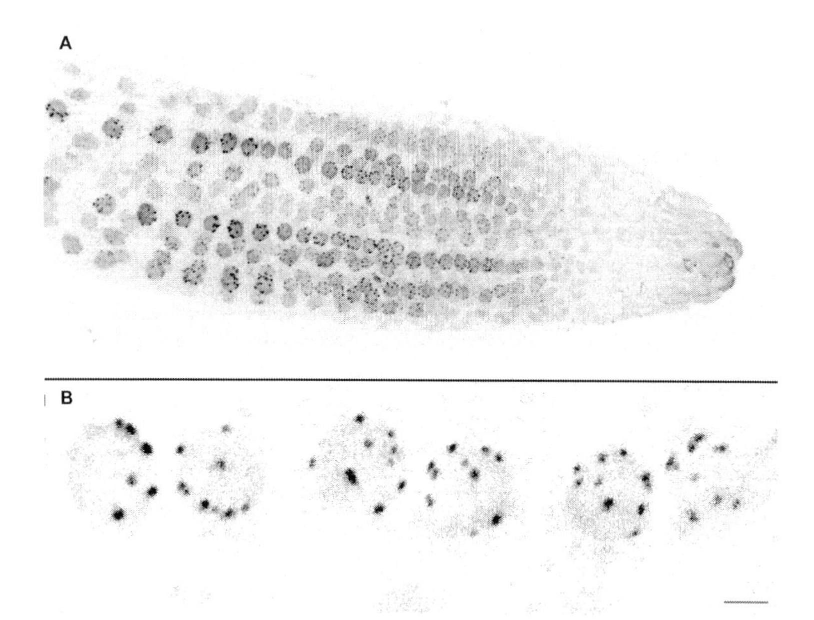

Fig. 24.2. Live cell imaging of centromere repeats in primary *Arabidopsis* root tips. For clarity, the fluorescent 180:GFP signals are shown as *dark spots*. (**A**) Overview of a complete root tip. (**B**) Detail of cells showing nuclei (*grey spheres*) with up to ten 180:GFP-labelled chromocenters. *Bar* represents 5 μm.

thoroughly with Milli-Q purified water, and drying them in a stream of filtered air. Thereby, a clean and hydrophilic surface of the coverslips is obtained.

2. Prepare 1% (w/v) PVA in PBS by gentle shaking over night. The next day, dilute stock solutions of 2% (solids) FluoSpheres and 1 μg/μL rEGFP by 10^5 and 10^6 times, respectively, in PBS/PVA. A volume of 50 μL of each preparation is then evenly spread on the cleaned coverslips using a spin coater. We achieve the best results using a two-step protocol: 10 s at 300 rpm and 1 min at 2,000 rpm. Spin coating the samples on coverslips imbeds and immobilizes the particles in a thin and hydrophobic film of PVA. The thickness of these films is 1 μm on average, as confirmed by confocal microscopy (7).
3. FluoSpheres are imaged on a confocal setup with the same settings used for imaging the PZF domains, i.e. laser intensity, scan speed, zoom settings, detector gain and offset, and filters need to be left unchanged. The same FluoSpheres are then recorded on the single-molecule microscopy setup. Settings for excitation intensity (~1 kW/cm^2 of the 488-nm line) and exposure (5 ms) are chosen to ensure bright diffraction-limited signals that do not exceed the detector's dynamic range (i.e. for a 16-bit recording, the brightest pixel

must not exceed a count value of $2^{16} = 65{,}536$). At the same time, these settings allow single-molecule sensitivity to be achieved for recordings of single rEGFP molecules, with sufficient signal to noise for subsequent quantifications of the signals.

3.4.2. Quantification of Fluorescent Signal Intensities in Digital Images

1. Since fluorescent signals of single FluoSpheres and rEGFPs are both diffraction limited (i.e. they do not exceed the spatial dimensions of the point spread function (PSF) of the respective microscope setup), they are quantified in the same way (*see* **Fig. 24.3**). The integrated intensity of the diffraction-limited spots is analysed by means of fitting a 2D Gaussian function (*see* equation [1] and **Fig. 24.3C**) to the each of the 2D intensity profiles.

$$f(x,y) = z_0 + 4\ln 2 \frac{A}{\pi\omega^2} e^{\left[-4\ln 2 \frac{(x-xc)^2}{\omega^2}\right]} e^{-\left[4\ln 2 \frac{(y-yc)^2}{\omega^2}\right]} \quad [1]$$

A custom Matlab script is used to automate this process. Each fluorescent spot exceeding a set intensity threshold is selected for a fit, which yields the local background (z_0),

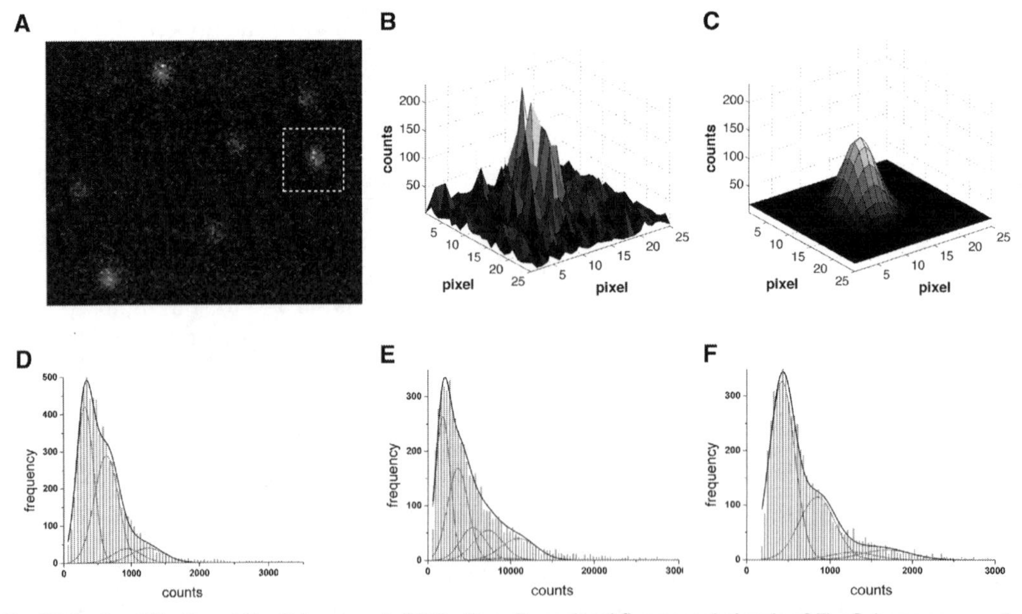

Fig. 24.3. Quantification of FluoSpheres and rEGFPs. Two-dimensional fluorescent signals of FluoSpheres as recorded in confocal images (**A**) are selected (**a**, *white box*, and **b**) and fit to a 2D Gaussian (**c** and equation [1]). A histogram of the obtained intensities (**D**) is fit to a sum of Gaussians (equation [2]), which reveals a mean intensity for single FluoSpheres of 310 counts. The same procedure for FluoSphere and rEGFP measurements on the single-molecule setup (**e** and **f**, respectively) yields mean intensities of 1,827 and 422 counts, respectively.

position (x_c, y_c), integrated intensity (A), and full width at half maximum peak height (FWHM, ω) for each spot.

2. To ensure that only in-focus FluoSpheres and rEGFPs are quantified, spots with a FWHM 30% below or above the width of the PSF of the respective setup are excluded from further analysis. A histogram of all remaining intensity values is then fit to a sum of Gaussians (*see* equation [2] and **Fig. 24.3D**).

$$f(x) = \sum_{1}^{i} A_i e^{-\frac{1}{2}\left[\frac{x - ix_c}{w\sqrt{i}}\right]^2} \qquad [2]$$

The peak position of the first Gaussian (x_c) indicates the mean intensity of single FluoSpheres or rEGFPs, respectively (*see* **Note 7**).

3. The axial dimension of the PSF on a confocal setup typically exceeds the lateral dimension by a factor of around 3, if the pinhole is set to a diameter of one Airy unit. Since the PZF domains are larger (1 μm) than the thickness of a confocal slice on our confocal setup (730 nm), their integrated intensities cannot be captured into a single image. Therefore, the integrated signal of all PZF:GFPs located in a domain needs to be determined from a z-series of scans (i.e. a 3D scan). In order not to overestimate the integrated signal by overlapping axial scans, the axial step size of the sequence (*see* equation [3]) needs to be adjusted to the thickness of a confocal slice.

$$\mathrm{W}_{\mathrm{FWHM}_z}\sqrt{\pi/2\ln 4} \qquad [3]$$

For each confocal slice of a centromeric region, the background signal outside the nucleus was subtracted from the integrated intensity of the region. Subsequently, all intensities of a single PZF domain are added up.

3.4.3. Calculation of the Number of PZF:GFPs Bound to a Centromeric Region/Domain

1. A division of the intensity (in counts) of single FluoSpheres on the single-molecule setup by the intensity of single rEGFPs measured on the same setup relates the intensities of the two fluorescent species. In our experiments, FluoSpheres had a brightness equivalent to 4.67 ± 0.12 rEGFP molecules.

2. The integrated intensity of a centromeric region is now related to the intensity of single FluoSpheres measured on the confocal setup. Multiplication with 4.67 ± 0.12 then yields the number of rEGFP molecules, which have the same

brightness as the domain, and provide a precise estimate for the number of PZF:GFPs that are bound to the domain (*see* **Note 8**).

4. Notes

1. It is of course possible to construct PZF domain for any other kind of repetitive targets, although not all ZF combinations will yield PZF domains with suitable affinity for the cognate DNA target sequences. Supposing that a biochemically suitable PZF domain is available, the chances for successful visualization obviously depend on several parameters. The exact number of 180-bp centromeric repeats in *Arabidopsis* chromosomes is estimated at 2,000 sites per centromere (1). Considering the clarity of signals in **Fig. 24.2**, one might expect that even one-tenth of the signal intensity would have sufficed for still obtaining a suitable picture. Hence, based upon other experiments performed in our laboratory and with other collaborators, we estimate that 200 targets are likely to form the lowest number of chromosomal binding sites that are still detectable by a PZF:GFP of this kind. Considering the physical spacing between the binding sites, this might vary between 40 and 200 bp. One way to meet and exceed this rather arbitrary threshold would be to co-express several PZF:GFPs that target different binding sites within the same repeating unit.
2. Addition of each ZF oligomer into the plasmid pSKN-SgrAI and analysis of the resulting plasmid by restriction digestion will take a minimum of 3 days.
3. Depending on oligomer quality and/or imperfections in annealing conditions, probably not all three-fingered PZF domains will possess the predicted sequence. It is therefore an absolute requirement to sequence the PZF-encoding domain prior to proceeding with Step 5 of **Section 3.1**. To avoid being stuck with a single imperfect sequence after having passed through the addition of ZF3, one could verify each step by sequence analysis; however, this is rather time consuming. In our hands, it is usually sufficient to work with two independent construction lines. Two different plasmids containing a restriction-verified ZF1 are used in parallel for the subsequent steps. This does not require much extra time, and the chance that two independent three-fingered PZFs are both imperfect is very limited.
4. Always use gloves and appropriate protective equipment such as a chemical hood when using phenol. For additional

details on the safe handling and extraction procedures using phenol/chloroform, *see* (8).

5. For additional details concerning electroporation of Agrobacterium cells, *see* (9). It is important to realize that plasmid carrying Agrobacterium cells should be grown at 28–29°C as plasmids can get unstable at higher temperatures.
6. For additional details concerning *Arabidopsis* transformation via the standard floral dip method, *see* (10).
7. A fit of the intensity histograms with multiple Gaussian is necessary, as two (or more) molecules or beads might either stick together or be too close to be resolvable as individual spots. Both cases lead to fluorescence signals with twice (or manifold) the intensity of the single object.
8. For a number of reasons, the calculated number of PZF:GFPs per domain is likely to underestimate the true number. Individual GFP molecules attached to the PZF can be incorrectly folded, can photo-bleach, or be in a so-called dark state during image acquisition (i.e. a process known as photo-blinking). In all three cases, a number of PZF:GFP molecules will not emit a fluorescent signal, leading to a lower integrated signal intensity of the PZF domain. In addition, the close proximity of GFP molecules in the domain may lead to processes such as inner filter effect (IFE) and photo-bleaching, both of which are difficult to quantify if the local density of GFP molecules is unknown. Our analysis therefore provides a good starting point to apply corrections for the above-mentioned processes.

Acknowledgements

We would like acknowledge support from the Netherlands Organisation for Scientific Research (NWO, grant 050-10-123) and Leiden University and Gerda Lammers for help with confocal microscopy.

References

1. Lindhout, B.I., Fransz, P., Tessadori, F., Meckel, T., Hooykaas, P.J., and van der Zaal, B.J. (2007) Live cell imaging of repetitive DNA sequences via GFP-tagged polydactyl zinc finger proteins. *Nucleic Acids Res.* *35*, e107.
2. Boisnard-Lorig, C., Colon-Carmona, A., Bauch, M., Hodge, S., Doerner, P.,

Bancharel, E., Dumas, C., Haseloff, J., and Berger, F. (2001) Dynamic analyses of the expression of the HISTONE::YFP fusion protein in *Arabidopsis* show that syncytial endosperm is divided in mitotic domains. *Plant Cell.* ***13***, 495–509.

3. Fang, Y. and Spector, D.L. (2005) Centromere positioning and dynamics in living Arabidopsis plants. *Mol Biol Cell.* ***16***, 5710–5718.
4. Matzke, A.J., Huettel, B., van der Winden, J., and Matzke, M. (2005) Use of two-color fluorescence-tagged transgenes to study interphase chromosomes in living plants. *Plant Phys.* ***139***, 1586–1596.
5. Robinett, C.C., Straight, A., Li, G., Willhelm, C., Sudlow, G., Murray, A., and Belmont, A.S. (1996) In vivo localization of DNA sequences and visualization of large-scale chromatin organization using lac operator/repressor recognition. *J Cell Biol.* ***135***, 1685–1700.
6. Neuteboom, L.W., Lindhout, B.I., Saman, I.L., Hooykaas, P.J., and van der Zaal, B.J. (2006) Effects of different zinc finger transcription factors on genomic targets. *Biochem Biophys Res Commun.* ***339***, 263–270.
7. Steinmeyer, R., Noskov, A., Krasel, C., Weber, I., Dees, C., and Harms, G.S. (2005) Improved fluorescent proteins for single-molecule research in molecular tracking and co-localization. *J Fluoresc.* ***15***, 707–721.
8. Sambrook, J., Russell, D.W., and Irwin, N. (2001) Molecular cloning: a laboratory manual. Cold Spring Harbor Laboratory Press, New York.
9. Mattanovich, D., Rüker, F., Machado, A.C., Laimer, M., Regner, F., Steinkellner, H., Himmler, G, and Katinger, H. (1989) Efficient transformation of *Agrobacterium* spp. by electroporation. *Nucleic Acids Res.* ***17***, 6747.
10. Clough, S.J. and Bent, A.F. (1998) Floral dip: a simplified method for *Agrobacterium*-mediated transformation of *Arabidopsis thalianat. Plant J.* ***16***, 735–743.

Chapter 25

Biophysical Analysis of the Interaction of Toxic Metal Ions and Oxidants with the Zinc Finger Domain of XPA

Andrea Hartwig, Tanja Schwerdtle, and Wojciech Bal

Abstract

Zinc-binding domains are frequently present in proteins required for maintaining genomic integrity. During recent years there has been accumulating evidence that they may be particularly sensitive targets for toxic metal ions, leading, for example, to DNA repair inhibition. Here, we present a broad set of methods for assessing the impact of metal ions on the zinc-binding domain of xeroderma group A protein, an essential factor in nucleotide excision repair. We envisage that these methods will be useful also for other proteins.

Key words: Xeroderma group A protein, toxicity, electrospray mass spectrometry, fluorescence spectroscopy, arsenic, cadmium.

1. Introduction

Zinc fingers (ZFs) and related structures are frequently found in transcription factors, DNA repair, and tumor-suppressor proteins and have been shown in many cases to be required for maintaining genomic integrity (1). Within these structures, the Zn(II) cation is complexed to four ligands, generally cysteines and/or histidines, thereby driving the folding of these protein domains that have been shown to be involved in DNA–protein or protein–protein interactions (2). The Zn(II) ions do not directly participate in these interactions, but rather stabilize the fold of the domain.

In recent years, accumulating evidence suggests that ZFs may be particularly sensitive targets for toxic metal ions, and that

J.P. Mackay, D.J. Segal (eds.), *Engineered Zinc Finger Proteins*, Methods in Molecular Biology 649,
DOI 10.1007/978-1-60761-753-2_25, © Springer Science+Business Media, LLC 2010

their functional inactivation may be responsible for altered gene expression, DNA repair disturbances as well as loss of tumor-suppressor functions, thus providing an important mechanism for metal-induced carcinogenicity (1, 3). Interactions have been shown on several levels, including the demonstration of zinc release and structural alterations of zinc-binding domains as well as inactivation of DNA repair processes and tumor-suppressor functions in cells. Thus, for example, trivalent inorganic and biomethylated arsenicals have been shown to release zinc from the ZF domain of the human repair protein XPA (XPAzf) (4) and $NiCl_2$ as well as $CdCl_2$ provoked conformational changes and altered sensitivity toward oxidizing agents in the same protein (5, 6).

Electrospray ionization mass spectrometry (ESI-MS) proved to be an excellent method for qualitative and quantitative studies of reactions of XPAzf with toxicologically relevant metals and oxidants, such as inorganic and organic arsenicals (7), hydrogen peroxide (8), and *S*-nitrosoglutathione (GSNO) (9). A single measurement may use as little as 50 pmol (~0.2 μg) of XPAzf. ESI-MS measurements allowed us to demonstrate that intrapeptide disulfides were the only products formed in the reaction of XPAzf with H_2O_2 and AsO_3^{3-} (7, 8), whereas GSNO formed transient *S*-nitroso derivatives that eventually released NO to form disulfides (9). Monomethylarsonous acid [MMA(III)] formed stable $AsCH_3$ adducts, in which one or two arsenic atoms were bonded to two Cys thiols each (7). The very mild ionization conditions available in ESI-MS enable detection of noncovalent interactions, such as Zn(II) coordination to XPAzf. As a result, the zinc release event can also be followed by ESI-MS (7–9). In general, ESI-MS is not a quantitative technique, because of ionization susceptibility differences, and thus the intensities of signals may be quite different even for closely related molecules. However, we demonstrated that modifications of XPAzf thiols, such as intramolecular disulfide formation, glutathionylation, S-nitrosylation, MMA adduct formation, and Zn(II) coordination, have little effect on either the molecular ion pattern or the signal intensity for XPAzf. As a result, ESI-MS can also be used as a tool for studying kinetics and mechanisms of reactions of XPAzf.

Fluorescence spectroscopy is another sensitive technique that has proven useful for studying interactions of metal ions with XPAzf. In this case, the fluorophores are provided by two Tyr residues in the protein, one of which is close to the N-terminus of the peptide, while the other is located in the middle of the loop formed upon the Zn(II) binding to four XPAzf thiols. Nevertheless, their contributions to the overall fluorescence of XPAzf cannot be separated easily. The binding of Zn(II) and Cd(II) ions resulted in an increase of fluorescence intensity, whereas that of Ni(II) and Co(II) had a strong opposite effect. The oxidation of zinc-bound XPAzf resulted in a decrease of fluorescence intensity compared to the Zn(II) complex, but the fluorescence was higher than that of the apopeptide. Thus, fluorescence spectroscopy can be used for studying zinc release, as well as competition between Zn(II) and Ni(II) or Co(II) and between Cd(II) and the latter two metals. However, it is blind to the Zn(II)/Cd(II) substitution, presumably due to structural similarities of zinc- and cadmium-bound structures (5, 6).

Within this contribution, we describe methods to elucidate interactions between toxic metal ions and the zinc-binding domain of XPA as a model for zinc-binding DNA repair proteins, including studies of zinc release as well as ESI-MS and fluorescence spectroscopy for structural investigations.

2. Materials

2.1. XPA Zinc Release

1. XPAzf: A commercially synthesized peptide with the sequence Ac-DYVICEECGKEFMDSYLMNHFDLPTCD NCRDADDKHK-NH_2 (e.g., Schafer-N or Alpha Diagnostic International). The peptide purity was determined to be higher than 98% by HPLC and its identity was confirmed by ESI-MS. Final concentration in the assay was 20 μM. *See* **Notes 1–3** for handling and control of XPAzf solutions.
2. Methylarsine oxide ($CH_3As(III)O$, ≥99% purity) was kindly provided by Prof. Dr. W. R. Cullen (University of British Columbia, Vancouver, Canada) and was used as the precursor to MMA(III). Sodium(meta)arsenite (≥99% purity) was purchased from Fluka Biochemika. High purity $Zn(NO_3)_2.6H_2O$, $ZnCl_2$, $Co(NO_3)_2.6H_2O$, $Cd(NO_3)_2.4H_2O$, and $NiCl_2.6H_2O$ were from Merck.
3. HEPES-NaOH buffer: 20 mM HEPES-NaOH (Fluka), pH 7.4. Store at room temperature.
4. Zinc buffer: 100 μM $ZnCl_2$ in HEPES-NaOH buffer. Store sterile at room temperature. Check for $Zn(OH)_2$ precipitation upon storage and discard if precipitate is formed.
5. 4-(2-Pyridylazo)-resorcinol monosodium salt (PAR, Riedel de-Haën): 2.5 mM working solution. Stable up to 1 month when stored in the dark (room temperature). Final concentration in the assay was 100 μM.

6. H_2O_2 (30%, p.a. grade, Merck): 1 M working solution. Prepare freshly. Final concentration in the assay was 10 mM.
7. Ellmans reagent (*see* **Note 2**).

2.2. HPLC Studies of Oxidative Reactions of XPAzf

1. Zinc Solution: 100 μM $ZnCl_2$ in high quality metal-free water (*see* **Note 3**). Store sterile at room temperature.
2. Zn(II)–XPAzf: XPAzf dissolved in 100 μM $ZnCl_2$ at 100 μM.
3. Sodium phosphate buffer: pH 7.4, final concentration in the reaction mixture is 50 mM.
4. H_2O_2 (30%, p.a. grade, Merck): 100 mM working solution in water (*see* **Note 3**), freshly prepared before each experiment.
5. Toxic metals: 10 mM metal ion stock solutions in water (*see* **Note 3**).
6. Mobile phase I and II: 15% methanol in water containing 0.1% TFA (I) and 100% methanol containing 0.1% TFA (II).

2.3. Mass Spectrometry Assays for Zinc Release and Covalent Modifications of Thiol Groups

1. Ammonium acetate buffer: 10 mM ammonium acetate with pH set to 7.4 using microlitre amounts of concentrated ammonia and acetic acid, when necessary. Should also be saturated with argon prior to the addition of XPAzf, in order to remove oxygen (*see* **Notes 3** and **4**).
2. Zinc solution: The use of Zn(II) acetate is preferred, but zinc chloride can be also used at final concentrations up to 10 μM. Store sterile at room temperature.
3. GSNO solution: *S*-Nitrosoglutathione can be prepared by mixing equal volumes of 220 mM GSH in water and 220 mM $NaNO_2$ in 0.1% TFA and incubating the resulting solution in the dark under nitrogen for 10 min (final concentration of approximately 100 mM GSNO). Use immediately after preparation. The outcome of the glutathione S-nitrosylation reaction can be monitored by HPLC (isocratic, 5% acetonitrile in water, 0.1% TFA, C_{18} column) to assure the purity of GSNO and absence of glutathione disulfide (GSSG), a typical product of GSH oxidation. Determine the concentration of GSNO at 336 nm (GSNO has a molar extinction coefficient of 900 $M^{-1}cm^{-1}$ at 336 nm) (10).

2.4. Fluorescence Spectroscopy

1. XPAzf solution in HEPES similar to that used in zinc release assay, except for final peptide concentrations, which can be lower than 10 μM, depending on the spectrofluorimeter sensitivity. Phosphate buffers can also be used.
2. Metal ion stock solutions in water (1 mM, *see* **Note 3**).

3. TCEP: 5 mM in water (can be added at a final concentration of 100 μM to help protect thiol groups; *see* **Note 1**).

3. Methods

3.1. XPA Zinc Release

1. Make up a solution of XPAzf peptide in 20 mM HEPES buffer. Check its concentration using Ellman's reagent (*see* **Note 2**).
2. Saturate the peptide with zinc to obtain the ZF structure (*see* **Note 5**).
3. Measure the release of zinc from the peptide (20 μM) spectrophotometrically by addition of 100 μM PAR after a 30-min incubation at 37°C in 20 mM HEPES-NaOH buffer (pH 7.4) with the respective metal compounds (*see* **Notes 6** and **7**). PAR forms a water soluble complex with zinc(II) with a 2:1 stoichiometry, thereby altering the absorbance spectrum of PAR.
4. Quantitate the zinc-PAR complex at 492 nm. For control experiments, add 10 mM H_2O_2 to release 100% of the zinc from the peptide.

3.2. HPLC Assay for Assessing Oxidation Sensitivity of XPAzf

1. Incubate Zn(II)–XPAzf (10 μM) in the absence or presence of the desired metals with 10, 100, or 1,000 μM H_2O_2 and analyse by HPLC after 0, 50, 100, 150, 200, and 250 min. For each toxic metal concentration to be analysed, a 1.25-mL solution is prepared and divided into six vials. For each time point, inject 100 μL of the reaction mixture into the HPLC (*see* **Notes 8** and **9**).
2. Carry out HPLC analysis using a system equipped with an autosampler and a gradient pump. XPAzf and its oxidation products can easily be separated on a 4.6 mm × 250 mm C_{18} analytical column (Luna C_{18}, 5 μM, Phenomenex) by a linear gradient from 15% methanol in water (0.1% TFA) to 100% methanol (0.1% TFA) within 30 min. Detection occurs by a fluorescence detector at an excitation wavelength of 260 nm and an emission wavelength of 304 nm (*see* **Note 10**).
3. Quantify the reaction products by comparing the peak areas of XPAzf at 0 h (control) with the peak areas at the respective incubation times.
4. Calculate kinetic parameters from substrate (XPAzf) decay using an appropriate kinetic equation (e.g., zeroth, first, or second order) and any general purpose data analysis software (e.g., Microsoft Excel, OriginLab Origin).

3.3. Electrospray Mass Spectrometry Assays for Zinc Release and Covalent Modifications of Thiol Groups

1. Make up solutions of 10 μM XPAzf in 10 mM ammonium acetate (*see* **Note 11**), containing appropriate amounts of metal ions and other reagents, as required (*see* **Note 12**). The minimal volume of the sample is ca. 20 μL for a single measurement.
2. Choose the mode of sampling. For slow reactions, aliquots of reaction mixtures are injected into the mass spectrometer at given intervals (8, 9). For faster reactions, or when there is enough material, the sample can be injected continuously using peristaltic pump; e.g., at 4 μL/min, and the sampling can be adjusted by choosing intervals for signal integration (7). Control the background Zn(II) in the mass spectrometer prior to actual measurements (*see* **Note 13**).
3. Assign the signals to individual reaction products on the basis of their theoretical m/z values and isotopic envelopes, which can be obtained from their atomic compositions using freely available software, such as ExPASy (www.expasy.ch).
4. Deconvolute overlapped signals numerically prior to integration using theoretical isotopic distributions (e.g., from ExPASy), as illustrated in **Fig. 25.1** for the formation of intramolecular disulfides (9).
5. Use integrals for individual peaks corresponding to specific reaction intervals to generate kinetic plots of substrate decay and/or product formation. Fit to an appropriate kinetic model, analogously to HPLC data.

3.4. Fluorescence Spectroscopy

1. Make up solutions of XPAzf (1–10 μM) in a desired buffer solution, e.g., 20 mM HEPES or 50 mM sodium phosphate (*see* **Notes 1**, **14,** and **15**).
2. Record the fluorescence spectra in 1-cm cuvettes using an excitation wavelength of 260 nm. Spectra can be recorded in the wavelength range of 270–400 nm, or alternatively the reaction can be followed at the emission maximum of 305 nm. **Figures 25.2** and **25.3** present examples of Zn(II) and Ni(II) titrations (5).
3. Identify an observable suitable for binding constant determinations (e.g., fluorescence intensity at a certain wavelength) and fit data to a model of complex formation to obtain conditional or relative binding constants. For example, the Ni(II) titration of XPAzf yielded a conditional binding constant of the Ni(II)–XPAzf complex and the subsequent Ni(II) titration of XPAzf + Zn(II) yielded the conditional binding constant for the Zn(II)–XPAzf complex (5).

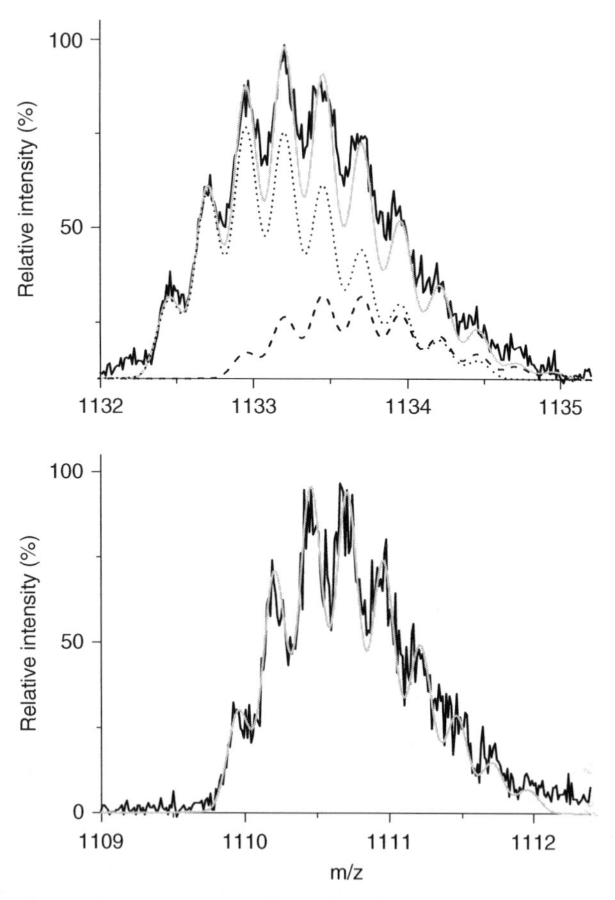

Fig. 25.1. Examples of isotopic analysis of composite ESI-MS signals. (*Top*) *black*: experimental peak; *grey*: best fit to the data, composed of 75% *dotted theoretical* $(AsCH_3)XPAzf(SS)$ and 25% *dashed theoretical* $(AsCH_3)XPAzf$; (*bottom*) *black*: experimental peak; *grey*: best fit composed of 100% $XPAzf(SS)_2$. Line widths of theoretical components (0.17 Da) were adjusted to fit the instrumental resolution of the spectra (8).

4. Notes

1. XPAzf, as other thiol peptides, is susceptible to air oxidation. It can be stored dry at –20°C for extended periods of time. Solutions of the apopeptide are unstable, unless prepared under strictly anaerobic conditions. In the latter case, aliquots of frozen solutions can be stored at –20°C for several days. Disulfides were the only products of air oxidation of XPAzf, and thus the apopeptide can be regenerated by reduction, followed by HPLC purification. Tris(2-carboxyethyl)phosphine (TCEP) is preferred as the reductant over DTT (11). Note that TCEP cannot be used

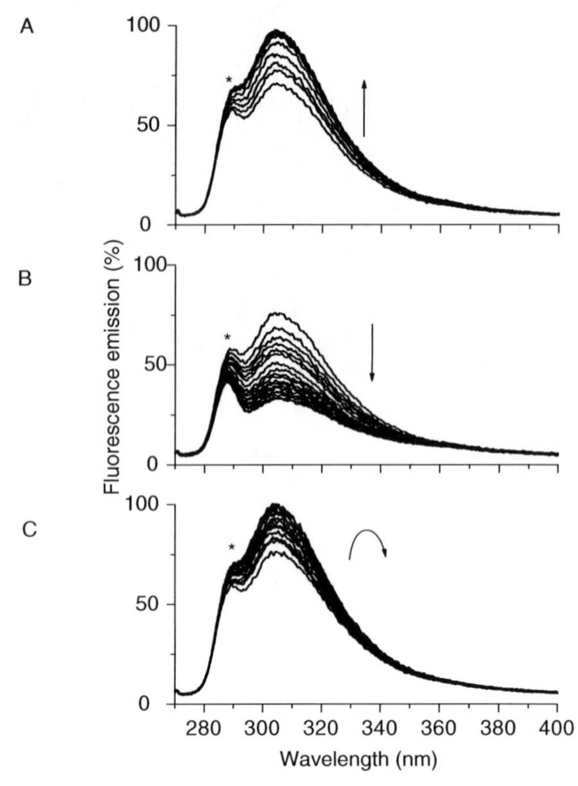

Fig. 25.2. Examples of fluorescence titrations of XPAzf in 50 mM sodium phosphate buffer containing 100 μM TCEP, pH 7.4. (**A**) Titration of 5.2 μM XPAzf with 1 mM Zn(II). (**B**) Titration of 7.0 μM XPAzf with 1 mM Ni(II). (**C**) Titration of 7.0 μM Zn(II)–XPAzf with 1 mM Ni(II). *Arrows* indicate the directions of spectral changes; *asterisks* mark the Raman band of water, which remained unchanged in the course of titrations (5).

together with phosphate buffer, as TCEP reacts with phosphate. If necessary, some assays not involving redox reactions can be performed in the presence of small amounts of TCEP, because it binds metal ions rather weakly (12). The solutions of XPAzf saturated with non-redox metal ions, such as Zn(II) or Cd(II) can be stored at 4°C for a day or two, but solutions of XPAzf with redox metals, such as Ni(II) or Co(II), cannot (5, 6). The amount of water and salt in the solid peptide may differ from sample to sample, depending on freeze-drying conditions. Therefore, the actual XPAzf concentration should be measured for each stock solution.

2. The Ellman's reagent [DTNB, 5,5′-dithiobis (2-nitrobenzoic acid)] assay for the thiol contents is recommended for the freshly made and stored stock solutions of XPAzf to obtain its actual concentration (corresponding to 0.25 × thiol amount from the assay). Make sure the Ellman's reagent for the thiol contents assay

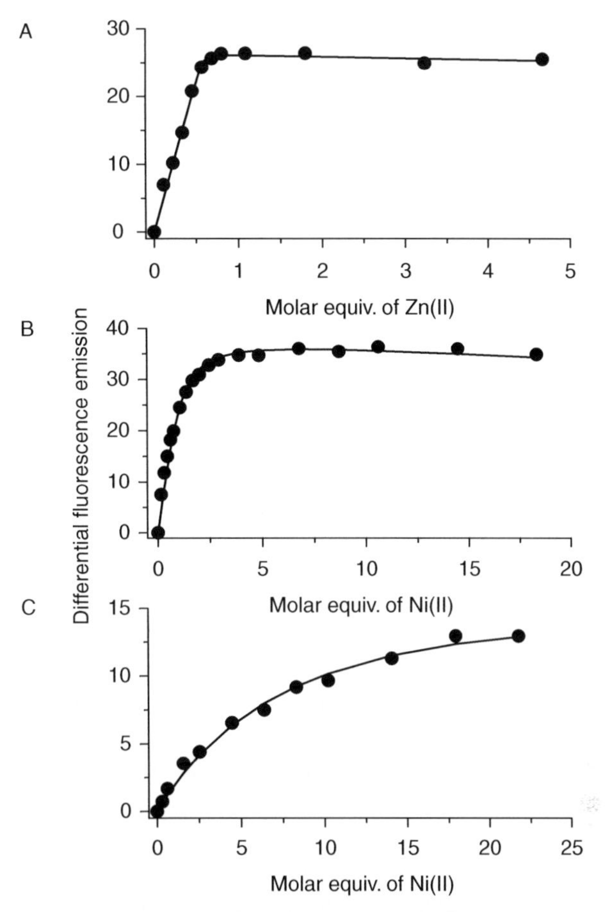

Fig. 25.3. Binding curves derived from data presented in **Fig. 25.2**, using emission at 305 nm (•), along with fits to apparent binding constants (*lines*); (**A**) Zn(II) titration of XPAzf, (**B**) Ni(II) titration of XPAzf, and (**C**) Ni(II) titration of Zn(II)–XPAzf (5).

is fresh. This reagent gets oxidized upon standing, even stored dry in a refrigerator. The usage of partially oxidized reagent may result in an overestimation of the reduced thiol concentration, *see* (13) and (14) for further details.

3. Prepare all solutions for XPAzf reactions using high-quality metal-free water, for example bidistilled water tested analytically for Zn(II) and other metals assayed, or analysis grade commercial water (e.g., from Baker). The requirements for water used to make HPLC buffers are less stringent, because their low pH prevents thiol oxidation. Regular bidistilled water is satisfactory for this purpose.
4. Minimize the contamination with zinc by using gloves especially in contact with the pipette tips or reaction tubes.
5. Check the zinc saturation of the peptide by titrating an aliquot of the peptide with PAR and compare the inten-

sity of the yellow complex with standard zinc solutions (2–6 mM $ZnCl_2$, dilute from the zinc buffer). If necessary, add more zinc buffer to the peptide.

6. Prepare a master mix of 53.1 μL buffer and 13.4 μL peptide. Add 0.7 μL of test substance or hydrogen peroxide solution at the wall of the reaction tube and centrifuge down. After incubation add 2.8 μL of PAR and put three aliquots of 20 μL each in a 384-well plate and measure the absorbance at 492 nm.
7. Be aware that PAR also reacts with other transition metals like copper or iron resulting in similar absorption spectra. These metals are unsuitable for the investigation on XPAzf in this assay and perturb zinc detection.
8. The metal complexes are fully dissociated in the acidic HPLC mobile phases and do not interfere with HPLC analysis.
9. Zn(II)–XPAzf is stable against air oxidation for at least 48 h. However, before each set of experiments analyze 100 μL of Zn(II)–XPAzf in water as a control.
10. Two oxidation products occur both at shorter retention times compared to XPAzf. MALDI-TOF measurements demonstrated that both the major product and the secondary product at shorter retention times had the same mass of 4,440 Da, which is two mass units less than the substrate XPAzf, thus indicating a formation of one intramolecular disulfide bridge (5, 6).
11. Volatile buffers are required by electrospray mass spectrometry, and even these need to be diluted, because higher buffer concentrations result in the loss of ESI-MS signal. Ammonium formate is an alternative ESI-MS buffer only at lower pH, as its buffering capacity at pH 7.4 is very poor. Inorganic ions (chloride, phosphate, etc.) should be strictly avoided.
12. Low molecular weight reagents, such as glutathione and its derivatives and arsenicals should be used only up to 100 μM, since higher concentrations of these ionic substances cause significant signal quenching.
13. Solutions and materials used in biochemistry and molecular biology are notoriously polluted with Zn(II); deionized or distilled water often contains micromolar levels of Zn(II) due to leaks from glass or plastic containers. Leaks of other metals, such as cadmium, may also occur. Residual Zn(II) from samples ran previously may be retained in tubes and capillaries of mass spectrometers. We often observed the appearance of substantial amounts of the Zn(II)–XPAzf

complex (30–50% in extreme cases) upon the injection of 10 μM solutions of apo-XPAzf. Therefore, the thorough purging of the mass spectrometer prior to XPAzf experiments is required. One way to do this is to inject a sample of apopeptide to serve as zinc sponge, prior to actual experiments.

14. The reduced state of XPAzf needs to be maintained strictly throughout the titrations.
15. Never prepare 1 mM stock solutions of metals in phosphate buffers, because colloidal precipitates of metal phosphates form slowly, resulting in uncertain metal ion dosage into the samples.

Acknowledgments

This research was supported by the Deutsche Forschungsgemeinschaft, grants no. HA 2372/3-4, SCHW 903/3-2, and by the Alexander von Humboldt-Stiftung subsistence grant to WB.

References

1. Hartwig, A. (2001) Zinc finger proteins as potential targets for toxic metal ions: differential effects on structure and function. *Antioxid Redox Signal.* **3**, 625–634.
2. Mackay, J.P. and Crossley, M. (1998) Zinc fingers are sticking together. *Trends Biochem Sci.* **23**, 1–4.
3. Witkiewicz-Kucharczyk, A. and Bal, W. (2006) Damage of zinc fingers in DNA repair proteins, a novel molecular mechanism in carcinogenesis. *Toxicol Lett.* **162**, 29–42.
4. Schwerdtle, T., Walter, I., and Hartwig, A. (2003) Arsenite and its biomethylated metabolites interfere with the formation and repair of stable BPDE-induced DNA adducts in human cells and impair XPAzf and Fpg. *DNA Repair (Amst).* **2**, 1449–1463.
5. Bal, W., Schwerdtle, T., and Hartwig, A. (2003) Mechanism of nickel assault on the zinc finger of DNA repair protein XPA. *Chem Res Toxicol.* **16**, 242–248.
6. Kopera, E., Schwerdtle, T., Hartwig, A., and Bal, W. (2004) Co(II) and Cd(II) substitute for Zn(II) in the zinc finger derived from the DNA repair protein XPA, demonstrating a variety of potential mechanisms of toxicity. *Chem Res Toxicol.* **17**, 1452–1458.
7. Piatek, K., Schwerdtle, T., Hartwig, A., and Bal, W. (2008) Monomethylarsonous acid destroys a tetrathiolate zinc finger much more efficiently than inorganic arsenite: mechanistic considerations and consequences for DNA repair inhibition. *Chem Res Toxicol.* **21**, 600–606.
8. Smirnova, J., Zhukova, L., Witkiewicz-Kucharczyk, A., Kopera, E., Oledzki, J., Wyslouch-Cieszynska, A., Palumaa, P., Hartwig, A., and Bal, W. (2007) Quantitative electrospray ionization mass spectrometry of zinc finger oxidation: the reaction of XPA zinc finger with H_2O_2. *Anal Biochem.* **369**, 226–231.
9. Smirnova, J., Zhukova, L., Witkiewicz-Kucharczyk, A., Kopera, E., Oledzki, J., Wyslouch-Cieszynska, A., Palumaa, P., Hartwig, A., and Bal, W. (2008) Reaction of the XPA zinc finger with S-nitrosoglutathione. *Chem Res Toxicol.* **21**, 386–392.
10. Stamler, J.S., Simon, D.I., Osborne, J.A., Mullins, M.E., Jaraki, O., Michel, T., Singel,

D.J., and Loscalzo, J. (1992) S-nitrosylation of proteins with nitric oxide: synthesis and characterization of biologically active compounds. *Proc Natl Acad Sci USA*. **89**, 444–448.

11. Han, J.C. and Han, G.Y. (1994) A procedure for quantitative determination of tris (2-carboxyethyl)phosphine, an odorless reducing agent more stable and effective than dithiothreitol. *Anal Biochem*. **220**, 5–10.
12. Krezel, A., Latajka, R., Bujacz, G.D., and Bal, W. (2003) Coordination properties of tris(2-carboxyethyl)phosphine, a newly introduced thiol reductant, and its oxide. *Inorg Chem*. **42**, 1994–2003.

Chapter 26

Preparation and Zinc-Binding Properties of Multi-Fingered Zinc-Sensing Domains

John H. Laity and Linda S. Feng

Abstract

Cys_2His_2 zinc finger proteins (ZFPs) adopt a highly conserved ββα-fold, which is stabilized by tetrahedral coordination of a central zinc ion by two cysteine and two histidine residues. Although the function of most zinc fingers is sequence-specific DNA binding, other roles such as zinc sensing have been identified. Here, methods are described to produce micromole quantities of zinc-sensing zinc finger domains from the Zap1 (two zinc fingers) and MTF-1 (six zinc fingers) transcription factors. Procedures are outlined to isolate recombinant zinc finger proteins from a bacterial expression system that generates both soluble intracellular (Zap1) and insoluble inclusion body (MTF-1) forms. Isolated proteins are reduced and subsequently HPLC purified at low pH and lyophilized, which generates proteins that are ideal for zinc-binding studies or metal substitution studies and stable during long-term storage. NMR and calorimetric methods are described to measure relative and individual zinc ion affinities in multi-fingered proteins, which is an essential step in unraveling the zinc-sensing mechanism of MTF-1 and Zap1.

Key words: Zinc fingers, zinc sensing, metal-binding, expression, HPLC purification, NMR, calorimetry, apo-protein.

1. Introduction

The divalent zinc ion is the second most abundant trace element in humans after iron. It is an essential nutrient, but also toxic if accumulated to excess. Organisms ranging from bacteria to mammals maintain intracellular zinc levels within a functional range through homeostatic mechanisms that can be regulated at the level of gene expression. Zn(II) homeostasis is regulated in part by two zinc finger (ZF) containing transcription factors, Zap1 (in yeast) (1) and MTF-1 (2) (mammals and insects). MTF-1

J.P. Mackay, D.J. Segal (eds.), *Engineered Zinc Finger Proteins*, Methods in Molecular Biology 649,
DOI 10.1007/978-1-60761-753-2_26,

and Zap1 sense and respond to intracellular levels of accessible or "labile" zinc at the level of coordination to a subset of Cys_2His_2 zinc fingers in each protein, and these events ultimately influence the expression of homeostatic proteins (3–5). Interestingly, MTF-1 activates the transcription of genes encoding sequestration and efflux transport (ZnT1) and sequestration (metallothionein) proteins in response to *excess* zinc, while Zap1 activates the transcription of genes encoding influx transporters in response to *low* intracellular zinc levels.

The classic Cys_2His_2 ZF forms a compact globular structure consisting of a small anti-parallel β-hairpin turn structure and a single helix that is stabilized by the central zinc ion and conserved hydrophobic residues (6, 7). The β-hairpin and helix structures donate two cysteine and two histidine ligands, respectively, for tetrahedral coordination to zinc. Studies of zinc finger proteins (ZFPs) have demonstrated that they are largely unstructured in the absence of zinc (8), which couples zinc-sensing functions to protein structure formation in MTF-1 and Zap1. While the canonical role ascribed to Cys_2His_2 ZFPs involves the regulation of cellular gene expression via highly specific interactions between zinc finger motifs and DNA (9), the non-canonical zinc-sensing roles carried out by MTF-1 and Zap1 ZFs has inspired the development of complementary nuclear magnetic resonance (NMR) and calorimetric zinc-binding methods to tease apart the unique mechanisms whereby zinc-sensing fingers respond at the level of zinc:polypeptide interactions. Here we describe detailed methodologies for an NMR-based approach to assess cooperative effects of larger multi-fingered domains with significant disparity in individual finger affinities such as MTF-1 (10). We also present isothermal titration calorimetry (ITC) approaches here, which provide more quantitative zinc stability measurements that report apparent (average) stabilities for zinc binding in multi-fingered proteins unless the affinities differ by several orders of magnitude. Cooperative finger–finger effects on the stabilization of zinc-binding affinities and relative finger affinities can be determined from ITC measurements of individual fingers in conjunction with complementary measurements of the intact multi-fingered domain (4).

The first reported zinc finger purification of the nine zinc finger TFIIIA transcription factor produced large quantities of protein, but the protein was isolated as an intact 7S particle in complex with the 5S RNA transcript from immature *Xenopus laevis* oocytes (11). Over the past 20+ years, numerous approaches have been reported to purify single- and multi-fingered ZFPs. Here we focus on methods specifically designed to obtain large quantities of recombinant zinc finger peptides suitable for isotopic enrichment, which are in a stable apo- and thiol-reduced form. These unstructured precursors are ideal for measurements of relative and quantitative individual zinc affinities using NMR and ITC and can also serve as targets for metal substitution studies. ZF peptides are prepared from bacterial expression of both soluble (Zap1) and insoluble (MTF-1) proteins. Reverse-phase HPLC is a highly advantageous approach, as the reduced apo-peptides do not bind zinc during the low-pH purification. Proteins are directly lyophilized without a dialysis step, stable for long-term storage, readily buffer matched with the ligand (zinc) buffer, and highly adaptable to small sample volumes, which greatly facilitates recording high-quality ITC measurements and NMR spectra.

2. Materials

All cell culture and laboratory reagent water are purified from a Milli-Q Synthesis A10 TOC (Millipore).

2.1. DNA Plasmid Transformation

1. Bacterial expression vectors: pET-21a(+) (Novagen) containing the wild-type zinc-sensing ZF domains of MTF-1 (pMZF1–6) (10, 12) and Zap1 (p21a_zf1-2_{WT}) (4). *See* **Note 1** for additional MTF1- and Zap1-derived zinc finger plasmids. Plasmids are available upon request by the authors.
2. Stock of frozen storage buffer (FSB)-prepared (13) frozen competent *Escherichia coli* BL21(DE3) cells stored at –80°C up to 1 year (*see* **Note 2**).
3. LB medium and LB agar plates supplemented with ampicillin (*see* **Note 3**).
4. Modified SOC media (13) consisting of LB broth supplemented with 20 mM glucose (49 mL of LB broth with 1 mL of 20% glucose).

2.2. Expression and Isolation of Recombinant ZFPs

1. Microfluidizer (e.g., M-110L, Microfluidics Corporation) and sonicator (e.g., Sonic Dismembrator 550 with microtip, Fisher Scientific).
2. Polyacrylamide gel electrophoresis system (e.g., Mini-Protean3, Bio-Rad).
3. Baffled culture flasks: Analytical 50- to 100-mL cultures use 250 mL size and 1 L cultures use 2,800-mL flasks.
4. 5× M9 salt stock solution: 6.4% (w/v) Na_2HPO_4, 1.5% (w/v) KH_2PO_4, 0.25% (w/v) NaCl. Autoclave (30 min, liquid cycle).
5. M9 minimal medium (13): 200 mL of 5× M9 salt stock solution added to 800 mL of autoclaved H_2O in a 2,800 mL flask. For standard "unlabeled" cultures, add 8 mL

(0.25 g/mL final concentration) of NH_4Cl and 10 mL of glucose (20% w/v final concentration) solutions, 0.2-μm filter sterilized (both solutions can be stored up to a year at room temperature). Uniform (U) stable ^{15}N isotopic enrichment of the media is used to produce protein for NMR studies by substituting 8 mL of freshly prepared U-[$^{15}NH_4Cl$] (0.25 g/mL final concentration) in place of unlabeled salt (*see* **Section 3.5**).

6. Trace element solutions (14, 15): Trace elements are made as 1,000× solutions that are 0.2-μm filter sterilized and frozen at –20°C in 2- or 4-mL aliquots. These solutions are added to the M9 minimal medium just prior to cell growth.
   a. 1,000× trace element master mix: 46 μM H_3BO_3, 0.17 μM $CoCl_2$, 0.8 μM $CuSO_4$, 1 mM $MgCl_2$, 0.9 μM $MnCl_2$, and 2 μM $ZnCl_2$.
   b. 1,000× $FeCl_3$ solution: 1 mM $FeCl_3$ made separately from the trace element master mix because it precipitates in the presence of the other trace elements.
   c. 1,000× $MgSO_4$ solution: 1 M $MgSO_4$, 0.2-μm filter sterilized and stored at room temperature.
   d. 10,000× $CaCl_2$ solution: 1 M $CaCl_2$, 0.2-μm filter sterilized, and stored at room temperature. Add 100 μL per liter of cell culture.
7. 200 mg/mL ampicillin: 0.2-μm filter sterilized into autoclaved, sealed tubes, and stored at –20°C in 1-mL aliquots.
8. 0.1 M isopropyl β-D-galactopyranoside (IPTG), 0.2 μm filter sterilized and freshly made the day of protein expression.
9. 100 mM phenylmethylsulphonyl fluoride (PMSF), prepared freshly in ethanol. Store PMSF powder desiccated at 4°C.
10. 50 mM dithiothreitol (DTT), added fresh to the degassed lysis buffer described below. Store DTT powder desiccated at –20°C.
11. Lysis buffer: Freshly helium degassed 50 mM 3-(*N*-morpholino)-propanesulfonic acid (MOPS), pH 7.0, 300 mM NaCl, and 0.5 % Triton X-100 (used in preparative samples only). After degassing and immediately before lysis, add DTT to 50 mM (*see* **Note 4**) and PMSF to 1 mM.
12. Polyacrylamide gel electrophoreisis (PAGE):
    a. Freshly cast discontinuous denaturing acrylamide gels are used, consisting of a 17% acrylamide/bis lower separating gel prepared in 0.375 M Tris–HCl, pH 8.8 (~4 mL poured per gel), and an upper 4% acrylamide/

bis stacking gel in 0.125 M Tris–HCl, pH 6.8 (1–1.5 mL per gel). Stock solutions for gels are as follows: 30% acrylamide/bis solution 29:1 (3.3% crosslinker), 1.5 M Tris–HCl, pH 8.8, and 0.5 M Tris–HCl, pH 6.8. Polymerization catalysts: freshly prepared 10% (w/v) ammonium persulfate (120 and 200 μL for separating gel and stacking gels, respectively) and tetramethylethylenediamine (20 and 24 μL for separating and stacking gels, respectively). Unused gels can be wrapped in plastic wrap and stored at 4°C for up to 1 week.

b. Running buffer: 25 mM Tris, 0.192 M glycine, 0.01% SDS.

c. 5× SDS loading dye: 0.01% (w/v) bromophenol blue, 20% SDS, 25% glycerol, 62.5 mM Tris–HCl, pH 6.8, and 5% (v/v) β-mercaptoethanol (BME) freshly added.

d. Coomassie blue gel staining solution: 0.25% (w/v) brilliant blue, 50% (v/v) methanol, 10% (v/v) acetic acid.

e. Gel destaining solution: 50% (v/v) methanol, 10% (v/v) acetic acid.

13. Wash buffer: Freshly helium degassed 25 mM MOPS, pH 7.0, 300 mM NaCl. Add DTT to 10 mM after degassing.

14. Solubilization buffer for insoluble recombinant proteins: 7 M guanidinium chloride, 25 mM MOPS, pH 7.0, 200 mM DTT. The amount of guanidine will be a significant fraction of the volume. To make 10 mL, resuspend 6.69 g of guanidine hydrochloride in 5 mL of helium degassed 25 mM MOPS, pH 7.0. Once it is dissolved, the buffer volume will be 10 mL and DTT can be added. Helium degassing prior to DTT addition is important because the DTT will not be quenched reducing the oxygen in the MOPS buffer, making it available to reduce the insoluble recombinant proteins (*see* **Note 4**).

2.3. Recombinant Apo-ZF Peptide HPLC Purification

1. High performance liquid chromatography (HPLC): Waters 626 LC (analytical/semi-prep) system with Waters Symmetry300 C18 or C4 (3.5 μm pore size) 4.6 mm × 75 mm columns, and Waters DeltaPrep4000 LC preparative system with C4 or C18 (300 Å pore size) 25 mm × 100 mm Waters Delta-Pak columns.

2. Reverse-phase HPLC buffers: stationary phase (buffer A) is 0.1% (v/v) trifluoroacetic acid, and mobile phase (buffer B) is 95% (v/v) acetonitrile, 0.1% (v/v) trifluoroacetic acid. Acetonitrile is HPLC grade.

3. Lyophilizer (e.g., Freezemobile 12-EL l from VirTis).

4. Computer-controlled dual-beam spectrophotometer (e.g., Cary100 Bio UV s from Varian Inc. with WinUV Scan v3.00 (182) software package). Also used in **Sections 3.4** and **3.5**.

2.4. Zinc Stability Measurements by Isothermal Titration Calorimetry

1. Microcal (Northhampton, MA) VP-ITC with VP2000 data acquisition software package (v1.4.12) and modified Origin7 SR4 (v7.0552-B552) analysis package.
2. ITC buffer: 200 mM MES (SigmaUltra grade, pH 6.9 matched buffers for ZF peptide (sample cell) and zinc ion solution (auto-pipette) ITC). All ITC reagents and buffers are freshly vacuum and helium degassed for 5 min each.
3. 0.1 M tri-(2-carboxyethyl)-phosphine hydrochloride (TCEP), freshly prepared in helium degassed 200 mM MES.
4. 26.7 mM $ZnSO_4$ stock solution: zinc sulfate heptahydrate (Sigma Ultra grade) as determined by atomic absorption (Varian AA400). Also used in **Section 3.5**.

2.5. NMR Measurements of Relative Zinc Affinities in MTF-1 Multi-Fingered Protein

1. High-resolution NMR spectrometer with cryogenically cooled ^1H, ^{15}N, ^{13}C triple resonance probe [e.g., Varian Inova 14.1 T (600 MHz ^1H)].
2. UNIX computer workstation running NMRPipe (16) and NmrView 5 (17) spectral processing and analysis packages, respectively.
3. NMR sample buffer: 20 mM MES-d_{13}, pH 6.9, 5% D_2O, 1 mM NaN_3, 0.2 mM 2,2-dimethyl-2-silapentane-5-sulfonic acid (DSS), with β-mercaptoethanol added after helium degassing to 2.0 mM. Sample buffer (without BME) can be stored in 1-mL aliquots at –80°C for up to 3 years.
4. 1.0 M NaOH and 1.0 M HCl solutions to adjust pH.
5. 5-mm diameter 535 PP NMR tube and cap (Wilmad-LabGlass), long-tip NMR pipette (e.g., Wilmad, catalog # 803A).

3. Methods

3.1. MTF-1 pMZF1–6 and Zap1 p21a_zf1-2$_{WT}$ Plasmid Transformation

1. Add 1 μL of freshly diluted 1.4 M BME (1:10 (v/v) dilution from pure into water) to a 100-μL aliquot of competent *E. coli* BL21(DE3) cells freshly thawed on ice in a 1.5-mL microcentrifuge tube. Flick tube with finger to mix and incubate for 5 min on ice.
2. Add ~50–500 ng (1–5 μL) of pMZF1–6 or p21a_zf1-2$_{WT}$ expression vector into cells. Gently mix with micropipette.

3. Incubate on ice for 0.5 h.
4. Heat shock cells at 37°C for 45 s.
5. Place cells on ice again for 2 min.
6. Add 400 μL of SOC, mix by inverting tube, and place on rotary shaker at 250 rpm for 1 h.
7. Depending on calibrated transformation efficiency of competent cells (*see* **Note 5**), plate 100–250 μL cells on LB agar plate supplemented with 200 μL/mL ampicillin using a rotary turntable and a sterile bent glass rod.
8. Invert plate and incubate at 37°C to form colonies (12–16 h).

3.2. Expression and Isolation of Recombinant MTF-1 and Zap1 Zinc-Sensing Domains

3.2.1. Pilot Expression Studies

Trials should always be carried out to determine *E. coli* BL21(DE3) expression yields and recombinant protein partitioning (soluble or insoluble) after transformations of newly created recombinant zinc finger DNA plasmids. The two-zinc finger Zap1 protein is isolated as a soluble intracellular protein (*see* **Fig. 26.1**); while the larger six-zinc finger MTF-1 protein is recovered from solubilized inclusion bodies (*see* **Note 6**). Pilot procedures applicable to most multi-domain ZFPs are outlined below.

1. Inoculate a single colony transformed with expression plasmid into 3–4 mL of LB medium supplemented with 200 μg/mL ampicillin in a sterile 15-mL conical tube. Place the tightly capped tube in a rotary 37°C incubator at 250 rpm overnight (8–16 h).
2. Gently pellet 2 mL of the inoculant at 5,000×*g* (e.g., 4,000 rpm in a Sorvall BIOshield rotor) for 10 min.
3. Decant the LB medium and use the cell pellet to inoculate 100 mL of M9 minimal medium supplemented with trace element solutions, NH_4Cl, and glucose solutions and 50 μg/mL ampicillin in a 250-mL baffled flask.
4. Incubate 100 mL of culture at 37°C and 250 rpm until the absorbance at 600 nm (OD_{600}) reaches 0.6–0.8 (typically 5–6 h, *see* **Note 6**), at which time IPTG is added to a final concentration of 1.0 mM. Prior to IPTG induction, remove 10 mL of culture and store on ice or, alternatively, pellet cultures, remove media, and freeze cell pellet at –20°C.
5. Grow cells overnight and take 10-mL samples at 1 h, 2 h, 3 h, 6 h, and overnight post-induction, store samples as

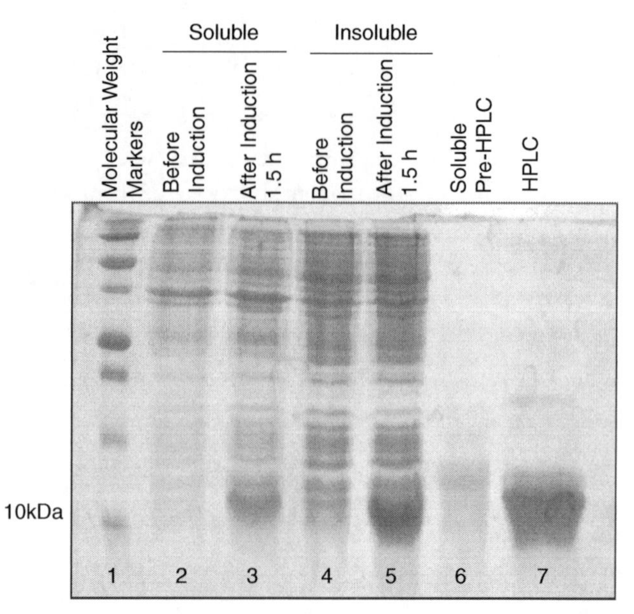

Fig. 26.1. SDS-PAGE analysis of soluble and insoluble fractions of *E. coli* BL21(DE3) transformed with p21a_zf1-2_{CA3} and HPLC purified. Zap1zf1-2_{CA3} is expressed predominantly in the soluble fraction. Each 10-μL lysate loaded in *lanes* 2–5 (from *left* to *right*) is derived from 200 μL of original cell culture. Soluble cell lysate adjusted to initial HPLC conditions (*lane* 6) was injected on C18 analytical column and the peak eluting at 38% buffer B was collected and lyophilized (*lane* 7).

described earlier. Process pre-induction and post-induction samples as described in steps 1–6 from **Section 3.2.2**. Measure the OD_{600} of the overnight culture to determine the point at which maximum cell density is obtained.

3.2.2. Assessment of ZF Expression Levels and Intracellular Partitioning

1. Gently pellet 10 mL of all cultures from **Section 3.2.1** at 4°C at 5,000×*g* (e.g., 4,000 rpm in a Sorvall BIOshield rotor) for 10 min.
2. Decant media and invert tube to drain briefly on a paper towel. Give tube a final tap to expel all media.
3. Resuspend cell pellet in 500 μL of lysis buffer.
4. Lyse resuspended cells by sonication: 3 × 10 pulses with a microtip sonicator at medium power setting, allowing the sample to ice and cool down between 10-pulse runs.
5. Separate soluble and insoluble fractions of lysate using a microcentrifuge at maximum speed for 5 min at 20,817×*g* (e.g., 14,000 rpm in an Eppendorf 5417C).
6. Transfer the soluble fraction to a clean microcentrifuge tube and resuspend the colloidal insoluble pellet in 500 μL of lysis buffer. Soluble and insoluble lysates can be stored at –20°C for up to 2 weeks.

7. Analyze the soluble and insoluble cell fractions by PAGE. Combine 10 μL of soluble or insoluble lysate with 5 μL of 5× SDS dye with BME and denature by heating to 95°C for 5 min. Load all samples on a 17% gel and run for 1 h. Typical power settings are: 200 V, 300 mA, and 300 W. Pre-induction control samples facilitate identification of recombinant proteins in post-induction samples. After electrophoresis, carefully extract gel from plates and place in Comassie blue stain bath (1 h to overnight). Transfer gel to destaining bath. Carrying out staining/destaining at 37°C and gentle shaking at 50 rpm speed up both processes considerably. Photograph or dry the gel for archival purposes (the gel can be wrapped and stored at 4°C for 1 week).
8. The results of the PAGE analysis (*see* **Fig. 26.1**) will clearly indicate whether expression yields are sufficient and whether the recombinant protein is predominantly expressed in soluble or insoluble forms (*see* **Note** 7). Even modest expression levels corresponding to 1–2 mg protein per liter will be plainly visible to the naked eye on the Coomassie stained gel.

3.2.3. Preparative Expression of MTF-1 and Zap1 Zinc-Sensing Domains

1. Inoculate a single colony into 20 mL of LB medium supplemented with 200 μg/mL ampicillin in a sterile 50-mL conical tube. Place tightly capped tube in a rotary incubator (250 rpm and 37°C) overnight (8–16 h).
2. Gently pellet all 20 mL of the inoculant (5,000×*g*, 10 min).
3. Decant the medium and use the cell pellet to inoculate 1 L of M9 minimal medium supplemented with trace element solutions and 50 μg/mL ampicillin in a 2.8-L baffled flask.
4. Incubate the 1-L culture at 37°C and 250 rpm until the OD_{600} reaches 0.6–0.8 (typically 5–6 h), at which time IPTG is added to a final concentration of 1.0 mM. For the six-zinc finger MTF-1 protein, overnight expression yields the best production of inclusion body protein, while the two-ZF Zap1 protein expression levels are optimal 1–2 h post-induction (*see* **Note 6**).

3.2.4. Harvest and Lysis of E. coli BL21(DE3) Cultures Expressing MTF-1 and Zap1 Proteins

1. Pellet the cells at the optimal post-induction time by centrifugation at 6,084×*g* using a refrigerated floor centrifuge (e.g., 6,000 rpm in a Sorvall SLA-3000 rotor) for 20 min at 4°C.
2. Decant medium and invert bottles on paper towels. At this point, cells can be lysed immediately or frozen in centrifuge bottles at –80°C and stored for up to a week.
3. Resuspend the cell pellets in a 1/20th volume of cell culture with lysis buffer chilled to 4°C. The presence of 50 mM

DTT at pH 7.0 will ensure soluble protein is completely reduced prior to low pH purification (*see* **Section 3.3.1**).

4. Lyse the cells with microfluidizer or cell sonication using large tip. Hold the lysate on ice. If lysing cells by sonication, lyse by running six rounds of a 3-min process time of 1 s on and 1 s off at medium power setting, allowing the sample to ice and cool down between 3-min process time runs.

3.2.5. Isolation of Soluble Zap1 two-ZFP

1. Adjust the lysate to initial conditions for reverse-phase HPLC by adding buffers A, B, and an additional 0.5% TFA until the pH matches the pH of the HPLC buffer reservoirs (pH 1.4–1.6, *see* **Section 3.3.2**). The solution will become cloudy, as some proteins will precipitate, although >90% of the Zap1 ZFP should remain soluble.
2. Remove 0.5 mL of lysis solution for gel and analytical HPLC screening, separate soluble from insoluble by maximum speed centrifugation for 5 min (e.g., 14,000 rpm on an Eppendorf 5417C, or 20,817×*g*), and filter supernatant through a 0.2-micron syringe filter. Analyze this sample using PAGE as in **Section 3.2.2** to confirm that the ZFP is predominantly in the soluble fraction. If the SDS sample buffer turns yellow from TFA (low pH) add 1 μL of 1.5 M Tris–HCl, pH 8.8 (the buffer should turn blue again).
3. Centrifuge the remaining lysate at 38,724×*g* (e.g., 18,000 rpm on a Sorvall SS-34 rotor) for 30 min.
4. Filter the soluble fraction through a 0.2-micron bottle top filter into clean bottle.
5. Store up to 24 h at 4°C until HPLC purification (*see* **Section 3.3.1**).

3.2.6. Isolation of MTF-1 Six-ZF Inclusion Body Protein

1. Take 0.5 mL of lysate for gel analysis, separating soluble from insoluble. Process this sample using PAGE as in **Section 3.2.2** to confirm that the ZFP is predominantly localized in inclusion bodies.
2. Centrifuge microfluidized lysate at 38,724×*g* for 30 min.
3. In the unlikely event that a significant quantity of soluble recombinant protein is evident from PAGE, decant the supernatant and store in a clean bottle at 4°C.
4. Wash the insoluble fraction with wash buffer by resuspension with a pipette.
5. Centrifuge again at 38,724×*g* for 30 min.
6. Decant the liquid. Repeat the wash and centrifugations steps.
7. Decant the second wash. A fuzzy white border around pellet indicates good expression.

8. Add 1–1.5 mL of solubilization buffer per liter of cell culture. Break up pellet with a spatula and alternate between agitation on a shaker and incubating on ice. Allow insoluble pellets to solubilize overnight (up to 3 days) at 4°C.
9. The solubilized protein will be a sticky, dark brown color. Centrifuge at 38,724×*g* for 1 h at 4°C to separate solubilized protein liquid from residual solid cell debris.
10. Transfer the solubilized liquid to a 15-mL conical tube. The residual solid may be green due to copper oxidation, which has no relevance to recombinant protein expression levels.
11. Measure the volume of the solubilized protein and dilute it 150-fold in HPLC buffer B (15% (v/v) acetonitrile with an additional 0.5% (v/v) TFA). Dilute by slowly adding solubilized protein to HPLC buffer in a beaker with stirring. Allow mixture to stir for 10–15 min.
12. Centrifuge the diluted protein at 16,900×*g* (e.g., 10,000 rpm in a Sorvall SLA-3000 rotor) for 1 h at 4°C.
13. Carefully decant the supernatant into a clean beaker immediately, as the centrifuged pellet of remaining cell debris can dislodge and clog up filter membrane in the next step.
14. Filter the solubilized protein solution through a 0.2-micron PVDF membrane bottle top filter and store at 4°C for HPLC.

3.3. HPLC Purification of Recombinant Apo-MTF-1 and Zap1 ZF Domains

3.3.1. Preparative HPLC Purification

Prior to the labor-intensive and costly preparative procedure outlined below, it is highly advisable to assess the quality and yield of recombinant protein preparations using analytical HPLC approaches (**Section 3.3.2**) with 100-500 μL samples of Zap1 (**Section 3.2.5**) and MTF-1 (**Section 3.2.6**) ZFPs.

1. Equilibrate a 25×100 mm preparative HPLC column with 5–7 column volumes of 20% buffer B (80% buffer A) using a flow rate of 30 mL/min. Optimal separation for MTF-1 and Zap1 recombinant proteins is achieved using C4 and C18 columns, respectively.
2. Load the protein solution from Zap1 (**Section 3.2.5**) or MTF-1 (**Section 3.2.6**) under equilibration conditions (20% buffer B, 30 mL/min). Monitor 220 and 280 nm absorbances simultaneously throughout purification (*see* **Note 9**). Collect the flow-through in a clean bottle until baseline absorbances are back down to preloading values.

3. Run a linear 20–50% buffer B gradient at 2% increase per min (*see* **Fig. 26.2b**). Recombinant proteins will elute at a different percentage of buffer B than is observed for analytical HPLC because of differences in flow rate, pump head volumes, and analytical/preparative column separation properties.
4. Collect any peaks eluting earlier than 30% with an absorbance greater than 0.1 in clean bottles or 50-mL centrifuge tubes, in case of premature elution of recombinant protein.
5. Start collecting all peaks with a fraction collector when the absorbance is 0.1 after 30% buffer B. The absorbance corresponding to the recombinant ZFP should be at least 10-fold larger than any other peaks in the range 30–70% buffer B. Note the buffer B percentage at peak maximum, as this

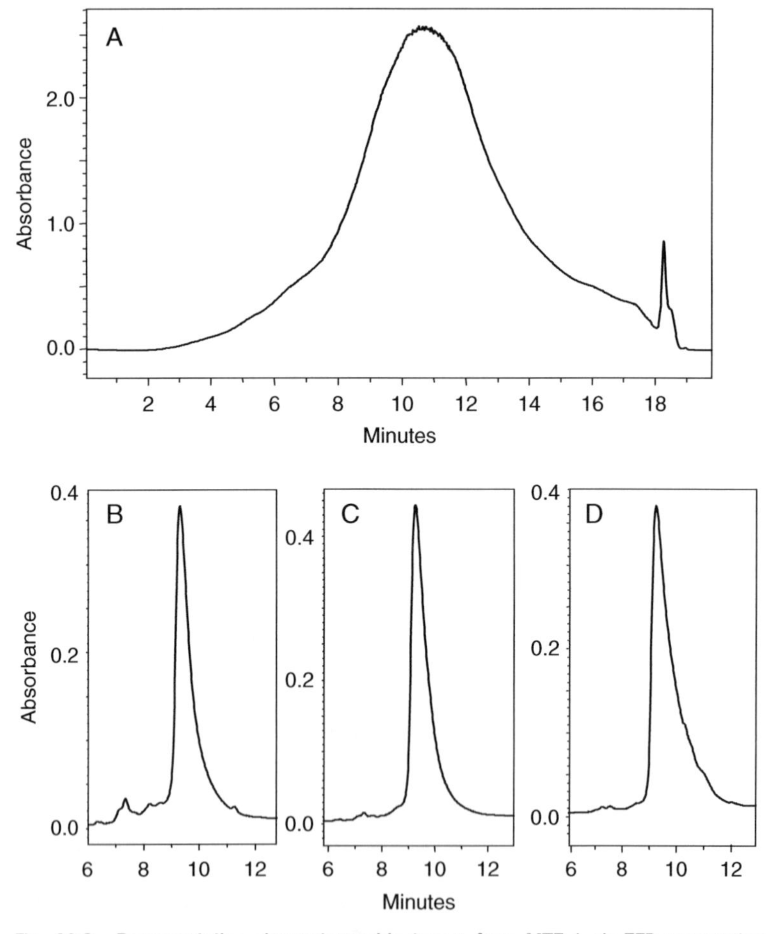

Fig. 26.2. Representative chromatographic traces from MTF-1 six-ZFP preparation monitored at 220 nm from (**a**) preparative HPLC purification and analytical HPLC screening of fractions corresponding to (**b**) ascending side, (**c**) maximum, and (**d**) descending side of the preparative peak from (**a**).

buffer composition will be used for a spectrophotometer "blank" during UV concentration determination of purified protein. Analyze fractions immediately (*see* **Section 3.3.2**) or store at 4°C for up to 48 h.

6. After reaching 50% buffer B or after the major peak has completely eluted, switch to 100% buffer B wash for 5–7 column volumes, followed by 5–7 column volumes of 100% pure acetonitrile (storage condition for C4 and C18 columns).

3.3.2. Analytical HPLC Analysis and Storage of Purified Apo-ZFPs

1. Equilibrate the analytical C4 or C18 column with 5 volumes of 20% buffer B at 1 mL/min.
2. Inject 20–100 μL of preparative fractions corresponding to the major peak and run the same linear 20–50% buffer B gradient at 2% increase per min at 1 mL/min, while monitoring at 220 and 280 nm. Representative ascending (*see* **Fig. 26.2B**), maximal (*see* **Fig. 26.2c**), and descending (*see* **Fig. 26.2d**) fractions from the prep peak (*see* **Fig. 26.2A**) are shown. Maximum absorbance ranges are typically 0.1–1 at 214 or 220 nm. Start with the fraction eluting at the preparative peak maxima as a first estimate of the most homogeneous recombinant protein sample. Then inject every other fraction on *each side* of the preparative peak center moving outward. Pool all pure fractions (>90–95% homogeneous).
3. Quantitate the pooled fractions of pure recombinant apo-ZFP and divide into aliquots sufficient for calorimetric (0.1–0.2 μmol, *see* **Section 3.4**) and NMR (0.35–0.45 μmol, *see* **Section 3.5**) zinc-binding studies. Mix buffer B with buffer A to the percentage where the preparative peak eluted (*see* step 5 of **Section 3.3.1**), for use as a buffer bank for UV spectrophotometry. Measure the absorbance of the pooled fractions at 280 nm. Calculate the concentration and quantity (concentration × volume) of purified ZFP using the Beer–Lambert's law:

$$C = A/\varepsilon L,$$

where C is calculated concentration, A is absorbance, and ε is the extinction coefficient at 280 nm (for MTF^{-1} six-ZFP, $\varepsilon = 11{,}180\ M^{-1}cm^{-1}$ (10, 12); for Zap1 two-ZFP, $\varepsilon = 11{,}375\ M^{-1}cm^{-1}$ (4)), and the cuvette pathlength L is 1 cm (*see* **Note 10**). Yields are typically 0.20–0.35 μmol of >95% pure ZFP per liter of cell culture.

4. Once the aliquot volumes have been calculated, transfer to 15-mL or 50-mL conical tubes (volume should be <60% of tube capacity).
5. Punch at least five holes in the cap of each tube with a needle and freeze at –80°C or flash freeze in liquid nitrogen.

6. Lyophilize frozen samples, which may take 2–3 days. Samples are thoroughly lyophilized when shaking them makes no sound of ice rattling in tube (appearance is like white cotton candy). If lyophilization is not complete, quickly re-freeze if necessary and put the sample back on the lyophilizer.
7. Pass argon or nitrogen gas at a *very low* flow rate through a hypodermic needle with a vent for a few min before sealing the cap with several layers of parafilm. Argon exchange displaces oxygen in the tube and creates a dry, stable, non-oxidative environment for long-term storage of the purified apo-ZFPs. *Warning*: a flow rate that is too fast will disperse the delicate lyophilized protein resulting in protein loss and contamination (rule of thumb; gas flow should be *barely detectable* at close range against your skin).
8. Store the protein with desiccant in jars or sealed freezer bags at –80°C. The protein is stable against oxidation and degradation for at least 6 months and usually several years.

3.4. ITC Zinc Stability Measurements of Zap1 Two-ZF Sensor Domain

1. Turn on the ITC instrument and initiate equilibration of the cell to 30°C.
2. Dissolve 0.1–0.2 μmol of lyophilized MTF-1 or Zap1 ZFP in 2.6 mL of ITC buffer with 500 μM TCEP.
3. Pellet any undissolved protein using a microcentrifuge at 20,817×*g* for 5 min, transferring the supernatant supernatant to a fresh tube.
4. Measure protein concentrations by UV absorbance (should be in the11–44 μM range, *see* **Note 10**).
5. Measure and adjust the pH of the protein solution to 6.90 ± 0.05 with 1 M NaOH and 1 M HCl solutions.
6. Rinse the calorimeter sample cell (1.43 mL) three times with ITC buffer.
7. Fill the sample cell with protein solution using the supplied clean and dry loading syringe. Store the remaining protein solution at 4°C.
8. Prepare 1 mL of a 0.5–2.0 mM $ZnSO_4$ solution from stock and adjust the pH of the diluted zinc solution to 6.90 ± 0.05. Choose a zinc concentration sufficient to achieve at least a 4:1 molar excess of zinc over protein for the Zap1 two-ZFP in 290 μL (*see* **Note 12**).
9. Load the ITC auto-pipette with 290 μL of zinc solution using the filling syringe and fill port, then execute a purge/refill command to eliminate any bubbles from pipette.

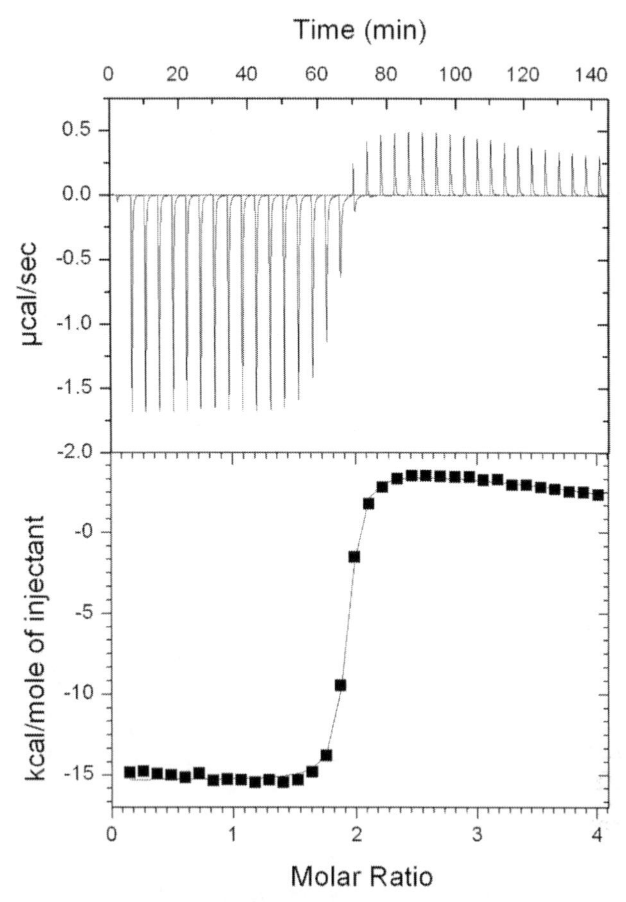

Fig. 26.3. ITC data recorded for the two-zinc finger Zap1 protein, where the *top panel* shows the differential heat change with time and the *bottom panel* shows integrated heats vs. molar ratio of Zn(II) to protein. Reproduced from Wang et al. (4) with permission from Elsevier Science.

10. Carefully position the titration syringe in the ITC sample cell.
11. Set up and start the ITC experiment using 2–5 µL zinc injections (*see* **Fig. 26.3** and **Note 13**).
12. Data analysis and fitting are done with the supplied vendor-specific Origin software package. Single-site model-fitting model is recommended to fit Zap1 two-ZF domain, which will result in an average stability constant (K_d) for the two ZF sites and an occupancy (N) of ~2 (4) (*see* **Note 14**).
13. Remove the protein solution from the sample cell using a clean dry loading syringe, microcentrifuge for 5 min at 20,817×g and accurately aliquot 100 µL of the supernatant for amino acid analysis.

3.5. NMR Measurement of Relative Zinc Affinities in the MTF-1 Six-ZF Domain

3.5.1. Zinc Titration and Recording of NMR Spectra

1. Resuspend 0.35–0.45 μM of lyophilized U-^{15}N-labeled MTF-1 six-ZFP in sufficient NMR sample buffer to reach a concentration of 0.30–0.35 mM (1.2–1.5 mL, *see* **Note 15**).
2. Adjust the pH to 6.90 ± 0.05 in 15-mL conical tube by 1-μL additions of 1 M NaOH or HCl as needed. Gently vortex the sample during additions.
3. Transfer sample to a 1.5-mL tube and centrifuge at 20,817×*g* for 5 min.
4. Measure the protein concentration by 2 successive UV absorbance measurements of protein solutions diluted 1:40 and 1:20, respectively, at 280 nm (*see* **Section 3.3.2**).
5. Transfer 550 μL of sample to a clean 5-mm diameter 535 PP NMR tube and seal with a cap. Retain the remaining sample at room temperature for later use.
6. Place the tube in the NMR instrument and record a ^{1}H-^{15}N-HSQC spectrum at 30°C after locking, tuning, shimming, and ^{1}H 90° pulse-width calibration with the following acquisition parameters (~90-min total ^{1}H-^{15}N-HSQC spectrum acquisition time) (10):
   a. Proton carrier frequency set to the H_2O frequency.
   b. $d1$ recycle delay = 1.2 s.
   c. 16 transients (nt) per t_1 increment.
   d. 512 complex points in the direct ^{1}H dimension.
   e. 128 complex t_1 increments (ni).
7. Eject the sample tube from the spectrometer and withdraw the sample from the NMR tube using a long-tip NMR pipette and combine with the remaining protein sample in a 15-mL conical tube.
8. Measure the total sample volume by transferring to a second clean tube using a 1-mL micropipette (two transfers will be needed).
9. Add enough zinc stock solution to the protein sample with simultaneous gentle vortex mixing to be 0.5 molar equivalents using the following equation:

$$C_1 V_1 = C_2 V_2$$

to solve for volume of zinc stock solution (V_2) addition to protein sample, where C_1 = (0.5 equivalents zinc)×(protein sample concentration), V_1=(protein sample volume), and C_2 = (zinc stock concentration). Repeat above **Sections** (steps 5–9 of **Section 3.4)** iteratively, adding 0.5–1.0 equivalents of zinc stock each time until NMR spectra have been recorded up to 8–9 total zinc equivalents (*see* **Note 16**).

3.5.2. Calculation of Relative Zinc Affinities for Each MTF-1 Finger

1. Retrieve MTF-1 six-ZFP ^{1}H and ^{15}N amide NMR chemical shifts from Biological Magnetic Resonance data Bank (BMRB) website (http://www.bmrb.wisc.edu/search/) using accession number 6275 (12).
2. Transfer the data directories for each NMR spectrum to a UNIX workstation and convert the Varian NMR binary "*fid*" file to NMRPipe (16) byte order by executing the following command inside < > brackets from a terminal window within the data directory: <*varian*> (*see* **Note 17**).
3. From the opened dialog window, adjust acquisition parameters to match those of the recorded ^{1}H-^{15}N-HSQC experiments (*see* **Note 18**) and then left mouse click on "*Save Script*" followed by "Quit" buttons (default filename is "*fid.com*").
4. Quit dialog window and execute <*fid.com*> from terminal window, which creates a "*filename.fid*" NMRPipe raw data file.
5. Create and execute <*nmrproc.com*> data processing file from the terminal window for each NMR experiment, which creates a NMRView 5 (17) compatible processed data file (*see* **Note 19**).
6. Open the NMRView 5 program from terminal window (<*installation path/nv5*>).
7. Open the "*filename.nv*" spectrum from NMRView 5 corresponding to six equivalents of zinc and draw in a graphics window using toolbar commands: "*Dataset>>Open Datasets*" and "*Windows>>Add>>Window Name*" enter "*filename*" then left mouse "*Create*" button, and right mouse in new spectral window commands: "*Attributes>>File*", select open spectral file then "*Attributes>>Draw*" (*see* **Fig. 26.4a**).
8. Adjust referencing of this spectrum to match BMRB-assigned values of MTF-1 ZF region by shifting the spectral window centers appropriately (use NMRView 5 toolbar commands "*Dataset>>Manage Dataset>>Reference*"), and then left click on "*WritePar*". A "*filename.par*" file will be created in same directory as "*filename.nv*" containing referencing data.
9. Generate a peak list from spectrum (in ^{1}H-^{15}N-HSQC spectral window, right mouse commands: "*Peak>>Pick*").

10. Export the peak list to text file using NMRView 5 toolbar commands: "Assign>>Peaks>>List" to select peak list name, followed by commands: "File>>Write List" toolbar command from Peaks dialog box.
11. Repeat above steps 7–10 **Section 3.5** for each ^{1}H-^{15}N-HSQC spectrum recorded at different zinc equivalent concentrations.
12. Import all peak lists single into an Excel (Microsoft Corp) spreadsheet file (*see* **Note 20**).
13. Copy ^{1}H, ^{15}N NMR chemical shifts and peak intensity columns from each file and to one new worksheet, keeping track of number of zinc equivalents corresponding to each set of data columns.
14. For each ^{1}H-^{15}N NMR chemical shift pair corresponding to an assigned residue from the BMRB list for the six-ZF MTF-1 protein, align the peak intensity data in one row of the spreadsheet (separate columns for each intensity) from lowest to highest zinc equivalence.
15. Calculate the fractional peak intensity for each set data corresponding to a single residue (*see* **Fig. 26.4B**) from the ratio of each peak intensity in the data set to the maximum peak intensity for that residue (*see* **Note 21**).
16. Average "per finger" titration curves which best differentiate relative zinc affinities for multi-fingered proteins are obtained from the average fractional intensities of all residues in regions of defined secondary structure (α-helix and two β-hairpins) within a given finger. An illustration of the qualitative results of this relative zinc affinity analysis of the MTF-1 ZFP is presented in **Fig. 26.4c, d** where it is clear that zinc finger 4, which is ~40% metal bound at 1 molar equivalent zinc per six-ZFP, has the largest relative affinity. The uncertainty of each titration point is calculated from the standard error of the average individual residue fractional peak intensity in each finger.

4. Notes

1. Other related ZF constructs purified using the same methods described here are summarized. Zap1 ZF-1 with linker (plasmid p21a_zf1L) and the two-ZF construct zf1-2_{CA3} (plasmid p21a_zf1-2_{CA3}) that had the non-canonical cysteines mutated to alanines (4). Canonical zinc-binding cysteines were mutated to alanines in ZFs 1 and 2 in the zf1-2_{F1KO} (plasmid p21a_zf1-2_{F1KO}) and zf1-2_{F2KO}

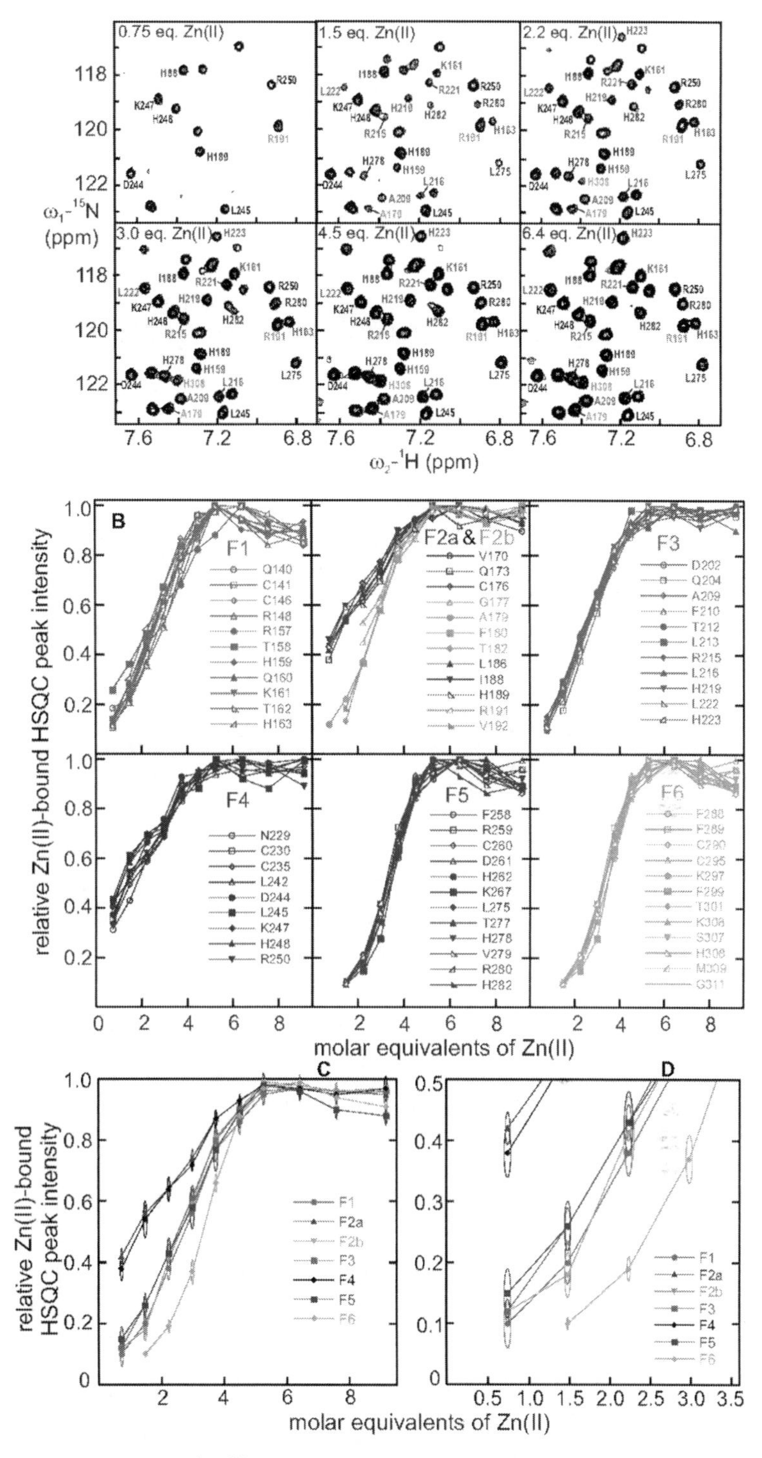

Fig. 26.4. (**a**) Region of the six-zinc finger MTF-1 ^{1}H-^{15}N-HSQC spectrum at different titration points. (**b**) Individual plots of zinc-dependent residue fractional ^{1}H-^{15}N-HSQC peak intensities. For finger 2, two populations intensities were observed, which are indicated as *light* and *dark gray*. (**c**) Average of per finger fractional ^{1}H-^{15}N-HSQC peak intensities from all residues in **b**. *Error arcs* around each data point represents the standard error from the averages. **(d)** Expanded early titration region from **c**. Reproduced from Potter et al. (10) with permission from the American Society for Biochemistry and Molecular Biology.

(plasmid p21a_zf1-2_{F2KO}) proteins, respectively. All these constructs were expressed as soluble protein in bacteria. The four N-terminal MTF-1 ZFP (plasmid pMZF1-4) was expressed in the insoluble fraction, while the two C-terminal MTF-1 ZFP (pMZF5-6) yielded soluble expression at 20°C.

2. For ZF domains with rare codons, *E. coli* BL21(DE3) RIL or RP from Stratagene can be used to improve recombinant protein expression levels. Other *E. coli* strains such as BL21 (DE3) pLysS and pLysE with more stringent expression control can also be used in cases of leaky expression or toxic recombinant proteins.
3. To ensure *E. coli* expression plasmid retention, a high-selection antibiotic concentration of 200 μg/mL is supplied to the LB medium and LB agar plates.
4. Water has a saturated dissolved oxygen concentration of 6.9 mg/L or 0.43 mM at 1 atmosphere and 37°C. Therefore, degassing buffers before adding reducing agents (i.e., DTT, BME, and TCEP) reduces protein instead of buffer. For ITC experiments, TCEP is preferable to DTT because DTT binds zinc with millimolar affinity (18), competing with lower affinity ZFPs present at micromolar concentrations in the ITC sample cell. In NMR experiments, the solubility of the six-ZF domain of MTF-1 was greatly reduced in the presence of TCEP. BME could be used for NMR studies since *relative* affinities rather than *absolute* affinities were determined, and the effects of competitive BME zinc binding are minimized by the higher protein concentrations (250–300 μM).
5. Transformation efficiency of competent *E. coli* cells is dependent on the number of colony forming units from a microgram of DNA. It is also dependent on the dilution factor and volume of culture plated. It is ideal to plate a volume of transformed culture that will yield at least five colonies but less than 500 colonies which would be difficult to pick a single colony for liquid culture inoculation.
6. In general (but not always), ZFPs of 1–3 domains are expressed as soluble proteins, while larger ZFr constructs tend to form inclusion bodies. For the 1–3 ZF constructs expressing soluble protein, inducing the cultures at an OD_{600} of 0.8 resulted in higher protein yields, while expression levels for larger ZF constructs that express as inclusion bodies were optimal when induced at an OD_{600} of 0.6.
7. If growth at 37°C does not result in good expression, repeat the pilot expression trial at 20°C. Cells grow

considerably slower at 20°C, and overnight growth of culture to reach mid-log phase induction point may be needed. In some cases, if a protein is expressed in the insoluble fraction at 37°C, it may partition into the soluble fraction at 20°C. Growth conditions may also be optimized by growing at 30°C or by growing the culture at 37°C and dropping the temperature to 20°C after induction. Pilot trials with different amounts of nitrogen and carbon per liter can also be carried out to optimize isotope usage. In general, 2 g/L for nitrogen and carbon is optimal for protein expression, although smaller ZFPs can have comparable expression with just 1 g/L of nitrogen and/or carbon. Expression analysis should be carried out empirically by PAGE and analytical HPLC.

8. ITC studies of individual MTF-1 ZF domains, which are typically challenging to produce from recombinant approaches due to intracellular degradation, are ongoing using synthetic peptides. The methods described here for HPLC purification and storage of recombinant protein are generally applicable to synthetic peptides with the following procedural recommendations: large-scale (100 mg), crude synthetic ZF preparations are supplied in lyophilized form; are resuspended in 2–5 mL of 50 mM MOPS, pH 7, 10% acetonitrile, and 100 μM TCEP; and allowed to incubate at room temperature for 1 h. Centrifuge peptide solution in a microcentrifuge at top speed, 20,817×*g* for 5 min. The peptide is now ready for preparative HPLC, although purifying synthetic peptides is typically more challenging because chemically similar truncated failure products have similar elution times to full-length peptide.

9. Simultaneous monitoring of the 220/280 nm ratio is a useful means to discriminate proteins from other possible species such as nucleic acids eluting from the HPLC column. The optimal detector for protein identification (but very expensive) is a photo diode array detector, since a full UV profile (i.e., 200–300 nm) can be monitored; this enables the discrimination of different proteins that have characteristic absorbance profiles.

10. A protein extinction coefficient at 280 nm is dependent on the number of tryptophans, phenylalanines, tyrosines, and cysteines in the sequence (19). If there are no tryptophans or tyrosines, as for many ZFs, the absorbance at 280 nm will be very low. We have found protein concentration measurements immediately after reverse-phase HPLC purification to be the most reliable. Ultimately, when very accurate extinction coefficients are needed, amino acid analysis should be carried out to calculate the number

of moles of protein in a solution of known volume and absorbance. Calibrated manual pipettes should be used to carefully measure volumes for amino acid analysis samples (5–100 μL).

11. After a good run, amino acid analysis using protein samples from before (apo) and after (zinc saturated) ITC is recommended to obtain a average protein concentration during the run. Some precipitation typically occurs during titration (<10%). This analysis yields the most accurate protein concentration measurement, in addition to a quantitative assessment of protein loss during experiment. Several ITC runs should be performed for a protein, with a minimum of three individual titrations for publication.

12. Zinc-sensing ZFs from Zap1 (4) and MTF-1 (unpublished results) have lower zinc affinities than typical Cys_2His_2 zinc fingers but still bind with an apparent $K_d < \mu M$. A good quality ITC titration curve will have at least 5–10 data points before and after zinc saturation and 1–3 points in the transition region in between (*see* **Fig. 26.3**). All of these points are essential for meaningful curve fitting analysis, so some experimentation with auto-pipette zinc concentrations and injection volumes in different experiments is always needed when studying new ZFPs. For a typical ITC run, sufficient zinc concentration and injections should be provided to achieve a 3–4:1 zinc:two-ZFP ratio. The midpoint of the saturation curve should be around 2:1 stoichiometry, although adjustment of calculated concentration after amino acid analysis may be needed before the stoichiometries are accurate (*see* **Notes 10** and **11**).

13. Before the start of an ITC run, the temperature, the sample, and the reference cells will be matched, after which the injection syringe will begin to spin for mixing (set to 290 rpm). The ITC then compensates for heat generated by mixing by reducing the power to heat the sample cell. In ~10–20 min, these steps will finish (indicated by baseline equilibration), after which titration of zinc into protein will automatically begin.

14. Two-site model fitting can be used for ZFs proteins with two (or possibly more) motifs if there are two populations of zinc-binding sites with very different affinities (>500 fold difference). The two-site model fails with similar affinity sites (very large uncertainties are obtained from the fitting process). For the two-ZF Zap1 protein, there were two distinct zinc-binding affinity populations, comprising two high-affinity ZF sites and three lower affinity sites contributed by three extra surface-exposed cysteines (4).

15. Preparing a twofold excess of protein solution over what is needed for recording an individual NMR spectrum eliminates problems associated with minor sample loss during initial protein concentration measurements using UV spectroscopy, solution transfer steps, and pH adjustments.
16. Be sure to measure total sample volume (for *V*1) after each titration. You also need to calculate the new protein concentration after each titration, which will be reduced 6–10% over the course of the entire experiment, due to dilution by zinc stock additions. For the first 4 equivalents, add in 0.5 equivalent steps and then increase to 1 equivalent step thereafter. We note that while the equation used to calculate volume of zinc stock for each titration is an approximation, the error is on the order of <1% and greatly simplifies the calculation. Other factors such as possible protein precipitation during titration are more likely potential sources of error and amino acid analysis of protein samples before and after successful titrations are recommended to assess any potential loss.
17. Other NMR data processing packages including, but not limited to, Sparky and CCPN are available. Most of these packages (including NMRPipe) can be used to process and analyze NMR data recorded from a variety of NMR spectrometers manufactured by Bruker, Varian, and JEOL.
18. Sample executable "*fid.com*" text file for NMR titration ("-yMODE" Varies from pulse sequence to pulse sequence):

```
#!/bin/csh
var2pipe -in./fid -noaswap \
-xN 1024 -yN 256 \
-xT 512 -yT 128 \
-xMODE Complex -yMODE Echo-AntiEcho \
-xSW 8396.306 -ySW 2200.000 \
-xOBS 599.732 -yOBS 60.777 \
-xCAR 4.773 -yCAR 118.272 \
-xLAB H1 -yLAB N15 \
-ndim 2 -aq2D States \
-out./filename.fid -verb -ov
sleep 5
```

19. Sample executable "*nmrproc.com*" text file for NMR titration spectrum (correct zero-order phase correction (PS – p0) will be needed for direct (t_2) dimension in line 6):

```
#!/bin/csh
nmrPipe -in filename.fid \
| nmrPipe -fn SP -off 0.3 -end 1.00 -pow 2 -c 1.0 \
# nmrPipe -fn ZF -auto \
| nmrPipe -fn FT -auto \
```

```
| nmrPipe -fn PS -p0 0.00 -p1 0.00 -di -verb \
| nmrPipe -fn EXT -left -sw \
| nmrPipe -fn TP \
| nmrPipe -fn SP -off 0.3 -end 1.00 -pow 1 -c 1.0 \
| nmrPipe -fn ZF -auto \
| nmrPipe -fn FT -auto \
| nmrPipe -fn PS -p0 0.00 -p1 0.00 -di -verb \
| nmrPipe -fn TP \
-ov -out filename.ft2
pipe2xyz -in filename.ft2 -out filename.nv –nv
```

20. A UNIX perl script "*gridsearch*" is available from the authors that concatenates and aligns peak intensities for each ^{1}H-^{15}N-HSQC peak corresponding to one residue from each zinc equivalent spectrum prior to import into Excel. This script greatly simplifies data analysis after peak export from NMRView 5.

21. Some broad and poorly dispersed peaks evident in low zinc equivalent titration spectra, which are not considered in the analysis, correspond to residues from apo-ZFPs. Maximum ^{1}H-^{15}N-HSQC peak intensity for most residues in the six-zinc finger MTF-1 protein occurs at six equivalents of zinc. For stronger and weaker affinity fingers, saturation may occur before or after six equivalents, respectively. After zinc saturation of a given finger-binding site, peak intensities tend to go down slightly.

References

1. Zhao, H. and Eide, D.J. (1997) Zap1p, a metalloregulatory protein involved in zinc-responsive transcriptional regulation in *Saccharomyces cerevisiae*, *Mol. Cell Biol.* **17**, 5044–5052.
2. Heuchel, R., Radtke, F., Georgiev, O., Stark, G., Aguet, M., and Schaffner, W. (1994) The transcription factor MTF-1 is essential for basal and heavy metal-induced metallothionein gene expression. *EMBO J.* **13**, 2870–2875.
3. Bird, A.J., McCall, K., Kramer, M., Blankman, E., Winge, D.R., and Eide, D.J. (2003) Zinc fingers can act as Zn2+ sensors to regulate transcriptional activation domain function. *EMBO J.* **22**, 5137–5146.
4. Wang, Z., Feng, L.S., Matskevich, V., Venkataraman, K., Parasuram, P., and Laity, J.H. (2006) Solution Structure of a Zap1 Zinc-responsive Domain Provides Insights into Metalloregulatory Transcriptional Repression in *Saccharomyces cerevisiae*. *J Mol Biol.* **357**, 1167–1183.
5. Li, Y., Kimura, T., Laity, J.H., and Andrews, G.K. (2006) The zinc-sensing mechanism of mouse MTF-1 involves linker peptides between the zinc fingers, *Mol. Cell Biol.* **26**, 5580–5587.
6. Lee, M.S., Gippert, G.P., Soman, K.V., Case, D.A., and Wright, P.E. (1989) Three-dimensional solution structure of a single zinc finger DNA-binding domain. *Science.* **245**, 635–637.
7. Laity, J.H. and Messerschmidt, A. (2004) Cys_2His_2 zinc fingers. In (A. Messerschmidt, W. Bode and M. Cygler eds.) *Handbook of metalloproteins.* John Wiley & Sons Press, West Sussex.
8. Frankel, A.D., Berg, J.M., and Pabo, C.O. (1987) Metal-dependent folding of a single zinc finger from transcription factor IIIA, *Proc. Natl Acad Sci USA.* **84**, 4841–4845.
9. Pavletich, N.P. and Pabo, C.O. (1991) Zinc finger-DNA recognition: crystal structure of a Zif268-DNA complex at 2.1 A. *Science.* **252**, 809–817.

10. Potter, B.M., Feng, L.S., Parasuram, P., Matskevich, V.A., Wilson, J.A., Andrews, G.K., and Laity, J.H. (2005) The six zinc fingers of metal-responsive element binding transcription factor-1 form stable and quasi-ordered structures with relatively small differences in zinc affinities. *J Biol Chem.* **280**, 28529–28540.
11. Miller, J., McLachlan, A.D., and Klug, A. (1985) Repetitive zinc-binding domains in the protein transcription factor IIIA from Xenopus oocytes. *EMBO J.* **4**, 1609–1614.
12. Potter, B.M., Knudsen, N.A., Feng, L.S., Matskevich, V., Wilson, J.A., Andrews, G.K., and Laity, J.H. (2005) NMR assignment of the six zinc fingers of MTF-1 in the free and DNA-bound states. *J. Biomol NMR.* **32**, 94.
13. Sambrook, J., Fritsch, E.F., and Maniatis, T. (1989) *Molecular cloning: a laboratory manual.* Vol. **3**, 2nd ed. Cold Spring Harbor Laboratory Press, New York.
14. Foster, M.P., Wuttke, D.S., Radhakrishnan, I., Case, D.A., Gottesfeld, J.M., and Wright, P.E. (1997) Domain packing and dynamics in the DNA complex of the N-terminal zinc fingers of TFIIIA. *Nat Struct Biol.* **4**, 605–608.
15. Laity, J.H., Chung, J., Dyson, H.J., and Wright, P.E. (2000) Alternative splicing of Wilms' tumor suppressor protein modulates DNA binding activity through isoform-specific DNA-induced conformational changes. *Biochemistry.* **39**, 5341–5348.
16. Delaglio, F., Grzesiek, S., Vuister, G.W., Zhu, G., Pfeifer, J., and Bax, A. (1995) NMRPipe: a multidimensional spectral processing system based on UNIX pipes, *J. Biomol NMR.* **6**, 277–293.
17. Johnson, B.A. and Blevins, R.A. (1994) NMRView: A computer program for the visualization and analysis of NMR data. *J Biomol NMR.* **4**, 603–614.
18. Cornell, N.W. and Crivaro, K.E. (1972) Stability constant for the zinc-dithiothreitol complex, *Anal. Biochem.* **47**, 203–208.
19. Gill, S.C. and von Hippel, P.H. (1989) Calculation of protein extinction coefficients from amino acid sequence data. *Anal Biochem.* **182**, 319–326.

Chapter 27

Using ChIP-seq Technology to Identify Targets of Zinc Finger Transcription Factors

Henriette O'Geen, Seth Frietze, and Peggy J. Farnham

Abstract

Half of all human transcription factors are zinc finger proteins and yet very little is known concerning the biological role of the majority of these factors. In particular, very few genome-wide studies of the in vivo binding of zinc finger factors have been performed. Based on in vitro studies and other methods that allow selection of high affinity-binding sites in artificial conditions, a zinc finger code has been developed that can be used to compose a putative recognition motif for a particular zinc finger factor (ZNF). Theoretically, a simple bioinformatics analysis could then predict the genomic locations of all the binding sites for that ZNF. However, it is unlikely that all of the sequences in the human genome having a good match to a predicted motif are in fact occupied in vivo (due to negative influences from repressive chromatin, nucleosomal positioning, overlap of binding sites with other factors, etc). A powerful method to identify in vivo binding sites for transcription factors on a genome-wide scale is the chromatin immunoprecipitation (ChIP) assay, followed by hybridization of the precipitated DNA to microarrays (ChIP-chip) or by high throughput DNA sequencing of the sample (ChIP-seq). Such comprehensive in vivo binding studies would not only identify target genes of a particular zinc finger factor, but also provide binding motif data that could be used to test the validity of the zinc finger code. This chapter describes in detail the steps needed to prepare ChIP samples and libraries for high throughput sequencing using the Illumina GA2 platform and includes descriptions of quality control steps necessary to ensure a successful ChIP-seq experiment.

Key words: Zinc fingers, chromatin immunoprecipitation, ChIP-seq, next generation sequencing.

1. Introduction

Half of all human transcription factors are zinc finger (ZF) proteins and yet very little is known concerning the biological role of the majority of these factors. The ZF transcription factors have

J.P. Mackay, D.J. Segal (eds.), *Engineered Zinc Finger Proteins*, Methods in Molecular Biology 649,
DOI 10.1007/978-1-60761-753-2_27, © Springer Science+Business Media, LLC 2010

evolved from a few ancestral genes, by diverging mainly in the number and sequence of their finger domains (1). Rapid expansion of the ZF family has occurred in the primate lineage and a large number of human ZF genes have no mouse ortholog, leading to speculation that ZF proteins could be important regulators of primate-specific transcription networks. If so, it is critical that we understand the DNA-binding properties of the ZF family and identify where different family members bind in the human genome. Various ZF codes have been developed that predict a binding motif for a particular ZNF (2–4). However, very few studies have performed in vivo analyses of ZNF-binding patterns and thus it is not clear if the binding codes developed using in vitro assays are physiologically relevant.

To understand the biological function of ZF transcription factors and to reveal the regulatory networks controlled by these factors, it is critical to identify all of the binding sites of a particular ZF factor that are present in the human genome. A powerful method to identify in vivo binding sites for transcription factors on a genome-wide scale is the chromatin immunoprecipitation (ChIP) assay, followed by hybridization of the precipitated DNA to microarrays (ChIP-chip) or by high throughput DNA sequencing of the sample (ChIP-seq). The ChIP assay uses formaldehyde to covalently cross-link proteins to their DNA substrates in living cells. This provides an opportunity to take a snapshot of DNA–protein interactions in a given cell type, using populations of cultured cells, subsets of cells taken at specific times of the cell cycle or development, or even cells taken directly from tissue samples. Initially, ChIP-chip was the method of choice for binding site analyses (5–10). However, multiple DNA microarrays are needed to cover the entire human genome, resulting in high costs for comprehensive studies. ChIP-seq technology, on the other hand, offers the possibility to identify transcription factor-binding sites in the entire genome in one sequencing run (11). In the beginning, ChIP samples were sequenced for the purpose of identifying binding sites after cloning individual immunoprecipitated fragments (12). Clearly this was an inefficient and costly method. Improvements in both cost and efficiency were gained when the Roche 454 DNA sequencer was used to analyze ChIP samples (13). However, 454 sequencing lacks the throughput necessary to cost-effectively identify the entire set of binding sites for a human transcription factor. In contrast, the Illumina GA2 sequencer generates over 10 times the number of DNA sequences at a fraction of the cost of sequencing with the 454 machine. It should be noted that the length of the reads from the Illumina sequencer are shorter than those obtained from the 454 machine. However, for the purposes of ChIP-seq, the short reads are sufficient to accurately map the enriched fragments to their genomic location. Applied Biosystems also produces a sequencer that provides high throughput sequencing of short tags. However, most of the published ChIP-seq papers to date have employed the Illumina system (14–17). This chapter describes in detail the steps needed to prepare ChIP samples and libraries for high throughput sequencing using the Illumina GA2 platform and includes descriptions of quality control steps necessary to ensure a successful ChIP-seq experiment (*see* **Fig. 27.1**).

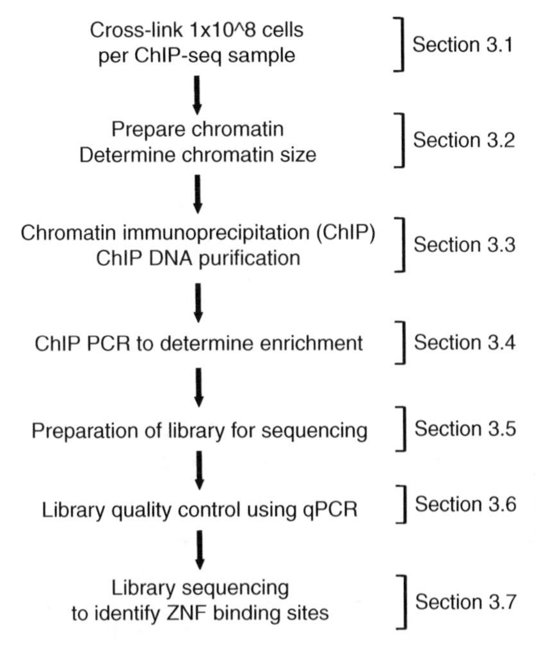

Fig. 27.1. ChIP-seq diagram summarizing experimental steps, including quality control procedures.

2. Materials

2.1. Cross-linking

1. Formaldehyde solution (37% w/w).
2. Glycine (electrophoresis grade).
3. Phosphate-buffered saline (PBS).

2.2. Chromatin Preparation

1. Protease inhibitor stock solutions (store in small aliquots at –20°C): 10 mg/mL aprotinin (in water), 10 mg/mL leupeptin (in water), 100 mM PMSF (in isopropanol).
2. Cell lysis buffer (store at room temperature): 5 mM PIPES pH 8, 85 mM KCl. Add igepal fresh each time to give a final concentration (f.c.) of 1% (10 μL/mL). Warm buffer in 37°C water bath and vortex briefly to help mixing of igepal.

After mixing has occurred, place buffer containing igepal on ice to allow solution to cool down and then add protease inhibitors [PMSF (10 μL/mL f.c.), aprotinin (1 μL/mL f.c.), and leupeptin (1 μL/mL f.c.)].

3. Nuclei lysis buffer (store at room temperature): 50 mM Tris–HCl, pH 8, 10 mM EDTA, and 1% (w/v) SDS. Place buffer on ice right before use to avoid precipitation of SDS and add protease inhibitors [PMSF (10 μL/mL f.c.), aprotinin (1 μL/mL f.c.), and leupeptin (1 μL/mL f.c.)] just prior to use.
4. Elution buffer (store at room temperature): 50 mM $NaHCO_3$, 1% (w/v) SDS.
5. DNase-free RNase A (10 mg/mL; Fermentas #EN0531).
6. QIAquick PCR purification kit (QIAGEN).
7. NanoDrop 1000 spectrophotometer is used to determine the concentration of double stranded DNA samples (*see* **Note 1**).
8. Type B Dounce homogenizer.

2.3. Chromatin Immunoprecipitation

1. Magnetic protein G beads (Cell Signaling Technology #9006) and magnetic rack (*see* **Note 2**).
2. Antibodies: Covance 8WG16 mouse monoclonal antibody against RNA polymerase II (Covance #MMS-1261R) is used as a positive control antibody for ChIP experiments, and normal/pre-immune control rabbit IgG (Alpha Diagnostic International #20009) is used as a negative control.
3. Protease inhibitor stock solutions (store in small aliquots at –20°C): 10 mg/mL aprotinin (in water), 10 mg/mL leupeptin (in water), 100 mM PMSF (in isopropanol).
4. IP dilution buffer (also known as RIPA buffer; store at 4°C): 50 mM Tris–HCl, pH 7.4, 150 mM NaCl, 1% (v/v) igepal, 0.25% (w/v) deoxycholic acid, 1 mM EDTA, pH 8. Add protease inhibitors [PMSF (10 μL/mL f.c.), aprotinin (1 μL/mL f.c.), and leupeptin (1 μL/mL f.c.)] just prior to use. This buffer is used to adjust the salt and SDS concentrations for the immunoprecipitation step.
5. IP wash buffer 1 (store at 4°C): same as IP dilution buffer, but no protease inhibitors are added.
6. IP wash buffer 2 (store at room temperature): 100 mM Tris–HCl, pH 9, 500 mM LiCl, 1% (v/v) igepal, 1% (w/v) deoxycholic acid.
7. IP wash buffer 3 (store at room temperature): 100 mM Tris–HCl, pH 9, 500 mM LiCl, 150 mM NaCl, 1% (v/v) igepal, 1% (w/v) deoxycholic acid.

8. Elution buffer (store at room temperature): 50 mM $NaHCO_3$, 1% (w/v) SDS.
9. 5 M NaCl.
10. DNase-free RNase A (10 mg/mL; Fermentas #EN0531).
11. QIAquick PCR purification kit (QIAGEN).

2.4. ChIP PCR Control

1. PCR reagents: 5 M betaine (stored at 4°C), 2 mM dNTP mix stored in small aliquots at –20°C, Taq polymerase such as GoTaq DNA polymerase (5 U/μL; Promega #M3005).
2. Oligonucleotide primers used for positive and negative control PCR reactions for standard PCR (for primer design, *see* **Note 3**):

   Positive control primers (to amplify a region which should be positive for binding of RNA Polymerase II):
   GAPDH-F: 5′-CACCGTCAAGGCTGAGAACG-3′
   GAPDH-R: 5′-ATACCCAAGGGAGCCACACC-3′

   Negative control primers (to amplify a region which should be negative for binding of RNA Polymerase II):
   ZNF333-F: 5′-CACAGGAGAGAAGCCCTACG-3′
   ZNF333-R: 5′-TCGCGCACTCATACAGTTTC-3′

2.5. Sequencing Library

1. End-It DNA END Repair Kit (Epicentre #ER0720).
2. Klenow (3′- to 5′-exo-minus) (5,000 U/mL; NEB #M0212S).
3. 100 mM dATP.
4. LigaFast DNA ligase (3 U/μL; Promega #M8221).
5. Oligo only kit for single end read sequencing containing genomic adapter oligo mix (100 μM) and genomic PCR primer 1.1 (25 μM) and PCR primer 2.1 (25 μM) (Illumina #FC-102-1003).
6. Phusion DNA polymerase (NEB #F531).
7. QIAquick PCR purification kit (QIAGEN) and MinElute PCR purification kit (QIAGEN).
8. QIAquick gel extraction kit (QIAGEN).
9. NanoDrop 1000 spectrophotometer is used to determine the concentration of double stranded DNA samples (*see* **Note 1**).

2.6. Pre-sequencing Control

1. Positive and negative control primers for qPCR (*see* **Note 3**).
2. SYBR-Green qPCR mix, such as SYBR Green JumpStart Taq ReadyMix for Quantitative PCR (SIGMA #S4438).

2.7. Sequencing

Sequencing reagents will be supplied by the sequencing facility.

3. Methods

3.1. Preparation of Cross-linked Cells

1. Cell cultures should be healthy and not density arrested prior to cross-linking. In general, 10^7 cells are used for a test ChIP assay and 10^8 cells are used per each ChIP-seq experiment. In a fume hood, add formaldehyde (37% stock) directly to tissue culture media to a final concentration of 1%. Incubate adherent cells on a gently shaking platform for 10 min at room temperature. Do not cross-link for longer periods since this may cause cells to form aggregates that do not sonicate efficiently.
2. Stop cross-linking reaction by adding glycine to a final concentration of 0.125 M. We use a 10× (1.25 M) stock solution. Continue to rock at room temperature for 5 min.
3. Pour off media; this media contains formaldehyde and should be treated as hazardous waste.
4. Rinse plates twice with ice-cold PBS and pour off wash solution. For 15 cm dishes, we add 20 mL of PBS per wash. Pipette gently so as not to dislodge cells. Swirl gently and discard. A few mL of PBS usually stay behind and can be used in step 5.
5. Using a cell scraper, transfer adherent cells from the culture dish to a 15 mL conical tube (or 50 mL tube for larger batches) on ice. Centrifuge the cross-linked cells at 430 rcf for 5 min at 4°C. Carefully aspirate supernatant so as to not lose cells. Cells may be used immediately for a chromatin preparation or snap frozen in liquid nitrogen and stored at –80°C.

3.2. Preparation of Chromatin

3.2.1. Cell Lysis and Chromatin Fragmentation

1. Thaw frozen cross-linked cells on ice. Resuspend cell pellet in freshly prepared ice-cold cell lysis buffer by pipetting. The final volume of cell lysis buffer should be sufficient so that there are no clumps of cells (~1 mL of Cell Lysis Buffer per 10^7 cells). Incubate on ice for 15 min.
2. Use a type B dounce homogenizer to break open the cells and release nuclei. Homogenize cells on ice with 20 strokes.
3. Centrifuge at 430 rcf for 5 min at 4°C.
4. Discard supernatant and resuspend nuclear pellet in nuclei lysis buffer. Use 200 μL of nuclei lysis buffer per 10^7 cells. Incubate on ice for 30 min.

5. Sonicate cells to achieve average chromatin length of 200–500 bp in a cold room and/or on ice (*see* **Note 4**). Larger chromatin fragments can negatively influence data quality and can lead to failure of the ChIP-seq experiment. Therefore, before processing large quantities of cells, sonication conditions should be optimized for each cell type (*see* **Note 5** and **Fig. 27.2**).

6. Transfer sonicated samples into an Eppendorf tube and centrifuge using a microcentrifuge at 10,000 rcf for 10 min at 4°C. Carefully transfer the supernatant (sonicated chromatin) to a new tube while avoiding cell debris. Chromatin can be snap frozen in liquid nitrogen and stored at –80°C or used immediately.

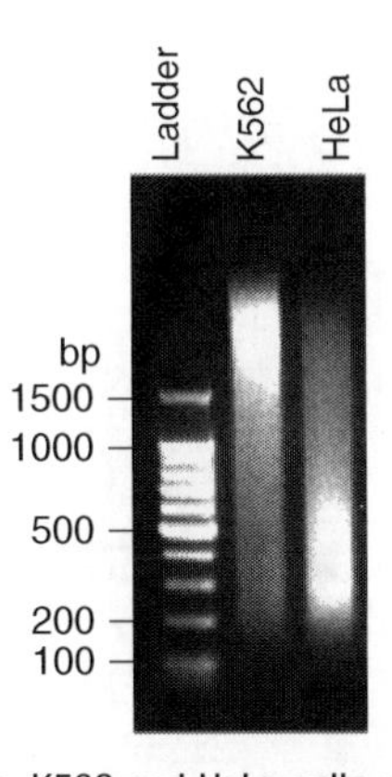

Fig. 27.2. Sonicated chromatin. K562 and HeLa cells were sonicated for 30 min (30 s pulses, 1.5 min pauses in between pulses). A 100 bp ladder was used to estimate chromatin size. The optimal fragment size of 200–500 bp is observed for HeLa cells, whereas K562 cells show an average chromatin size larger than 1.5 kb. Before proceeding, the sonication conditions should be optimized for K562 cells.

3.2.2. Determination of Chromatin Size and Concentration

1. Take an aliquot of a chromatin sample prepared as described above. We typically use 20 μL of chromatin, which is the equivalent to 10% input from 10^7 cells.

2. Add elution buffer to a total volume of 100 μL and add 12 μL of 5 M NaCl to give a final salt concentration of 0.54 M. Boil samples in a water bath for 15 min to reverse cross-links.

3. Allow the sample to cool down, add 1 μL of DNase-free RNase A, and incubate for 20 min at 37°C. This step is important because the presence of RNA results in a false estimation of chromatin size.

4. Purify DNA using a PCR purification kit, then elute DNA in 50 μL water. Measure chromatin concentration by NanoDrop. The concentration can be used to calculate the chromatin yield. Approximately 50–100 μg of chromatin can be isolated from 10^7 cells.

5. Run 1 μg of chromatin on a 1.5% agarose gel to visualize the average size of the chromatin (*see* **Fig. 27.2**). If necessary, adjust sonication conditions by repeating steps 1–5 (e.g., add more pulses if the majority of the chromatin is bigger than 600 bp).

3.3. Chromatin Immunoprecipitation

1. Thaw chromatin preparation from **Section 3.2.1** on ice. This can take hours depending on the volume. Measure the volume and divide chromatin as needed; 200 μL of chromatin (equivalent to 10^7 cells) is used per ChIP. Multiple ChIPs may need to be performed to obtain sufficient material for Illumina GA2 sequencing; the number required will vary between different antibodies. Typically 10 ChIP assays are performed for one ChIP-seq experiment (10^7 cells/assay; this requires 10^8 cells total, which is ~2 mL of chromatin).
2. The protein/chromatin complexes are captured using antibodies specific to the protein of interest (*see* **Note 6**). We also recommend that you include a positive and a negative control antibody to monitor the success of the ChIP experiment. An RNA polymerase II antibody is used as a positive control and normal rabbit IgG is used as a negative control. These control ChIPs are especially important when no target genes are known for the protein of interest. Therefore, in a typical experiment you will have three antibodies (against RNA polymerase II, IgG, and a ZF transcription factor). Approximately 10^7 cells (~200 μL of chromatin) are sufficient for the positive and negative control antibody ChIP assays because these samples will not be used for library preparation and sequencing, but will only serve as controls in PCR confirmations (*see* **Section 3.4**).
3. Save a 10% equivalent from 10^7 cells (~20 μL of chromatin); this sample is the input sample. Store the 10% input sample at –20°C until the next day when cross-links of ChIP samples are reversed.
4. For each ChIP from 10^7 cells, dilute chromatin to 1 mL with ice-cold IP dilution buffer containing protease inhibitors. When performing multiple ChIP assays of a given antibody for ChIP-seq, two ChIPs can be combined into a 2 mL tube.
5. Add the appropriate antibody to the samples. Although the amount of antibody should be optimized by immunoprecipitation experiments (*see* **Note 6**), typically between 1 and 5 μg of antibody is sufficient per 10^7 cells. For example, we use 2.4 μg of the Covance 8WG16 mouse monoclonal antibody against RNA polymerase II for 10^7 cells.

6. Incubate for 12–16 h on a rotating platform at 4°C.
7. Add 30 μL of magnetic protein G beads to each ChIP sample from 10^7 cells to capture antibody/chromatin complexes, and incubate on a rotating platform for 2 h at 4°C (*see* **Note 7**).
8. At room temperature, allow the beads to settle for 1 min in a magnetic separation rack. Carefully remove and discard the supernatant without disturbing the magnetic beads.
9. At room temperature, pool ChIPs obtained using the same antibody into one tube for washes. To pool, resuspend the beads of one tube in 1 mL of IP wash buffer 1, then add this suspension to the next tube. Repeat until all identical ChIPs are pooled into one tube. Do not discard the tubes; repeat the entire process with 0.5 mL of IP wash buffer 1 to rinse and retain as much bead material as possible.
10. Wash magnetic beads from the control ChIPs and pooled ChIPs two times with 1 mL of IP wash buffer 1. For each wash, take the tubes out of the magnetic rack, mix by pipetting, and collect as in step 8. Discard all wash solutions. Efficient washing is critical to reduce background. Avoid cross-contamination of samples and loss of magnetic beads.
11. Wash magnetic beads three times with IP wash buffer 2 (take tubes out of the magnetic rack and mix by pipetting and collect as in step 8). Discard all wash solutions.
12. Wash once with the higher stringency IP wash buffer 3. Discard wash solution.
13. Elute antibody/chromatin complexes from magnetic beads by adding 100 μL of elution buffer per tube. Shake on a vortexer for 30 min, setting 4 (on a scale of 1–10). Repeat this step for pooled ChIP samples, resulting in a total volume of 200 μL. The second elution is not necessary for ChIPs from 10^7 cells.
14. Allow beads to settle for 1 min in a magnetic separation rack. Carefully transfer the supernatant containing antibody/chromatin complexes to a fresh tube.
15. Add 12 μL of 5 M NaCl per 100 μL elution buffer to give a final concentration of 0.54 M NaCl.
16. At this point, thaw the input sample from the previous day. Bring the volume to 100 μL with elution buffer and add 12 μL of 5 M NaCl.
17. Incubate all samples in a 67°C water bath for 4 h to overnight to reverse the formaldehyde cross-links.
18. Allow samples to cool down, add 1 μL of RNaseA, and incubate at 37°C for 20 min.

19. Purify DNA with a PCR clean up kit, using one column per sample. Elute each sample with 50 μL of water.

3.4. ChIP Control Assays

The success of the ChIP experiment can be assessed by standard PCR using primers designed to amplify a known binding site for the positive control antibody and for the ZF protein antibody (*see* **Note 8**). IgG serves as a negative control antibody and should not show significant enrichment at any region of the genome if the IP washes were satisfactory. Examples of control ChIP experiments are shown in **Fig. 27.3**. Alternatively, qPCR can be used to determine ChIP enrichment. If this method is used, then the protocol outlined in **Section 3.6** should be followed.

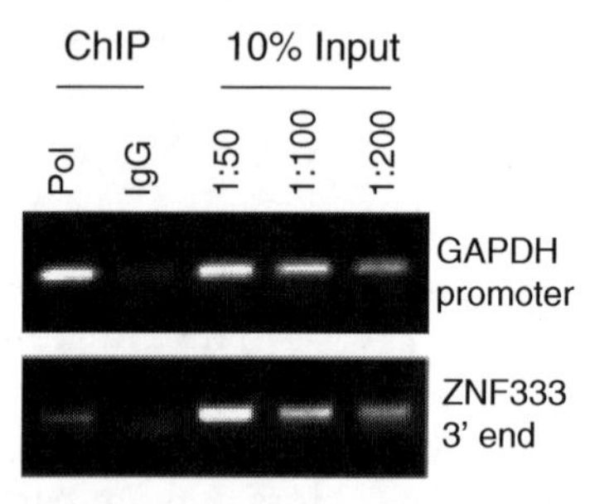

Fig. 27.3. PCR validation of a ChIP experiment. An antibody to RNA polymerase II (Pol) is used as a positive control and normal rabbit IgG serves as a negative control. Success of the ChIP experiment is evaluated by comparing the amplified PCR fragments to dilutions of the 10% input sample. Enrichment of Pol at the active *GAPDH* promoter is equivalent to 0.2% of input (1:50 dilution of 10% input), while IgG shows no significant enrichment. Neither Pol nor IgG is enriched at the negative control region (3′-end of *ZNF333*). These results are typical for a successful RNA polymerase II ChIP experiment.

Example for a standard PCR reaction: ChIP samples from 10^7 cells are used undiluted, whereas ChIP samples from 10^8 cells are diluted 1:10 before PCR (1 μL of ChIP sample + 9 μL of water). The 10% input sample should be diluted 1:50, 1:100, and 1:200. Set up individual reactions for each set of positive and negative control primer pairs.

4 μL	5× GoTaq PCR buffer
4 μL	5 M betaine
2 μL	2 mM dNTPs
1 μL	10 μM forward primer
1 μL	10 μM reverse primer
1 Unit	GoTaq polymerase
1.5 μL	ChIP sample or diluted input sample (1:50, 1:100, 1:200)
20 μL	Total reaction volume

Amplify using the following PCR protocol:

3 min at 95°C

35 cycles: 30 s at 95°C, 30 s at 60°C, 30 s at 72°C

5 min at 72°C
Hold at 10°C

The PCR reactions are run on a 1.5% agarose gel. Enrichment is evaluated relative to input. Strong enrichments give signals comparable to the signal obtained using the 1:50 dilution of 10% input sample (equivalent to 0.2% enrichment, *see* **Fig. 27.3**).

3.5. Preparation of the Sequencing Library

The library protocol is based on the Illumina Sample Preparation Kit for Genomic DNA with some modifications. This protocol describes the preparation of libraries from ChIP DNA that are compatible with the Illumina sequencing platforms.

3.5.1. End Repair

End repair is performed using the "End-It DNA End Repair Kit" from Epicentre. This step ensures that all DNA fragments are converted to 5′-phosphorylated blunt-ended DNA. ChIP DNA from **Section 3.3** can be concentrated to 34 μL using a speed vac, but avoid over-drying since this causes loss of DNA.

34 μL	ChIP DNA from **Section 3.3**
5 μL	10× end repair buffer
5 μL	10 mM ATP
5 μL	2.5 mM dNTP Mix
1 μL	end repair enzyme mix
50 μL	Total reaction volume

Incubate at room temperature for 45 min and purify DNA using a PCR purification kit (such as QIAquick PCR purification kit). Elute in 34 μL of the kit elution buffer.

3.5.2. Addition of an 'A' Base to the 3′-End of DNA Fragments

Prepare 1 mM dATP aliquots from 100 mM dATP stock. Store aliquots of 11 μL at –20°C. 1 mM dATP solution should not be refrozen.

34 μL	DNA from **Section 3.5.1**
5 μL	10× Klenow buffer
10 μL	1 mM dATP
5 units	Klenow fragment (3′- to 5′-exo-minus)
50 μL	Total reaction volume

Incubate for exactly 30 min at 37°C. Purify DNA using a PCR purification kit (such as MinElute PCR purification kit). Elute in 12 μL of the kit elution buffer.

3.5.3. Ligation of Adapters to DNA Fragments

The genomic adapter oligo mix from Illumina is diluted 1:10 in water before use.

12 μL	DNA from **Section 3.5.2**
15 μL	2× DNA ligase buffer
1 μL	1:10 dilution of genomic adapter oligo mix
2 μL	LigaFast DNA ligase
30 μL	Total reaction volume

Incubate for 15 min at room temperature. Purify DNA using a PCR purification kit (such as QIAquick PCR purification kit). Elute in 25 μL of the kit elution buffer.

3.5.4. Amplification of Adapter Modified DNA Fragments and Gel Purification

24 μL DNA from **Step 3.5.3**
25 μL 2× Phusion DNA polymerase
0.5 μL Illumina Genomic PCR primer 1.1
0.5 μL Illumina Genomic PCR primer 2.1
50 μL Total reaction volume

Amplify using the following PCR protocol:

30 s at 98°C
15 cycles: 10 s at 98°C, 30 s at 65°C, 30 s at 72°C
5 min at 72°C
Hold at 4°C

Purify and concentrate DNA using a PCR purification kit (such as MinElute PCR purification kit). Elute in 10 μL of the kit elution buffer. Run purified DNA product on a 2% agarose gel next to a 50 bp or 100 bp ladder. Excise DNA between 150 and 500 bp. This step ensures removal of unused primers and adapters and selection of proper fragment size for sequencing. Purify DNA using a gel extraction kit. Elute DNA in 30 μL of the kit elution buffer and measure DNA concentration by NanoDrop. At this point, follow instructions of the sequencing facility for sample submission.

3.6. Pre-sequencing Control Assays: Enrichment Check Using Positive/Negative Control Primers

To verify that a library has maintained a specific enrichment of ChIP target sites, perform qPCR on the ChIP-seq library using both target and negative control primer pairs (*see* **Note 9**). An input library derived from the 10% input sample serves as a control to normalize the qPCR data to determine the relative enrichment of a given target (*see* **Note 10**).

3.6.1. Prepare Input Library

Using 50–200 ng of input chromatin extract, construct an input library as described in **Section 3.5**.

3.6.2. Real-Time Quantitative PCR (qPCR)

Analyze the ChIP-seq library as well as an input library for reference. Prepare a master reaction mix for each library with triplicate reactions per primer set. Add extra reagents for 10% of the total number of reagents to account for loss of volume. Add 15 μL of reaction mix to each PCR reaction well.

2μL 2 ng library from **Section 3.5.4** and **Section 3.6.1**
3.5 μL Nuclease-free H_2O
7.5 μL 2× SYBR Green mix (containing polymerase)
2 μL 5 μM target primer mix
20 μL Total reaction volume

Amplify using the following PCR protocol:

3 min at 95°C
40 cycles of 30 s at 95°C then 30 s at 60°C

Include a 70–95°C melting curve at the end of the qPCR program, reading all points or every 0.2°C.

3.6.3. Determine Enrichment

Analyze the qPCR results by first manually determining the cycle threshold for each reaction across the plate within the linear range of the amplification curve. Using Microsoft Excel, calculate the average cycle threshold for each triplicate reaction of each sample. Create a column on the spreadsheet and copy the corresponding averaged cycle thresholds for an input library and ChIP-seq library so that they are adjacent for a specific primer set inside the column. In an adjacent column use the following formula to determine the 'relative DNA amount' for each sample $2^{(\max(A\$1{:}A\$6)-A1)}$, where A$1:A$6 is the list of average sample cycle thresholds for all the samples analyzed in the column (in this example there are six samples, the first at cell A1 and so on). Divide the relative DNA amount of the ChIP-seq library by the relative DNA amount of the input library for a given primer set. The resultant quotient corresponds to the enrichment value of a target in the library over the input library. The enrichment value for a target primer set should be at least 20-fold greater than the enrichment of a negative control primer set. The enrichment value for a negative control site should be near one, signifying no enrichment for this negative control. To further account for any bias when comparing the ChIP-seq library to the input library, determine the fold enrichment using the data from the negative control primers. This is accomplished by dividing the relative DNA amount of each sample for a target primer set by the corresponding value for a negative control primer set. The resulting quotient represents a normalized DNA amount. Divide the normalized DNA amount for the ChIP-seq sample by the corresponding normalized DNA amount for the input sample to determine the fold enrichment of a target site in the ChIP-seq library (*see* **Fig. 27.4**). If the positive control primers show greater than 20-fold enrichment over the negative control primers, then the sample can be provided to a sequencing facility (e.g., http://genomecenter.ucdavis.edu/dna_technologies) for high throughput sequencing.

3.7. Library Sequencing

An estimated 10–20 million uniquely mapped tags should be sufficient to identify all binding sites for a site-specific factor (which is approximately two lanes using an Illumina GA2 machine). Ideally, these reads should come from two independent ChIP samples, with the binding sites identified in each

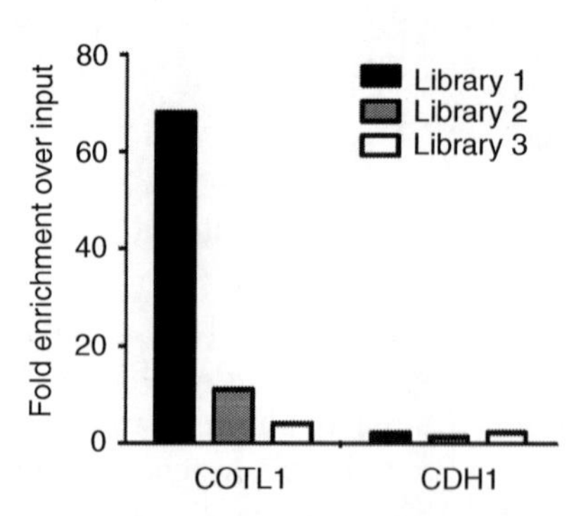

Fig. 27.4. Confirmation of ChIP targets in ChIP-seq libraries. Three independent ChIP-seq libraries were prepared from K562 cells using an antibody to ZNF263 (17). The relative enrichments of a target site (*COTL1*) and a negative control site (*CDH1*) were determined by comparing qPCR signals from the ZNF263 ChIP-seq libraries to the qPCR signals from an input library. Based on the enrichment values, it would be expected that library 1 would give robust ChIP-seq peaks, library 2 would give modest ChIP-seq peaks, and library 3 would not produce peaks (*see* **Fig. 27.5**).

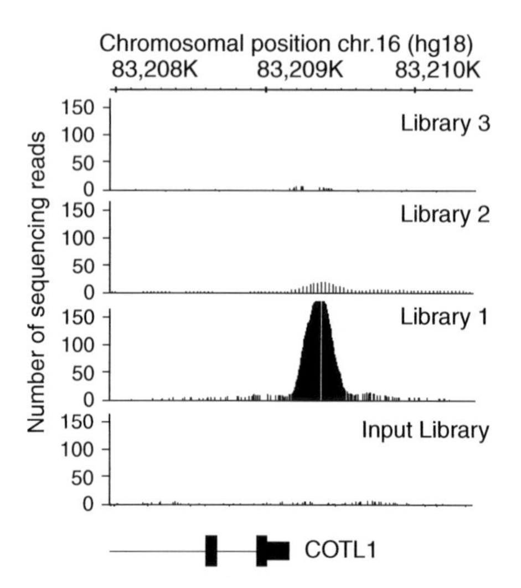

Fig. 27.5. Comparison of ChIP-seq data. ChIP-seq data from three independent ZNF263 ChIP-seq libraries that were analyzed by qPCR (*see* **Fig. 27.4**) and from an input library were visualized using the Integrated Genome Browser. The positive ZNF263 target site (*COTL1*) is shown. The *COTL1* gene is located on chromosome 16. The 5′-end starts at 83,209,170 (shown in the schematic) and the gene extends to the left and ends at 83,156,705 (not shown); all sequences are relative to the hg18 human genome data set on the UCSC browser. Unique sequencing reads (4–5 million) were analyzed for libraries 1–3, and 12 million unique sequencing reads were analyzed for the input library. The peak heights can be compared by the number of sequencing reads plotted on the *y*-axis. The number of reads observed at the binding site in the *COTL1* promoter correlates with the qPCR enrichment data shown in **Fig. 27.4**.

replicate having at least a 60% overlap. After the sequencing is performed, the short tags (~25–50 nt) are mapped to the human genome, the tags that map uniquely to only one location in the genome are selected, and then the unique tags are extended to the average size of the library fragments (~200 nt) and binned into consecutive bins running the length of each chromosome. The binned data can be visualized using the UCSC browser (http://www.genome.ucsc.edu) or the Affymetrix Integrated Genome Browser (http://www.affymetrix.com/partners_programs/-programs/developer/tools/download_igb.affx) (*see* **Fig. 27.5**). Target sites can be identified using a variety of peak calling methods (14, 15, 18–22).

4. Notes

1. This instrument is invaluable for measuring low DNA concentrations (e.g., 10 ng/uL) and for small sample volumes (as little as 1 μL of sample can be measured).
2. Do not use magnetic beads that have been blocked with foreign DNA, such as Herring sperm or Salmon sperm DNA. This may result in sequencing of the blocking DNA, resulting in lower quality ChIP-seq data. Although protein G binds antibodies from a variety of species (rabbit, mouse, goat, etc.) with high affinity, magnetic protein A beads can be used if desired.
3. Use the primer design program Primer3 (23) or another suitable program to design the target and control primers, making the product length 90–150 bp. If possible, design primers for at least two targets as well as for two negative control regions.
4. Chromatin can be sheared by sonication or digested by micrococcal nuclease. Although both methods work well, this protocol is based on sonication. An alternative method using micrococcal nuclease is available from cell signaling technology (http://www.cellsignal.com).
5. We use a BioRuptor on high setting for sonication. Wear hearing protection! Volumes between 0.5 and 2 mL are sonicated in 15 mL tubes, whereas volumes between 0.1 and 0.3 mL are sonicated in 1.5 mL Eppendorf tubes. The pulse duration, intensity, and number will vary depending on the sonicator, the extent of cross-linking, and cell type. Ideally the least amount of input energy that gives satisfactory fragmentation should be used. To optimize sonication conditions, it is recommended to take 20 μL samples of chromatin in 5 min intervals. We commonly sonicate for

30 min (pulses of 30 s at setting high, with 1.5 min pauses between pulses).

6. Antibody efficiency can vary significantly between different batches, resulting in variation of the quality of the resultant ChIP-seq data. It is therefore important to record antibody details, such as catalog number, lot number, batch of affinity purification, etc. It is helpful to test that the antibody works in a regular immunoprecipitation (IP) first before performing a ChIP assay (*see* **Fig. 27.6**). Ideally, the test immunoprecipitation is carried out in the same buffer as the ChIP assay. After immunoprecipitation, the sample is run on an SDS-PAGE gel and western blotted using the same antibody. The major band on the blot should be the correct size of the protein of interest. Antibodies that work in a test immunoprecipitation will most likely work in a ChIP assay. This standard immunoprecipitation assay is also a good way to optimize the amount of antibody that should be used for a specific number of cells. It should also be noted that antibodies that specifically recognize only one ZF protein are difficult to develop due to the high similarity of different ZFs to each other and to the low expression level of most ZFs. If the IP-western blot does not show a single major band (or if no antibody to a particular ZF is available), then we recommend expressing a tagged version of the ZF of interest. Transient expression of the tagged ZF protein, as well as stably transfected cell lines, can be used. However, in either case, the expression level of the tagged ZF protein should be determined by western blot analysis. The ChIP assay can then be performed with a commercially available antibody specific to the tag used, such as an HA or Flag tag, that has been previously shown to work in immunoprecipitations.

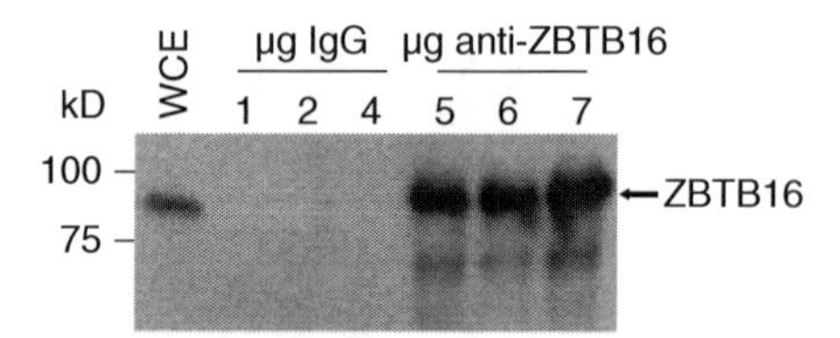

Fig. 27.6. IP-western blot analysis of the zinc finger protein ZBTB16. ZBTB16 was immunoprecipitated from LNCaP nuclear extracts using 1, 2, and 4 μg of either control IgG or anti-ZBTB16 antibody. ZBTB16 protein was detected in whole cell extract (WCE) prior to immunoprecipitation and in immunoprecipitations using ZBTB16 antibody, but not IgG. Based on the success of the antibody in these standard immunoprecipitations, it is expected that the antibody would be appropriate for use in ChIP assays.

7. This protocol is based on magnetic protein A/G beads because of their ease of use, but protein A/G

agarose beads as well as StaphA cells are also commonly used to capture the antibody/protein/DNA complexes. Some antibodies only give good ChIP enrichments when used in combination with StaphA cells. A detailed StaphA ChIP protocol can be found at http://www.genomecenter.ucdavis.edu/farnham. ChIP samples obtained from StaphA cells can be used for ChIP-seq, but high enrichment of the protein of interest by the antibody and proper technique to avoid carryover of StaphA cells is critical for the success of the experiment. Carryover of StaphA cells, in combination with low ChIP enrichment, results in amplification of StaphA DNA in the library preparation step, which ultimately leads to sequencing of StaphA DNA. This can significantly lower the quality of ChIP-seq data.

8. Unfortunately, the in vivo binding specificities of very few ZF proteins have been characterized and therefore binding sites are not known for most members of this family. When no target gene is known for the ZF protein of interest, the RNA polymerase II ChIP sample can be used to determine if the ChIP experiment itself was successful. Enrichment of RNA polymerase II can be observed at active promoters, such as the *GAPDH* promoter or the promoter for the large subunit of RNA polymerase II. In contrast, RNA polymerase II should not be significantly enriched at silenced promoters or at regions outside of a core promoter region; we routinely use a primer set to the 3′-end of the *ZNF333* gene as a negative control for RNA polymerase II.
9. Before testing the library, it is important to determine that the primers work for the SYBR Green-based real-time PCR assay. To do this, use input DNA and run a melting curve following the real-time PCR reaction conditions. View the dissociation curve to ensure that the desired amplicon was detected, as seen by a single peak.
10. An input library is also critical for determining a baseline genome for identification of binding sites. For each cell type, ~10–20 million sequenced tags of an input library are required. The same input library used to determine enrichment can be used for sequencing.

References

1. Emerson, R.O. and Thomas, J.H. (2008) Adaptive evolution in zinc finger transcription factors. *PLOS Genet.* **5**, e1000325.
2. Liu, J. and Stormo, G.D. (2008) Context-dependent DNA recognition code for C2H2 zinc-finger transcription factors. *Bioinformatics.* **24**, 1850–1857.

3. Cho, S.Y., Chung, M., Park, M., Park, S., and Lee, Y.S. (2008) ZIFIBI: prediction of DNA binding sites for zinc finger proteins. *BBRC.* **369**, 845–848.
4. Segal, D.J. and Barbas, C.F., 3rd (1999) Design of novel sequence-specific DNA-binding proteins. *Curr Opin Chem Biol.* **4**, 34–39.
5. Weinmann, A.S., Yan, P.S., Oberley, M.J., Huang, T.H.-M., and Farnham, P.J. (2002) Isolating human transcription factor targets by coupling chromatin immunoprecipitation and CpG island microarray analysis. *Genes Dev.* **16**, 235–244.
6. Johnson, D.S., Li, W., Gordon, D.B., Bhattacharjee, A., Curry, B., Ghosh, J., Brizuela, L., Carroll, J.S., Brown, M., Flicek, P., Koch, C.M., Dunham, I., Bieda, M., et al. (2008) Systematic evaluation of variability in ChIP-chip experiments using predefined DNA targets. *Genome Res.* **18**, 393–403.
7. Kirmizis, A., Bartley, S.M., Kuzmichev, A., Margueron, R., Reinberg, D., Green, R., and Farnham, P.J. (2004) Silencing of human polycomb target genes is associated with methylation of histone H3 lysine 27. *Genes Dev.* **18**, 1592–1605.
8. Kim, T.H., Barrera, L.O., Zheng, M., Qu, C., Singer, M.A., Richmond, T.A., Wu, Y., Green, R.D., and Ren, B. (2005) A high-resolution map of active promoters in the human genome. *Nature.* **436**, 876–880.
9. Cawley, S., Bekiranov, S., Ng, H.H., Kapranov, P., Sekinger, E.A., Kampa, D., Piccolboni, A., Sementchenko, V., Cheng, J., Williams, A.J., Wheeler, R., Wong, B., Drenkow, J., Yamanaka, M., Patel, S., Brubaker, S., Tammana, H., Helt, G., Struhl, K., and Gingeras, T.R. (2004) Unbiased mapping of transcription factor binding sites along human chromosomes 21 and 22 points to widespread regulation of noncoding RNAs. *Cell.* **116**, 499–509.
10. Carroll, J.S., Liu, X.S., Brodsky, A.S., Li, W., Meyer, C.A., Szary, A.J., Eeckhoute, J., Shao, W., Hestermann, E.V., Geistlinger, T.R., Fox, E.A., Silver, P.A., and Brown, M. (2005) Chromosome-wide mapping of estrogen receptor binding reveals long-range regulation requiring the forkhead protein FoxA1. *Cell.* **122**, 33–43.
11. Hoffman, B.G. and Jones, S.J. (2009) Genome-wide identification of DNA-protein interactions using chromatin immunoprecipitation coupled with flow cell sequencing. *J Endocrinol.* **201**, 1–13.
12. Weinmann, A.S., Bartley, S.M., Zhang, M.Q., Zhang, T., and Farnham, P.J. (2001) The use of chromatin immunoprecipitation to clone novel E2F target promoters. *Mol Cell Biol.* **21**, 6820–6832.
13. Loh, Y.-H., Wu, Q., Chew, J.-L., Vega, V.B., Zhang, W., Chen, X., Bourque, G., George, J., Leong, B., Liu, J., Wong, K.-Y., Sung, K.W., Lee, C.W., Zhao, X.D., Chiu, K.P., Lipovich, L., Kuznetsov, V.A., Robson, P., Stanton, L.W., Wei, C.L., Ruan, Y., Lim, B., and Ng, H.H. (2006) The Oct4 and Nanog transcription network regulates pluripotency in mouse embryonic stem cells. *Nat Genet.* **38**(4), 431-440, on line March 5, 2006.
14. Robertson, G., Hirst, M., Bainbridge, M., Bilenky, M., Zhao, Y., Zeng, T., Euskirchen, G., Bernier, B., Varhol, R., Delaney, A., Thiessen, N., Griffith, O.L., He, A., Marra, M., Snyder, M., and Jones, S. (2007) Genome-wide profiles of STAT1 DNA association using chromatin immunoprecipitation and massively parallel sequencing. *Nat Methods.* **4**, 1–7.
15. Johnson, D.S., Mortazavi, A., Myers, R.M., and Wold, B. (2007) Genome-wide mapping of in vivo protein-DNA interactions. *Science.* **316**, 1497–1502.
16. Cuddapah, S., Jothi, R., Schones, D.E., Roh, T.-Y., Cui, K., and Zhao, K. (2009) Global analysis of the insulator CTCF in chromatin barrier regions reveals demarcation of active and repressive domains. *Genome Res.* **19**, 24–32.
17. Frietze, S., Lan, X., Jin, V.X., and Farnham, P.J. (2010) Genomic targets of the KRAB and SCAN domain-containing zinc finger protein 263. *J Biol Chem.* **285**, 1393–1403.
18. Blahnik, K.R., Dou, L., O'Geen, H., McPhillips, T., Xu, X., Cao, A.R., Iyengar, S., Nicolet, C.M., Ludäscher, B., Korf, I., and Farnham, P.J. (2010) Sole-Search: an integrated analysis program for peak detection and functional annotation using ChIP-seq data. *Nucleic Acids Res.* **38**, e13.
19. Fejes, A.P., Robertson, G., Bilenky, M., Varhol, R., Bainbridge, M., and Jones, S.J.M. (2008) FindPeaks 3.1: a tool for identifying areas of enrichment from massively parallel short-read sequencing technology. *Bioinformatics.* **24**, 1729–1730.
20. Xu, H., Wei, C.-L., Lin, F., and Sung, W.-K. (2008) An HMM approach to genome-wide identification of differential histone modification sites from ChIP-seq data. *Bioinformatics.* **24**, 2344–2349.
21. Zhang, Y., Liu, T., Meyer, C.A., Eeckhoute, J., Johnson, D.S., Bernstein, B.E., Nussbaum, C., Myers, R.M., Brown, M., Li, W.,

and Liu, X.S. (2008) Model-based analysis of ChIP-Seq (MACS). *Genome Biol.* **9**, R137.
22. Jothi, R., Cuddapah, S., Barski, A., Cui, K., and Zhao, K. (2008) Genome-wide identification of in vivo protein-DNA binding sites from ChIP-seq data. *Nucleic Acids Res.* **36**, 5221–5231.
23. Rozen, S. and Skaletsky, H.J. (2000) Primer3 on the WWW for general users and for biologist programmers. In: (Krawetz, S., and Misener, S., Eds.), Bioinformatics methods and protocols: methods in molecular biology, pp. 365–386. Humana Press, Totowa.

Chapter 28

Crystallization of Zinc Finger Proteins Bound to DNA

Nancy C. Horton and Chad K. Park

Abstract

A method is presented for determining conditions for the cocrystallization of zinc finger proteins with DNA. The method describes steps beginning with protein expression, through purification, design of DNA for cocrystallization, and the conditions to screen for cocrystallization.

Key words: Crystallization, zinc finger, DNA, X-ray crystal structure, hanging drop vapor diffusion.

1. Introduction

Here we present a method used in our laboratory for the cocrystallization of zinc finger proteins (ZFPs) and other DNA-binding proteins with DNA. The crystallization experiments described are performed using the hanging drop vapor diffusion method of crystallization (1). Important parameters include the initial choice of expression vector, purification of the ZFP, design of DNA for cocrystallization, and the solution conditions to be used in screening for cocrystallization. The expression vector choice is important, since relatively large quantities (milligrams) of protein are needed, which must be purified to a very high level. If initial expression of the protein is strong, then fewer purification steps are needed. In addition, the expression vector can add sequences useful for purification and can even be cleaved from the ZFP efficiently. We have not tested all of the available expression vectors, but suggest the expression vector that has been useful in our work.

J.P. Mackay, D.J. Segal (eds.), *Engineered Zinc Finger Proteins*, Methods in Molecular Biology 649,
DOI 10.1007/978-1-60761-753-2_28, © Springer Science+Business Media, LLC 2010

The design of DNA to be used in cocrystallization experiments is critical. Several variants should be tested, which are varied both in length and in the nature of the ends of the DNA molecules. We provide a summary of DNA constructs that have been used successfully in other ZFP/DNA crystals, as well as in our own. We have found that each different DNA construct leads to different crystal qualities including morphology, unit cell, space group, crystallization conditions, and the resolution of diffraction from the crystal. Since the variation on the cocrystallization DNA is so important, we have found that far less screening of crystallization conditions (i.e., salt, pH, precipitating agent) is required. We suggest two screens, each requiring only 20 individual conditions to be tested, that have been particularly successful in our lab.

2. Materials

In general, all chemicals should be of ACS reagent grade or ≥99.0% purity.

2.1. Protein Expression

1. Expression vector (e.g., pMAL-c4X, New England Biolabs, for protein expression).
2. Competent cells (for protein expression, e.g., BL21 *Escherichia coli*).
3. Luria-Bertani medium (LB) for growth: in 1 L total volume, 10 g tryptone, 5 g yeast extract, 10 g NaCl, adjust to final pH of ≈7 with NaOH, autoclave 30 min, and add sterile filtered glucose to 0.2% (w/v).
4. 50 mg/mL ampicillin in double distilled H_2O (dd H_2O), sterile filtered (0.2 μm), stored at –20°C.
5. LB/ampicillin plates (for transformations and growth): LB with 15 g/L agar, autoclaved, cooled until glass is comfortable to touch, then add 50 μg/mL final concentration of sterile ampicillin and pour into Petri dishes.
6. 0.3 mM isopropyl-β-D-1-thiogalactoside (IPTG), sterile filtered.
7. 60% glycerol (sterile filtered through 0.2 micron filter).
8. Equipment: Petri dishes, fermenter with 18 L capacity (optional).

2.2. Protein Purification

1. Breaking buffer (BB): 100 mM Tris–HCl, pH 7.4, 800 mM NaCl, 1 mM EDTA (ethylenediaminetetraacetic acid), 1 mM PMSF (phenylmethanesulfonyl fluoride), 1 mM benzamidine, 10 mM BME (β-mercaptoethanol). Can be stored at 4°C for up to 1 week.

2. Amylose buffer A (ABA): 20 mM Tris–HCl, pH 7.5, 90 μM $ZnCl_2$, 150 mM KCl, 10 mM BME. Can be stored at 4°C for up to 1 week.
3. Amylose buffer B (ABB): ABA with 1.5 M KCl. Can be stored at 4°C for up to 1 week.
4. Amylose buffer C (ABC): ABA with 10 mM maltose.
5. Cleavage buffer: 20 mM Tris–HCl, pH 8.0, 100 mM NaCl, 2 mM $CaCl_2$, 10 mM BME.
6. Zinc buffer A (ZBA): 20 mM Tris–HCl, pH 7.5, 90 μM $ZnCl_2$, 1 mM $MgCl_2$, 90 mM KCl, 10 mM BME. Store at 4°C for up to 1 week.
7. Enzymes: Factor Xa protease (New England Biolabs) or appropriate protease.
8. Resins: Amylose resin (agarose coupled, New England Biolabs) and optionally heparin resin (e.g., HiPrep 16/10 Heparin FF, GE Healthcare; Mono Q HR 16/10, GE Healthcare).
9. Standard equipment and reagents for SDS-PAGE and staining with Coomassie blue.
10. Other equipment: Sonicator (e.g., Branson Sonifier 450), chromatography columns, dialysis tubing, dialysis clips, Centricon ultrafiltration devices (Millipore).

2.3. Crystallization of ZFPs with DNA

1. 18.2 MΩ cm water and glassware (*see* **Note 1**).
2. Synthetic oligonucleotides for cocrystallization (reverse-phase cartridge purified or better).
3. ZBA (*see* step 6 of **Section 2.2**).
4. 1.0 M sodium acetate: titrate 1 M acetic acid with 1 M NaOH to pH 4.5.
5. 1.0 M sodium citrate: titrate 1 M citric acid with NaOH to pH 5.5.
6. 1.0 M imidazole: titrate 1 M imidazole with HCl to pH 6.5.
7. 1.0 M 4-(2-hydroxyethyl)-1-piperazineethanesulfonic acid (HEPES): titrate 1 M HEPES with NaOH to pH 7.5.
8. 1.0 M Tris–HCl: titrate 1 M Tris base with HCl to pH 8.5.
9. 0.02% NaN_3 in 1.0 M NaCl. Note that NaN_3 is toxic; use gloves and appropriate precautions when handling.
10. 50% w/v PEG 4 K in water (the highest purity PEG 4 K is available from Hampton Research). Add 0.02% NaN_3 (toxic, use gloves when handling) to prevent fungal growth.
11. 4.0 M ammonium sulfate, freshly filtered before use.

12. Crystallization materials and equipment: Linbro or VDX crystallization plates (Hampton Research), glass cover slips (22 mm, siliconized), silicone vacuum grease, crystallization incubator or styrofoam boxes, cryoloops, seeding pipette, three-well glass plate.

3. Methods

3.1. Protein Expression

1. Select and clone your gene of interest into an appropriate vector (*see* **Note 2**).
2. Perform expression tests prior to the large-scale growth (*see* **Note 3**).
3. For the large-scale growth, inoculate 3 mL of LB containing 50 μg/mL ampicillin using a single colony from a fresh (less than 1 month) plate of the cells from a glycerol stock (*see* **Note 2**), and grow the culture overnight at 37°C with shaking. The next day, we use the 3 mL to inoculate 1 L LB containing 50 μg/mL ampicillin and grow the culture overnight at 37°C with shaking. The following day, the 1 L of culture is used to inoculate 18 L of LB containing 50 μg/mL ampicillin, 0.2% glucose, and 90 μM $ZnCl_2$ in a fermenter, which is then grown at 37°C with aeration. If a fermenter is not available, multiple 1 L cultures can be grown in 2.8 L Fernbach flasks.
4. Induce using the conditions optimized in step 2. For Aart, induce expression with 0.3 mM IPTG when the OD at 600 nm reaches 0.6. The cultures are grown overnight, and the next day the cells are pelleted by centrifugation at 5,000 rpm. The cells are stored by first dropping spoonfuls into a Dewar containing liquid nitrogen, then placing the frozen cells in a closed plastic bottle, and storing at –80°C.

3.2. Protein Purification

3.2.1. Prepare Cell-Free Lysate

1. Thaw, if necessary, and mix cells from the large growth into 4 mL breaking buffer (BB) per 1 g of cells (*see* **Note 4**). Keep the cells in a beaker on ice. Break the cells with sonication (e.g., a Branson Sonifier 450 set at 50%, output 4, with 20 pulses per set).
2. Mix the cells between sets and take a 20 μL aliquot after each set. After five sets, spin the aliquots at 10,000×*g* for 10 min

in a microfuge. Use 5 μL of the supernatant diluted in 1 mL of BB to read the absorbance over the range 220–330 nm.

3. Repeat the sonication sets until the absorbance at 260 nm of the centrifuged aliquots reaches a constant value, indicating no further cell breakage. This often takes as many as 20 sonication sets or more.

4. Separate the cell debris from the cell-free lysate by centrifuging at 30,000×*g* (16,000 rpm in a Sorval SS-34 rotor) for 30 min.

3.2.2. Amylose Resin Chromatography

This first chromatographic step utilizes the ability of the MBP fused to the ZFP to bind to amylose, which is coupled to agarose beads. Note that the agarose-coupled amylose resin has a finite lifetime that is reduced with each purification (amylases from *E. coli* reduce the binding capacity of the resin, and the added glucose to the growth medium helps to reduce their expression).

1. Use 1 mL of resin per 2 mL of cell-free lysate. (The manufacturer notes that 3 mg of MBP binds 1 mL of resin. For very strong MBP-ZFP expression systems, more resin may be required.)

2. The resin must be rinsed with 5–10 column volumes (CV) of water to remove the storage buffer (usually 20% ethanol or 0.02% sodium azide), then equilibrated with 5–10 CV of amylose buffer A (ABA).

3. Load the sample onto the resin. This can be done in a beaker or by flowing lysate over the resin in a column. Incubation times from 1 h to overnight at 4°C have been used.

4. If using a beaker for batch binding your protein, load the resin/lysate slurry into a column. Wash column with 5 CV of the ABA, 3 CV of ABB (containing high salt to wash away any copurifying DNA), and then 2 CV of ABA.

5. The protein is then eluted with 2–3 CV of ABC (containing 10 mM maltose). At this point the protein solution may be stored at −20°C as a 50% v/v glycerol solution. Subsequent steps will need to be prefaced by dialysis since protease cleavage is not appreciable in 50% glycerol.

3.2.3. Cleavage

1. The protein should be dialyzed into cleavage buffer or concentrated and cleavage buffer added (4 volumes cleavage buffer to 1 volume of eluted ZFP solution). For MBP–Aart fusion protein, 12,000 MWCO dialysis tubing is used.

2. Factor Xa is added (approximately 1 μg of factor Xa to 50 μg of fusion protein). Fusion protein concentration is estimated from measurement of the optical density of the solution at 280 nm. Shake at room temperature for 16 h. Check for complete cleavage by SDS-PAGE.

3.2.4. Separation of Zinc Finger from Uncleaved Fusion Protein, Protease, and MBP

1. There are several methods of separating cleaved ZFP from the protease, MBP fusion protein, and uncleaved ZFP. The pMal vector information describes ion exchange chromatography. The simplest method that has successfully recovered ZFP in our hands is to use amylose chromatographic resin to bind the MBP and uncleaved ZFP leaving the cleaved product to flow through the column.
2. Dialyze the ZFP exhaustively against ABA to remove the maltose prior to the amylose chromatography. Also, wash the amylose resin free of maltose and any contaminants.
3. Load the cleaved protein mixture onto the amylose resin (*see* **Section 3.2.2**). Typically, we use 1 h incubations of solution and resin on ice.
4. Collect the solution that flows through the resin as the cleaved ZFP. This solution will also contain the protease. However, we have found that at the low concentrations of use, the protease is not visible by Coomassie-stained SDS-PAGE. The resin can be regenerated by eluting off the MBP and uncleaved ZFP with maltose.
5. Analyze the proteins by SDS-PAGE followed by staining with Coomassie blue. Concentrations can be determined by UV/Vis absorbance at 280 nm. Such a purification in our lab resulted in the protein shown in **Fig. 28.1** lane 4. Crystallization requires very pure protein, and the laboratory standard of purity is defined as the absence of any additional visible bands on a Coomassie-stained SDS-PAGE gel

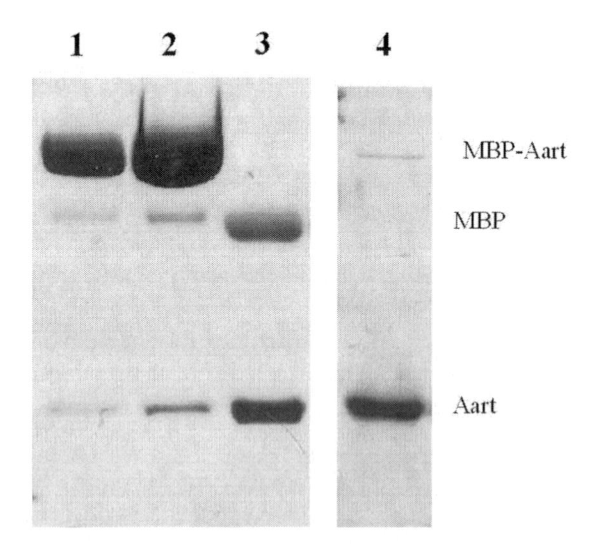

Fig. 28.1. Cleavage and repurification of Aart. *Lanes* 1–2: partial cleavage of MBP–Aart; *Lane* 3: complete cleavage of MBP–Aart into MBP and Aart; *Lane* 4: Aart following purification of cleaved protein from *Lane* 3 using amylose resin chromatography.

overloaded with the protein (25–50 μg, depending on the gel thickness, the protein should form a large, dark band). Although one additional band is visible in the protein of **Fig. 28.1** in lane 4 (most likely the uncleaved fusion protein), this protein sample did in fact produce usable crystals. In general, though, the ZFP should be of the highest purity in order to increase the chances of crystallization (*see* **Note 5** for modifications of purification protocol).

3.2.5. Storage of ZFP

1. Dialyze into ZBA with 50% glycerol, aliquot into 1 mg aliquots in 1.5 mL tubes.
2. Concentrate the ZFP if necessary (with centrifugal concentrators to >1 mg/mL) before dialyzing into glycerol containing buffer.
3. Flash freeze in liquid nitrogen and store at –80°C.

3.3. Cocrystallization of ZFP with DNA

3.3.1. Design DNA for Cocrystallization

1. Design DNA oligonucleotide containing the zinc finger recognition sequence with additional nucleotides or base pairs both upstream and downstream (*see* **Note 6**).
2. We typically design and screen at least three different DNA duplexes, although screening up to 10 or more in order to obtain crystals with the desired diffraction qualities is not unusual.
3. In **Fig. 28.2**, the zinc finger recognition sequence is shown in the large black rectangle and does not vary from one construct to the other. The small black boxes are the nucleotides (base pairing) that can be varied in length and composition. We typically use random sequences here, and do not usually vary the sequence from one oligonucleotide construct to another. The overhanging nucleotides are shown in small grey boxes, with G on the top strand and C on the bottom, making base pairing overhangs. The overhang need not

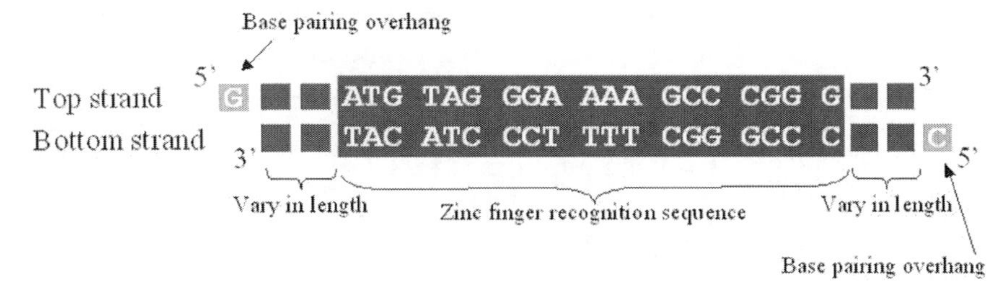

Fig. 28.2. Design of DNA for cocrystallizaton. Bases with *black* background indicate the zinc finger recognition sequence. The four *squares* on each side of the recognition sequence represent variable length regions. The *grey* 'G' and 'C' indicate the possible inclusion of GC overhangs.

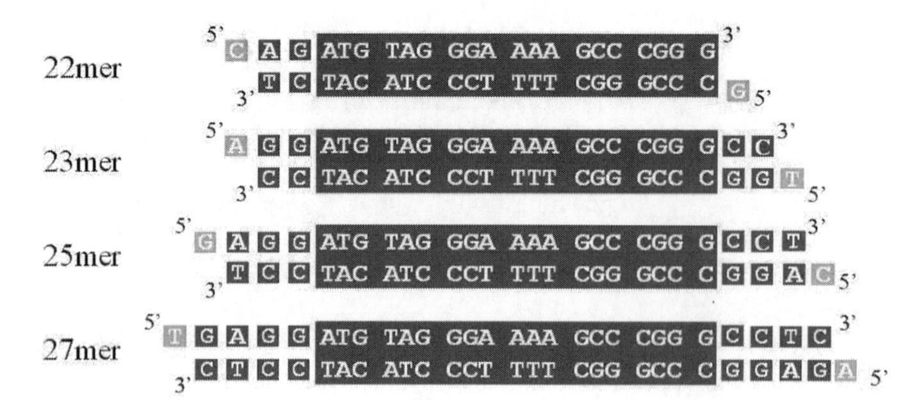

Fig. 28.3. DNA used in cocrystallization screens with Aart, a six-finger zinc finger protein. Colors follow the definitions in **Fig. 28.2**.

be only on the 5′-end, nor base paring, and crystals have been formed with non base pairing overhangs, as well as 3′-overhangs. We generally use 5′-base pairing overhangs as well as blunt duplexes without any overhangs.

4. For cocrystallization with Aart, we screened the DNA duplexes shown in **Fig. 28.3** (2).
5. The 22mer was designed based on the crystal structure of a three-finger ZFP bound to DNA that crystallized with two copies in the asymmetric unit (mimicking a six finger:DNA complex) (2). Only the 22mer gave decent-sized crystals that were eventually used in the structure solution of Aart bound to DNA (3). The 23mer, 25mer, and 27mer produced only microcrystals with Aart. As this example shows, the cocrystallization process is empirical and screening many DNA constructs increases the chances of obtaining usable crystals.
6. The DNA oligonucleotides are prepared synthetically, and many companies sell DNA oligonucleotides with sequences to order. The top and bottom strands are ordered separately. Most companies that sell synthetic oligonucleotides also offer the option of purification; either by gel or by reverse-phase HPLC (RP-HPLC). Purification by RP-HPLC is best, although we have obtained usable crystals from DNA constructs that have been only desalted after synthesis. However, especially for the initial screen, the oligonucleotides should be purified by RP-HPLC. A method for purification has been published (4).

3.3.2. Prepare ZFP for Crystallization

1. Remove an aliquot (~1 mg protein) from the –80°C and place on ice.
2. Dialyze a sample of 5 mL or less against four changes of 250 mL of dialysis buffer (ZBA). Use dialysis tubing

with an appropriate molecular weight cutoff (Aart = 21,000 Da, therefore we use tubing with a cutoff of 12,000–14,000 Da). Dialyze at 4°C with gentle stirring for at least 4 h per 250 mL. Try not to let the dialysis go beyond 24 h and seal the dialysis with a minimal amount of air above the buffer. Parafilm, or preferably jars that seal tight, can be used.

3. After dialysis, the ZFP should be concentrated to 10 mg/mL. This can be accomplished using a Centricon ultrafiltration device with an appropriate molecular weight cutoff (*see* **Note 7**).
4. First, rinse the Centricon unit with ZBA by centrifuging at 1,000×*g* for 15 min at 4°C. After the spin, remove the buffer.
5. Add the ZFP solution to the Centricon unit and centrifuge at 3,000–4,000 rpm for 15 min at 4°C. After the spin, mix the ZFP solution using a pipette, as the protein can adhere to the filter and block the pores. More protein solution can be added, and the process repeated, until the volume of the protein solution gives a predicted concentration of 10 mg/mL.
6. Measure the concentration using absorbance at 280 nm: blank the spectrophotometer with dialysis buffer (ZBA) and measure the full spectrum from 220 to 330 nm. The protein should produce a peak near 280 nm with absorbance of 0 at 330 nm. If the absorbance at 330 nm is significant, then the protein may contain particulates that scatter light and interfere with the absorption measurement. To remove the particulates, centrifuge the sample at 10,000×*g* for 10 min, transfer to a new tube, and repeat the absorbance measurement.
7. The extinction coefficient, ε, for the zinc finger can be calculated from the amino acid sequence using the following formula (5): ε = (number of Tyr) × ε(Tyr) + (number of Trp) × ε(Trp), where ε(Tyr) = 1,490 L/mole/cm and ε(Trp) = 5,500 L/mole/cm (for proteins in water measured at 280 nm). The calculation of extinction coefficients, molecular weights, and other parameters base on amino acid sequence can be carried out at Expasy: http://www.expasy.org/tools/protparam.html
8. The concentration is calculated from Beer's Law: $C = A/\varepsilon b$, where A is absorbance at 280 nm, ε is the extinction coefficient, and b is the path length in the spectrophotometer, usually 1 cm. The concentration of the ZFP should be at least 10 mg/mL if possible, but can be higher, up to 30 mg/mL.

3.3.3. Prepare DNA for Cocrystallization

1. Wear gloves when handling DNA to prevent nuclease contamination of the DNA sample. All buffers must be prepared carefully using sterile ddH_2O water handled with gloves.
2. The purified synthetic DNA to be used in cocrystallization is dissolved in the dialysis buffer used for the ZFP (ZBA or ZBA with 400–600 mM NaCl) to a final concentration of at least 10 mg/mL.
3. Measure the concentration of the DNA by absorbance at 260 nm, with ZBA buffer as the background blank, and measuring from 220 to 330 nm. The DNA should give a peak at 260 nm with very little absorbance at 330 nm. If significant absorbance occurs at 330 nm, centrifuge the sample at $10{,}000\times g$ for 10 min, transfer to a new tube, and repeat the measurement.
4. Convert the absorbance to concentration using the extinction coefficient calculated from the sequence using the following formula (6, 7): ε = (number of *dA*)× (15,400) + (number of *dC*)× (7,400) + (number of *dG*)× (11,500) + (number of *dT*)× (8,700) in L/mole/cm.
5. Anneal the DNA by mixing equimolar quantities of top and bottom strands for each construct and heating to 98°C for 5 min, followed by cooling to room temperature or 4°C over several hours. This can be performed in a thermocycler, or using a heat block set to 98°C and turned off after the addition of the samples, then covered with some insulating material (e.g., test tube racks or freezer bags equilibrated to room temperature).
6. The annealed DNA should be stored at –20°C. Note that repeated freeze–thawing of the DNA can cause un-annealing of the DNA, and it can be preferable to reanneal just before use.

3.3.4. Mix ZFP with DNA

1. Label 1.5 mL tubes for each of the different DNA constructs to be screened. The initial screen will require a total volume of 30 μL per tray, but since some volume is typically lost to the sides of the tubes, 35 μL should be prepared for each tray.
2. Calculate the quantity of DNA to add to the ZFP solution. We use excess duplex DNA to ZFP (i.e., a molar ratio of 1.5:1 DNA:ZFP), since in our experience some proportion of the DNA is not competent for binding by the ZFP, and the excess DNA is less detrimental to crystallization than is the excess protein.
3. Mix the ZFP solution with the DNA and note any precipitate. If precipitation occurs, the concentration of NaCl

should be increased. In this case, add additional NaCl from a 1 M stock to 450, 500, 550, or 600 mM, watching for the disappearance of the precipitate after each addition. Do not add more NaCl than is necessary to solubilize the precipitate.

4. The concentration of the ZFP in the final solution should be 5 mg/mL or higher. If it is lower than 5 mg/mL due to dilution with NaCl or DNA, it should be concentrated as described in **Section 3.3.2**.

5. Note that the actual concentration of the ZFP cannot be measured reliably at this point using absorbance at 280 nm, since the DNA in the mixture will dominate the absorption spectrum. We concentrate the zinc finger/DNA samples to a volume to give a predicted concentration of the zinc finger of at least 5 mg/mL.

3.3.5. Crystallization Screen Number 1: PEG vs. pH

1. The experiments are conducted in 24-well plates. Use 20 of the 24 wells in the screen, leaving the first well of each row empty. The lip of each well must be greased with silicon vacuum grease (alternatively, trays can be purchased pre-greased).
2. The easiest way to grease the lips is by using a 5 mL syringe filled with the grease, and applying a uniform 2 mm coating of the grease to the entire lip of each well. To fill the 5 mL syringe with the vacuum grease, heat the vacuum grease tube in a flask of hot water first, then squeeze into the barrel of the 5 mL syringe before adding the plunger.
3. The screen is composed of varying pH in each column (pH 4.5, 5.5, 6.5, 7.5, 8.5), and varying % PEG 4 K in each row (25, 20, 15, 10%). Each well also contains 150 mM NaCl.
4. **Table 28.1** gives an example of the components to be added to the wells of a crystallization plate for such a screen. Use a 1 mL pipette to mix the contents of each well by pumping in and out of the pipette tip until Schlieren mixing lines (visible lines in the solutions due to differences in local densities) are no longer visible.
5. Each drop should be prepared independently to minimize drop drying. Place 1.5 μL of the ZFP/DNA mixture onto the drop and 1.5 μL of the well solution (*see* **Note 8** for variations). Place the cover slip with the drop side down over the well and be sure that the grease seals the cover slip completely over the well. Prepare the remaining drops similarly.
6. Place the finished tray (labeled with a unique number) in the either a crystallization incubator or a styrofoam box at room temperature (to help limit temperature fluctuations).

Table 28.1
Crystallization screen number 1: PEG vs. pH

Well	50% PEG 4K (μL)	1 M buffer	1 M NaCl (μL)	ddH_2O (μL)
A2	500	100 μL sodium acetate pH 4.5	150	250
A3	500	100 μL sodium citrate pH 5.5	150	250
A4	500	100 μL imidazole pH 6.5	150	250
A5	500	100 μL HEPES pH 7.5	150	250
A6	500	100 μL Tris–HCl pH 8.5	150	250
B2	400	100 μL sodium acetate pH 4.5	150	350
B3	400	100 μL sodium citrate pH 5.5	150	350
B4	400	100 μL imidazole pH 6.5	150	350
B5	400	100 μL HEPES pH 7.5	150	350
B6	400	100 μL Tris–HCl pH 8.5	150	350
C2	300	100 μL sodium acetate pH 4.5	150	450
C3	300	100 μL sodium citrate pH 5.5	150	450
C4	300	100 μL imidazole pH 6.5	150	450
C5	300	100 μL HEPES pH 7.5	150	450
C6	300	100 μL Tris–HCl pH 8.5	150	450
D2	200	100 μL sodium acetate pH 4.5	150	550
D3	200	100 μL sodium citrate pH 5.5	150	550
D4	200	100 μL imidazole pH 6.5	150	550
D5	200	100 μL HEPES pH 7.5	150	550
D6	200	100 μL Tris–HCl pH 8.5	150	550

We typically use a crystallization incubator set at 17°C (*see* **Note 9**).

7. Prepare a sheet for each crystallization tray that will allow the recording of observations for each well. The sheet should also contain information on the contents of each drop and well and the temperature and dates of setup.
8. Record observations about once a week. It is unusual to find diffraction quality crystals in the first screen. However, important observations include the conditions under which precipitation occurs; crystallization requires that the complex be reduced in solubility, so if all drops are clear, then the protein and/or the precipitating agent concentrations must be increased. In addition, note any conditions where any crystalline type of precipitate forms.
9. Follow-up experiments should include finer gradations of precipitating agent (i.e., PEG 4 K) between the concentrations where some and no precipitation occurs. In addition, follow-up screens should include varying the salt (i.e.,

NaCl) concentration vs. PEG, at pH values that give any crystalline precipitate. Varying the drop size and ratio of precipitant to ZFP:DNA complex (i.e., 1.0 μL/1.5 μL, and 1.5μL/1.0 μL), as well as testing the effect of the addition of additives, is recommended for improving crystal size and morphology (*see* **Section 3.5.6**).

10. Repeating the initial screen but placing the tray at 4°C is also recommended, as results at 4°C can often be very different than those at 17°C or room temperature.

3.3.6. Crystallization Screen Number 2: Ammonium Sulfate vs. pH

This screen varies pH across the tray as above, but instead of the PEG and NaCl, only ammonium sulfate is used.

1. To each column, add 100 μL of appropriate 1.0 M buffer (using the buffers from screen number 1, pH 4.5, 5.5, 6.5, 7.5, 8.5).
2. To each row, add ammonium sulfate: row A, add 550 μL of 4.0 M ammonium sulfate (2.2 M final); row B, add 500 μL of 4.0 M (2.0 M final); row C, add 450 μL of 4.0 M (1.8 M final); row D, add 400 μL of 4.0 M (1.6 M final).
3. Use ddH_2O to bring each well to 1 mL and mix using a 1 mL pipette.
4. Prepare the drops as above, using 1.5 μL of each well solution in the hanging drop and 1.5 μL of the ZFP:DNA solution.
5. Place the tray in the desired temperature (room temperature, 4 and 17°C are recommended) and observe the results after 1 week. If no precipitation occurs even at the highest precipitant concentration, then the concentration of the ZFP/DNA complex and/or the ammonium sulfate concentration must be increased. Remember, crystals form under conditions where the ZFP/DNA complex is slowly coming out of solution, so conditions that produce slow precipitation are desired. Heavy immediate precipitation is undesirable and indicates that the precipitant and/or ZFP/DNA concentration are too high.

3.3.7. Additional Screens

1. Many additional crystallization screens are commercially available (Hampton Research, Qiagen). Crystallization is an empirical process and large amounts of the ZFP/DNA solution may be required to test the large number of potential precipitating agents.
2. We have found that crystals are usually obtained with the PEG vs. pH screen, and that varying the DNA construct used in the crystallization provides a greater chance of success in crystallization. In addition, the purity of the ZFP should be at least 99% or higher for the greatest chance of success.

3.3.8. Optimization of Cocrystallization Conditions

1. If large single crystals that diffract to high resolution are obtained from the initial screens, then optimization need not be performed. However, typically the crystals form as showers in the drop and do not become large enough for data collection. Additionally, crystals may be undesirable in morphology, such as thin needles or plates, or growing as stacks of small crystals.
2. To improve the size and morphology of the crystals, the crystallization conditions can be varied including varying the salt (i.e., NaCl or ammonium sulfate) concentration by 10–50 mM at a time, co-varying with 1% increments in the PEG 4 K concentration in the case of screen number 1, while keeping the other parameters constant (i.e., the pH).
3. Testing the effect of 0.2 unit increments in pH can also lead to improved crystallization conditions.
4. The largest crystals will grow when one or only a few nucleations occur in the drop, allowing the ZFP:DNA complex to be added to these nucleations, rather than forming additional small crystals. Precipitant concentrations (and/or ZFP:DNA concentrations) that are too high will lead to multiple nucleations, but if too low then no nucleation will occur. Therefore, conditions must be found where the ZFP:DNA complex loses solubility slowly as the vapor diffusion process results in concentration of components of the drop.
5. If optimizing conditions using PEG 4 K, other molecular weight PEG can be tested, as well as modified PEG such as monomethyl ether derivations (available from Hampton Research).
6. Additives to be tested include salts (commonly ammonium chloride, ammonium acetate, ammonium phosphate, ammonium sulfate, sodium chloride, potassium chloride, calcium chloride, sodium potassium tartarate, sodium acetate, magnesium chloride, sodium citrate, and lithium sulfate. Essentially any water soluble salt can be tested), detergents, different molecular weight PEG, modified PEG (i.e., monomethyl ethers), and organics (e.g., methylpentanediol, *tert*-butanol, and DMSO).
7. We have found that 10 mM $NaNO_3$ can often alter crystal morphology from stacks of plates to large single crystals.
8. The final conditions for Aart crystals using a hanging drop vapor diffusion method at 17°C in for at least 1 week are: Drop: 1.5 μL of Aart/DNA (Aart 10 mg/mL, 2:1 molar ratio zinc finger) and 1.5 μL of 22% PEG 4 K,0.1 M sodium citrate pH 5.5, 0.1 M ammonium acetate. Well: 22% PEG

4 K, 0.1 M citrate pH 5.5, 0.1 M ammonium acetate, 0.2 M NaCl.

9. If crystals still do not grow to a large enough size, seeding can be attempted.

3.3.9. Freezing Crystals for Storage and for Data Collection

1. The crystals must be exchanged into a suitable cryoprotectant that protects the crystals from the destruction of water ice crystal formation, but does not dissolve the ZFP:DNA crystal. We use a cryoprotectant that contains 30% glycerol in addition to components that mimic the crystallization solution, but with slightly higher concentrations to offset the dissolving effect of glycerol. For example, if crystals grow in 15% PEG 4 K, 0.1 M citrate, pH 5.5, 0.1 M ammonium acetate, and 0.3 M NaCl, the cryoprotectant is made of 30% glycerol, 20% PEG 4 K, 0.1 M citrate, pH 5.5, 0.2 M ammonium acetate, and 0.3 M NaCl.
2. The cryoprotectant is added to the crystal in its drop using very small volumes at first (3 μL), followed by mixing with the pipette. Eventually larger volumes are added (10 μL).
3. Finally, much of the volume of the drop is removed and 100% cryoprotectant added to ensure complete exchange into the cryoprotectant. The whole process of exchange into cryoprotectant should be performed relatively quickly (i.e., 5 min) to minimize any negative effects on the crystal.
4. The crystal is then scooped up in a cryoloop and placed in liquid nitrogen in a cryotube. Keep in liquid nitrogen until data collection.

4. Notes

1. Prepare solutions to be used in the crystallization screen with Milli-Q grade water (ddH_2O, resistivity of 18.2 MΩ cm). Rinse all glassware three times and wear sterile gloves to prevent nuclease contamination. Store at room temperature in closed containers unless otherwise noted.
2. Expression vectors for expression of the ZFP in *E. coli* can be obtained from commercial (Invitrogen, New England Biolabs) or academic/noncommercial sources (Dana-Farber/Harvard Cancer Center DNA Resource Core, or Addgene Inc.). Vectors vary in expression level, method of control (e.g., the lac repressor/IPTG), and the presence or absence of coding sequences for tags that facilitate purification. The tags may or may not have linker sequences allowing their cleavage from the ZFP using a specific

protease. We have had success with the pMAL-c2X expression vector from New England Biolabs. This expresses a factor Xa-cleavable, N-terminal maltose-binding protein (MBP) prior to the multiple cloning site. With appropriate subcloning, the protease site can be placed adjacent to the N-terminus of the ZFP, cleaving away all non-zinc finger residues. The MBP–Aart construct contained 11 additional amino acid residues C-terminal to the protease cleavage site; however, these did not appear to interfere with cocrystallization (3, 8). Other versions of the pMAL-c4 vector have protease cleavage sites for the proteases enterokinase or Genenase I (New England Biolabs). Another variant of the expression vector, pMAL-p4X, contains a signal sequence on the MBP that targets the protein for the periplasmic space, which can be helpful if toxicity is encountered with the recombinant protein. However, the expression from this version can be much less from the cytoplasmic version (pMAL-c4X). We found good expression of the Aart ZFP in the pMAL-c4X vector (at least 1 mg of purified protein from 1 L of cell culture). Note that pMAL-c2X is an older version of the pMAL-c4X, the difference is the incorporation of mutations into MBP in the newer vectors that are reported to increase amylose resin binding (New England Biolabs). Since ZFPs are susceptible to oxidation at the cysteine residues that ligate the zinc ion, reducing agents such as BME or TCEP must always be present, and New England Biolabs indicates that factor Xa, enterokinase, and Genenase I are compatible with 10 mM BME. Alternatively, the TEV protease cleavage site, ENLYFQF|S (| indicates cleavage site), can be placed just upstream of the ZFP-coding sequence in any given expression vector (9).

Once the expression vector is made, it should be stored at –20 or –80°C in TE (10 mM Tris–HCl, pH 7.5, 1 mM EDTA). Transformation into *E. coli* can be performed with competent cells (available from various commercial sources or made in the lab) following standard heat shock protocols. Store all expression vectors as glycerol stocks after transformation. To make glycerol stocks, select a single colony with which to inoculate 3 mL of LB containing 50 μg/mL ampicillin and grow overnight. Mix 1.5 mL of the bacterial culture with 0.5 mL of 60% sterile glycerol in a cryotube. Mix well and flash freeze by dropping the cryotube into liquid nitrogen. Store at –80°C.

3. For the expression test, select a single colony with which to inoculate 3 mL of LB containing 50 μg/mL ampicillin and grow overnight. Use all 3 mL of the confluent culture

to inoculate 100 mL of LB containing 50 μg/mL ampicillin in a 500 mL or larger flask. Important parameters to vary are (a) the OD at 600 nm at which induction is performed (e.g., 0.3, 0.6, and 1.0 OD), (b) the incubation temperature after induction, (typically 37°C or 17°C), and (c) the length of time after induction (2 h, 4 h, overnight). Expression is assayed by SDS-PAGE of cell pellets boiled in loading buffer.

4. The high NaCl concentration (800 mM) in the breaking buffer (BB) is useful to prevent precipitation of the ZFP after cell lysis and to reduce the amount of DNA copurifying with the zinc finger. We have also successfully purified the MBP–Aart fusion protein using BB containing 50 mM KCl, although the solubility of the Aart ZFP after cleavage from MBP was found to be limiting in KCl concentrations below 150 mM.

5. More involved purifications have also been performed in our lab, involving chromatography with heparin and Mono Q resins. The heparin chromatography (HiPrep 16/10 Heparin FF) is used with ZBA as buffer A and a linear gradient to 50% buffer B: (ZBA + 2 M KCl) over 10 CV. We found that Aart elutes early in the gradient, along with any uncleaved fusion protein, but MBP flows through the column during loading, providing good separation of Aart from MBP. The Mono Q step was used with the same buffers A and B above. We found that Aart flowed through the column during loading. However, the Mono Q step was useful in that it retained DNA. Use of a high salt wash (1.5 M KCl) in the first amylose chromatographic step eliminates the copurifying DNA. It should also be noted that resins exist that capture proteases, such as enterokinase (trypsin inhibitor agarose, Sigma) and factor Xa (anti-factor Xa-agarose conjugate, Sigma). It should be noted that precipitates of DTT with Zn^{2+} can form, and therefore TCEP-HCl (tris(2-carboxyethyl) phosphine hydrochloride) or BME is preferred over DTT as the reducing agent in buffers. In addition, the ZFP may require more NaCl (300–750 mM) to remain soluble following cleavage from MBP. Finally, use of the α-methylglucopyranoside in place of maltose has been reported to improve the second round of amylose resin chromatography, as it is more easily separated from MBP during dialysis (Dr. Dave Waugh, NCI, personal communication). Maltose binds MBP tightly inhibiting MBP binding to the amylose resin, and therefore must be removed by dialysis prior to amylose resin chromatography.

6. The number of base pairs to be added is a parameter that will be varied in the cocrystallization experiments.

Cocrystallization appears most facilitated with 0–4 additional base pairs at either end of the recognition sequence (*see* **Table 28.2**). In addition, an additional base at the 5′-end of each strand, which is not base paired, can be added to some of the designed oligonucleotides. These 'overhanging' nucleotides can be designed to base pair in an intermolecular fashion, potentially allowing for neighboring duplexes to base pair to each other in the crystal, as commonly occurs in crystals of DNA, with or without bound protein.

9. An important issue to be determined in working with a ZFP is the solubility of the protein, particularly after mixing with DNA. We have found that the solubility of Aart is greatly reduced after mixing with DNA, and 400–600 mM NaCl is necessary to keep the complex soluble. The additional NaCl may be added to the ZFP just prior to adding the DNA, or alternatively, it may be added to the ZBA used for dialysis. Initially, we added the NaCl prior to mixing with DNA, but more recently we have added it to the ZBA we use for dialysis of the Aart protein.

10. Variations on screen number 1. If the ZFP:DNA mixture contains a relatively high concentration of NaCl (400 mM) to solubilize the complex, it may be desirable to add a similar amount of NaCl to each well. We have found that in the hanging drop vapor diffusion method, higher concentrations of NaCl in the drop relative to that in the well can lead to drop dilution, rather than concentration, and that this effect can lead to destabilization of crystals over time (or the failure to produce crystals). To counteract this effect, we have found it useful to raise the NaCl concentration in the well to that in the ZFP:DNA solution (i.e., 400 mM), but to prepare the drop from a solution without the NaCl, and otherwise mimicking the well components. For example, in a well at pH 7.5 containing HEPES, the well would be composed of 25% PEG 4 K, 400 mM NaCl, 100 mM HEPES, and the drop prepared from an equal volume of ZFP:DNA solution (in buffer and 400 mM NaCl) and an equal volume of a solution containing 25% PEG 4 K and 100 mM HEPES.

11. Many published ZFP:DNA structures make use of an anaerobic chamber for crystallization setup. Anaerobic conditions will reduce the amount of oxidation of the ZFP although are not always necessary, as we were able to obtain unoxidized ZFP:DNA crystals for data collection without use of an anaerobic chamber (3).

Table 28.2
Crystallization DNA and conditions of other ZFP:DNA structures

References	Number of fingers	DNA (black indicates target site, grey indicates flanking sequence, underline indicates overhangs)	Crystallization conditions
(10)	3	5′-A GCGTGGGCG T -3′ 3′- CGCACCCGC A T-5′	Hanging drop vapor diffusion method in anaerobic chamber. Drop composed of equal volume protein: DNA solution (1 mM protein, 1 mM DNA, 450–750 mM NaCl, 125 mM bis–tris propane HCl (BTP) pH 8.0) and well solution: 10% PEG 400, 350–650 mM NaCl, 25 mM BTP pH 8.0
(11)	3	5′- GAC GCTATAAAA GGA G-3′ 3′-C CTG CGATATTTT CCT -5′	Hanging drop vapor diffusion method: 1 mM protein + 1.1 mM DNA, 30–35% PEG 400, 100–200 mM NaCl, 10–50 mM BTP pH 8.0
(12)	2	5′-C TAATA AGGAT AACGTCCG -3′ 3′- ATTAT TCCTA TTGCAGGC T-5′	Batch method: 0.3 mM 1:1 complex, 20 mM MES pH 6, 5–20 mM NaCl, 1.75–3.5 mM spermine
(13)	4	5′- AGGGT CTCCATTTT GAAGCG -3′ 3′- TCCCA GAGGTAAAA CTTCGC -5′	Hanging drop vapor diffusion at 4°C. Drop: equal volumes 1.56:1 molar ratio of peptide:DNA (in 20 mM HEPES pH 7.5, 100 mM NaCl, 100 mM ammonium acetate, 5 mM $MgCl_2$, 10 mM DTT, 100 μM zinc acetate at 0.9–1.2 mM ZFP) and well solution: 100 mM HEPES, 100 mM NaCl, 5 mM $MgCl_2$, 25% PEG monomethyl ether 5 K, 5 mM $MgCl_2$
(14)	6	5′-A CGGGC CTGGTTAGTACCTGGATGGGAGAC C -3′ 3′- GCCCG GACCAATCATGGACCTACCCTCTG G T-5′	Hanging drop vapor diffusion at 18°C. Protein:DNA complex reconstituted in 0.25–0.75 M NaCl. Drop: 165 mM NaCl, 35 mM sodium acetate, 3.2 mM DTT, 9.2% glycerol, 1.8 mM NaN_3, 1.8 mM cadaverine-2 HCl, 5.5 mM Tris pH 8, 22.5% PEG 4 K
(15)	3	5′-A T GAGGCAGAA CT -3′ 3′- A CTCCGTCTT GA T-5′	Hanging drop vapor diffusion: 1 mM protein, 1:1.2 ZFP:DNA, NaCl added to 230 mM to solubilize, pH 7.5, filtered, well: 25 mM BTP pH 7.6–8.0, 10 mM $MgCl_2$, 80–100 mM NaCl, 22–24% PEG 1,500. Prepared in anaerobic chamber but placed outside chamber at 20°C

(continued)

Table 28.2 (continued)

References	Number of fingers	DNA (black indicates target site, grey indicates flanking sequence, underline indicates overhangs)	Crystallization conditions
(16)	2×2	5′- A TGGGCGCGCCCA T -3′ 3′- T ACCCGCGCGGGT A -5′	Hanging drop vapor diffusion method in anaerobic chamber. 1.1 mM protein dimer: 1.5 mM DNA, NaCl added to 0.4 mM to solubilize. Drops had equal volumes protein:DNA and well solution: 13–20% PEG 4K, 50–150 mM NaCl, 10 mM $MgCl_2$, 50 mM MES pH 6.2
(17)	5	5′-A CGTG GACCACCCA AGACGAA -3′ 3′- GCAC CTGGTGGGT TCTGCTT T-5′	Hanging drop in anaerobic chamber: 0.5 mM protein:0.6 mM DNA, 60–100 mM $MgCl_2$, 20–25% PEG 400, 50 mM BTP pH 7.0
(18)	3	5′-A GCGTGGGCGT -3′ 3′- CGCACCCGCA T-5′	Hanging drop in anaerobic chamber. Equal volume per drop: complex in 200 mM NaCl, 50 mM MES pH 6.2, well: 27.5–35% PEG 3350, 0–200 mM NaCl, 100 mM Tris, pH 8.5
(19)	3	5′-A GCGTGGGCGG -3′ 3′- CGCACCCGCC T-5′	Hanging drop in anaerobic chamber. Equal volume per drop: complex in 200 mM NaCl, 50 mM MES pH 6.2, well: 10–15% PEG 1450, 300–500 mM NaCl, 25 mM BTP pH 8.0
(20)	5	5′-A GCGTGGGCGT -3′ 3′- CGCACCCGCA T-5′	Hanging drop vapor diffusion method in anaerobic chamber: 3.2 μL protein:DNA complex (0.095 mM ZFP, 0.0532 mM DNA, in 10 mM BTP pH 8.0, 1 mM $MgCl_2$, and 1.5 molar zinc finger equivalents of $CoCl_2$) over a well of 600 mM ammonium acetate
(4)	2×3	5′-G TG GCGTGGGCG GCGTGGGCG T -3′ 3′- AC CGCACCCGC CGCACCCGC A C-5′	Hanging drop vapor diffusion method in anaerobic chamber: drops were prepared by mixing the protein (0.7 mM):DNA (0.28 mM) complex with an equal volume of well solution (4–15% PEG 400, 0–200 mM ammonium acetate, 100 mM bis–tris-propane pH 7.0)

References

1. McPherson, A. (2009) Introduction to macromolecular crystallography. 2nd ed. John Wiley & Sons, Hoboken.
2. Peisach, E. and Pabo, C.O. (2003) Constraints for zinc finger linker design as inferred from X-ray crystal structure of tandem Zif268-DNA complexes. *J Mol Biol.* **330**, 1–7.
3. Segal, D.J., Crotty, J.W., Bhakta, M.S., Barbas, C.F., 3rd, and Horton, N.C. (2006) Structure of Aart, a designed six-finger zinc finger peptide, bound to DNA. *J Mol Biol.* **363**, 405–421.
4. Aggarwal, A.K. (1990) Crystallization of DNA binding proteins with oligodeoxynucleotides., *Methods: A Companion to Methods. Enzymology.* **1**, 83–90.
5. Gill, S.C. and von Hippel, P.H. (1989) Calculation of protein extinction coefficients from amino acid sequence data. *Anal Biochem.* **182**, 319–326.
6. Cantor, C.R., Warshaw, M.M., and Shapiro, H. (1970) Oligonucleotide interactions. 3. Circular dichroism studies of the conformation of deoxyoligonucleotides. *Biopolymers.* **9**, 1059–1077.
7. Warshaw, M.M. and Cantor, C.R. (1970) Oligonucleotide interactions. IV. Conformational differences between deoxy- and ribodinucleoside phosphates. *Biopolymers.* **9**, 1079–1103.
8. Crotty, J.W., Etzkorn, C., Barbas, C.F., Segal, D.J., and Horton, N.C. (2005) Crystallization and preliminary X-ray crystallographic analysis of Aart, a designed six-finger zinc-finger peptide, bound to DNA. *Acta Cryst.* **F61**, 573–576.
9. Blommel, P.G. and Fox, B.G. (2007) A combined approach to improving large-scale production of tobacco etch virus protease. *Protein Expr Purif.* **55**, 53–68.
10. Pavletich, N.P. and Pabo, C.O. (1991) Zinc finger-DNA recognition: crystal structure of a Zif268-DNA complex at 2.1 A. *Science.* **252**, 809–817.
11. Wolfe, S.A., Grant, R.A., Elrod-Erickson, M., and Pabo, C.O. (2001) Beyond the "recognition code": structures of two Cys2His2 zinc finger/TATA box complexes. *Structure.* **9**, 717–723.
12. Fairall, L., Schwabe, J.W., Chapman, L., Finch, J.T., and Rhodes, D. (1993) The crystal structure of a two zinc-finger peptide reveals an extension to the rules for zinc-finger/DNA recognition. *Nature.* **366**, 483–487.
13. Houbaviy, H.B., Usheva, A., Shenk, T., and Burley, S.K. (1996) Cocrystal structure of YY1 bound to the adeno-associated virus P5 initiator. *Proc Natl Acad Sci USA.* **93**, 13577–13582.
14. Nolte, R.T., Conlin, R.M., Harrison, S.C., and Brown, R.S. (1998) Differing roles for zinc fingers in DNA recognition: structure of a six-finger transcription factor IIIA complex. *Proc Natl Acad Sci USA.* **95**, 2938–2943.
15. Kim, C.A. and Berg, J.M. (1996) A 2.2 A resolution crystal structure of a designed zinc finger protein bound to DNA. *Nat Struct Biol.* **3**, 940–945.
16. Wang, B.S., Grant, R.A., and Pabo, C.O. (2001) Selected peptide extension contacts hydrophobic patch on neighboring zinc finger and mediates dimerization on DNA. *Nat Struct Biol.* **8**, 589–593.
17. Pavletich, N.P. and Pabo, C.O. (1993) Crystal structure of a five-finger GLI-DNA complex: new perspectives on zinc fingers. *Science.* **261**, 1701–1707.
18. Elrod-Erickson, M., Benson, T.E., and Pabo, C.O. (1998) High-resolution structures of variant Zif268-DNA complexes: implications for understanding zinc finger-DNA recognition. *Structure.* **6**, 451–464.
19. Miller, J.C. and Pabo, C.O. (2001) Rearrangement of side-chains in a Zif268 mutant highlights the complexities of zinc finger-DNA recognition. *J Mol Biol.* **313**, 309–315.
20. Wolfe, S.A., Grant, R.A., and Pabo, C.O. (2003) Structure of a designed dimeric zinc finger protein bound to DNA. *Biochemistry.* **42**, 13401–13409.

Chapter 29

Beyond DNA: Zinc Finger Domains as RNA-Binding Modules

Josep Font and Joel P. Mackay

Abstract

Over the last 25 years, we have learned that many structural classes of zinc-binding domains (zinc fingers, ZFs) exist and it has become clear that the molecular functions of these domains are by no means limited to the sequence-specific recognition of double-stranded DNA. For example, ZFs can act as protein recognition or RNA-binding modules, and some domains can exhibit more than one function. In this chapter we describe the progress that has been made in understanding the role of ZF domains as RNA-recognition modules, and we speculate about both the prevalence of such functions and the prospects for creating designer ZFs that target RNA.

Key words: Zinc fingers, RNA-binding domain, protein interaction domain, protein design.

1. Introduction

As the result of work from many different laboratories since the mid 1980s, we now know that at least 20 different structural classes of protein domains exist that use one or more zinc ions to stabilize their structures. Of these, by far the most numerous and most intensively studied are the classical zinc fingers (classical ZFs) that were originally discovered by Sir Aaron Klug in 1985 (1–3). A wealth of data has led us to a quite sophisticated understanding of the mechanism through which classical ZFs can act as double-stranded DNA recognition modules, and it is likely that many of the >12,000 classical ZFs predicted in the human genome (according to Pfam) will bind DNA in this manner. It is also clear from these numbers that many classical ZFs are *un*likely to be DNA-binding domains. For example, whereas classical ZF proteins most commonly recognize DNA using an array of two

J.P. Mackay, D.J. Segal (eds.), *Engineered Zinc Finger Proteins*, Methods in Molecular Biology 649,
DOI 10.1007/978-1-60761-753-2_29, © Springer Science+Business Media, LLC 2010

or more contiguous ZFs (although proteins such as GAGA (4) constitute exceptions), many of the classical ZFs in the human genome are found as isolated ZF domains. We have speculated that these domains might in many cases turn out to be protein-recognition motifs (5).

Examples of both classical ZFs and other classes of zinc-binding domains have now been demonstrated to act as protein- or RNA-binding domains and the structural basis for a number of these activities is now known. These data reveal that a great deal of diversity both in the structures of ZF domains and in the mechanisms through which they recognize molecular partners. Here we outline this diversity, focusing on both classical (**Sections 2**, **3**, **4** and **5**) and non-classical (**Sections 6**, **7** and **8**) ZF domains that exhibit RNA-binding activity.

In recent years, the potential of ZFs as molecular tools has begun to be realized. Their stable, modular structure has permitted the design of classical ZF sequence variants that can recognize many different DNA sequences. The assembly of arrays of these variants and their fusion to either transcriptional regulatory domains or endonucleases has led to the creation of customized gene regulatory proteins or site-specific genomic scissors, as demonstrated throughout this volume. The discovery of new functions for ZFs raises the possibility of designing new RNA-targeting ZF modules. We briefly discuss the prospects for this eventuality.

2. TFIIIA

Despite the historical focus on the ability of ZFs to recognize DNA, the first ZFs to be identified, in the transcription factor TFIIIA, were discovered for their ability to bind to the double-stranded 5S ribosomal RNA particle in *Xenopus* oocytes (1). Through this binding activity, they regulate the transport and storage of the 5S particle. TFIIIA contains nine classical ZFs, and fingers 4–6 (F4–6) are the most important for this RNA-binding activity (6, 7). The X-ray structure of these ZFs bound to a minimal 5S rRNA (*see* **Fig. 29.1A**) (8) revealed that F4 and F6 make base-specific contacts, whereas F5 makes exclusively non-sequence-specific interactions with the phosphate backbone via basic side chains. The converse situation exists when TFIIIA binds to DNA (at an internal control region within the 5S RNA gene (9)); it is F1–3 and F5 that make sequence-specific contacts, whereas F4 and F6 appear to act as spacers.

It is also notable that F4 and F6 recognize bases that are not base paired, but rather exist in 'flipped-out' conformations in loop

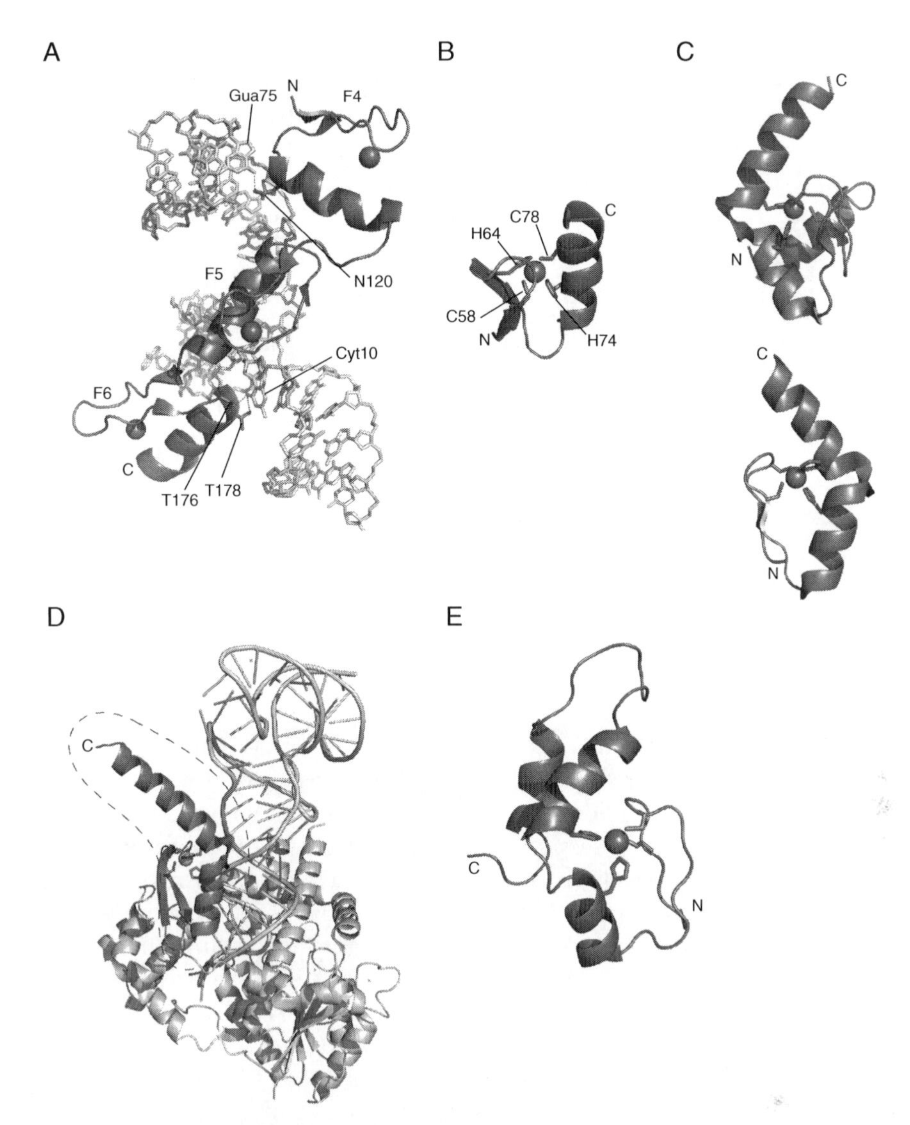

Fig. 29.1. Structures of classical ZFs involved in RNA recognition. (**A**) TFIIIA bound to a minimal 5S rRNA (8) (PDB code 1un6). Sequence-specific hydrogen bonds made by N120, T176, and T178 are shown. Almost no other sequence-specific contacts are observed in the complex. (**B**) The ZF from U11-48 K (24) (PDB code 2vy4), which displays a CHHC zinc coordination pattern. (**C**) The two ZF domains from dsRBP-ZFa (31) (PDB code 1zu1). F1 is shown *above* and F2 *below*. Note the additional N-terminal helix present in F1. (**D**) IPTase bound to tRNA (33) (PDB code 3eph). The ZF domain is circled. (**E**) The U1C ZF (35) (PDB code 2vrd). Note the additional C-terminal α-helix.

structures. The N-terminal end of the α-helix in each ZF points directly toward the RNA, in contrast to the customary orientation of the helix of a classical ZF in DNA complexes, where each α-helix sits along the major groove *(see*, for example, 10). Despite these differences, the residues in F4 and F6 that mediate contacts with RNA (–2, –1, +1, and +2 for F4 and –1, +1, and +2 for F6) overlap almost entirely with the regular DNA-contact residues in classical ZF:DNA complexes (–1, +2, +3, and +6). Thus, roughly the same surface of the classical ZF has been adapted to bind two different target structures in two different contexts.

It has also been demonstrated that a second *Xenopus* classical ZF protein, the nine ZF p43, can bind to 5S RNA (11). Both similarities and differences exist between this protein and TFIIIA. Although both proteins recognize the same RNA target with a similar affinity and using an array of classical ZFs, it is ZF4–6 from TFIIIA that recognizes RNA, which contrasts with the observation that ZF1–4 of p43 is both necessary and sufficient for high-affinity binding. Mutagenesis has been used to map a number of the important residues for RNA recognition, but a structure of the p43:5S RNA complex is currently lacking. A structure for this complex will be an important step toward understanding the diversity of classical ZF:RNA interactions.

3. WT1

The Wilm's tumor 1 protein (WT1) contains a C-terminal array of four closely spaced classical ZFs, which are separated by the short linker sequence (TGEKP) that is widely conserved in DNA-binding classical ZFs. The protein is essential for normal urogenital development (12) and it has been shown that a number of splice isoforms of the protein exist (13), including one in which three amino acids (KTS) are inserted in the linker between F3 and F4. The –KTS form can act as a transcriptional regulator, binding to double-stranded DNA (dsDNA) sequences in the promoters of a number of putative target genes. The structure of a four-ZF construct bound to dsDNA (14) shows that recognition is mediated by F2–F4, which make interactions of the canonical type with the DNA major groove; in fact, F2–F4 are 65% identical to F1–F3 of Zif268 (2) and the two proteins make all of the same contacts to DNA. In contrast, the +KTS isoform cannot bind to the same DNA sites. Both –KTS and +KTS forms can bind with high affinity to both cellular RNA species (15, 16) and to combinatorially selected aptamers (17, 18). Functional data suggest that the +KTS form acts upstream of the sex-determining factor SRY (19), and this isoform has been shown to colocalize with the splicing machinery in nuclear speckles (20).

In a situation reminiscent of *Xenopus* p43, deletion of either F1 or F4 significantly reduces binding to the RNA aptamers (which are predicted to contain substantial secondary structure), as do mutations in F2 or F3 (21). The residues for which point mutations caused the most significant decreases in binding affinity were again located at or immediately preceding the α-helices:

arginines at the –1 and +6 positions in F2 and the –1 arginine in F3. It was also demonstrated that these residues are important for DNA recognition by WT1. Taken together, these data hint that a common mode might exist for the binding of classical ZFs to structured RNAs, and it will be very interesting to see whether this is indeed the case once more structural data are available.

4. A CHHC ZF in U11-48 K

Recently, a single ZF has been characterized in U11-48 K (22), a component of the minor spliceosome (which is responsible for splicing so-called U12 introns (23)). This domain has an unusual CHHC zinc ligation topology and belongs to a small family that includes the TRM13 methyltransferases and the gametocyte-specific factors (22). Despite the difference in coordination geometry, the structure of the U11-48 K ZF closely resembles the classical ZF fold (24) (*see* **Fig. 29.1B**), and fluorescence aniostropy-binding data reveal that the domain recognizes a specific dsRNA sequence corresponding to the 5′-splice site at a human U12-type intron–exon boundary. Binding to ssRNA with the same sequence, or to RNA oligonucleotides with unrelated sequences, is much weaker.

5. dsRBP-ZFa (Matrin) ZFs

The *Xenopus* protein dsRBP-ZFa was identified from a screen designed specifically to detect dsRNA-binding proteins (25). dsRBP-ZFa contains seven ZF domains with sequences that resemble the consensus for classical ZFs. Significantly though, all of the fingers are connected by rather long linkers; these sequences are 34–44 residues in length and none contain the TGEKP sequences commonly associated with dsDNA-binding activity. Gel shift data demonstrated that the full-length protein binds with reasonably high affinity to both dsRNA and an RNA–DNA hybrid, although not to dsDNA or ssRNA. Strikingly, no clear sequence specificity was observed, although this issue was not exhaustively examined. It is worth noting in passing that the transcription factor Sp1 was shown more than 10 years ago to bind a RNA–DNA hybrid with the same affinity as a dsDNA target (26). Curiously though, rather little has come so far of this provocative finding, due in part perhaps to difficulties in finding a clear function for such an activity.

Although the function of dsRBP-ZFa has not been resolved, the ZFs have strong homology to ZFs in the mammalian proteins Wig-1 (27) and JAZ (28); these proteins each contain four widely spaced ZFs (the linkers, which have lower homology, range up to ~70 residues in length) and both have been implicated in the regulation of apoptosis (29, 30).

The solution structures of ZFs 1 and 2 from dsRBP-ZFa (31) reveal folds that resemble overall the classical ZF, with the most notable difference being that the α-helix in the dsRBP-ZFa fingers is roughly twice as long and contains a substantial kink in the middle, located between the two zinc-ligating histidine residues (*see* **Fig. 29.1C**). The structure of the first finger is also elaborated with a short N-terminal helix and a third β-strand. Although no RNA-binding data were provided by Moller et al., Andreeva and Murzin (32) subsequently noted that the dsRBP-ZFa fold is observed in the C-terminal zinc-binding domain of the enzyme isopentenyltransferase (IPTase) (33), a protein that binds to and covalently modifies transfer RNAs. As shown in **Fig. 29.1D**, the N-terminal portion of the long helix packs against the minor groove of a double-stranded portion of the tRNA, whereas the N-terminal part of the C-terminal portion of the helix makes contacts with the major groove (somewhat reminiscent of the TFIIIA:RNA interaction). All of the ZF:RNA contacts observed are between protein side chains and the phosphate backbone, and therefore do not appear to impose any sequence specificity. These residues are largely conserved between IPTase and the proteins dsRBP-ZFa, Wig-1, and JAZ, hinting that these proteins might recognize nucleic acids in a similar fashion, especially given the lack of observed sequence specificity for dsRBP-ZFa.

U1C is a component of the U1 riboprotein complex that in turn forms part of the major spliceosome. It has been demonstrated that, at least in yeast, U1C can recognize 5′-splice site type sequences in vitro (34). U1C contains a single matrin-type ZF domain, the structure of which has been determined (35) (*see* **Fig. 29.1E**). However, in vitro analysis using recombinant human U1C failed to reveal direct binding to RNA (35), so it is currently unclear exactly what role the U1C ZF plays in spliceosomal activity.

Although sequence differences have been noted between dsRBP-ZFa and proteins such as U1C (31), all of these ZFs can be probably be considered to be members of the matrin class of ZF domains. Relatively little is known about this class of domains, which are named for the nuclear matrix protein matrin 3 – a protein that has been implicated in RNA processing (36). There may, however, be more than 40 proteins in humans that contain these domains, and more information is required to define the functional and structural diversity of the class.

6. Nucleocapsid-Type Zinc Knuckles

Nucleocapsid (NC) proteins are essential for the correct packaging of retroviral genomes during viral replication. In almost all cases, the NC protein contains one or two copies of a very small CCHC zinc finger motif known as a zinc knuckle. This motif has the consensus sequence C-X_2-C-X_4-H-X_4-C and forms a compact structure (*see*, for example, 37 and references therein) that binds to several stem-loop structures within a specific sequence element (known as the Ψ-site) in the HIV-1 genome. The structure of the double zinc knuckle from HIV-1 NC bound to stem-loop 3 (SL3) (38) shows that recognition of two of the tetraloop guanine residues, one by each zinc knuckle, dominates the interaction (*see* **Fig. 29.2A**). Each knuckle has a cleft that accommodates the guanine, making van der Waals contacts on both faces of each base. Several hydrogen bonds are also made to the Watson–Crick edge of each guanine, providing specificity for guanine over adenine. An additional short α-helix preceding the first knuckle also sits in the major groove of the RNA, making largely nonsequence-specific contacts with the RNA backbone. A subsequent structure of the same two-ZF protein bound to stem-loop 3 (SL3) showed that the same two guanines in the tetraloop are targeted by the protein (37), despite the fact that they lay on the opposite surface of the SL3 hairpin compared to their position in SL2. In this case, the N-terminal helix cannot lie in the major groove, and instead sits along the RNA backbone, allowing a phenylalanine

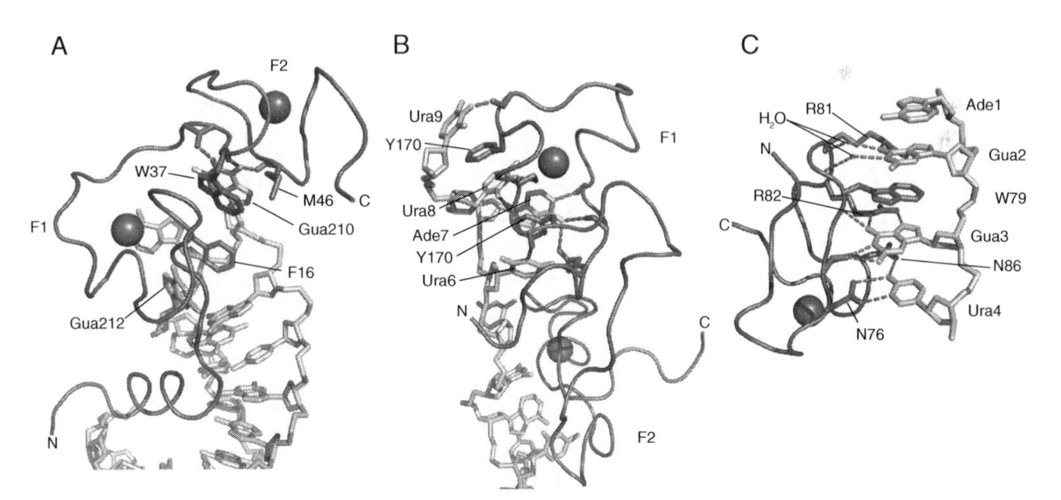

Fig. 29.2. Structures of non-classical ZFs that recognize RNA. (**A**) The double-zinc knuckle from the HIV-1 nucleocapsid protein bound to the SL3 recognition element (38) (PDB code 1a1t). (**B**) Tis11d bound to an AU-rich element (50) (PDB code 1rgo). (**C**) ZRANB2 bound to a GGU sequence of ssRNA (65) (PDB code 3g9y).

side chain in this helix to contact the guanine base to which the more N-terminal of the knuckles is bound.

It has also been demonstrated that NC can play a role in the conversion of the viral RNA genome into dsDNA during the early stages of viral replication. The protein binds to stem-loop structures in the initiation sites (primer-binding sites, PBS) of ssDNA intermediates that are formed during replication (39). This binding event alters the conformation of the stem-loop, promoting strand transfer reactions that are required for replication. The structure of an NC:PBS complex (40) revealed that NC binds to the ssDNA structure in a similar fashion overall to the ssRNA interactions described above, with the exception that the N-terminal knuckle takes a thymine in its binding cleft rather than a guanine.

7. CCCH Zinc Fingers: Tis11d and Friends

A number of proteins in organisms ranging from yeast to mammals contain up to five (but typically one or two) CCCH ZFs that display a C-X_{5-8}-C-X_{4-5}-C-X_3-H consensus sequence. The most widely studied is tetratrisprolin (TTP or Tis11/ZFP36/Nup475), which regulates the expression of inflammatory proteins such as tumor necrosis factor alpha (TNF-α) (41) by binding through its ZFs to AU-rich elements in the 3′-untranslated regions of the target mRNA and promoting degradation of the message (42). Other mammalian proteins in the family include the closely related ZFP36L1, ZFP36L2 (Tis11d), and ZFP36L3, as well as the four proteins MCPIP1-4, which have also been implicated in immune regulation (43). The 30-kDa subunit of cleavage and polyadenylation-stimulating factor (CPSF-30, (44), which is essential for both 3′-cleavage and polyadenylation of mRNAs (45), contains five CCCH ZFs (as well as a C-terminal zinc knuckle) and binds to U-rich sequences found near the cleavage site (45). The immune regulator roquin (46, 47) and the antiviral protein ZAP (48) also contain CCCH ZFs and both have been associated with binding to mRNA to control message stability.

Tis11d is important for normal hematopoiesis (49) and, like TTP, recognizes AU-rich elements. The three-dimensional structure of the Tis11d double-ZF domain bound to the single-stranded RNA (ssRNA) sequence 5′-UUAUUUAUU-3′ (*see* **Fig. 29.2B**) revealed that the Watson–Crick edges of the bases were recognized by specific hydrogen bonds but, unusually, that these hydrogen bonds were mediated almost exclusively by the protein backbone (rather than side chain functional groups) (50).

Two stacking interactions are also observed in each ZF; in each case an aromatic sidechain sits between two RNA bases, making contacts that are important for the stability of the complex (51). Given the similarity of the folds of the other CCCH-ZF proteins described herein and the observation that most of the protein–RNA interactions are mediated by the backbone, it is likely that a similar recognition mechanism will be observed for other family members as well.

An exception appears to be the *Drosophila* protein muscleblind (and its human orthologues MBNL1-3), which have been implicated in alternative splicing (52, 53) and, in the case of the human proteins, in the disease myotonic dystrophy (54, 55). These proteins contain two to four CCCH ZFs with the consensus $C\text{-}X_7\text{-}C\text{-}X_6\text{-}C\text{-}X_3\text{-}H$ and have been shown to bind with high affinity to both CUG repeats and sequences such as AGUCU, but surprisingly only in the context of double-stranded RNA. Given that these domains are likely to have the same fold as the other CCCH ZFs, it will be of great interest to learn what sequence and/or structural features specify dsRNA recognition.

8. ZRANB2

Another small family of ZF has been recently demonstrated to recognize ssRNA in a sequence-specific manner. RanBP2-type zinc fingers are found throughout Eukarya and have been implicated in several different functions. For example, RanBP2 is a component of the nuclear pore complex and ZFs in this protein recognize RanGTP/GDP (56), whereas the RanBP2 ZF in Npl4 mediates an interaction with ubiquitin (57). We showed that the two RanBP2 ZFs from the splicing factor ZRANB2 are able to each recognize a GGU motif through a combination of hydrogen bonding and aromatic stacking interactions (*see* **Fig. 29.2C**). In particular, two arginine residues make double-headed hydrogen bonds to recognize the two guanine bases, in a manner that resembles the arginine–guanine recognition that is well known in the recognition of dsDNA by classical ZFs (58). The intercalation of a tryptophan side chain between the two guanines is also a notable feature of the structure that provides significant stabilization to the complex.

Comparison of the sequences of human RanBP2 ZFs reveals that at least six other proteins contain ZFs that are likely to have similar RNA-binding activity; i.e., translocated in liposarcoma (TLS or FUS) (59), Ewing's sarcoma (EWS) (60), RNA-binding module 5 (RBM5) (LUCA-15) (61), RBM10 (62), RNA-binding protein 56 RBP56 (TAF15 or $TAF_{II}68$) (63), and testes-expressed 13a (Tex13a) 64). Subtle sequence changes between these domains and the ZRANB2 ZFs suggest that specificity differences might exist within the family.

9. Conclusions and Prospects

Our understanding of the RNA-binding properties of zinc-binding domains is in some ways still catching up to our appreciation of ZF:DNA interactions, but it is already becoming clear that the more diverse conformational properties of RNA are reflected in the diversity of observed recognition modes. An interesting observation is that the interactions made by classical ZFs (including the dsRBP-ZFa-type ZFs described in **Section 5**) appear to exhibit relatively little sequence specificity, in contrast to classical ZF:DNA interactions. Far more work is required to determine how general this observation is and what (if any) are its functional ramifications. Given the large number of classical ZFs for which no function is yet experimentally ascribed (probably 90% or so), it is also worth noting that there is every possibility that RNA recognition is a major function of this class of ZFs that has yet to be fully explored.

Other structural classes of ZFs have evolved to bind RNA by quite different mechanisms – using, for example, hydrogen bonds made predominantly by either backbone (CCCH ZFs) or side chain (ZRANB2) functional groups. The RanBP2 and CCCH ZF families are much smaller than the classical ZFs and it will be interesting to see whether this observation translates to a smaller diversity of sequences that can be recognized by such domains. This might particularly be the case for the CCCH domains, given the degree to which the backbone appears to drive specificity.

Given the explosion in the use of designer ZF proteins to target a chosen DNA sequence, we wait with bated breath to see whether ZF domains might be similarly tailored to recognize a wide range of RNA targets. The ability to create designer RNA-binding proteins on demand could be of enormous benefit, both in the analysis of RNA function and for applications in the area of gene regulation. Again, the diversity of RNA structure will play a role here and this feature (together with the technical problems associated with working with RNA) has presumably contributed to the relative lack of real progress to date in the design of RNA-binding proteins in general. However, ZFs are now known that recognize both ssRNA and dsRNA, and structural information is emerging for both of these types of interactions. It seems likely that we will see an increase in activity in this area in the near future.

References

1. Miller, J., McLachlan, A.D., and Klug, A. (1985) Repetitive zinc-binding domains in the protein transcription factor IIIA from Xenopus oocytes. *EMBO J.* **4**, 1609–1614.
2. Pavletich, N.P. and Pabo, C.O. (1991) Zinc finger-DNA recognition: crystal structure of a Zif268-DNA complex at 2.1 A. *Science.* **252**, 809–817.
3. Segal, D.J. and Barbas, C.F., 3rd (2001) Custom DNA-binding proteins come of age: polydactyl zinc-finger proteins. *Curr Opin Biotechnol.* **12**, 632–637.
4. Omichinski, J.G., Pedone, P.V., Felsenfeld, G., Gronenborn, A.M., and Clore, G.M. (1997) The solution structure of a specific GAGA factor-DNA complex reveals a modular binding mode. *Nat Struct Biol.* **4**, 122–132.
5. Mackay, J.P. and Crossley, M. (1998) Zinc fingers are sticking together. *Trends Biochem. Sci.*. **23**, 1–4.
6. Zang, W.Q. and Romaniuk, P.J. (1995) Characterization of the 5 S RNA binding activity of Xenopus zinc finger protein p43. *J Mol Biol.* **245**, 549–558.
7. Zang, W.Q., Veldhoen, N., and Romaniuk, P.J. (1995) Effects of zinc finger mutations on the nucleic acid binding activities of Xenopus transcription factor IIIA. *Biochemistry.* **34**, 15545–15552.
8. Lu, D., Searles, M.A., and Klug, A. (2003) Crystal structure of a zinc-finger-RNA complex reveals two modes of molecular recognition. *Nature.* **426**, 96–100.
9. Pelham, H.R. and Brown, D.D. (1980) A specific transcription factor that can bind either the 5S RNA gene or 5S RNA. *Proc Natl Acad Sci USA.* 77, 4170–4174.
10. Nolte, R.T., Conlin, R.M., Harrison, S.C., and Brown, R.S. (1998) Differing roles for zinc fingers in DNA recognition: structure of a six-finger transcription factor IIIA complex. *Proc Natl Acad Sci USA.* **95**, 2938–2943.
11. Joho, K.E., Darby, M.K., Crawford, E.T., and Brown, D.D. (1990) A finger protein structurally similar to TFIIIA that binds exclusively to 5S RNA in Xenopus. *Cell.* **61**, 293–300.
12. Pritchard-Jones, K., Fleming, S., Davidson, D., Bickmore, W., Porteous, D., Gosden, C., Bard, J., Buckler, A., Pelletier, J., Housman, D., et al (1990) The candidate Wilms' tumour gene is involved in genitourinary development. *Nature.* **346**, 194–197.
13. Hastie, N.D. (2001) Life, sex, and WT1 isoforms – three amino acids can make all the difference. *Cell.* **106**, 391–394.
14. Stoll, R., Lee, B.M., Debler, E.W., Laity, J.H., Wilson, I.A., Dyson, H.J., and Wright, P.E. (2007) Structure of the Wilms tumor suppressor protein zinc finger domain bound to DNA. *J Mol Biol.* **372**, 1227–1245.
15. Caricasole, A., Duarte, A., Larsson, S.H., Hastie, N.D., Little, M., Holmes, G., Todorov, I., and Ward, A. (1996) RNA binding by the Wilms tumor suppressor zinc finger proteins. *Proc Natl Acad Sci USA.* **93**, 7562–7566.
16. Morrison, A.A., Venables, J.P., Dellaire, G., and Ladomery, M.R. (2006) The Wilms tumour suppressor protein WT1 (+KTS isoform) binds alpha-actinin 1 mRNA via its zinc-finger domain. *Biochem Cell Biol.* **84**, 789–798.
17. Bardeesy, N. and Pelletier, J. (1998) Overlapping RNA and DNA binding domains of the wt1 tumor suppressor gene product. *Nucleic Acids Res.* **26**, 1784–1792.
18. Zhai, G., Iskandar, M., Barilla, K., and Romaniuk, P.J. (2001) Characterization of RNA aptamer binding by the Wilms' tumor suppressor protein WT1. *Biochemistry.* **40**, 2032–2040.
19. Bradford, S.T., Wilhelm, D., Bandiera, R., Vidal, V., Schedl, A., and Koopman, P. (2009) A cell-autonomous role for WT1 in regulating Sry in vivo. *Hum Mol Genet.* **18**, 3429–3438.
20. Larsson, S.H., Charlieu, J.P., Miyagawa, K., Engelkamp, D., Rassoulzadegan, M., Ross, A., Cuzin, F., van Heyningen, V., and Hastie, N.D. (1995) Subnuclear localization of WT1 in splicing or transcription factor domains is regulated by alternative splicing. *Cell.* **81**, 391–401.
21. Weiss, T.C. and Romaniuk, P.J. (2009) Contribution of individual amino acids to the RNA binding activity of the Wilms' tumor suppressor protein WT1. *Biochemistry.* **48**, 148–155.
22. Andreeva, A. and Tidow, H. (2008) A novel CHHC Zn-finger domain found in spliceosomal proteins and tRNA modifying enzymes. *Bioinformatics.* **24**, 2277–2280.
23. Burge, C.B., Padgett, R.A., and Sharp, P.A. (1998) Evolutionary fates and origins of U12-type introns. *Mol Cell.* **2**, 773–785.
24. Tidow, H., Andreeva, A., Rutherford, T.J., and Fersht, A.R. (2009) Solution structure of the U11-48 K CHHC zinc-finger domain that specifically binds the 5′ splice site of U12-type introns. *Structure.* **17**, 294–302.
25. Finerty, P.J., Jr. and Bass, B.L. (1997) A Xenopus zinc finger protein that specifically

binds dsRNA and RNA-DNA hybrids. *J Mol Biol.* **271**, 195–208.
26. Shi, Y. and Berg, J.M. (1995) Specific DNA-RNA hybrid binding by zinc finger proteins. *Science.* **268**, 282–284.
27. Varmeh-Ziaie, S., Okan, I., Wang, Y., Magnusson, K.P., Warthoe, P., Strauss, M., and Wiman, K.G. (1997) Wig-1, a new p53-induced gene encoding a zinc finger protein. *Oncogene.* **15**, 2699–2704.
28. Yang, M., May, W.S., and Ito, T. (1999) JAZ requires the double-stranded RNA-binding zinc finger motifs for nuclear localization. *J Biol Chem.* **274**, 27399–27406.
29. Higashi, Y., Asanuma, M., Miyazaki, I., Haque, M.E., Fujita, N., Tanaka, K., and Ogawa, N. (2002) The p53-activated gene, PAG608, requires a zinc finger domain for nuclear localization and oxidative stress-induced apoptosis. *J Biol Chem.* **277**, 42224–42232.
30. Yang, M., Wu, S., Su, X., and May, W.S. (2006) JAZ mediates G1 cell-cycle arrest and apoptosis by positively regulating p53 transcriptional activity. *Blood.* **108**, 4136–4145.
31. Moller, H.M., Martinez-Yamout, M.A., Dyson, H.J., and Wright, P.E. (2005) Solution structure of the N-terminal zinc fingers of the Xenopus laevis double-stranded RNA-binding protein ZFa. *J Mol Biol.* **351**, 718–730.
32. Andreeva, A. and Murzin, A.G. (2008) A fortuitous insight into a common mode of RNA recognition by the dsRNA-specific zinc fingers. *Proc Natl Acad Sci USA.* **105**, E.
33. Zhou, C. and Huang, R.H. (2008) Crystallographic snapshots of eukaryotic dimethylallyltransferase acting on tRNA: insight into tRNA recognition and reaction mechanism. *Proc Natl Acad Sci USA.* **105**, 16142–16147.
34. Du, H. and Rosbash, M. (2002) The U1 snRNP protein U1C recognizes the 5′ splice site in the absence of base pairing. *Nature.* **419**, 86–90.
35. Muto, Y., Pomeranz Krummel, D., Oubridge, C., Hernandez, H., Robinson, C.V., Neuhaus, D., and Nagai, K. (2004) The structure and biochemical properties of the human spliceosomal protein U1C. *J Mol Biol.* **341**, 185–198.
36. Zeitz, M.J., Malyavantham, K.S., Seifert, B., and Berezney, R. (2009) Matrin 3: Chromosomal distribution and protein interactions. *J Cell Biochem.* **108**, 125–133.
37. Amarasinghe, G.K., De Guzman, R.N., Turner, R.B., Chancellor, K.J., Wu, Z.R., and Summers, M.F. (2000) NMR structure of the HIV-1 nucleocapsid protein bound to stem-loop SL2 of the psi-RNA packaging signal. Implications for genome recognition. *J Mol Biol.* **301**, 491–511.
38. De Guzman, R.N., Wu, Z.R., Stalling, C.C., Pappalardo, L., Borer, P.N., and Summers, M.F. (1998) Structure of the HIV-1 nucleocapsid protein bound to the SL3 psi-RNA recognition element. *Science.* **279**, 384–388.
39. Johnson, P.E., Turner, R.B., Wu, Z.R., Hairston, L., Guo, J., Levin, J.G., and Summers, M.F. (2000) A mechanism for plus-strand transfer enhancement by the HIV-1 nucleocapsid protein during reverse transcription. *Biochemistry.* **39**, 9084–9091.
40. Bourbigot, S., Ramalanjaona, N., Boudier, C., Salgado, G.F., Roques, B.P., Mely, Y., Bouaziz, S., and Morellet, N. (2008) How the HIV-1 nucleocapsid protein binds and destabilises the (–)primer binding site during reverse transcription. *J Mol Biol.* **383**, 1112–1128.
41. Taylor, G.A., Carballo, E., Lee, D.M., Lai, W.S., Thompson, M.J., Patel, D.D., Schenkman, D.I., Gilkeson, G.S., Broxmeyer, H.E., Haynes, B.F., and Blackshear, P.J. (1996) A pathogenetic role for TNF alpha in the syndrome of cachexia, arthritis, and autoimmunity resulting from tristetraprolin (TTP) deficiency. *Immunity.* **4**, 445–454.
42. Lai, W.S., Carballo, E., Strum, J.R., Kennington, E.A., Phillips, R.S., and Blackshear, P.J. (1999) Evidence that tristetraprolin binds to AU-rich elements and promotes the deadenylation and destabilization of tumor necrosis factor alpha mRNA. *Mol Cell Biol.* **19**, 4311–4323.
43. Liang, J., Wang, J., Azfer, A., Song, W., Tromp, G., Kolattukudy, P.E., and Fu, M. (2008) A novel CCCH-zinc finger protein family regulates proinflammatory activation of macrophages. *J Biol Chem.* **283**, 6337–6346.
44. DuBois, R.N., McLane, M.W., Ryder, K., Lau, L.F., and Nathans, D. (1990) A growth factor-inducible nuclear protein with a novel cysteine/histidine repetitive sequence. *J Biol Chem.* **265**, 19185–19191.
45. Barabino, S.M., Ohnacker, M., and Keller, W. (2000) Distinct roles of two Yth1p domains in 3′-end cleavage and polyadenylation of yeast pre-mRNAs. *EMBO J.* **19**, 3778–3787.
46. Vinuesa, C.G., Cook, M.C., Angelucci, C., Athanasopoulos, V., Rui, L., Hill, K.M., Yu, D., Domaschenz, H., Whittle, B., Lambe, T., Roberts, I.S., Copley, R.R., Bell, J.I., Cornall, R.J., and Goodnow, C.C. (2005) A RING-type ubiquitin ligase family member

required to repress follicular helper T cells and autoimmunity. *Nature.* **435**, 452–458.

47. Yu, D., Tan, A.H., Hu, X., Athanasopoulos, V., Simpson, N., Silva, D.G., Hutloff, A., Giles, K.M., Leedman, P.J., Lam, K.P., Goodnow, C.C., and Vinuesa, C.G. (2007) Roquin represses autoimmunity by limiting inducible T-cell co-stimulator messenger RNA. *Nature.* **450**, 299–303.
48. Guo, X., Carroll, J.W., Macdonald, M.R., Goff, S.P., and Gao, G. (2004) The zinc finger antiviral protein directly binds to specific viral mRNAs through the CCCH zinc finger motifs. *J Virol.* **78**, 12781–12787.
49. Stumpo, D.J., Broxmeyer, H.E., Ward, T., Cooper, S., Hangoc, G., Chung, Y.J., Shelley, W.C., Richfield, E.K., Ray, M.K., Yoder, M.C., Aplan, P.D., and Blackshear, P.J. (2009) Targeted disruption of Zfp36l2, encoding a CCCH tandem zinc finger RNA-binding protein, results in defective hematopoiesis. *Blood.*
50. Hudson, B.P., Martinez-Yamout, M.A., Dyson, H.J., and Wright, P.E. (2004) Recognition of the mRNA AU-rich element by the zinc finger domain of TIS11d. *Nat Struct Mol Biol.* **11**, 257–264.
51. Lai, W.S., Kennington, E.A., and Blackshear, P.J. (2002) Interactions of CCCH zinc finger proteins with mRNA: non-binding tristetraprolin mutants exert an inhibitory effect on degradation of AU-rich element-containing mRNAs. *J Biol Chem.* **277**, 9606–9613.
52. Ranum, L.P. and Cooper, T.A. (2006) RNA-mediated neuromuscular disorders. *Annu Rev Neurosci.* **29**, 259–277.
53. Vicente, M., Monferrer, L., Poulos, M.G., Houseley, J., Monckton, D.G., O'dell, K.M., Swanson, M.S., and Artero, R.D. (2007) Muscleblind isoforms are functionally distinct and regulate alpha-actinin splicing. *Differentiation.* **75**, 427–440.
54. Fardaei, M., Rogers, M.T., Thorpe, H.M., Larkin, K., Hamshere, M.G., Harper, P.S., and Brook, J.D. (2002) Three proteins, MBNL, MBLL and MBXL, co-localize in vivo with nuclear foci of expanded-repeat transcripts in DM1 and DM2 cells. *Hum Mol Genet.* **11**, 805–814.
55. Miller, J.W., Urbinati, C.R., Teng-Umnuay, P., Stenberg, M.G., Byrne, B.J., Thornton, C.A., and Swanson, M.S. (2000) Recruitment of human muscleblind proteins to (CUG)(n) expansions associated with myotonic dystrophy. *EMBO J.* **19**, 4439–4448.
56. Higa, M.M., Alam, S.L., Sundquist, W.I., and Ullman, K.S. (2007) Molecular characterization of the Ran-binding zinc finger domain of Nup153. *J Biol Chem.* **282**, 17090–17100.
57. Alam, S.L., Sun, J., Payne, M., Welch, B.D., Blake, B.K., Davis, D.R., Meyer, H.H., Emr, S.D., and Sundquist, W.I. (2004) Ubiquitin interactions of NZF zinc fingers. *EMBO J.* **23**, 1411–1421.
58. Jamieson, A.C., Wang, H., and Kim, S.H. (1996) A zinc finger directory for high-affinity DNA recognition. *Proc Natl Acad Sci USA.* **93**, 12834–12839.
59. Crozat, A., Aman, P., Mandahl, N., and Ron, D. (1993) Fusion of CHOP to a novel RNA-binding protein in human myxoid liposarcoma. *Nature.* **363**, 640–644.
60. Delattre, O., Zucman, J., Plougastel, B., Desmaze, C., Melot, T., Peter, M., Kovar, H., Joubert, I., de Jong, P., Rouleau, G., et al (1992) Gene fusion with an ETS DNA-binding domain caused by chromosome translocation in human tumours. *Nature.* **359**, 162–165.
61. Gure, A.O., Altorki, N.K., Stockert, E., Scanlan, M.J., Old, L.J., and Chen, Y.T. (1998) Human lung cancer antigens recognized by autologous antibodies: definition of a novel cDNA derived from the tumor suppressor gene locus on chromosome 3p21.3. *Cancer Res.* **58**, 1034–1041.
62. Nagase, T., Seki, N., Tanaka, A., Ishikawa, K., and Nomura, N. (1995) Prediction of the coding sequences of unidentified human genes. IV. The coding sequences of 40 new genes (KIAA0121-KIAA0160) deduced by analysis of cDNA clones from human cell line KG-1. *DNA Res.* **2**(167–74), 99–210.
63. Morohoshi, F., Arai, K., Takahashi, E.I., Tanigami, A., and Ohki, M. (1996) Cloning and mapping of a human RBP56 gene encoding a putative RNA binding protein similar to FUS/TLS and EWS proteins. *Genomics.* **38**, 51–57.
64. Wang, P.J., McCarrey, J.R., Yang, F., and Page, D.C. (2001) An abundance of X-linked genes expressed in spermatogonia. *Nat Genet.* **27**, 422–426.
65. Loughlin, F.E., Mansfield, R.E., Vaz, P.M., McGrath, A.P., Setiyaputra, S., Gamsjaeger, R., Chen, E.S., Morris, B.J., Guss, J.M., and Mackay, J.P. (2009) The zinc fingers of the SR-like protein ZRANB2 are single-stranded RNA-binding domains that recognize 5′ splice site-like sequences. *Proc Natl Acad Sci USA.* **106**, 5581–5586.

Index

A

B

J.P. Mackay, D.J. Segal (eds.), *Engineered Zinc Finger Proteins*, Methods in Molecular Biology 649,
DOI 10.1007/978-1-60761-753-2,

C

D

E

F

G

H

I

J

K

L

M

R

S

T

U